BIRKHÄUSER

AF213200

Rock Physics and **Natural Hazards**

Edited by
Sergio Vinciguerra
Yves Bernabé

Birkhäuser
Basel · Boston · Berlin

Reprint from Pure and Applied Geophysics
(PAGEOPH), Volume 166 (2009) No. 5 – 7

Editors
Sergio Vinciguerra
Sezione di Sismologia e
Tettonofisica
Istituto Nazionale di Geofisica e
Vulcanologia (INGV)
Via di Vigna Murata 605
00143 Roma
Italy
Email: vinciguerra@ingv.it

Yves Bernabé
Dept. Earth, Atmosphere
& Planetary Sciences
Massachussets Institute of Technology
77, Massachussets Ave.
Cambridge, MA 02139
USA
Email: yvb@mit.edu

Library of Congress Control Number: 2009930008

Bibliographic information published by Die Deutsche Bibliothek:
Die Deutsche Bibliothek lists this publication in the Deutsche Nationalbibliografie; detailed bibliographic data is available in the Internet at <http://dnb.ddb.de>

ISBN 978-3-0346-0121-4 Birkhäuser Verlag AG, Basel · Boston · Berlin

Basel · Boston · Berlin
P.O. Box 133, CH-4010 Basel, Switzerland
Part of Springer Science+Business Media
Printed on acid-free paper produced from chlorine-free pulp. TCF ∞
Cover graphic: Based on a graphic from the article "Axial and Radial Permeability Evolutions of Compressed Sandstones: End Effects and Shear-band Induced Permeability Anisotropy" by J. Dautriat, N. Gland, J. Guelard, A. Dimanov and J. L. Raphanel.

Printed in Germany

ISBN 978-3-0346-0121-4

e-ISBN 978-3-0346-0122-1

9 8 7 6 5 4 3 2 1

www.birkhauser.ch

Contents

Transport Properties Modelling and Poro-elasticity

Shear Induced Anisotropy

Geothermics and Reservoir Characterization

Pure appl. geophys. 166 (2009) 737–739
0033–4553/09/050737–3
DOI 10.1007/s00024-009-0507-0

Pure and Applied Geophysics

Introduction

SERGIO VINCIGUERRA[1] and YVES BERNABÉ[2]

Natural hazards events such as earthquakes or volcanic eruptions involve activation of coupled thermo-hydro-chemo-mechanical processes in rocks. The 7th Euro-Conference of Rock Physics and Geomechanics, sponsored by the Italian Istituto Nazionale di Geofisica e Vulcanologia (INGV), the French Centre National de Recherche Scientifique (CNRS) and Exxon-Mobil, was held September 25 to 30, 2007, in Erice (Italy), to explore how rock physics experiments and models can facilitate understanding and constraint of natural hazards' mechanisms, and, to foster cross-disciplinary collaborations. The Erice Conference was organized around six sessions: 1) fault healing, sealing and rupture process, 2) rock deformation and coupled processes in rocks, 3) fault structure and mechanics, from laboratory to exploratory drilling of major faults, 4) fracture and flow in volcanic and geothermal areas, and physical properties of volcanic rocks, 5) reservoir characterization and stimulation, 6) scaling laws, applications to rupture mechanics and fluid flow. The present topical issue originated from the desire to assemble in a single volume unpublished contributions to the Erice Conference. This volume presents new laboratory data, theoretical and numerical rock physics models and field observations relevant to the study of natural hazards.

The bulk of this topical issue is devoted to questions related to rock failure and earthquake source mechanisms. BIZZARRI presents a review of various competing mechanisms which control dynamic failure during large earthquakes and attempts to infer a realistic earthquake source model from them. GROB *et al.* advance the fact that, despite the huge reduction in scale, laboratory experiments sometimes display phenomena strikingly similar to field observations and can therefore provide insight into natural phenomena. Here, the temporal and spatial distributions of events during interface crack propagation suggest that dynamic fracture propagation is controlled by elastic interactions between structural heterogeneities. At the laboratory scale, NASSERI *et al.* explore the relationship between fracture toughness and roughness in thermally (up to 850°C) treated Westerly granite, while FORTIN *et al.*, STANCHITS *et al.* and BAUD *et al.* investigate the relationship between competing deformation micro-mechanisms, as measured and

[1] Istituto Nazionale di Geofisica e Vulcanologia, Sezione di Roma1, Via di Vigna Murata 605, 00143 Rome, Italy.
[2] Earth, Atmospheric and Planetary Sciences Department, Massachusetts Institute of Technology, Cambridge, MA 02139, U.S.A.

characterized by acoustic emission activity and microstructural observations, and strain localization in sandstones and carbonates. The effect of a number of factors such as confining pressure, porosity, and the presence or absence of water is investigated. Pressure solution is another important example of a deformation mechanism involving coupled chemical and mechanical processes. BERNABÉ *et al.* present microscale numerical simulations of pressure solution creep, and attempt to interpret them via a commonly used, generic creep law. This creep law could be, in turn, incorporated into models of fault gouge healing during the seismic cycle. Mechanical and transport properties of rocks are known to be coupled in many field situations. In order to help constrain this type of problem, NASSERI *et al.* propose an effective medium-based inversion scheme and employ it to infer the mechanical properties and permeability of thermally cracked Westerly granite under isostatic compression from experimental data. The transport properties of rocks are, of course, primarily controlled by the geometry of the pore space. PAPE and CLAUSER propose a novel model linking the Nuclear Magnetic Resonance T1 and T2 distributions to permeability in rocks with fractal pore space. Contrary to previous models, which assume that pores are isolated, the new model includes the effect of proton diffusion coupling between connected pores. BERNABÉ revisits AC flow of a compressible fluid through a single deformable pipe and heterogeneous pipe networks. These AC flow waves may occur in open fractures, possibly causing resonance effects such as might be involved in volcanic tremors. Shear deformation in the Earth, ductile and/or brittle, is often associated with anisotropy of the rock physical properties. CIRRINCIONE *et al.* document microstructural evidence of this process in naturally deformed leuco-gneiss from an Alpine shear zone while CHEN *et al.* characterize the anisotropy of permeability in cores from the Chelungpu fault and assess the implications of their findings to fault models and weakening mechanisms inferred for the Chi-Chi Earthquake. At the laboratory scale, DAUTRIAT *et al.* and LOUIS *et al.* investigate stress-induced anisotropy and its possible control by the transport properties and pre-existing fabric, respectively. Finally, rock physics techniques and models can be applied or tested in geotechnical applications such as geothermal energy recovery or oil/gas production. The work of ZIMMERMANN *et al.* and BLÖCHER *et al.* stress the importance of hydro-mechanical coupling for optimal reservoir management. PAPE *et al.* estimate rock permeability at the well-bore and core scale, based on Nuclear Magnetic Resonance T1 and T2 measurements. Although most studies in this topical issue are not directly devoted to natural hazards, they present rock physics data and models that can be applied to the rock mechanisms and processes occurring during events such as earthquakes or volcanic eruptions.

Acknowledgements

The guest editors are thankful to the contributors for their commitment and enthusiasm and to Renata Dmowska for providing editorial guidance. They also recognize the efforts of the colleagues who conscientiously performed critical reviews of the papers: M. Balme, P. Baud, P. Benson, Y. Bernabé, P. Besuelle, A. Bizzarri, M. Cocco, C. David,

G. Dresen, D. Faulkner, J. Fortin, P. Glover, J-P. Gratier, Y. Guéguen, B. Haimson, M. Heap, H. Kern, I. Main, A. Maineult, C. Marone, P. Meredith, A. Niemeijer, D. Olgaard, T. Reuschlé, F. Rossetti, J. Rudnicki, A. Schubnel, S. Shapiro, S. Stanchits, T.-f. Wong, W. Zhu.

Pure appl. geophys. 166 (2009) 741–776
0033–4553/09/050741–36
DOI 10.1007/s00024-009-0494-1

Pure and Applied Geophysics

What Does Control Earthquake Ruptures and Dynamic Faulting? A Review of Different Competing Mechanisms

ANDREA BIZZARRI

Abstract—The fault weakening occurring during an earthquake and the temporal evolution of the traction on a seismogenic fault depend on several physical mechanisms, potentially concurrent and interacting. Recent laboratory experiments and geological field observations of natural faults revealed the presence, and sometime the coexistence, of thermally activated processes (such as thermal pressurization of pore fluids, melting of gouge and rocks, material property changes, thermally-induced chemical environment evolution), elasto-dynamic lubrication, porosity and permeability evolution, gouge fragmentation and wear, etc. In this paper, by reviewing in a unifying sketch all possible chemico–physical mechanisms that can affect the traction evolution, we suggest how they can be incorporated in a realistic fault governing equation. We will also show that simplified theoretical models that idealistically neglect these phenomena appear to be inadequate to describe as realistically as possible the details of breakdown process (i.e., the stress release) and the consequent high frequency seismic wave radiation. Quantitative estimates show that in most cases the incorporation of such nonlinear phenomena has significant, often dramatic, effects on the fault weakening and on the dynamic rupture propagation. The range of variability of the value of some parameters, the uncertainties in the relative weight of the various competing mechanisms, and the difference in their characteristic length and time scales sometime indicate that the formulation of a realistic governing law still requires joint efforts from theoretical models, laboratory experiments and field observations.

Key words: Rheology and friction of the fault zones, constitutive laws, mechanics of faulting, earthquake dynamics, computational seismology.

1. Introduction

Contrary to other ambits of physics, seismology presently lacks knowledge of *exact* physical law which governs natural faults and makes the understanding of earthquakes feasible from a deterministic point of view. In addition to the ubiquitous ignorance of the initial conditions (i.e., the initial state) of the seismogenic region of interest, we also ignore the equations that control the traction evolution on the fault surface; this has been recently recognized as one of the grand challenges for seismology (http://www.iris.edu/hq/lrsps/seis_lrp_12_08_08.pdf, December 2008). Indirect information comes from theoretical and numerical studies, which, under some assumptions and hypotheses, try to reproduce real–world events and aim to infer some constraints from a systematic

Istituto Nazionale di Geofisica e Vulcanologia, Sezione di Bologna, Italy. E-mail: bizzarri@bo.ingv.it comparison of synthetics with observations. Much information arises from laboratory experiments, which, on the other hand, suffer from the technical impossibility of reproducing (in terms of confining stress *and* sliding velocity) the conditions typical of a fault during its coseismic failure. Finally, recent observations of mature faults (drilling projects and field studies) open new frontiers on the structure of faults and their composition, although at the same time they raise questions regarding physical processes occurring during faulting.

It has been widely established that earthquakes are complex at all scales, both in the distributions of slip and of stress drop on the fault surface. After the pioneering studies by DAS and AKI (1977a), AKI (1979) and DAY (1982), the first attempts to model complexities were the "asperity" model of KANAMORI and STEWART (1976) and the "barrier" model proposed by DAS and AKI (1977b). Subsequently, in his seminal paper, MADARIAGA (1979) showed that the seismic radiation from these models would be very complex and, more recently, BAK *et al.* (1987) suggested that complexity must be due to the spontaneous organization of a fault that is close to criticality. The slip complexity on a fault may arise from many different factors. Just for an example: (*i*) effects of heterogeneous initial stress field (MAI and BEROZA, 2002; BIZZARRI and SPUDICH, 2008 among many others); (*ii*) interactions with other active faults (STEACY *et al.*, 2005 and references therein; BIZZARRI and BELARDINELLI, 2008); (*iii*) geometrical complexity, non–planarity, fault segmentation and branching (POLIAKOV *et al.*, 2002; KAME *et al.*, 2003; FLISS *et al.*, 2005; BHAT *et al.*, 2007); and (*iv*) different and competing physical mechanisms occurring within the cosesimic temporal scale, such as thermal pressurization, rock melting, mechanical lubrication, ductile compaction of gouge, inelastic deformations, etc.

In this paper we will focus on the latter aspect, which has been the object of increasing interest in recent years. Starting from the key question raised by SCHOLZ and HANKS (2004), it is now clear that in the tribological community, as well as in the dynamic modelling community, "the central issue is whether faults obey simple friction laws, and if so, what is the friction coefficient associated with fault slip" (see also BIZZARRI and COCCO, 2006d and references therein). We will discuss in the progression of the paper how these phenomena can be incorporated in a governing equation and in most cases we will show the significant effect on rupture propagation of such an inclusion.

As an opposite of a fracture criterion — which is a condition that specifies in terms of energy (PARTON and MOROZOV, 1974) or maximum frictional resistance (REID, 1910; BENIOFF, 1951) whether there is a rupture at a given fault point and time on a fault — a governing (or constitutive) law is an analytical relation between the components of stress tensor and some physical observables. Following the Amonton' s Law and the Coulomb–Navier criterion, we can relate the magnitude τ of the shear traction **T** to the effective normal stress on the fault σ_n^{eff} through the well known relation:

$$\tau = \|\mathrm{T}\| = \mu\sigma_n^{eff}, \tag{1}$$

μ being the (internal) friction coefficient and

$$\sigma_n^{eff} = \sigma_n - p_{fluid}^{wf} \tag{2}$$

(TERZAGHI *et al.*, 1996). In equation (1) an additional term for the cohesive strength C_0 of the contact surface can also appear on the right–hand side (C_0 = 1–10 MPa; RANALLI, 1995);

Table 1

List of main symbols used in the paper

Symbol	Meaning
$\tau = \|\mathrm{T}\|$	Fault traction, equations (1) and (37)
τ_{ac}	Average local shear strength of asperities contacts, equation (16)
μ	Friction coefficient
σ_n^{eff}	Effective normal stress, equations (2) and (32)
σ_n	(Reference) normal stress of tectonic origin
p_{fluid}^{wf}	Pore fluid pressure on the fault (i.e., in the middle of the slipping zone); see also equation (13) for the solution of the thermal pressurization problem
$p_{fluid}^{(lub)}$	Lubrication fluid pressure, equation (30)
$p^{(lub)}$	Approximated expression for lubrication fluid pressure, equations (32) and (36)
u	Modulus of fault slip
u_{tot}	Total cumulative fault slip
v	Modulus of fault slip velocity
Ψ	Scalar state variable for rate– and state–dependent governing laws, equation (39)
$2w$	Slipping zone thickness; see also equations (28), (29), (33) and (34)
T	Temperature field
T^{wf}	Temperature on the fault (i.e., in the center of the slipping zone), equation (8)
T_{weak}	Absolute weakening temperature of asperity contacts
κ	Thermal conductivity
ρ_{bulk}	Cubic mass density of the bulk composite
C_{bulk_p}	Specific heat of the bulk composite at constant pressure
$\chi = \kappa/\rho_{bulk}C_{bulk_p}$	Thermal diffusivity
$c \equiv \rho_{bulk}C_{bulk_p}$	Heat capacity for unit volume of the bulk composite
α_{fluid}	Volumetric thermal expansion coefficient of the fluid
β_{fluid}	Coefficient of the compressibility of the fluid
ω	Hydraulic diffusivity, equation (10)
k	Permeability
η_{fluid}	Fluid viscosity; see also equation (19)
Φ	Porosity, see also equations (21) to (25)
A_{ac}	Asperity (i.e., real) contacts area
A_m	Macroscopic (i.e., nominal) area in contact
D_{ac}	Average diameter of asperity contacts
E_a	Activation energy, equations (19) and (39)
V_a	Activation volume, equation (40)
R	Universal gas constant, equations (19) and (39)
k_B	Boltzmann constant, equation (40)
h	Material hardness, equation (40)

in equation (2) σ_n is the normal stress (having tectonic origin) and p_{fluid}^{wf} is the pore fluid pressure on the fault. For convenience, we list in Table 1 the main symbols used in this paper.

Once the boundary conditions (initial conditions, geometrical settings and material properties) are specified, the value of fault friction τ controls the metastable rupture nucleation, the further (spontaneous) propagation (accompanied by stress release, seismic wave excitation and stress redistribution in the surrounding medium), the healing of slip and finally the arrest of the rupture (i.e., the termination of the seismogenic phase of the rupture), which precedes the restrengthening interseismic stage. With the only exception of slow nucleation and restrengthening, all the above–mentioned phases of the rupture process are accounted for in fully dynamic models of an earthquake, provided that the exact analytical form of the fault strength is given. The inclusion of all the previously–mentioned physical processes that can potentially occur during faulting is a clear requisite of a realistic fault governing law. In the light of this, equation (1) can be rewritten in a more verbose form as follows (generalizing equation (3.2) in BIZZARRI and COCCO, 2005):

$$\tau = \tau(w_1 O_1, w_2 O_2, \ldots, w_N O_N) \tag{3}$$

where $\{O_i\}_{i=1,\ldots,N}$ are the physical observables, such as cumulative fault slip (u), slip velocity modulus (v), internal variables (such as state variables, $\mathbf{\Psi}$; RUINA, 1983), etc. (see BIZZARRI and COCCO, 2005 for further details). Each observable can be associated with its evolution equation, which is coupled to equation (3).

It is unequivocal that the *relative* importance of each process (represented by the weights $\{w_i\}_{i=1,\ldots,N}$) can change depending on the specific event we consider; it is therefore very easily expected that not all independent variables O_i will appear in the expression of fault friction for all natural faults. Moreover, each phenomenon is associated with its own characteristic scale length and duration and is controlled by certain parameters, some of which are sometimes poorly constrained by observational evidence. The difference in the length (and time) scale parameters of each chemico–physical process potentially represents a theoretical complication in the effort to include different mechanisms in the governing law, as we will discuss in the following of the paper.

2. *The Fault Structure*

In Figure 1 there is a sketch representing the most widely accepted model of a fault, which is considered in the present paper. It is essentially based on the data arising from numerous field observations and geological evidence (EVANS and CHESTER, 1995; CHESTER and CHESTER, 1998; LOCKNER *et al.*, 2000; HEERMANCE *et al.*, 2003; SIBSON, 2003; BILLI and STORTI, 2004). Many recent investigations focussing on the internal structure of fault zones reveal that coseismic slip on mature fault often occurs within an ultracataclastic, gouge–rich and possibly clayey zone (the foliated fault core), generally having a thickness of the order of a few centimeters (2–3 mm in small faults with 10 cm of slip in

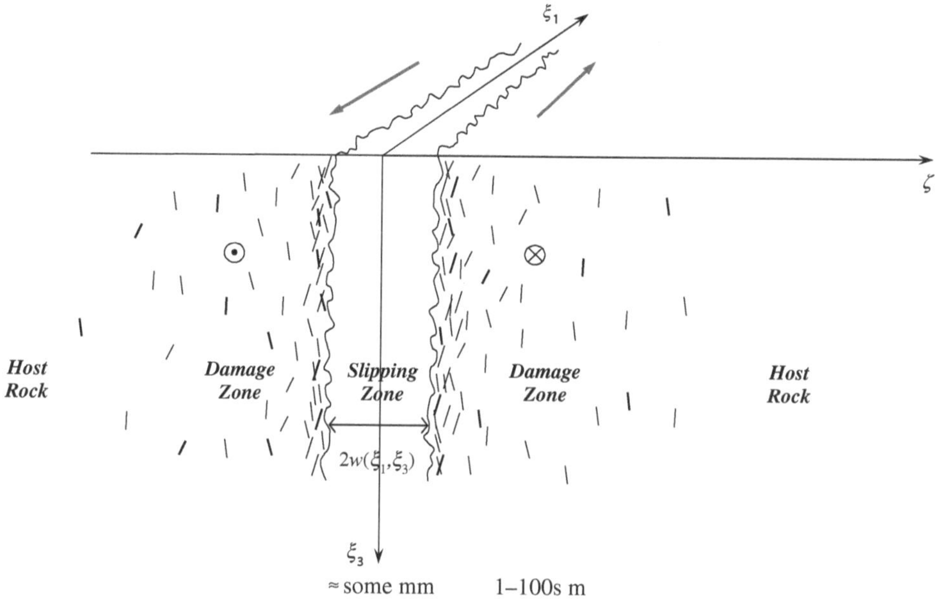

Figure 1
Sketch representing the fault structure described in section 2, as suggested by geological observations (e.g., CHESTER and CHESTER, 1998; LOCKNER *et al.*, 2000; SIBSON, 2003). The slipping zone of thickness $2w$ is surrounded by highly fractured damage zone and finally by the host rocks.

the Sierra Nevada; 10–20 cm in the Punchbowl fault). The fault core, which typically is parallel to the macroscopic slip vector, is surrounded, with an abrupt transition (CHESTER *et al.*, 2004), by a cataclastic damage zone, which can extend up to hundreds of meters. This is composed of highly fractured, brecciated and possibly granulated materials and it is generally assumed to be fluid–saturated. The degree of damage diminishes to the extent as we move far from the ultracataclastic fault core. Outside the damage zone is the host rock, composed of undamaged materials (e.g., WILSON *et al.*, 2003).

The above–mentioned observations (see also BEN–ZION and SAMMIS, 2003; CASHMAN *et al.*, 2007) tend to suggest that slip is accommodated along a single, nearly planar surface, the prominent slip surface (pss) — sometimes called principal fracture surface (pfs) — which generally has a thickness of the order of millimeters. TCHALENKO (1970) described in detail the evolution, for increasing displacement, of localized Riedel zones and conjugate set of Riedel shears, existing around the yield stress, into a concentration of rotated Riedel, P and Y shear zones within a relatively narrow and contiguous tabular zone. When the breakdown process is realized (i.e., the traction is degraded down to its kinetic, or residual, level), the fault structure reaches a mature stage and the slip is concentrated in one (or sometime two) pss, which can be in the middle or near one border of the fault core (symmetric or asymmetric disposition, respectively; see

Sibson, 2003). Moreover, field observations from exhumed faults indicate that fault zones grow in width by continued slip and evolve internally as a consequence of grains size reduction (e.g., Engelder, 1974). We will discuss a possible way to incorporate such a variation in a numerical model of earthquake rupture in section 5. As we will see in the advancement of the paper, the fault zone width, which is a key parameter for many phenomena described in this paper, is difficult to quantify even for a single fault (Rathbun and Marone, 2009) and exhibits an extreme variation along the strike direction. In extreme cases where damage is created off–fault (as in the laboratory experiments of Hirose and Bystricky, 2007) and the definition itself of the fault zone width is not straightforward. One possible (conservative) approach is to assume that the fault zone width is spatially uniform and temporally constant (see next sections 3.1 and 3.2) and to test the effects on the numerical results of the different widths, exploring the range of variability suggested by natural observations. In this way we can take into account physical constraints and, at the same time, we could try to address new laboratory experiments in order to validate the accuracy of theoretical and numerical approaches.

3. *Thermal Effects*

3.1. Temperature Changes

In this section we will focus on the thermally–activated processes. It is well known that when contacting surfaces move relative to each other the friction existing between the two objects converts kinetic energy into thermal energy, or heat. Indicating with q the rate of frictional heat generation ($[q] = \mathrm{W/m^3}$) and neglecting state changes, the temperature in a point of a thermally isotropic medium is the solution of the heat conduction equation

$$\frac{\partial}{\partial t}T = \chi\left(\frac{\partial^2}{\partial \xi_1^2} + \frac{\partial^2}{\partial \zeta^2} + \frac{\partial^2}{\partial \xi_3^2}\right)T + \frac{1}{c}q, \tag{4}$$

where (ξ_1, ξ_3) is a fault point, ζ denotes the spatial coordinate normal to the fault (see Fig. 1), χ is the thermal diffusivity ($\chi = \kappa/\rho_{bulk}C_{bulk_p}$, where κ is the thermal conductivity, ρ_{bulk} is the cubic mass density of the bulk composite and C_{bulk_p} is the specific heat of the bulk composite at constant pressure) and $c \equiv \rho_{bulk}C_{bulk_p}$ is the heat capacity for unit volume of the bulk composite. Following Bizzarri and Cocco (2006a; the reader can refer to that paper for a comprehensive discussion concerning numerical values of the parameters appearing in the model), considering that in the coseismic temporal scale the 1–D (normal to fault plane) approximation of the thermal conduction problem is acceptable and that the temperature in a fault point mainly depends on the fault slip velocity and traction time histories in that point, we have:

$$T^{w}(\xi_1,\zeta,\xi_3,t) = T_0 + \int_0^t \mathrm{d}t' \int_{-\infty}^{+\infty} \mathrm{d}\zeta' q(\xi_1,\zeta',\xi_3,t') K_\chi(\xi_1,\zeta-\zeta',\xi_3,t-t') \quad (5)$$

where $T_0 \equiv T(\xi_1, \zeta, \xi_3, 0)$, i.e., the host rock temperature prior to faulting, and K_χ is the Green's kernel of the conduction equation (expressed by equation (A1), with $h = 1$, of BIZZARRI and COCCO, 2006a).

We can express q considering that the rate of frictional heat generation within the slipping zone (see Fig. 1) can be written as the product of the shear stress τ and the shear strain rate. According to CARDWELL *et al.* (1978), FIALKO (2004) and BIZZARRI and COCCO (2004, 2006a, 2006b), we assume here that the shear strain rate is constant within the slipping zone. In laboratory experiments MAIR and MARONE (2000) have shown that this hypothesis might be adequate, but we want to remark that, in general, the slip velocity profile may be nonlinear across the slipping zone. It follows that the shear strain rate becomes the ratio of the total slip velocity v over the thickness of the slipping zone $2w$. Referring to section 2, as a first approximation of the reality we can regard $2w$ indifferently as the width of the fault core (which includes the pss), the ultracataclastic shear zone or the gouge layer. At the present state of the art we do not have a sufficiently accurate mathematical model to distinguish between these structures which are identified basically from a microstructural point of view. Therefore we will refer to $2w$ as the thickness of the slipping zone and of the gouge layer. Assuming that all the work spent to allow the fault sliding is converted into heat (PITTARELLO *et al.*, 2008), we can write the quantity q in equation (4) as (BIZZARRI and COCCO, 2004, 2006a):

$$q(\xi_1,\zeta,\xi_3,t) = \begin{cases} \dfrac{\tau(\xi_1,\xi_3,t)\upsilon(\xi_1,\xi_3,t)}{2w(\xi_1,\xi_3)}, & t>0, |\zeta| \le w(\xi_1,\xi_3) \\ 0, & |\zeta| > w(\xi_1,\xi_3) \end{cases} \quad (6)$$

where $2w$ explicitly depends on the on–fault coordinates; indeed, there is experimental evidence (e.g., KLINGER *et al.*, 2005) that the slipping zone thickness can change along strike and dip, even on a single fault. In section 5 we also will discuss possible temporal changes of $2w$. By using (6), equation (5) can be solved analytically (BIZZARRI and COCCO, 2004, 2006a, c; see also CARDWELL *et al.*, 1978; FIALKO, 2004):

$$T^{w}(\xi_1,\zeta,\xi_3,t) = T_0 + \frac{1}{4cw(\xi_1,\xi_3)} \int_0^{t-\varepsilon} \mathrm{d}t' \left\{ \mathrm{erf}\left(\frac{\zeta + w(\xi_1,\xi_3)}{2\sqrt{\chi(t-t')}}\right) - \mathrm{erf}\left(\frac{\zeta - w(\xi_1,\xi_3)}{2\sqrt{\chi(t-t')}}\right) \right\}$$
$$\times \tau(\xi_1,\xi_3,t')\upsilon(\xi_1,\xi_3,t') \quad (7)$$

erf(.) being the error function $\left(\mathrm{erf}(z) \underset{\mathrm{df}}{=} \frac{2}{\sqrt{\pi}} \int_0^z e^{-x^2} \mathrm{d}x\right)$ and ε an arbitrarily small, positive, real number (see BIZZARRI and COCCO, 2006a for technical details). It is clear from equation (7) that temperature changes involve the damage zone as well as the slipping zone. In the center of the slipping zone (namely in $\zeta = 0$, which can be regarded as the idealized (or virtual mathematical) fault plane), equation (7) reduces to:

$$T^{w^f}(\xi_1, \xi_3, t) = T_0^f + \frac{1}{2cw(\xi_1, \xi_3)} \int_0^{t-\varepsilon} dt' \operatorname{erf}\left(\frac{w(\xi_1, \xi_3)}{2\sqrt{\chi(t-t')}}\right) \tau(\xi_1, \xi_3, t') \upsilon(\xi_1, \xi_3, t') \quad (8)$$

T_0^f being the initial temperature distribution on the fault plane (i.e., $T_0^f \equiv T(\xi_1, 0, \xi_3, 0)$). Examples of temperature rises due to frictional heat are shown for different values of the slipping zone thickness in Figure 2. For a typical earthquake event, if the thickness of the slipping zone is extremely thin ($w \leq 1$ mm), the increase of temperature is significant: for a half meter of slip, and for a slipping zone 1 mm thick, the temperature change might be of the order of 800°C (FIALKO, 2004; BIZZARRI and COCCO, 2006a,b) and can still be sufficiently large to generate melting of gouge materials and rocks. We will discuss this issue in section 3.4.

3.2. *Thermal Pressurization of Pore Fluids*

The role of fluids and pore pressure relaxation on the mechanics of earthquakes and faulting is the subject of an increasing number of studies, based on a new generation of laboratory experiments, field observations and theoretical models. The interest is motivated by the fact that fluids play an important role in fault mechanics: They can affect the earthquake nucleation and earthquake occurrence (e.g., SIBSON, 1986; ANTONIOLI *et al.*, 2006), can trigger aftershocks (NUR and BOOKER, 1972; MILLER *et al.*, 1996; SHAPIRO *et al.*, 2003 among many others) and can control the breakdown process through the so–called thermal pressurization phenomenon (SIBSON, 1973; LACHENBRUCH, 1980; MASE and SMITH, 1985, 1987; KANAMORI and HEATON, 2000; ANDREWS, 2002; BIZZARRI and COCCO, 2004, 2006a, b; RICE, 2006). In this paper we will focus on the coseismic time scale, however we want to remark that pore pressure can also change during the interseismic period, due to compaction and sealing of fault zones (BLANPIED *et al.*, 1995; SLEEP and BLANPIED, 1992).

Temperature variations caused by frictional heating (equation (7) or (8)) heat both rock matrix and pore fluids; thermal expansion of fluids is paramount, since thermal expansion coefficient of water is greater than that of rocks. The stiffness of the rock matrix works against fluid expansion, causing its pressurization. Several *in situ* and laboratory observations (LOCKNER *et al.*, 2000) show that there is a large contrast in permeability (k) between the slipping zone and the damage zone: in the damage zone k might be three orders of magnitude greater than that in the fault core (see also RICE, 2006). Consequently, fluids tend to flow in the direction perpendicular to the fault. Pore pressure changes are associated to temperature variations caused by frictional heating, temporal changes in porosity and fluid transport through the equation:

$$\frac{\partial}{\partial t} p_{fluid} = \frac{\alpha_{fluid}}{\beta_{fluid}} \frac{\partial}{\partial t} T - \frac{1}{\beta_{fluid} \Phi} \frac{\partial}{\partial t} \Phi + \omega \frac{\partial^2}{\partial \zeta^2} p_{fluid} \quad (9)$$

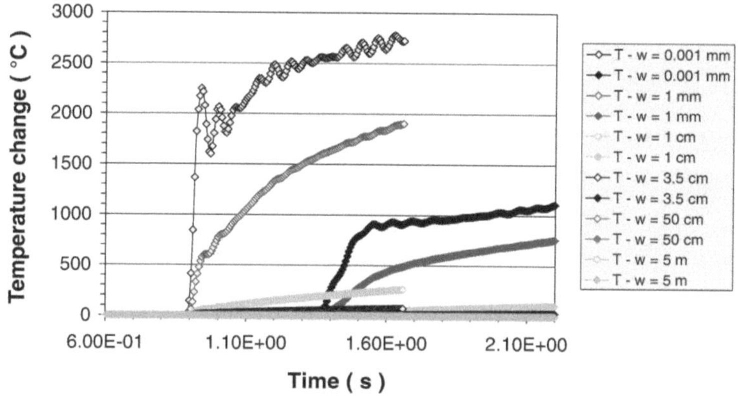

Figure 2
Temperature rises w. r. to $T_0^f = 100°C$ for different values of the slipping zone thickness $2w$ on a vertical strike-slip fault obeying the linear slip–weakening law (IDA, 1972). Temperatures time histories are calculated from equation (8) at the hypocentral depth (6200 m) and at a distance along the strike of 2750 m from the hypocenter. Solid symbols refer to a configuration with strength parameter $S = 1.5$; empty ones are for $S = 0.8$. From BIZZARRI and COCCO (2006a).

where α_{fluid} is the volumetric thermal expansion coefficient of the fluid, β_{fluid} is the coefficient of the compressibility of the fluid and ω is the hydraulic diffusivity, expressed as (e.g., WIBBERLEY, 2002):

$$\omega \equiv \frac{k}{\eta_{fluid}\Phi\beta_{fluid}}, \tag{10}$$

η_{fluid} being the dynamic fluid viscosity and Φ the porosity[1]. The solution of equation (9), coupled with the heat conduction equation, can be written in the form:

$$p^w_{fluid}(\xi_1,\zeta,\xi_3,t) = p_{fluid_0} + \frac{\alpha_{fluid}}{\beta_{fluid}} \int_0^t dt' \int_{-\infty}^{+\infty} d\zeta' \left\{ \frac{q(\xi_1,\zeta',\xi_3,t')}{\omega-\chi} \cdot \right.$$
$$\left. (-\chi K_\chi(\xi_1,\zeta-\zeta',\xi_3,t-t') + \omega K_\omega(\xi_1,\zeta-\zeta',\xi_3,t-t')) \right\} \tag{11}$$

From equation (11) it emerges that there is a coupling between temperature and pore fluid pressure variations; in the case of constant diffusion coefficients, rearranging terms of equation (11), we have that $p^w_{fluid}(\xi_1,\zeta,\xi_3,t) - p_{fluid_0} = -\frac{\chi}{\omega-\chi}\frac{\alpha_{fluid}}{\beta_{fluid}}\,(T^w(\xi_1,\ \zeta,\ \xi_3,t) - T_0) + \frac{\omega}{\omega-\chi}\frac{\alpha_{fluid}}{\beta_{fluid}} \cdot \int_0^t dt' \int_{-\infty}^{+\infty} d\zeta' \{\frac{q(\xi_1,\zeta',\xi_3,t')}{\omega-\chi} K_\omega\ (\xi_1\zeta - \zeta',\xi_3,t-t')\}$.

[1] Equation (9) is derived under the assumption that permeability (k), dynamic density (η_{fluid}) and cubic mass density of the fluid are spatially homogeneous. The quantity $\Phi\beta_{fluid}$ in equation (10) is an adequate approximation of the storage capacity, $\beta_c = \Phi(\beta_{fluid} - \beta_{grain}) + (\beta_{bulk} - \beta_{grain})$, because the compressibilities of mineral grain (β_{grain}) and bulk composite (β_{bulk}) are negligible w. r. to β_{fluid}.

For the specific heat source in (6), equation (11) becomes (BIZZARRI and COCCO, 2006b):

$$p^{w}_{fluid}(\xi_1,\zeta,\xi_3,t) = p_{fluid_0} + \frac{\gamma}{4w(\xi_1,\xi_3)}\int_0^{t-\varepsilon} dt'\left\{ -\frac{\chi}{\omega-\chi}\left[\mathrm{erf}\left(\frac{\zeta+w(\xi_1,\xi_3)}{2\sqrt{\chi(t-t')}}\right) - \mathrm{erf}\left(\frac{\zeta-w(\xi_1,\xi_3)}{2\sqrt{\chi(t-t')}}\right)\right] + \frac{\omega}{\omega-\chi}\left[\mathrm{erf}\left(\frac{\zeta+w(\xi_1,\xi_3)}{2\sqrt{\omega(t-t')}}\right) - \mathrm{erf}\left(\frac{\zeta-w(\xi_1,\xi_3)}{2\sqrt{\omega(t-t')}}\right)\right]\right\}\left\{\tau(\xi_1,\xi_3,t')\upsilon(\xi_1,\xi_3,t'). - \frac{2w(\xi_1,\xi_3)}{\gamma}\frac{1}{\beta_{fluid}\Phi(t')}\frac{\partial}{\partial t'}\Phi(\xi_1,\zeta,\xi_3,t')\right\} \tag{12}$$

which is simplified to

$$p^{w_f}_{fluid}(\xi_1,\xi_3,t) = p^{f}_{fluid_0} + \frac{\gamma}{2w(\xi_1,\xi_3)}\int_0^{t-\varepsilon} dt'\left\{ -\frac{\chi}{\omega-\chi}\mathrm{erf}\left(\frac{w(\xi_1,\xi_3)}{2\sqrt{\chi(t-t')}}\right). + \frac{\omega}{\omega-\chi}\mathrm{erf}\left(\frac{w(\xi_1,\xi_3)}{2\sqrt{\omega(t-t')}}\right)\right\}\left\{\tau(\xi_1,\xi_3,t')\upsilon(\xi_1,\xi_3,t') - \frac{2w(\xi_1,\xi_3)}{\gamma}\frac{1}{\beta_{fluid}\Phi(t')}\frac{\partial}{\partial t'}\Phi(\xi_1,0,\xi_3,t')\right\} \tag{13}$$

in the middle of the slipping zone. In previous equations p_{fluid_0} the initial pore fluid pressure (i.e., $p_{fluid_0} \equiv p_{fluid}(\xi_1,\zeta,\xi_3,0)$) and $\gamma \equiv \alpha_{fluid}/(\beta_{fluid}c)$. In (12) and (13) the term involving Φ accounts for compaction or dilatation and it acts in competition with respect to the thermal contribution to the pore fluid pressure changes. Additionally, variations in porosity will modify, at every time instant (see equation (10)), the arguments of error functions which involve the hydraulic diffusivity.

As a consequence of equations (1) and (2), it follows from equation (13) that variations in pore fluid pressure lead to changes in fault friction. In fact, in their fully dynamic, spontaneous, *truly* 3–D (i.e., not mixed–mode as in ANDREWS, 1994) earthquake model BIZZARRI and COCCO (2006a, b) demonstrated that the inclusion of fluid flow in the coseismic process strongly alters the dry behavior of the fault, enhancing instability, even causing rupture acceleration to super–shear rupture velocities for values of strength parameter ($S \underset{\mathrm{df}}{=} \frac{\tau_u-\tau_0}{\tau_0-\tau_f}$; DAS and AKI, 1977a, b) which do not allow this transition in dry conditions (see also BIZZARRI and SPUDICH, 2008). For extremely localized slip (i.e., for small values of slipping zone thickness) or for low value of hydraulic diffusivity, the thermal pressurization of pore fluids increases the stress drop, causing a near complete stress release (see also ANDREWS, 2002), and changes the shape of the slip–weakening curve and therefore the value of the so–called fracture energy. In Figure 3 we report slip–

weakening curves obtained in the case of Dieterich–Ruina law (LINKER and DIETERICH, 1992) for different vales of $2w$, ω and ${\alpha_{DL}}$[2]. It has been emphasized (BIZZARRI and COCCO, 2006a, b) that in some cases it is impossible to determine the equivalent slip–weakening distance (in the sense of OKUBO, 1989 and COCCO and BIZZARRI, 2002; see also TINTI *et al.*, 2004) and the friction exponentially decreases as recently suggested by several authors (ABERCROMBIE and RICE, 2005; MIZOGUCHI *et al.*, 2007 and SUZUKI and YAMASHITA, 2007).

If we are interested in considering temporal windows longer than those typical of coseismic ruptures, we emphasize that equations (7) and (12) can be directly applicable not only in a perfectly elastic model, but also in cases accounting for a more complex rheology, where stress tensor components explicitly depend on variations of temperature (ΔT) and pore fluid pressure (Δp_{fluid}) fields (e.g., BOLEY and WEINER, 1985):

$$\sigma_{ij} = 2Ge_{ij} + \lambda e_{kk}\delta_{ij} - \frac{2G + 3\lambda}{3}\alpha_{soild}\Delta T\delta_{ij} - \left(1 - \frac{2G + 3\lambda}{3}\beta_{soild}\right)\Delta P_{fluid}\delta_{ij} \qquad (14)$$

where G is the rigidity, λ is the first Lamé' s constant, $\{e_{ij}\}$ is the deformation tensor and Einstein' s convention on repeated indices is assumed. In (14) α_{solid} is the volumetric thermal expansion coefficient of the solid phase and β_{solid} is the compressibility of the solid devoid of any cavity, expressing the pressure necessary to change its volume (i.e., the interatomic spacing). The third term on the right–hand side in equation (14) accounts for the thermal strain of an elastic body, due to thermal expansion.

We finally recognize that in the solution of the thermal pressurization problem presented above we neglected the advection of pore fluids (approximation which has been shown to be valid if permeability is lower than 10^{-16} m^2 (LACHENBRUCH, 1980; MASE and SMITH, 1987; LEE and DELANEY, 1987)). In of generality an additional term will appear in the right–hand side of equation (4), which becomes

$$\frac{\partial T}{\partial t} = \chi \frac{\partial^2 T}{\partial \zeta^2} + \frac{1}{c} q - \frac{k}{\eta_{fluid}} \frac{\partial T}{\partial \zeta} \frac{\partial p_{fluid}}{\partial \zeta} \qquad (15)$$

This will add a direct coupling between temperature and pore fluid pressure, and therefore analytical solutions (7) and (12) are no longer valid.

3.3. *Flash Heating*

Another physical phenomenon associated with frictional heat is the flash heating (TULLIS and GOLDSBY, 2003; HIROSE and SHIMAMOTO, 2003; PRAKASH and YUAN, 2004; RICE, 2006; HAN *et al.*, 2007; HIROSE and BYSTRICKY, 2007) which might be invoked to explain the reduction of the friction coefficient μ from typical values at low slip rate

[2] The parameter α_{DL} controls the coupling of the pore fluid pressure and the evolution of the state variable. It typically ranges between 0.2 and 0.56 (LINKER and DIETERICH, 1992). A null value of α_{DL} means that the evolution equation for the state variable does not depend on the effective normal stress and that the pore fluid affects only the expression of the fault friction τ.

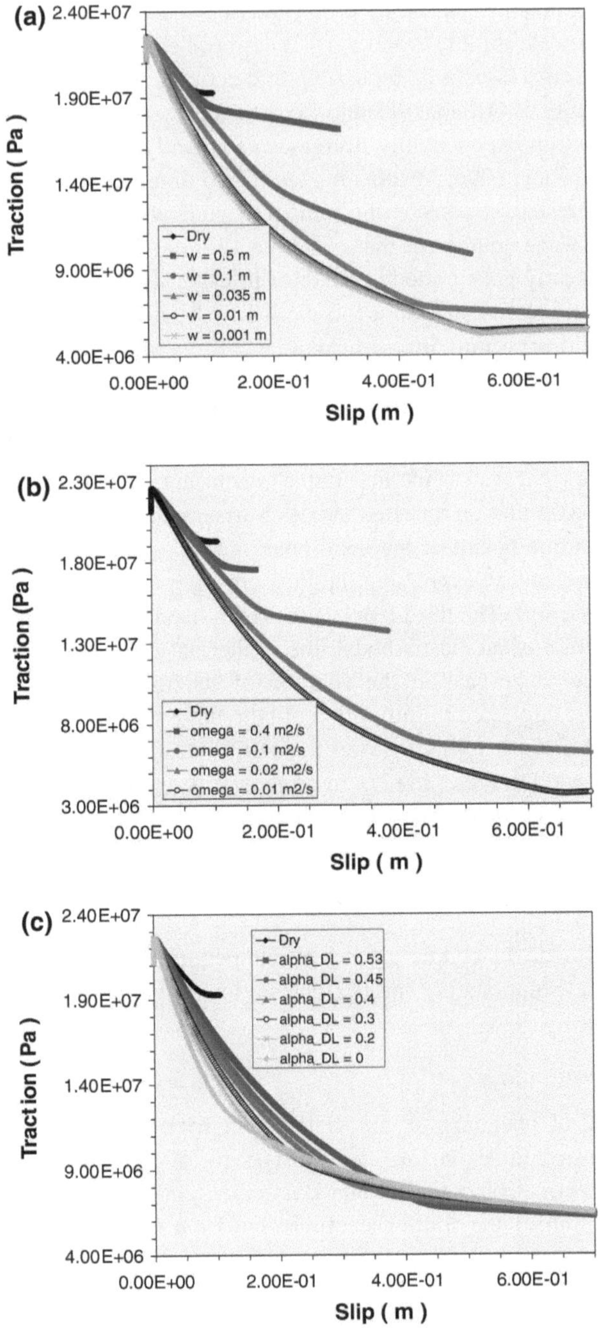
(a)
Traction (Pa)
Slip (m)
2.40E+07
1.90E+07
1.40E+07
9.00E+06
4.00E+06
0.00E+00
2.00E-01
4.00E-01
6.00E-01
Dry
w = 0.5 m
w = 0.1 m
w = 0.035 m
w = 0.01 m
w = 0.001 m
(b)
Traction (Pa)
Slip (m)
2.30E+07
1.80E+07
1.30E+07
8.00E+06
3.00E+06
Dry
omega = 0.4 m2/s
omega = 0.1 m2/s
omega = 0.02 m2/s
omega = 0.01 m2/s
(c)
Traction (Pa)
Slip (m)
2.40E+07
1.90E+07
1.40E+07
9.00E+06
4.00E+06
Dry
alpha_DL = 0.53
alpha_DL = 0.45
alpha_DL = 0.4
alpha_DL = 0.3
alpha_DL = 0.2
alpha_DL = 0

◀ Figure 3
Slip–weakening curves for a rupture governed by the LINKER and DIETERICH (1992) friction law and considering thermal pressurization of pore fluids (see section 3.2 for further details). Solutions are computed at a distance along strike of 1300 m from the hypocenter. (a) Effect of different slipping zone thickness, $2w$. (b) Effect of different hydraulic diffusivities, ω. (c) Effect of different couplings between pore fluid pressure and state variable. In all panels blue diamonds refer to a dry fault, without fluid migration. From BIZZARRI and COCCO (2006b).

($\mu = 0.6$–0.9 for most all rock types; e.g., BYERLEE, 1978) down to $\mu = 0.2$–0.3 at seismic slip rate. It is assumed that the macroscopic fault temperature (T^{wf}) changes substantially more slowly than the temperature on an asperity contact (since the asperity (i.e., real) contact area, A_{ac}, is smaller than the macroscopic (nominal) area in contact, A_m), causing the rate of heat production at the local contact to be higher than the average heating rate of the nominal area. In the model, flash heating is activated if the sliding velocity is greater than the critical velocity

$$v_{fh} = \frac{\pi\chi}{D_{ac}}\left(c\frac{T_{weak} + T^{wf}}{\tau_{ac}}\right)^2 \tag{16}$$

where τ_{ac} is the local shear strength of an asperity contact (which is far larger than the macroscopic applied stress; $\tau_{ac} \sim 0.1\ G =$ few GPa), D_{ac} ($\sim$few μm) its diameter and T_{weak} ($\sim$several 100 s of °C, near the melting point) is a weakening temperature at which the contact strength of an asperity begin to decrease. Using the parameters of RICE (2006) we obtain that v_{fh} is of the order of several centimeters per second; we remark that v_{fh} changes in time as does macroscopic fault temperature T^{wf}. When fault slip exceeds v_{fh} the analytical expression of the steady–state friction coefficient becomes:

$$\mu_{fh}^{ss}(v) = \mu_{fh} + \left[\mu_* - (b-a)\ln\left(\frac{v}{v_*}\right) - \mu_{fh}\right]\frac{v_{fh}}{v}, \tag{17}$$

being μ_*, μ_{fh} and v_* reference values for friction coefficient and velocity, respectively, and b and a the dimensionless constitutive parameters of the Dieterich–Ruina governing equations (DIETERICH, 1979; RUINA, 1983). For $v < v_{fh}$, on the contrary, $\mu^{ss}(v)$ retains the classical form $\left(\text{i.e., } \mu_{lv}^{ss}(v) = \left[\mu_* - (b-a)\ln\left(\frac{v}{v_*}\right)\right]\right)$.

We note that thermal pressurization of pore fluids and flash heating are inherently different mechanisms because they have a different length scale: the former is characterized by a length scale on the order of few micron (D_{ac}), while the length scale of the latter phenomenon is the thermal boundary layer ($\delta = \sqrt{2\chi t_d}$, where t_d is the duration of slip, of the order of seconds), which is $\sim$ mm up to a few cm. Moreover, while thermal pressurization affects the effective normal stress, flash heating causes changes only in the analytical expression of the friction coefficient at high slip rates. In both cases the evolution equation for the state variable is modified: by the coupling of variations in σ_n^{eff} for the first phenomenon, by the presence of additional terms involving v_{fh}/v in the latter. In a recent paper, Noda *et al.* (2009) integrate both flash heating and thermal pressurization in a single constitutive framework, as described above. Their results are properly constrained to the scale of typical laboratory experiments. When we try to apply the model to real–world events, we would encounter the well–known problem of scaling the values of the parameters of the rate– and state–dependent friction law (in whom the two phenomena are simultaneously incorporated) from laboratory–scale to real faults. This, which is particularly true for the scale distance for the evolution of the state variable, would additionally raise the problem to properly resolve, from a numerical point of view, the different spatial scales of flash heating and thermal pressurization.

3.4. Gouge and Rocks Melting

As pointed out by Jeffreys (1942), McKenzie and Brune (1972) and Fialko and Khazan (2005), melting should probably occur during coseismic slip, typically after rocks comminution. Rare field evidence for melting on exhumed faults (i.e., the apparent scarcity of glass or pseudotachylytes, natural solidified friction melts produced during coseismic slip) generates scepticism for the relevance of melt in earthquake faulting (Sibson and Toy, 2006; Rempel and Rice, 2006). However, several laboratory experiments have produced melt, when typical conditions of seismic deformation are attained (Spray, 1995; Tsutsumi and Shimamoto, 1997; Hirose and Shimamoto, 2003). Moreover, as mentioned in section 3.1, it has been demonstrated by theoretical models that for thin slipping zones (i.e., $2w/\delta < 1$) melting temperature T_m can easily be exceeded in dynamic motion (Fialko, 2004; Bizzarri and Cocco, 2006a, b). Even if performed at low (2–3 MPa) normal stresses, the experiments of Tsutsumi and Shimamoto (1997) demonstrated significant deviations from the predictions obtained with the usual rate– and state–friction laws (e.g., Dieterich, 1979; Ruina, 1983). Fialko and Khazan (2005) suggested that fault friction simply follows the Coulomb–Navier equation (1) before melting and the Navier–Stokes constitutive relation $\tau = \eta_{melt} \frac{v}{2w_{melt}}$ after melting ($2w_{melt}$ being the thickness of the melt layer).

Nielsen *et al.* (2008) theoretically interpreted the results from high velocity (i.e., with $v > 0.1$ m/s) rotary friction experiments on India Gabbro and derived the following relation expressing the fault traction in steady–state conditions (i.e., when $\partial T/\partial t = 0$ in (4)) when melting occurs:

$$\tau = \sigma_n^{eff^{1/4}} \frac{K_{NEA}}{\sqrt{R_{NEA}}} \sqrt{\frac{\ln\left(\frac{2v}{v_m}\right)}{\frac{2v}{v_m}}} \qquad (18)$$

where K_{NEA} is a dimensional normalizing factor, R_{NEA} is the radius of the contact area (i.e., the radius of sample; $R_{NEA} \approx 10 - 20$ mm in their lab experiments) and v_m is a characteristic slip rate ($v_m \leq 14$ cm/s).

3.5. Additional Effects of Temperature

3.5.1 Material property changes. An additional complication in the model described above can also arise from the dependence of properties of the materials on temperature. For the sake of simplicity, we neglect the variations of rigidity, volume and density of fault fluids and the surrounding medium due to temperature and pressure, even if they might be relevant[3]. We will focus our attention on the rheological properties of the fault. It is well known that the yield strength depends on temperature and that at high temperatures and pressures failure can result in ductile flow, instead that brittle failure (e.g., RANALLI, 1995). Moreover, dynamic fluid viscosity also strongly depends on temperature, through the Arrhenious Law (also adopted in cases where fluid is represented by melted silicates; see FIALKO and KHAZAN, 2005 and references therein):

$$\eta_{fluid} = K_0 e^{\frac{E_a}{RT}} \tag{19}$$

where E_a is the activation energy, R is the universal gas constant ($E_a/R \approx 3 \times 10^4$ K) and K_0 is a pre–exponential reference factor (which in some cases also might be temperature–dependent; typically is $K_0 = 1.7 \times 10^3$ Pa s for basalts and $K_0 = 2.5 \times 10^6$ Pa s for granites). This explicit dependence on the absolute temperature T will add an implicit dependence on temperature in the hydraulic diffusivity (see equation (10)), which in turn strongly affects the pore pressure evolution, as mentioned in section 3.2. As for an example, for a temperature change from 100°C to 300°C we might expect a variation in η_{fluid} of about – 83%, which in turn translates in to a dramatic change in ω (nearly 500% for typical parameters; see Table 1 in BIZZARRI and COCCO, 2006a). We note that such a continuous variation in ω can be easily incorporated in a thermal pressurization model, simply by coupling (19) with (10) and (13).

Furthermore, fluid compressibility increases with increasing temperature (WIBBERLEY, 2002) and this causes additional temporal variations in hydraulic diffusivity. Finally, also thermal conductivity changes with absolute temperature, according to the following equation:

$$\kappa = K_1 + \frac{K_2}{T + 77} \tag{20}$$

where $K_1 = 1.18$ J/(kg K) and $K_2 = 474$ J/kg (CLAUSER and HUENGES, 2000). For the parameters used in BIZZARRI and COCCO (2006a), equation (20) implies a decrease of 16%

[3] For instance, a temperature change of 200°C can lead to a variation of the order of 30% in fluid density (since $\frac{d\rho_{fluid}}{\rho_{fluid}} = -\alpha_{fluid} dT$).

in thermal diffusivity χ for a temperature rise from 100°C to 300°C and this will directly affect the evolution of the pore fluid pressure (see again equation (13)).

3.5.2 Chemical environment changes. It is known that fault friction can be influenced also by chemical environment changes. Without any clear observational evidence, OHNAKA (1996) assumed that the chemical effects of pore fluids (as that of all other physical observables), are negligible compared to that of fault slip. On the contrary, chemical analyses of gouge particles formed in high velocity laboratory experiments by HIROSE and BYSTRICKY (2007) showed that dehydration reactions (i.e., the release of structural water in serpentine) can take place. Moreover, recent experiments on Carrara marble performed by HAN *et al.* (2007) using a rotary–shear, high–velocity friction apparatus demonstrated that thermally activated decomposition of calcite (into lime and CO_2 gas) occurs from a very early stage of slip, in the same temporal scale as the ongoing and enhanced fault weakening. Thermal decomposition weakening may be a widespread chemico–physical process, since natural gouges commonly are known to contain sheet silicate minerals. The latter can decompose, even at lower temperatures than that for calcite decomposition, and can leave geological signatures of seismic slip (HAN *et al.*, 2007), different from pseudotachylytes. Presently, there are no earthquake models in which chemical effects are incorporated within a governing equation. We believe that efforts will be directed to this goal in the near future.

4. Porosity and Permeability Changes

4.1. Porosity evolution

As pointed out by BIZZARRI and COCCO (2006a, b), values of permeability (k), porosity (Φ) and hydraulic diffusivity (ω) play a fundamental role in controlling the fluid migration and the breakdown processes on a seismogenic fault. During an earthquake event frictional sliding tends to open (or dilate) cracks and pore spaces (leading to a decrease in pore fluid pressure), while normal traction tends to close (or compact) cracks (therefore leading to a pore fluid pressure increase). Stress readjustment on the fault can also switch from ineffective porosity (i.e., closed, or non–connected, pores) to effective porosity (i.e., catenary pores), or *vice versa*. Both ductile compaction and frictional dilatancy cause changes to k, Φ and therefore to ω. It is clear from equation (13) that this leads to variations to p_{fluid}^{wf}.

Starting from the theory of ductile compaction of MCKENZIE (1984) and assuming that the production rate of the failure cracks is proportional to the frictional strain rate and combining the effects of the ductile compaction, SLEEP (1997) introduced the following evolution equation for the porosity:

$$\frac{\mathrm{d}}{\mathrm{d}t}\phi = \frac{\upsilon\beta_{cp}\mu_*}{2w} - \frac{\sigma_n^{eff^n}}{C_\eta(\phi_{sat} - \phi)^m} \tag{21}$$

where β_{cp} is a dimensionless factor, C_η is a viscosity parameter with proper dimensions, n is the creep power law exponent and m is an exponent that includes effects of nonlinear rheology and percolation theory[4]. Equation (21) implies that porosity cannot exceed a saturation value ϕ_{sat}.

As noted by SLEEP and BLANPIED (1992), frictional dilatancy also is associated with the formation of new voids, as well as with the intact rock fracturing (i.e., with the formation of new tensile microcracks; see VERMILYE and SCHOLZ, 1988). In fact, it is widely accepted (e.g., OHNAKA, 2003) that earthquakes result in a complex mixture of frictional slip processes on pre–existing fault surfaces and shear fracture of initially intact rocks. This fracturing will cause a change in porosity; fluid within the fault zone drains into these created new open voids and consequently decreases the fluid pressure. The evolution law for the porosity associated with the new voids is (SLEEP, 1995):

$$\frac{\mathrm{d}}{\mathrm{d}t}\phi = \frac{\upsilon\beta_{ov}\mu}{2w}, \tag{22}$$

where the factor β_{ov} is the fraction of energy that creates the new open voids[5].

SLEEP (1997) also proposed the following relation that links the increase of porosity to the displacement: $\frac{\partial}{\partial u}\Phi = \frac{\Phi\alpha_{fluid}\tau}{2wc}$. This leads to an evolution law for porosity:

$$\frac{\mathrm{d}}{\mathrm{d}t}\Phi = \frac{\Phi\alpha_{fluid}\tau}{2wc} \tag{23}$$

Finally, SEGALL and RICE (1995) proposed two alternative relations for the evolution of Φ. The first mimics the evolution law for state variable in the Dieterich–Ruina model (BEELER *et al.*, 1994 and references therein):

$$\frac{d}{dt}\Phi(\xi_1,\zeta,\xi_3,t) = -\frac{\upsilon}{L_{SR}}\left[\Phi(\xi_1,\zeta,\xi_3,t) - \varepsilon_{SR}\ln\left(\frac{c_1\upsilon + c_2}{c_3\upsilon + 1}\right)\right], \tag{24}$$

where ε_{SR} ($\varepsilon_{SR} < 3.5 \times 10^{-4}$ from extrapolation to natural fault of SEGALL and RICE, 1995) and L_{SR} are two parameters representing the sensitivity to the state variable evolution (in the framework of rate– and state–dependent friction laws) and a characteristic length–scale, respectively, and $\{c_i\}_{i\,=\,1,2,3}$ are constants ensuring that Φ is in the range [0,1]. In principle, ε_{SR} can decrease with increasing effective normal stress, although present by we have no detailed information about this second–order effect.

The second model, following SLEEP (1995), postulates that Φ is an explicit function on the state variable Ψ:

[4] The exponent m in equation (19) can be approximated as (KRAJCINOVIC, 1993): $m \cong 2 + 0.85(n + 1)$.

[5] It is important to remark that in (22) an absolute variation of the porosity is involved, while in (21) a relative change is involved.

$$\Phi(\xi_1, \zeta, \xi_3, t) = \Phi_* - \varepsilon_{SR} \ln\left(\frac{\Psi v_*}{L_{SR}}\right). \tag{25}$$

Φ_* being a reference value, assumed to be homogeneous over the entire slipping zone thickness. Considering the latter equation, coupled with (13), and assuming as SEGALL and RICE (1995) that the scale lengths for the evolution of porosity and state variable are the same, BIZZARRI and COCCO (2006b) showed that even if the rupture shape, the dynamic stress drop and the final value of σ_n^{eff} remain unchanged w. r. to a corresponding simulated event in which a constant porosity was assumed, the weakening rate is not constant for increasing cumulative slip. Moreover, they showed that the equivalent slip–weakening distance becomes meaningless. This is clearly visible in Figure 4, where we compare the solutions of the thermal pressurization problem in cases of constant (black curve) and variable porosity (gray curve).

All the equations presented in this section clearly state that porosity evolution is concurrent with the breakdown processes, since it follows the evolution of principal variables involved in the problem (v, τ, σ_n^{eff}, Ψ). However, in spite of the above–mentioned profusion of analytical relations (see also SLEEP, 1999), porosity is one of the biggest unknowns in the fault structure and presently available evidence from the laboratory, and from geological observations as well, does not allow us to discriminate between different possibilities. Only numerical experiments performed by coupling one of the equations (21) to (25) with (13) can show the effects of different assumptions and suggest what is the most appropriate. Quantitative results will be reported at a later date and they will plausibly give useful indications for the design of laboratory experiments.

4.2. Permeability Changes

As mentioned above, changes in hydraulic diffusivity can be due not only to the time evolution of porosity, but also to variations of permeability. k is known to suffer large variations with type of rocks and their thermo–dynamical state (see for instance TURCOTTE and SCHUBERT, 1982) and moreover local variation of k has been inferred near the fault (JOURDE *et al.*, 2002). Several laboratory results (BRACE *et al.*, 1968; PRATT *et al.*, 1977; BRACE, 1978; HUENGES and WILL, 1989) supported the idea that k is an explicit function of σ_n^{eff}. A reasonable relation (RICE, 1992) is:

$$k = k_0 e^{-\frac{\sigma_n^{eff}}{\sigma_*}}, \tag{26}$$

where k_0 is the permeability at zero effective normal stress and σ_* is a constant (between 5 and 40 MPa). For rocks in the center of the Median Tectonic Line fault zone (Japan) WIBBERLEY and SHIMAMOTO (2005; their Figure 2a) found the same relation with $k_0 = 8.71 \times 10^{-21}$ m^2 and $\sigma_* = 30.67$ MPa. For typical changes in σ_n^{eff} expected during coseismic ruptures (see for instance Figures 3c and 5d of BIZZARRI and COCCO, 2006a) we can estimate an increase in k at least of a factor 2 within the temporal scale of the

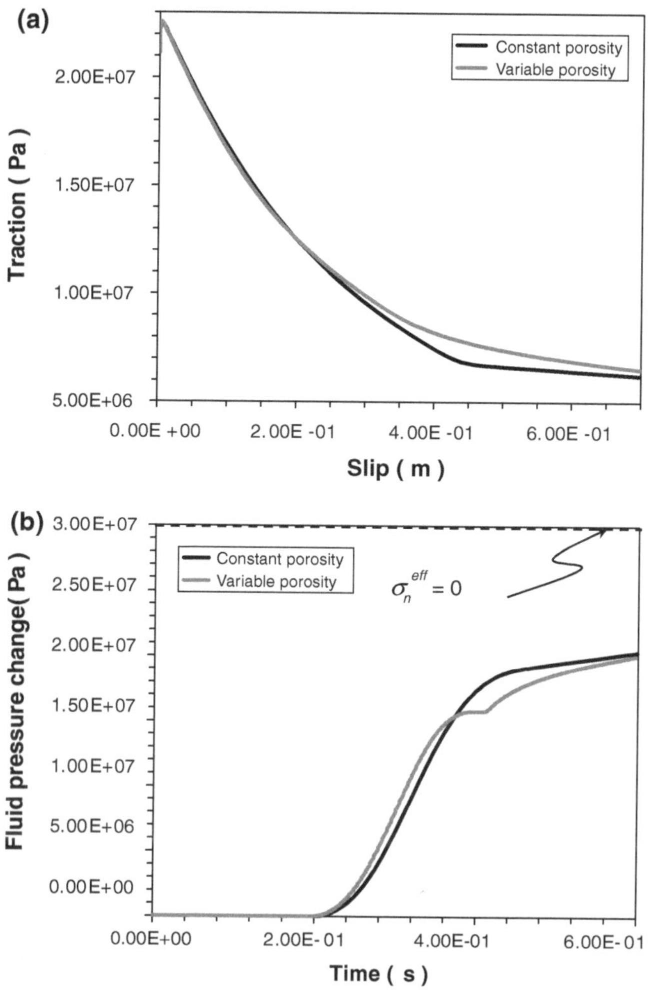

Figure 4
Comparison between solutions of the thermal pressurization problem in the case of constant (black curve) and variable porosity (as described by equation (25); gray curve). (a) Traction vs. slip curve, emphasizing that the equivalent slip–weakening distance becomes meaningless in the case of variable porosity. (b) Traction evolution of the effective normal stress. The relative minimum in σ_n^{eff} is the result of the competition between the two terms in equation (13), the thermal contribution and the porosity contribution. From Bizzarri and Cocco (2006b).

dynamic rupture. In principle, this can counterbalance the enhancement of instability due to the fluid migration out of the fault. This is particularly encouraging because seismological estimates of the stress release (almost ranging from about 1 to 10 MPa; Aki, 1972; Hanks, 1977) do not support the evidence of a nearly complete stress drop, as predicted by numerical experiments of thermal pressurization (ANDREWS, 2002; BIZZARRI and COCCO, 2006a, b).

Another complication may arise from the explicit dependence of permeability on porosity and on grain size d. Following one of the most widely accepted relation, the Kozeny–Carman equation (KOZENY, 1927; CARMAN, 1937, 1956), we have:

$$k = K_{KC}\frac{\Phi^3}{1-\Phi^2}d^2, \tag{27}$$

where $K_{KC} = 8.3 \times 10^{-3}$. Gouge particle refinement and temporal changes in Φ, such as that described in equations (21) to (25), affect the value of k.

As in the case of porosity evolution, permeability changes also occur during coseismic fault traction evolution and equations (26) or (27) can be easily incorporated in the thermal pressurization model (i.e., coupled with equation (12)).

5. *Gouge evolution and Gelation*

Numerous number of laboratory and geological studies on mature faults (among the others TULLIS and WEEKS, 1986) emphasized that, during sliding, a finite amount of wear is progressively generated by abrasion, fragmentation and pulverization of rocks (see also SAMMIS and BEN–ZION, 2007). Further slip causes comminution (or refinement) of existing gouge particles, leading to a net grain size reduction and finally to the slip localization onto discrete surface (see also section 2). There is also evidence that the instability of natural faults is controlled by the presence of granular wear products (see for instance MARONE *et al.*, 1990). POWER *et al.* (1988) suggested that natural rock surface roughness leads to a linear relationship between wear zone width and cumulative slip:

$$2w \cong 2K_{PEA}u \tag{28}$$

($K_{PEA} = 0.01$[6]), which in principle complicates the solution of the thermal pressurization problem, because it will insert an implicit temporal dependence in the slipping zone thickness. Temporal variations of $2w$ are also expected as a consequence of thermal erosion of the host rocks (e.g., JEFFREYS, 1942; MCKENZIE and BRUNE, 1972). More plausibly, in natural faults the variations in $2w$ described by equation (28) are appreciable if we consider the entire fault history, and not within the coseismic temporal scale. In equation (28) u has therefore to be regarded as the total fault slip after each earthquake event (u_{tot_n}); in light of this, equation (28) is in agreement with the empirical relation found by SCHOLZ (1987). The increasing value of this cumulative slip leads to a net increase in $2w$ and this further complicates the constraints on this important parameter.

Inversely, MARONE and KILGORE (1993) and MAIR and MARONE (1999) experimentally found that

[6] This value agrees with ROBERTSON (1983).

$$\frac{\Delta w}{\Delta Log(v_{load})} = K_{MK}, \tag{29}$$

where v_{load} is the loading velocity in the laboratory apparatus and K_{MK} is a constant depending on the applied normal stress. As an example, a step in v_{load} from 1 to 10 mm/s will raise $2w$ by 8 μm at $\sigma_n = 25$ MPa. Unfortunately, the extrapolation to natural fault conditions is not trivial.

Another interesting and non–thermal mechanism related to the gouge particle comminution is the silica gel formation (GOLDSBY and TULLIS, 2002), phenomenon which is likely restricted to faults embedded in silica–rich host rocks (e.g., granite). The water in pores, liberated through fracturing during sliding, can be absorbed by the fine particles of silica and wear debris produced by grain fragmentation and refinement, and this will cause the formation of moisture of silica gel (see for instance ILER, 1979). Net effects of the gouge gelation are pore–fluid pressure variation (due to water absorption) and mechanical lubrication of fault surface (due to the gel itself; see next section for more details). At the actual state of the art we do not have sufficient information to write equations describing the gouge evolution. Therefore additional investigations are needed, both in the laboratory and in the field, enabling inclusion of thermal erosion, grain evolution and gouge gelation within a constitutive model.

6. *Mechanical Lubrication and Gouge Acoustic Fluidization*

6.1. *Mechanical Lubrication*

An important effect of the presence of pore fluids within the fault structure is represented by the mechanical lubrication (SOMMERFELD, 1950; SPRAY, 1993; BRODSKY and KANAMORI, 2001; MA *et al.*, 2003). In the model of BRODSKY and KANAMORI (2001) an incompressible fluid obeying the Navier–Stokes equation's flows around the asperity contacts of the fault, without leakage, in the direction perpendicular to the fault surface[7]. In the absence of elastic deformations of the rough surfaces, the fluid pressure in the lubrication model is:

$$p_{fluid}^{(lub)}(\xi_1) = p_{res} + \frac{3}{2}\eta_{fluid}V\int_0^{\xi_1}\frac{w^* - w(\xi_1')}{(w(\xi_1'))^3}d\xi_1', \tag{30}$$

where p_{res} is the initial reservoir pressure (which can be identified with quantity p_{fluid}^{wf} of equation (13)), V is the relative velocity between the fault walls ($2v$ in our notation), $w^* \equiv w(\xi_1^*)$, where ξ_1^* is such that $\left.\frac{dp_{fluid}^{(lub)}}{d\xi_1}\right|_{\xi_1=\xi_1^*} = 0$, and ξ_1 maps the length of the lubricated

[7] This is a consequence (HAMROCK, 1994) of the fact that the lubricated zone is much wider than long.

zone $L^{(lub)}$[8]. Qualitatively, $L^{(lub)}$ is equal to the total cumulative fault slip u_{tot}. Interestingly, simple algebra illustrates that if the slipping zone thickness is constant along the strike direction, the lubrication pore fluid pressure is always equal to p_{res}.

The net result of the lubrication process is that the pore fluid pressure is reduced by an amount equal to the last member of equation (30). This in turn can be estimated as

$$P^{(\mathrm{lub})} \cong 12\eta_{fluid}\upsilon \frac{ru_{tot}^2}{(\langle 2w \rangle)^3} \tag{31}$$

where r is the aspect ratio constant for roughness ($r = 10^{-4}$–10^{-2}; POWER and TULLIS, 1991) and $\langle\ 2w\ \rangle$ is the average slipping zone thickness. Therefore equation (2) is rewritten as:

$$\sigma_n^{eff} = \sigma_n - p_{fluid}^{wf} - P^{(lub)}. \tag{32}$$

The fluid pressure can also adjust the fault surface geometry, since

$$2w(\xi_1) = 2w_0(\xi_1) + u^{(\mathrm{lub})}(\xi_1), \tag{33}$$

where $2w_0$ is the initial slipping zone profile and $u^{(lub)}$ is elasto–static displacement caused by lubrication (see equation (8) in BRODSKY and KANAMORI, 2001). Equation (33) can be approximated as

$$\langle 2w \rangle = \langle 2w_0 \rangle + \frac{P^{(\mathrm{lub})}L}{E} \tag{34}$$

E being the Young' s modulus. $u^{(lub)}$ is significant if $L^{(lub)}$ (or u_{tot}) is greater than a critical length, defined as (see also MA *et al.*, 2003):

$$L_c^{(\mathrm{lub})} = 2\langle 2w_0 \rangle \left(\frac{\langle 2w_0 \rangle E}{12\eta_{fluid}\upsilon r} \right)^{\frac{1}{3}}. \tag{35}$$

otherwise the slipping zone thickness does not widen. If $u_{tot} > L_c^{(lub)}$ then $P^{(lub)}$ is the positive real root of the following equation

$$P^{(\mathrm{lub})} \left(\langle 2w_0 \rangle + \frac{P^{(\mathrm{lub})} u_{tot}}{E} \right)^3 - 12\eta_{fluid}\upsilon r u_{tot}^2 = 0. \tag{36}$$

It is clear from equation (32) that lubrication contributes to reduce the fault traction (and therefore to increase the fault slip velocity, which in turn further increases $P^{(lub)}$, as stated in equation (31)). Moreover, if lubrication increases the slipping zone thickness, then it will reduce asperity collisions and the contact area between the asperities (which in turn will tend to decrease $P^{(lub)}$, as still stated in equation (31)).

It is generally assumed that when effective normal stress vanishes then material interpenetration and/or tensile (i.e., mode I) cracks (YAMASHITA, 2000;

[8] The length of the lubricated zone $L^{(lub)}$ is defined as the length over which $p_{fluid}^{(lub)}$ returns to p_{res} level.

DALGUER *et al.*, 2003) develop, leading to the superposition during an earthquake event of all three basic modes of fracture mechanics (ATKINSON, 1987; ANDERS and WILTSCHKO, 1994; PETIT and BARQUINS, 1988). An alternative mechanism that can occur when σ_n^{eff} falls to zero, if fluids are present in the fault zone, is that the frictional stress of contacting asperities described by the Amonton' s Law (1) becomes negligible w. r. to the viscous resistance of the fluid and the friction can be therefore expressed as

$$\tau = \frac{\langle 2w \rangle}{u_{tot}} P^{(\text{lub})} \tag{37}$$

which describes the fault friction in the hydrodynamic regime. Depending on the values of u_{tot} and v, in equation (37) $P^{(lub)}$ is alternatively expressed by (31) or by the solution of (36). For typical conditions ($\langle 2w_0 \rangle = 1$ mm, $E = 5 \times 10^4$ Pa, $v = 1$ m/s, $u_{tot} = 2$ m, $r = 10 \times 10^{-3}$ m), if the lubricant fluid is water ($\eta_{fluid} = 1 \times 10^{-3}$ Pa s), then $u_{tot} < L_c^{(lub)}$ and (from equation (31)) $P^{(lub)} \cong 4.8 \times 10^4$ Pa. Therefore the lubrication process is negligible in this case and the net effects of the fluid presence within the fault structure will result in thermal pressurization only. On the contrary, if the lubricant fluid is a slurry–formed form the mixture of water and refined gouge ($\eta_{fluid} = 10$ Pa s), then $u_{tot} > L_c^{(lub)}$ and (from equation (36)) $P^{(lub)} \cong 34.9$ MPa, which can be a significant fraction of tectonic loading σ_n. In this case hydro–dynamical lubrication can coexist with thermal pressurization: In a first stage of the rupture, characterized by the presence of ample aqueous fluids, fluids can be squeezed out of the slipping zone due to thermal effects. In a next stage of the rupture, the gouge, rich of particles, can form the slurry with the remaining water; at this moment thermal pressurization is not possible but lubrication effects will become paramount. This is an example of how two different physical mechanisms can be incorporated in a single frictional model.

6.2. *Gouge Acoustic Fluidization*

Another phenomenon which involves the gouge is its acoustic fluidization (MELOSH, 1979, 1996): acoustic waves (high–frequency vibrations) in incoherent rock debris can momentarily decrease the overburden pressure (i.e., normal stress) in some regions of the rocks, allowing sliding in the unloaded regions. It is known that vibrations can allow granular material to flow like liquids and acoustically fluidized debris behaves like a Newtonian fluid. From energy balance considerations, MELOSH (1979) established that acoustic fluidization will occur within a fluidized thickness

$$w_{af} \leq \frac{\tau e u_{tot} v_S^2}{K_M \rho_{rock} g^2 H^2}, \tag{38}$$

where e is the seismic efficiency (typically $e \approx 0.5$), v_S is the S–waves velocity, K_M is a constant ($K_M = 0.9$), ρ_{rock} the cubic mass density of rocks, g is the acceleration of gravity and H is the overburden thickness. For typical conditions, MELOSH (1996) found $w_{af} \leq 20$ m, which can be easily satisfied in mature faults (see also section 2). We notice that this phenomenon does not require the initial presence of fluid within the fault structure and can be regarded as an alternative w. r. to the above–mentioned processes of thermal pressurization in the attempt to explain the weakness of the faults. Unfortunately, it is difficult to detect evidence of fluidization of gouge in exhumed fault zones because signatures of this process generally are not preserved.

7. *Normal Stress Changes and Inelastic Deformations*

7.1. *Bi–material Interface*

Traditional and pioneering earthquake models (see for instance BRACE and BYERLEE, 1966; SAVAGE and WOOD, 1971) simply account for the reduction of the frictional coefficient from its static value to the kinetic frictional level, taking the effective normal stress constant over the duration of the process. Subsequently, WEERTMAN (1980) suggested that a reduction in σ_n during slip between dissimilar materials can influence the dynamic fault weakening. Considering an asperity failure occurring on a bi–material, planar interface separating two uniform, isotropic, elastic half spaces, HARRIS and DAY (1997; their equation (A12)) analytically demonstrated that σ_n can change in time. On the other hand, a material property contrast is not a rare phenomenon in natural faults: LI *et al.* (1990), HOUGH *et al.* (1994) and LI and VIDALE (1996) identified certain strike–slip faults in which one side is embedded in a narrow, fault parallel, low–velocity zone (the width of a few hundred meters). Simultaneously several authors (LEES, 1990; Michael and EBERHART–PHILLIPS, 1991; MAGISTRALE and SANDERS, 1995) inferred the occurrence of significant velocity contrasts across faults, generally less than 30% (e.g., TANIMOTO and SHELDRAKE, 2002).

Even if RENARDY (1992) and ADAMS (1995) theoretically demonstrated that Coulomb frictional sliding is unstable if it occurs between materials with different properties, there is not a general consensus regarding the importance of the presence of bi–material interface on natural earthquakes (BEN–ZION, 2006 vs. ANDREWS and HARRIS, 2005 and HARRIS and DAY, 2005). More recently, DUNHAM and RICE (2008) showed that spatially inhomogeneous slip between dissimilar materials alters σ_n^{eff} (with the relevant scale over which poroelastic properties are to be measured being of the order of the hydraulic diffusion length, which is mm to cm for large earthquakes). Moreover, it is known that the contrast in poroelastic properties (e.g., permeability) across faults can alter both σ_n and p_{fluid} (while the elastic mismatch influences only σ_n).

7.2. *Inelastic Deformations*

Calculations by POLIAKOV *et al.* (2002) and RICE *et al.* (2005) suggested that when the rupture tip passes by, the damage zone is inelatically deformed, even if the primary shear is confined to a thin zone (see Fig. 1). These inelastic deformations, that are a consequence of the high stress concentration near the tip, are likely interacting with the stress evolution and energy flow of the slipping process; therefore they are able to modify the magnitude of traction on the fault. In particular, TEMPLETON and RICE (2008) numerically showed that the accumulation of off–fault plastic straining can delay or even prevent the transition to super–shear rupture velocities. In his numerical simulations of a spontaneously growing 2–D, mode II crack having uniform stress drop, ANDREWS (2005) demonstrated that the energy absorbed off the fault is proportional to the thickness of the plastic strain zone, and therefore to rupture propagation distance[9]. He also showed that the energy loss in the damage zone contributes to the so–called fracture energy, which in turn determines the propagation velocity of a rupture. YAMASHITA (2000), employing a finite–difference method, and DALGUER *et al.* (2003), using a discrete element method, modelled the generation of off–fault damage as the formation of tensile cracks, which is in agreement with the high velocity friction experiments by HIROSE and BYSTRICKY (2007).

8. Discussion

8.1. The Prominent Effects of Temperature on Fault Weakening

In the previous sections we have emphasized that one of the most important effects on fault friction evolution is that of temperature, which might cause pressurization of pore fluids present in the fault structure (see section 3.2), flash heating (section 3.3), melting (see section 3.4), material property changes (see section 3.5.1) and chemical reactions (see section 3.5.2). Assuming that the microscopic processes controlling the direct effect of friction and its decay are thermally activated and follow an Arrhenius relationship, CHESTER (1994, 1995; see also BLANPIED *et al.*, 1995) directly incorporated the temperature dependence in the analytical expression of the governing law, proposing a rate–, state– and temperature–dependent version of the Dieterich' s Law (DIETERICH, 1979):

$$\tau = \left[\mu_* - a\ln\left(\frac{\upsilon_*}{\upsilon}\right) + \frac{E_a^{(a)}}{R}\left(\frac{1}{T} - \frac{1}{T_*}\right) + b\Psi\right]\sigma_n^{eff}$$
$$\frac{\mathrm{d}}{\mathrm{d}t}\Psi = -\frac{\upsilon}{L}\left[\Psi - \ln\left(\frac{\upsilon_*}{\upsilon}\right) + \frac{E_a^{(b)}}{R}\left(\frac{1}{T} - \frac{1}{T_*}\right)\right] \tag{39}$$

where $E_a^{(a)}$ and $E_a^{(b)}$ are the activation energies for the direct and evolution effect, respectively ($E_a^{(a)}$ = 60–154 kJ/mol; $E_a^{(b)}$ = 43–168 kJ/mol), and T_* is a reference (i.e., initial) temperature. As in the model of Linker and DIETERICH (1992), a variation in the

[9] In his model, the stress components are first incremented elastically through time. If Coulomb yield criterion is violated, then they are reduced by a time–dependent factor, accounting for Maxwellian visco–plasticity, so that the maximum shear stress resolved over all orientations is equal to the yield stress.

third independent variable (T in this case) will cause explicit and implicit (through the evolution equation for Ψ) changes to the fault friction. We also recall here that a direct effect is also controlled by temperature through the relation (NAKATANI, 2001)

$$a = \frac{k_B T_{ac}}{V_a h}, \tag{40}$$

where k_B is the Boltzmann constant ($k_B = 1.38 \times 10^{-23}$ J/K), T_{ac} is the absolute temperature of contacts, V_a is the activation volume ($V_a \sim 10^{-29}$ m^3) and h is the hardness ($h = A_m \sigma_n / A_{ac}$; $h \sim$ GPa), which in turn may decrease with T_{ac} due to plastic deformation of contacts. Additionally, nonuniform temperature distributions in contacting bodies can lead to thermo–elastic deformations, which in turn might modify the contact pressure and generate normal vibrations and dynamic instabilities, the so–called frictionally–excited thermo–elastic instability (TEI; e.g., BARBER, 1969).

8.2. Scaling Problems and Related Issues

It has been mentioned that each nonlinear dissipation process that can potentially act during an earthquake rupture has its own distance and time scales, which can be very different from one phenomenon to another. The difference in scale lengths, as well as the problem of the scale separation, can represent a limitation in the attempt to simultaneously incorporate *all* the mechanisms described in this paper in a single constitutive model. In the paper we have seen that thermal pressurization (section 3.2) can coexist with mechanical lubrication (section 6.1) as well as with porosity (section 4.1) and permeability evolutions (section 4.2). The same holds for flash heating and thermal pressurization (NODA *et al.*, 2009). This simultaneous incorporation ultimately leads to numerical problems, often severe, caused by the need to properly resolve the characteristic distances and times of each separate process. The concurrent increase in computational power and the development of new numerical algorithms can definitively assist us in this effort.

Table 2 reports a synoptic view of the characteristic length scales for the processes described in this paper. Two important lengths (see for instance BIZZARRI *et al.*, 2001), that are not negligible w. r. to the other scale lengths involved in the breakdown process, are the breakdown zone length (or size, X_b) and the breakdown zone time (or duration, T_b). They quantify the spatial extension, and time duration, of the cohesive zone, respectively; in other words they express the amount of cumulative fault slip, and the elapsed time, required to the friction to drop, in some way, from the yield stress down to the residual level.

Another open problem is related to the difficulty in moving from the scale of the laboratory (where samples are of the order of several meters) up to the scale of real faults (typically several kilometers long). Many phenomena described in the paper have been measured in the lab: This raises the problem of how to scale the values of the parameters of the inferred equations to natural faults. Both geological observations and improve-

Table 2

Synoptic view of the characteristic distances and scale lengths of the processes described in the paper

Process	Characteristic distance Scale length	Typical value range
Macroscopic decrease of fault traction from yield stress to residual level	d_0	~ few mm in the lab(a), (b)
	X_b	~ 100 of m(c)
Temporal evolution of the state variable in the framework of the rate– and state–dependent friction laws	L	~ few μm in the lab(b)
Thermal pressurization (section 3.2)	$2w$	$\leq$ 1 cm(d)
	$\delta = \sqrt{2\chi t_d}$	~ few cm
Flash heating (section 3.3)	D_{ac}	~ few μm
Gouge and rocks melting (section 3.4)	$2w_{melt}$	~ 100 of μm in the lab
Porosity evolution (section 4.1):		
– equations (18) to (20)	$2w$	$\leq$ 1 cm(d)
– equations (21) and (22)	L_{SR}	assumed to be equal to L
Gouge acoustic fluidization (section 6.2)	w_{af}(e)	20 m

(a) See also OHNAKA (2003)
(b) These estimates refer to an idealized fault of zero thickness (i.e., bare surfaces)
(c) Estimates from numerical 3–D fault models of BIZZARRI and COCCO (2005). In each fault node, the breakdown zone time, T_b, is associated to X_b through the local rupture velocity v_r via: $T_b = X_b/v_r$
(d) From geological estimates SUPPE (1985) and CHESTER and CHESTER (1998)
(e) Acoustic fluidization is effective for thicknesses lower than w_{af} (theoretical estimates by Melosh, 1996)

ments in laboratory machines are necessary ingredients in our understanding of earthquake source physics and our capability to reproduce it numerically.

9. Conclusions

Earthquake models can help us in the attempt to understand how a rupture starts to develop and propagate, how seismic waves travel in the Earth crust and how high frequency radiation can affect a site on the ground. Since analytical, closed–form analytical solutions of the fully dynamic, spontaneous rupture problem do not exist (even in homogeneous conditions), it is clear that accurate and realistic numerical simulations are the only available path. In addition to the development of robust and capable computer codes, the solution of the elasto-dynamic problem requires the introduction of a fault governing law, which ensures that the energy flux is bounded at the crack tip and prevents the presence of singularities of solutions at the rupture front. At the present state of the art, unfortunately, among the different possibilities presented in the literature, the most appropriate form of the physical law that governs fault during its seismic cycle is not agreed upon. This is particularly true for the traction evolution within the coseismic temporal scale, where the stress release and the consequent emission of seismic waves are realized.

In this paper we have presented and discussed a large number of physical mechanisms that can potentially take place during an earthquake event. These phenomena are macroscopic, in that fundamental variables (i.e., the physical observables) describing them have to be regarded as macroscopic averages (see also OHNAKA, 2003). As a result, the fault friction does not formally describe the stress acting on each single asperity, but the macroscopic average of the stress acting within the slipping zone. The same holds for the sliding velocity: v is the macroscopic slip rate and not the microscopic one, at solid–solid contacts. Unlikely, a link between the microphysics of materials, described in terms of lattice or atomic properties, and the macrophysical description of friction, obtained from stick–slip laboratory experiments, is missing. On the other hand, we cannot expect to be able to mathematically (either deterministically or statistically) describe the evolution of all the surface asperities and of all micro cracks in the damage zone.

Recent laboratory experiments and geological investigations have clearly shown that different dissipative processes can lead to the same steady state value of friction. In the simple approximation which considers only one single event on an isolated fault, some authors claim that the slip dependence is paramount (e.g., OHNAKA, 2003). On the other hand, the explicit dependence of friction on sliding velocity (e.g., DIETERICH, 1986) is unquestionable, even at high slip rates (e.g., TSUTSUMI and SHIMAMOTO, 1997). In fact, in the literature there is a major debate (see for instance BIZZARRI and COCCO, 2006d) concerning the most important dependence of fault friction. Actually, the problem of what is (are) the dominant physical mechanism(s) controlling the friction evolution (i.e., the quantitative estimate of the weights w_i in equation (3)) is still unsolved.

We believe that seismic data presently available are not sufficient to clarify what specific mechanism is operating (or dominant) during a specific earthquake event. The inferred traction evolution on the fault, as retrieved from seismological records (e.g., IDE and TAKEO, 1997; GUATTERI and SPUDICH, 2000), gives us only some general information about the average weakening process on an idealized mathematical fault plane. Moreover, it is affected by the unequivocal choice of the source time function adopted in kinematic inversions and by the frequency band limitation in data recording (SPUDICH and GUATTERI, 2004) and sometimes could be inconsistent with dynamic ruptures (PAGE *et al.*, 2005). On the other side, we have seen in previous sections that the insertion of additional physical and chemical mechanisms in the analytical expression of a fault governing equation cannot be neglected *a priori*. The first reason is that, compared to results obtained by adopting a simplified or idealized constitutive relation (where effective normal stress is assumed to be constant and where, for instance, only slip or slip rate dependence is postulated), numerical experiments from models in which additional physical mechanisms are accounted for show a significant, often dramatic, change in the dynamic stress drop (and therefore in the resulting ground motions), in the distance over which it is realized, in the so–called fracture energy and in the total scalar seismic moment. The second reason is that, as we have shown in the paper, the inclusion of different mechanisms in some cases requires a modification of its usual, or classical, analytical expression (recall the effects of gouge and rocks melting and those of hydrodynamic lubrication).

As a future perspective, it would be intriguing—and at the same time extraordinarily challenging—to try to compare synthetics, both on the fault and out of the fault, obtained by assuming that one particular physical mechanism is paramount w. r. to the others, in order to look for some characteristic signatures and specific features in the solutions and eventually attempt to envisage such features in real seismological data.

Moreover, we are inclined to think that only a multidisciplinary approach to source mechanics, which combines results from accurate theoretical models, advanced laboratory experiments, systematic field observations and data analyses, can eventually lead to the formulation of a realistic fault governing model and can assist us in the ambitious aim of describing as realistically as possible the complexity of the breakdown process during an earthquake rupture. In this paper we have underlined that some different, nonlinear, chemico–physical processes can potentially cooperate, interact, or even compete one with each other. We have shown that in most cases we are able to write equations describing them (sometimes with constrained parameters), and we have explicitly indicated how they can be incorporated into a fault governing model. In order to reproduce the complexity and the richness of the inelastic and dissipative mechanisms occurring on a fault surface during an earthquake event, a "classical" constitutive relation appears to be currently inadequate and the above–mentioned multidisciplinary approach would be painstakingly considered for years to come.

Acknowledgements

The Euro–Conference on Rock Physics and Geomechanics "Natural hazards: thermo–hydro–mechanical processes in rocks" held in Erice in September 2007 motivated this work and the Director of the School, Enzo Boschi, and the Directors of the course, Sergio Vinciguerra and Yves Bernabé, are kindly acknowledged. Moreover, I benefitted from insightful discussions with Renata Dmowska, Eric Dunham, Yann Klinger, Laurent Montési, Jim Rice, Paul Segall and Paul Spudich. Comments by Emily Brodsky concerning mechanical lubrication are also greatly appreciated. I lastly acknowledge Chris Marone and Massimo Cocco for their reviews.

References

Abercrombie, R. E. and Rice, J. R. (2005), *Can observations of earthquake scaling constrain slip weakening?*, Geophys. J. Int. *162*, 406–424.

Adams, G. G. (1995), *Self-excited oscillations of two elastic half–spaces sliding with a constant coefficient of friction*, J. Appl. Mech. *62*, 867–871.

Aki, K. (1972), *Earthquake mechanism*, Tectonophys. *13*, 423–446.

Aki, K. (1979), *Characterization of barriers on an earthquake fault*, J. Geophys. Res. *84*, B11, 6140–6148.

Anders, M. H. and Wiltschko, D. V. (1994), *Microfracturing, paleostress and the growth of faults*, J. Struct. Geol. *16*, 795–815.

Andrews, D. J. (1994), *Dynamic growth of mixed–mode shear cracks*, Bull. Seism. Soc. Am. *84*, 4, 1184–1198.

Andrews, D. J. (2002), *A fault constitutive relation accounting for thermal pressurization of pore fluid*, J. Geophys. Res. *107*, B12, 2363, doi: 10.1029/2002JB001942.

Andrews, D. J. (2005), *Rupture dynamics with energy loss outside the slip zone*, J. Geophys. Res. *110*, B01307, doi: 10.1029/2004JB003191.

Andrews, D. J. and Harris, R. A. (2005), *The wrinkle–like slip pulse is not important in earthquake dynamics, Geophys.* Res. Lett. *32*, L23303, doi:10.1029/2005GL023996.

Antonioli, A., Belardinelli, M. E., Bizzarri, A., and Vogfjord, K. S. (2006), *Evidence of instantaneous dynamic triggering during the seismic sequence of year 2000 in south Iceland*, J. Geophys. Res. *111*, B03302, doi: 10.1029/2005JB003935.

Atkinson, B. K. *Fracture Mechanics of Rock*, (Academic, San Diego, CA. 1987).

Bak, P., Tang, C., and Wiesenfled, K. (1987), *Self–organized criticality: An explanation of the 1/f noise*, Phys. Rev. Lett. *59*, 381–384.

Barber, J. R. (1969), *Thermoelastic instabilities in the sliding of conforming solids*, Proc. Roy. Soc. (London) *A312*, 381–394.

Beeler, N. M., Tullis, T. E. and Weeks, J. D. (1994), *The roles of time and displacement in the evolution effect in rock friction*, Geophys. Res. Lett. *21*, 18, 1987-1990.

Ben–Zion, Y. (2006), *A comment on "Material contrast does not predict earthquake rupture propagation direction" by R. A. Harris and S. M Day*. Geophys. Res. Lett. *33*, L13310, doi:10.1029/2005GL025652.

Ben–Zion, Y. and Sammis, C. G. (2003), *Characterization of fault zones*, Pure Appl. Geophys. *160*, 677–715.

Benioff, H. (1951), *Earthquakes and rock creep Part I: Creep characteristics of rocks and the origin of aftershocks*, Bull. Seism. Soc. Amer. *41*, 31–62.

Bhat, H. S., Dmowska, R., King, G. C. P., Klinger, Y., and Rice, J. R. (2007), *Off–fault damage patterns due to supershear ruptures with application to the 2001 M_w 8.1 Kokoxili (Kunlun) Tibet earthquake*, J Geophys Res. *112*, B06301, doi: 10.1029/2006JB004425.

Billi, A. and Storti, F. (2004), *Fractal distribution of particle size in carbonate cataclastic rocks from the core of a regional strike–slip fault zone*, Tectonophys. *384*, 115–128.

Bizzarri, A. and Belardinelli, M. E. (2008), *Modelling instantaneous dynamic triggering in a 3–D fault system: Application to the 2000 June South Iceland seismic sequence*, Geophys. J. Intl. *173*, 906–921, doi: 10.1111/j.1365-246X.2008.03765.x.

Bizzarri, A. and Cocco, M. (2004), *Thermal pressurization in 3–D dynamic spontaneous rupture models with cohesive zone*, EOS Trans. AGU, *85*, No. 47, Fall Meet. Suppl., Abstract T23A–0572.

Bizzarri, A. and Cocco, M. (2005), *3D dynamic simulations of spontaneous rupture propagation governed by different constitutive laws with rake rotation allowed*, Annals of Geophys. *48*, 2, 277–299.

Bizzarri, A. and Cocco, M. (2006a), *A thermal pressurization model for the spontaneous dynamic rupture propagation on a three–dimensional fault: 1. Methodological approach*, J. Geophys. Res. *111*, B05303, doi: 10.1029/2005JB003862.

Bizzarri, A. and Cocco, M. (2006b), *A thermal pressurization model for the spontaneous dynamic rupture propagation on a three–dimensional fault: 2. Traction evolution and dynamic parameters*, J. Geophys. Res. *111*, B05304, doi: 10.1029/2005JB003864.

Bizzarri, A. and Cocco, M. (2006c), *Correction to "A thermal pressurization model for the spontaneous dynamic rupture propagation on a three–dimensional fault: 1. Methodological approach"*, J. Geophys. Res. *111*, B11302, doi: 10.1029/2006JB004759.

Bizzarri, A. and Cocco, M. (2006d), *Comment on "Earthquake cycles and physical modeling of the process leading up to a large earthquake"*, Earth Planets Space *58*, No. 11, 1525–1528.

Bizzarri, A., Cocco, M., Andrews, D. J., and Boschi, E. (2001), *Solving the dynamic rupture problem with different numerical approaches and constitutive laws*, Geophys. J. Int. *144*, 656-678.

Bizzarri, A. and Spudich, P. (2008), *Effects of supershear rupture speed on the high-frequency content of S waves investigated using spontaneous dynamic rupture models and isochrone theory*, J. Geophys. Res. *113*, B05304, doi: 10.1029/2007JB005146.

Blanpied, M. L., Lockner, D. A., and Byerlee, J. D. (1995), *Frictional slip of granite at hydrothermal conditions*, J. Geophys. Res. *100*, B7, 13,045–13,064.

Boley, B. A. and Weiner, J. H. *Theory of Thermal Stresses* (Krieger, Malabar 1985).

BRACE, W. F. (1978), *A note on permeability changes in geologic material due to stress*, Pure Appl. Geophys. *116*, 627–633.

BRACE, W. F. and BYERLEE, J. D. (1966), *Stick–slip as a mechanism for earthquakes*, Science *153*, 3739, 990–992.

BRACE, W. F., WALSH, J. B., and FRANGOS, W. T. (1968), *Permeability of granite under high pressure*, J. Geophys. Res. *73*, 6, 2225–2236.

BRODSKY, E. E. and KANAMORI, H. (2001), *Elastohydrodynamic lubrication of faults*, J. Geophys. Res. *106*, B8, 16,357–16,374.

BYERLEE, J. D. (1978), *Friction of rocks*, Pure Appl. Geophys. *116*, 615–626.

Cardwell, R. K., Chinn, D. S., Moore, G. F., and Turcotte, D. L. (1978), *Frictional heating on a fault zone with finite thickness*, Geophys J. Roy. Astr. Soc. *52*, 525–530.

CARMAN, P. C. (1937), *Fluid flow through granular beds*, Trans. Inst. Chem. Eng. 15,150.

CARMAN, P.C. *Flow of Gases through Porous Media* (Butterworths Scientific Publications, London 1956).

CASHMAN, S. M., BALDWIN, J. N., CASHMAN, K. V., SWANSON, K., and CRAWFORD, K. (2007), *Microstructures developed by coseismic and aseismic faulting in near-surface sediments, San Andreas fault, California*, Geology *35*, 611–614, doi: 10.1130/G23545A.

CHESTER, F. M. (1994), *Effects of temperature on friction: Constitutive equations and experiments with quartz gouge*, J. Geophys. Res. *99*, 7247–7262.

CHESTER, F. M. (1995), *A rheologic model for wet crust applied to strike–slip faults*, J. Geophys. Res. *100*, B7, 13,033–13,044.

CHESTER, F. M. and CHESTER, J. S. (1998) *Ultracataclasite structure and friction processes of the Punchbowl fault, San Andreas system, California*, Tectonophys. *295*, 199–221.

CHESTER, F. M., CHESTER, J. S., KIRSCHNER, D. L., SCHULZ, S. E., and EVANS, J. P. *Structure of large-displacement, strike-slip fault zones in the brittle continental crust*. In *Rheology and Deformation in the Lithosphere at Continental Margins* (eds. G. D. Karner, B. Taylor, N. W. Driscoll, and D. L. Kohlstedt) pp. 223–260 (Columbia Univ. Press, New York 2004).

CLAUSER, C. and HUENGES, E. (2000), *Thermal conductivity of rocks and minerals*, In *Rock Physics and Phase Relations*, AGU Ref. Shelf, vol. 3 (ed. T. J. Ahrens), pp. 105–126 (AGU, Washington, D. C. 2000).

COCCO, M. and BIZZARRI, A. (2002), *On the slip–weakening behavior of rate– and state–dependent constitutive laws*, Geophys. Res. Lett. *29*, 11, doi: 10.1029/2001GL01399.

DALGUER, L. A., IRIKURA, K., and RIERA, J. D. (2003), *Simulation of tensile crack generation by three–dimensional dynamic shear rupture propagation during an earthquake*, J. Geophys. Res. *108*, B3, 2144, doi:10.1029/2001JB001738.

DAY, S. M. (1982), *Three-dimensional finite difference simulation of spontaneous rupture: the effect of nonuniform prestress*, Bull. Seism. Soc. Am. *72*, 1881–1902.

DAS, S. and AKI, K. (1977a), *A numerical study of two–dimensional spontaneous rupture propagation, Geophys*, J. Roy. Astr. Soc. *50*, 643–668.

DAS, S. and AKI, K. (1977b), *Fault plane with barriers: a versatile earthquake model*, J. Geophys. Res. *82*, 36, 5658–5670.

DIETERICH, J. H. (1979), *Modeling of rock friction: 1. Experimental results and constitutive equations*, J. Geophys. Res. *84*, 2161–2168.

DIETERICH, J. H. (1986), *A model for the nucleation of earthquake slip*, In *Earthquake Source Mechanics, Geophysical Monograph, 37, Maurice Ewing Series*, 6 (ed. S. DAS, J. BOATWRIGHT and H. SCHOLZ), (Am. Geophys. Union, Washington D. C. 1986) pp. 37–47.

DUNHAM, E. M. and RICE, J. R. (2008), *Earthquake slip between dissimilar poroelastic materials*, J. Geophys. Res. *113*, B09304, doi: 10.1029/2007JB005405.

ENGELDER, J. T. (1974), *Cataclasis and the generation of fault gouge*, Geol. Soc. Amer. Bull. *85*, 1515–1522.

EVANS, J. P. and CHESTER, F. M. (1995), *Fluid–rock interaction in faults of the San Andreas system: Inferences from San Gabriel fault rock geochemistry and microstructures*, J. Geophys. Res. *100*, 13,007–13,020.

FIALKO, Y. A. (2004), *Temperature fields generated by the elastodynamic propagation of shear cracks in the Earth*, J. Geophys. Res. *109*, No. B01303, doi: 10.1029/2003JB002497.

FIALKO, Y. and KHAZAN, Y. (2005), *Fusion by the earthquake fault friction: stick or slip?* J. Geophys. Res. *110*, B12407 doi.org/10.1029/2005JB003869.

Fliss, S., Bhat, H. S., Dmowska, R., and Rice, J. R. (2005), *Fault branching and rupture directivity*, J. Geophys. Res. *110*, B06312, doi: 10.1029/2004JB003368.

Goldsby, D. L. and Tullis, T. E. (2002), *Low frictional strength of quartz rocks at subseismic slip rates*, Geophys. Res. Lett. *29*, 17, doi: 10.1029/2002GL015240.

Guatteri, M. and Spudich, P. (2000), *What can strong-motion data tell us about slip-weakening fault-friction laws?* Bull. Seism. Soc. Am. *90*, 1, 98-116.

Hamrock, B. J. (1994), *Fundamentals of Fluid Film Lubrication* (McGraw Hill, New York 1994).

Han, R., Shimamoto, T., Hirose, T., Ree, J.–H., and Ando, J.–I. (2007), *Ultralow friction of carbonate faults caused by thermal decomposition*, Science *316*, 878–881.

Hanks, T. C. (1977), *Earthquake stress drop, ambient tectonic stresses, and stresses that drive plate motion*, Pure Appl. Geophys. *115*, 441–458.

Harris, R. A. and Day, S. M. (1997), *Effects of a low-velocity zone on a dynamic rupture*, Bull. Seismol. Soc. Am. *87*, 1267–1280.

Harris, R. A. and Day, S. M. (2005), *Material contrast does not predict earthquake rupture propagation direction*, Geophys. Res. Lett. *32*, L23301, doi:10.1029/2005GL023941.

Heermance, R., Shipton, Z. K., and Evans, J. P. (2003), *Fault structure control on fault slip and ground motion during the 1999 rupture of the Chelungpu Fault, Taiwan*, Bull. Seismol. Soc. Am. *93*, 3, 1034–1050.

Hirose, T. and Bystricky, M. (2007), *Extreme dynamic weakening of faults during dehydration by coseismic shear heating*, Geophys. Res. Lett. *34*, L14311, doi:10.1029/2007GL030049.

Hirose, T. and Shimamoto, T. (2003), *Fractal dimension of molten surfaces as a possible parameter to infer the slip–weakening distance of faults from natural pseudotachylytes*, J. Struct. Geol. *25*, 1569–1574.

Hough, S. E., Ben–Zion., Y., and Leary, P. (1994), *Fault–zone waves observed at southern Joshua Tree earthquake rupture zone*, Bull. Seism. Soc. Am. *84*, 761–767.

Huenges, E., and Will, G. *Permeability, bulk modulus and complex resistivity in crystalline rocks*, In *Fluids Movements – Element Transport and the Composition of the Deep Crust,* (Ed. D. Bridgwater) pp. 361–375, (Kluwed, Dordrecht 1989).

Ida, Y. (1972), *Cohesive force across the tip of a longitudinal-shear crack and Griffith' s specific surface energy*, J. Geophys. Res. *77*, 20, 3796–3805.

Ide, S. and Takeo, M. (1997), *Determination of constitutive relations of fault slip based on seismic wave analysis*, J. Geophys. Res. *102*, B12, 27,379-27,391.

Iler, R. K. *The Chemistry of Silica* (John Wiley and Sons, New York 1979), 866 pp

Jeffreys, H. (1942), *On the mechanics of faulting*, Geological Magazine *79*, 291–295.

Jourde, H., Flodin, E. A., Aydin, A., Durlofsky, L. J. and Wen, X. H. (2002), *Computing permeability of fault zones in eolian sandstone from outcrop measurements*, AAPG Bull. *86*, 1187–1200.

Kame, N., Rice, J. R., and Dmowska, R. (2003), *Effects of pre–stress state and rupture velocity on dynamic fault branching*, J. Geophys. Res. 108, B5, 2265, doi: 10.1029/2002JB002189.

Kanamori, H. and Stewart, G. S. (1976), *Mode of strain release along Gibbs fracture zone, Mid-Atlantic ridge,* Phys. Earth and Plan. Interiors, *11* (4), 312–332.

Kanamori, H. and Heaton, T. H. (2000), *Microscopic and macroscopic physics of earthquakes*, GeoComplexity and the Physics of Earthquakes–Geophysical Monograph *120*, 147–163.

Klinger, Y., Xu, X., Tapponnier, P., Van der Woerd, J., Laserre, C., and King, G. (2005), *High–resolution satellite imagery mapping of the surface rupture and slip distribution of the $M_w \sim 7.8$, November 14, 2001 Kokoxili earthquake* (Kunlun fault, Northern Tibet, China), Bull. Seis. Soc. Am. *95*, 5, 1970–1987.

Kozeny, J. (1927), *Uber kapillare Leitung des wassers in Böden*, Sitzungsber. Akad. Wiss. Wien Math. Naturwiss. Kl., Abt. 2a, 136,271–306 (In German).

Krajcinovic, D. (1993), *Scaling law for the blue cheese model of damaga*, Phys. Lett. A *173*, 433–437.

Lachenbruch, A. H. (1980), *Frictional heating, fluid pressure, and the resistance to fault motion*, J. Geophys. Res. *85*, B11, 6097–6122.

Lee, T. C. and Delaney, P. T. (1987), *Frictional heating and pore pressure rise due to a fault slip*, Geophys. J. Roy. Astr. Soc. *88*, 3, 569–591.

Lees, J. M. (1990), *Tomographic P–wave velocity images of the Loma Prieta earthquake asperity*, Geophys. Res. Lett. *17*, 1433–1436.

LI, Y.–G., LEARY, P., AKI, K., and MALIN, P. (1990), *Seismic trapped modes in the Oroville and San Andreas fault zones*, Science *249*, 763–766.

LI, Y.–G. and VIDALE, J. E. (1996), *Low–velocity fault–zone guided waves: numerical investigations of trapping efficiency*, Bull. Seism. Soc. Am. *86*, 371–378.

LINKER, M. F. and DIETERICH, J. H. (1992), *Effects of variable normal stress on rock friction: observations and constitutive equations*, J. Geophys. Res. *97*, B4, 4923-4940.

LOCKNER, D., NAKA, H., TANAKA, H., ITO, H., and IKEDA, R. (2000), *Permeability and strength of core samples from the Nojima fault of the 1995 Kobe earthquake*. In *Proc. International Workshop on the Nojima Fault Core and Borehole Data Analysis* (eds. H. Ito, K. Fujimoto, H. Tanaka, and D. Lockner), pp. 147–152, U.S. Geol. Surv. Open File Rep., 00–129.

MA, K.-F., BRODSKY, E. E., MORI, J., JI, C., SONG, T.-R. A., and KANAMORI, H. (2003), *Evidence for fault lubrication during the 1999 Chi–Chi, Taiwan, earthquake (M_w 7.6)*, G. Res. Lett. ***30***, 5, 1244, doi: 10.1029/2002GL015380.

MADARIAGA, R. (1979), *On the relation between seismic moment and stress drop in the presence of stress and strength heterogeneity*, J. Geophys. Res. *84*, B5, 2243–2250.

MAGISTRALE, H. and SANDERS, C. (1995), *P–wave image of the Peninsular Ranges batholith, southern California*, Geophys. Res. Lett. *22*, 2549–2552.

MAI, P. M. and BEROZA, G. C. (2002), *A spatial random field model to characterize complexity in earthquake slip*, J. Geophys. Res. *107*, B11, 2308, doi: 10.1029/2001JB000588.

MAIR, K. and MARONE, C. (1999), *Friction of simulated fault gauge for a wide range of slip velocities and normal stresses*, J. Geophys. Res. *104* (B12), 28, 899–28, 914.

MARONE, C. and KILGORE, B. (1993), *Scaling of the critical slip distance for seismic faulting with shear strain in fault zones*, Nature *362*, 618-621.

MARONE, C., RALEIGH, C. B., and SCHOLZ, C. H. (1990), *Frictional behavior and constitutive modeling of simulated fault gouge*, J. Geophys. Res. *95*, B5, 7007–7025.

MASE, C. W. and SMITH, L. (1985), *Pore–fluid pressures and frictional heating on a fault surface*, Pure Appl. Geophys. *92*, 6249–6272.

MASE, C. W. and SMITH, L. (1987), *Effects of frictional heating on the thermal, hydrologic, and mechanical response of a fault*, J. Geophys. Res. *92*, B7, 6249–6272.

MCKENZIE, D. P. (1984), *The generation and compaction of partially molten rock*, J. Petrol. *25*, 713–765.

MCKENZIE, D. and BRUNE, J. (1972), *Melting on fault planes during large earthquakes*, Geophys. J. R. Astron. Soc. *29*, 65–78.

MELOSH, H. J. (1979), *Acoustic fluidization: A new geologic process?* J. Geophys. Res. *84*, B13, 7513–7520.

MELOSH, H. J. (1996), *Dynamical weakening of faults by acoustic fluidization*, Nature *397*, 601–606.

MICHAEL, A. J. and EBERHART–PHILLIPS, D. (1991), *Relations among fault behavior, subsurface geology, and three–dimensional velocity models*, Science *253*, 651–654.

MILLER, S. A., NUR, A., and OLGAARD, D. L. (1996), *Earthquakes as a coupled shear stress–high pore pressure dynamical system*, Geophys. Res. Lett. *23*, 197–200.

MIZOGUCHI, K., HIROSE, T., SHIMAMOTO, T., and FUKUYAMA, E. (2007), *Reconstruction of seismic faulting by high–velocity friction experiments: An example of the 1995 Kobe earthquake*, Geophys. Res. Lett. *34*, L01308, doi: 10.1029/2006GL027931.

NAKATANI, M. (2001), *Conceptual and physical clarification of rate– and state–dependent friction law: Frictional sliding as a thermally activated rheology*, J. Geophys. Res. *106*, 13,347–13,380, doi: 10.1029/2000JB900453.

NIELSEN, S., DI TORO, G., HIROSE, T., and SHIMAMOTO, T. (2008), *Frictional melt and seismic slip*, J. Geophys. Res. *113*, B01308, doi: 10.1029/2007JB005122.

NODA, H., DUNHAM, E. M., and RICE, J. R. (2009), *Earthquake ruptures with thermal weakening and the operation of major faults at low overall stress levels*, submitted to J. Geophys. Res.

NUR, A. and BOOKER, J. (1972), *Aftershocks caused by pore fluid flow?* Science *175*, 885–887.

OHNAKA, M. (2003), *A constitutive scaling law and a unified comprehension for frictional slip failure, shear fracture of intact rocks, and earthquake rupture*, J. Geophys. Res. *108*, B2, 2080, doi: 10.1029/2000JB000123.

Ohnaka, M. (1996), *Nonuniformity of the constitutive law parameters for shear rupture and quasistatic nucleation to dynamic rupture: a physical model of earthquake generation processes*, Proc. Natl. Acad. Sci. USA, *93*, 3795–3802.

Okubo, P. G. (1989), *Dynamic rupture modeling with laboratory-derived constitutive relations*, J. Geophys. Res. *94*, B9, 12,321-12,335.

Page, M. T., Dunham, E. M., and Carlson, J. M. (2005), *Distinguishing barriers and asperities in near–source ground motion,* J. Geophys. Res. *110*, B11302, doi:10.1029/2005JB003736.

Parton, V. Z. and Morozov, E. M. *Elastic-Plastic Fracture Mechanics* (Mir Publishers Moscow, 1974), 427 pp.

Petit, J. P. and Barquins, M. (1988), *Can natural faults propagate under mode II conditions?* Tectonics *7*, 1243–1256.

Pittarello, L., Di Toro, G., Bizzarri, A., Pennacchioni, G., Hadizadeh, J., and Cocco, M. (2008), *Energy partitioning during seismic slip in pseudotachylyte–bearing faults (Gole Larghe Fault, Adamello, Italy)*, Earth Planet Sci. Lett. *269*, 131–139, doi: 10.1016/j.epsl.2008.01.052.

Poliakov, A. N. B., Dmowska, R., and Rice, J. R. (2002), *Dynamic shear rupture interactions with fault bends and off–axis secondary faulting*, J. Geophys. Res. *107*, doi:10.1029/2001JB000572.

Power, W. L. and Tullis, T. E. (1991), *Euclidean and fractal models for the description of rock surface roughness*, J. Geophys. Res. *93*, 415–424.

Power, W. L., Tullis, T. E., and Weeks, D. J. (1988), *Roughness and wear during brittle faulting*, J. Geophys. Res. *93*, B12, 15,268–15,278.

Prakash, V. and Yuan, F. (2004), *Results of a pilot study to investigate the feasibility of using new experimental techniques to measure sliding resistance at seismic slip rates*, EOS Trans. AGU *85*, 47, Fall Meet. Suppl., Abstract T21D-02.

Pratt, H. R., Black, A. D., Brace, W. F., and Swolfs, H. (1977), *Elastic and transport properties an in–situ jointed granite*, Int. J. Rock Mech. Min. Sci. *14*, 35–45.

Ranalli, G. *Rheology of the Earth*, 2nd Ed. (Chapman and Hall, London 1995).

Rathbun, A. P. and Marone, C. (2009), *Effect of strain localization on frictional behaviour of sheared granular materials*, submitted to J. Geophys. Res.

Reid, H. F. *The Mechanics of the Earthquake*, Rept. State Earthquake Inv. Comm., The California Earthquake of April 18, 1906 (Washington D. C. 1910).

Rempel, A. and Rice, J. (2006), *Thermal pressurization and on set of melting in fault zones*, J. Geophys. Res. *111*, B09314, doi:10.1029/2006JB004314.

Renardy, Y. (1992), *Ill–posedness at the boundary for elastic solids sliding under Coulomb–friction*, J. Elasticity *27*, 3, 281–287.

Rice, J. R. *Fault stress states, pore pressure distributions, and the weakness of the San Andreas Fault.* In *Fault Mechanics and Transport Properties in Rocks (the Brace Volume)*, (eds. by B. Evans and T.-F. Wong), (*Academic*, San Diego 1992) pp. 475–503.

Rice, J. R. (2006), *Heating and weakening of faults during earthquake slip*, J. Geophys. Res. *111*, B5, B05311, doi: 10.1029/2005JB004006.

Rice, J. R., Sammis, C. G. and Parsons, R. (2005), *Off–fault secondary failure induced by a dynamic slip–pulse*, Bull. Seism. Soc. Am. *95*, 1, 109–134, doi: 10.1785/0120030166, 2005.

Robertson, E. C. (1983), *Relationship of fault displacement to gouge and breccia thickness*, Mining Engin. *35*, 10, 1426–1432.

Ruina, A. L. (1983), *Slip instability and state variable friction laws*, J. Geophys. Res. *88*, B12, 10,359–10,370.

Sammis, C. G. and Ben–Zion, Y. (2007), *The mechanics of grain-size reduction in fault zones*, submitted to J. Geophys. Res.

Savage, J. C. and Wood, M. D. (1971), *The relation between apparent stress and stress drop*, Bull. Seismol. Soc. Am. *61*, 1381–1388.

Scholz, C. H. (1987), *Wear and gouge formation in brittle fracture*, Geology *15*, 493–495.

Scholz, C. H. and Hanks, T. C. *The strength of the San Andreas fault: A discussion.* In *Rheology and Deformation of the Lithosphere at Continental Margins* (eds. G. D. Karner, B. Taylor and D. Kohlstedt) (Columbia University Press, New York 2004).

Shapiro, S. A., Patzig, R., Rothert, E., and Rindshwentner, J. (2003), *Triggering of seismicity by pore–pressure perturbations: Permeability–related signature of the phenomenon*, Pure Appl. Geophys. *160*, 1051–1066.

Segall, P. and Rice, J. R. (1995), *Dilatancy, compaction, and slip instability of a fluid–infiltrated fault*, J. Geophys. Res. *100*, 101, 22,155–22,171.

Sibson, R. H. (1973), *Interaction between temperature and pore–fluid pressure during earthquake faulting – A mechanism for partial or total stress relief*, Nature *243*, 66–68.

Sibson, R. H. (1986), *Brecciation processes in fault zones: Inferences from earthquake rupturing*, Pure Appl. Geophys. *124*, 169–175.

Sibson, R. H. (2003), *Thickness of the seismic slip zone*, Bull. Seism. Soc. Am. *93*, 3, 1169–1178.

Sibson, R. and Toy, V. *The habitat of fault-generated pseudotachylyte: Presence vs. absence of friction melt*. In *Radiated Energy and the Physics of Earthquake Faulting* (eds. R. Abercrombie, A. McGarr, H. Kanamori, and G. Di Toro), (AGU, Washington, D. C. 2006).

Sleep, N. H. (1995), *Ductile creep, compaction, and rate and state dependent friction within major fault zones*, J. Geopys. Res. *100*, B7, 13,065–13,080.

Sleep, N. H. (1997), *Application of a unified rate and state friction theory to the mechanics of fault zones with strain localization*, J. Geophys. Res. *102*, B2, 2875-2895.

Sleep, N. H. (1999), *Rate– and state–dependent friction of intact rock and gouge*, J. Geophys. Res. *104*, B8, 17,847-17,855.

Sleep, N. H. and Blanpied, M. L. (1992), *Creep, compaction and the weak rheology of major faults*, Nature *359*, 687–692.

Sommerfeld, A. *Mechanics of Deformable Bodies* (Academic Press, San Diego, CA 1950).

Spray, J. G. (1993), *Viscosity determinations of some frictionally generated silicate melts: Implications for fault zone rheology at high strain rates*, J. Geophys. Res. *98*, 8053–8068.

Spray, J. (1995), *Pseudotachylyte controversy; fact or friction?* Geology *23*, 1119–1122.

Spudich, P. and Guatteri, M. (2004), *The effect of bandwidth limitations on the inference of earthquake slip–weakening distance from seismograms*, Bull. Sesim. Soc. Am. *94*, 6, 2028–2036.

Suppe, J. *Principles of Structural Geology* (Prentice–Hall, Englewood Cliffs, N. J. 1985) 537 pp.

Suzuki, T. and Yamashita, T. (2007), *Understanding of slip–weakening and –strengthening in a single framework of modeling and its seismological implications*, Geophys. Res. Lett. *34*, L13303, doi: 10.1029/2007GL030260.

Tanimoto, T. and Sheldrake, K. P. (2002), *Three–dimensional S–wave velocity structure in Southern California*, Geophys. Res. Lett. *29*, 8, 1223, doi:10.1029/2001GL013486.

Tchalenko, J. S. (1970), *Similarities between shear zones of different magnitudes*, Bull. Geol. Soc. Am. *81*, 1625–1640.

Templeton, E. L. and Rice, J. R. (2008), *Off–fault plasticity and earthquake rupture dynamics: 1. Dry materials or neglect of fluid pressure changes*, J. Geophys. Res. *113*, B09306, doi: 10.1029/2007JB005529.

Terzaghi, K., Peck, R. B., and Mesri, G. Soil Mechanics in Engineering Practice, 3rd ed. (John Wiley, Hoboken, N. J. 1996).

Tinti, E., Bizzarri, A., Piatanesi, A., and Cocco, M. (2004), *Estimates of slip weakening distance for different dynamic rupture models*, Geophys. Res. Lett. *31*, L02611, doi: 10.1029/2003GL018811.

Tsutsumi, A. and Shimamoto, T. (1997), *High–velocity frictional properties of gabbro*, Geophys. Res. Lett. *24*, pp.699–702.

Tullis, T. E. and Goldsby, D. L. (2003), *Flash melting of crustal rocks at almost seismic slip rates*, EOS Trans. AGU *84*, 46, Fall Meet. Suppl., Abstract S51B–05.

Tullis, T. E. and Weeks, J. D. (1986), *Constitutive behavior and stability of frictional sliding of granite*, Pure Appl. Geophys. *124*, 3, 383-414.

Turcotte, D. L. and Schubert, G. *Geodynamics* (John Wiley and Sons, New York, USA. 1982).

Vermilye, J. M. and Scholz, C. H. (1988), *The process zone: A microstructural view of fault growth*, J. Geophys. Res. *103*, 12,223–12,237.

Yamashita, T. (2000), *Generation of microcracks by dynamic shear rupture and its effects on rupture growth and elastic wave radiation*, Geophys. J. Int. *143*, 395– 406.

Weertman, J. (1980), *Unstable slippage across a fault that separates elastic media of different elastic constants*, J. Geophys. Res. *85*, B3, 1455-1461.

WIBBERLEY, C. A. J. (2002), *Hydraulic diffusivity of fault gauge zones and implications for thermal pressurization during seismic slip*, Earth Planets Space, *54* (11), 1153–1171.
WIBBERLEY, C. A. J. and SHIMAMOTO, T. (2005), *Earthquake slip-weakening and asperities explained by thermal pressurization*, Nature *436* (4), 689–792, doi: 10.1038/nature03901.
WILSON, J. E., CHESTER, J. S., and CHESTER, F. M. (2003), *Microfracture analysis of fault growth and wear processes, Punchbowl Fault, San Andreas system, California*, J. Struct. Geol. *25*, 1855–1873.

(Received March 27, 2008, revised October 3, 2008, accepted March 2, 2009)
Published Online First: May 23, 2009

Pure appl. geophys. 166 (2009) 777–799
0033–4553/09/050777–23
DOI 10.1007/s00024-004-0496-z

Pure and Applied Geophysics

Quake Catalogs from an Optical Monitoring of an Interfacial Crack Propagation

M. GROB,[1] J. SCHMITTBUHL,[1] R. TOUSSAINT,[1] L. RIVERA,[1] S. SANTUCCI,[2,3] and K. J. MÅLØY[3]

Abstract—Using an experimental setup which allows to follow optically the propagation of an interfacial crack front in a heterogeneous medium, we show that the fracture front dynamics is governed by local and irregular avalanches with large velocity fluctuations. Events defined as high velocity bursts are ranked in catalogs with analogous characteristics to seismicity catalogs: time of occurence, epicenter location and energy parameter (moment). Despite differences in the fracturing mode (opening for the experiments and shear rupture for earthquakes), in the acquisition mode and in the range of time scales, the distributions of moment and epicenter jumps in the experimental catalogs obey the same scaling laws with exponents similar to the corresponding distributions for earthquakes. The record-breaking event analysis also shows very strong similarities between experimental and real seismicity catalogs. The results suggest that the dynamics of crack propagation is controlled by the elastic interactions between microstructures within the material.

Key words: Crack front dynamics, asperity, microseismicity catalogs, analog model.

1. Introduction

In observational seismology, attempts to characterize earthquake ruptures have motivated numerous studies from the very early stages (REID, 1910). The description of small earthquakes is often reduced to a source-point description given by a set of simple parameters: time of occurrence, source location, seismic moment or magnitude, possibly focal mechanism (CLINTON *et al.*, 2006). Despite this crude information, organized in well-known seismic catalogs, these simple parameters have provided extremely useful earthquake features like the Gutenberg-Richter magnitude-frequency relationship (GUTENBERG and RICHTER, 1944). A complementary approach is to analyze the properties

[1] Institut de Physique du Globe de Strasbourg, UMR 7516 CNRS, EOST/Université de Strasbourg, 5 rue René Descartes, F-67084 Strasbourg Cedex, France.
[2] Physics of Geological Processes, University of Oslo, PO Box 1048, Blindern, N-0316 Oslo, Norway.
[3] Physics Institut, University of Oslo, PO Box 1048, Blindern N-0316 Oslo, Norway.

of interactions between earthquakes (BAK *et al.*, 2002; KAGAN, 2002; ZIV *et al.*, 2003; CORRAL, 2004; DAVIDSEN and PACZUSKI, 2005; BIZZARRI and BELARDINELLI, 2008). It has been shown for instance that the spatiotemporal evolution of earthquakes in a specific region exhibits a fractal pattern resulting from a dynamical process and expressing a hierarchical organization of events in space and time (MARSAN *et al.*, 2000; DAVIDSEN *et al.*, 2006).

Seismic inversions of slip history during large earthquakes have been an important step in the description of the rupture process (WALD *et al.*, 1991; BOUCHON *et al.*, 2000). The complete fault region associated to the earthquake is then considered. Slip history inversions provide important features of the rupture propagation: fault geometry, rupture length, rupture speed, time rise, rupture width, etc. However, they rely on strong assumptions like simple fault models for kinematic inversions (KIKUCHI and KANAMORI, 1991) or simple friction processes for dynamic inversion (IDE and TAKEO, 1997; FUKUYAMA *et al.*, 2003) and subsequently provide low resolution or speculative images of the rupture complexity. Statistics from these inversions however allow more precise statistical relationships such as the ones linking the seismic moment M_0 to the three geometric quantities characterizing the rupture: Its length L, its width W and the average slip on the fault u (KANAMORI and ANDERSON, 1975; WELLS and COPPERSMITH, 1994; MAI and BEROZA, 2000). In general, all these relationships can be approximated by a power law and exponents are found to be mostly universal (SCHOLZ, 2002; KAGAN, 1999). New tools to account for the full complexity of the earthquake rupture are required and under development (MAI and BEROZA, 2002; AOCHI and FUKUYAMA, 2002; TINTI *et al.*, 2005; RUBIN and AMPUERO, 2007).

Fracture complexity has also been addressed in the mechanical community, particularly since the pioneer work of MANDELBROT *et al.* (1984). In particular, scaling properties of fractures in heterogeneous materials were discovered (BOUCHAUD, 1997). For instance, the roughness of fracture surfaces in heterogeneous media exhibits a self-affine morphology. The associated roughness exponent is found to be very robust for different materials such as steel or aluminium alloy (MÅLØY *et al.*, 1992; BOUCHAUD *et al.*, 1993) or fault rocks (SCHMITTBUHL *et al.*, 1993, 1995). Experiments were also carried out to investigate the interactions between the crack front and the material heterogeneities (DAGUIER *et al.*, 1995). In a simplified two-dimensional configuration, studies of the crack front that is constrained geometrically in a plane, both theoretically (SCHMITTBUHL *et al.*, 2003a, 2003b) and experimentally (SCHMITTBUHL and MÅLØY, 1997; DELAPLACE *et al.*, 1999) were performed. The front line morphology is shown to exhibit scaling invariances (SCHMITTBUHL and MÅLØY, 1997; DELAPLACE *et al.*, 1999; SANTUCCI *et al.*, 2007). More recent studies were performed to describe the crack front dynamics (SCHMITTBUHL *et al.*, 2001; SANTUCCI *et al.*, 2006; MÅLØY *et al.*, 2006). These works show that the crack propagation is controlled by the pinning and depinning of the front owing to local asperities. The fracture front dynamics is then governed by local and irregular avalanches of varying sizes and velocities.

In this article, we propose a link between the fracture dynamics analyzed at the laboratory scale and the earthquake dynamics extracted from a subset of the Southern California catalog (SHLK catalog). The core of the study is the construction of laboratory-scale catalogs of events or quakes from an optical monitoring of an interfacial crack front propagation in a heterogeneous medium. Using a transparent material (Plexiglas), the experimental setup allows to follow visually at high spatial resolution and continuously in time, the rupture process. The first part of the paper describes the sample preparation and how the experimental setup works. The next section explains the image processing of the raw crack front pictures in order to obtain the parameters needed to build the quake catalogs made of several thousands of events. Since we aim at comparing our experimental results to natural seismicity data, we show in a following section the results of statistical tools typically applied to natural seismicity data when applied to the experimental data. Finally a discussion of the reliability of analogies or differences between experimental and natural data is proposed.

2. An Interfacial Rupture Experiment

2.1. Sample Preparation

The samples are built from two types of transparent polymethylmethacrylate (PMMA) plates: The main plate is 16-cm long, 14-cm large and 1-cm thick; the others are a set of 17-cm long, 0.4-cm thin and 1-cm wide bands. One surface of each plate is first sand-blasted with glass beads of diameter ø = 200 μm or 500 μm. The thick plate is then assembled to a set of four thin plates (see Fig. 1(a)), sand-blasted surfaces facing each other, and placed under a normal homogeneous load, so that the remaining air is expelled from the contact area when the PMMA sandwich is loaded. The loading frame for the sample preparation is made of two parallel aluminium plates between which the Plexiglas plates sandwich is placed. Twelve screws are used to exert a normal load on the sample (see Fig. 1(b)). The torque applied when tightening the screws is equal to 10 N.m.

To anneal the plate assembly, the whole block (sample + press) is put in an oven at 205°C during 30 minutes, which allows the Plexiglas plates to adhere. The thermal annealing process produces a cohesive interface along the former boundary between the two plates which is weaker than the bulk PMMA (the critical stress intensity factor K_c, also called fracture toughness, for the growth of our interfacial crack is around 40 $kPa.m^{1/2}$, more than 55 times smaller than in the bulk). Thus this sample preparation constrains the block to break along an interfacial fracture during mode I crack. Figure 1 summarizes the sample preparation procedure.

An important feature of the interface preparation is the sand-blasting procedure that induces a random topography on the PMMA surface to be annealed and accordingly controls the local toughness during annealing. Another consequence of the sand-blasting technique is that the transparency of the Plexiglas plates is lost: light scatters because of

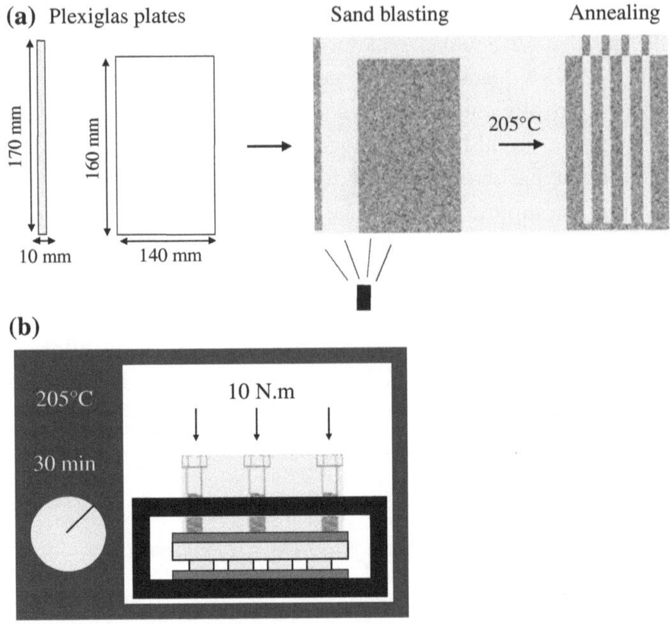

Figure 1
(a) Sample preparation procedure. Plexiglas plates are sand-blasted and sintered together to form a new single block. Sandblasting creates roughness fluctuations on the plate surface that control the local toughness during the annealing process. It also induces a loss of transparency by unpolishing the surfaces. Transparency is however recovered thanks to the sintering process. (b) Scheme of the loading system: In white, the PMMA sample; in gray, the two aluminium plates; in black, the solid frame with the screws to apply a torque $N =$ 10 N.m. The whole block is put in an oven at 205°C during 30 minutes.

the microstructures introduced by the sand-blasting. Nonetheless after annealing, the newly formed block recovers its transparency since contrast of the refraction index along the interface disappears.

2.2. *Setup*

The thick PMMA plate is clamped to a stiff aluminium frame. The thin plates are pulled apart thanks to a rod that presses on the excess length of the thin Plexiglas bands. A diagram of the setup is displayed in Figure 2. The loading translation stage is controlled by a stepping motor that applies a continuous descent of the rod at a speed in the range: 100–200 $\mu m.s^{-1}$ (in our configuration, the mean crack velocity is proportional to the speed of the rod motion). The normal displacement imposed to the lower Plexiglas plate induces a stable mode I crack propagation along the artificially introduced weak plane of the sample. The crack propagation is stable since the total energy of the system decreases with the crack advance, owing to the plate geometry (it would not be the case

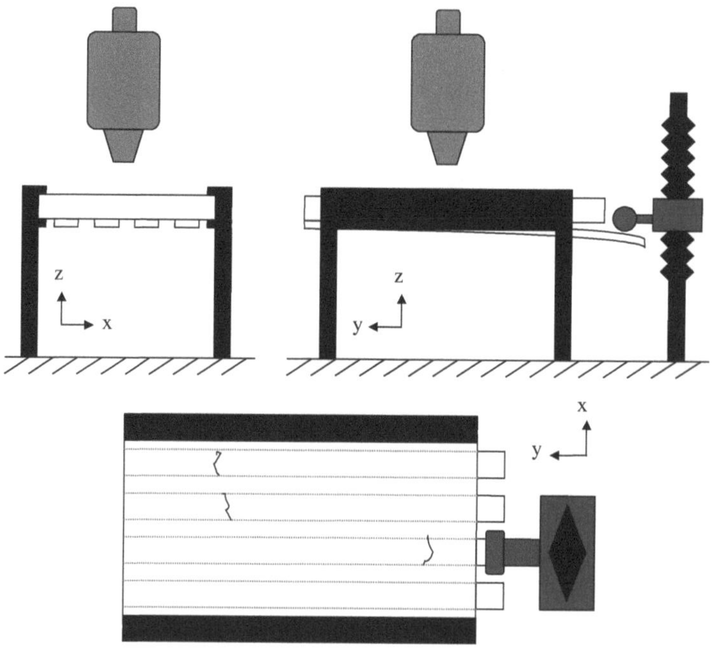

Figure 2
Sketch of the experimental setup (*top left:* Front view; *top right:* Side view; *bottom:* View from above). The thick plate of the sample (in white) is clamped into a rigid aluminium frame (in black). A cylindrical rod imposes a displacement of the thin Plexiglas band. The crack propagates in the y direction and is observed from above using a high speed camera mounted on a microscope (in light gray).

for a crack in an infinite elastic medium). Indeed the energy per unit of front length can be found from simple beam theory (LAWN, 1993):

$$U_E = \frac{Ed^3h^2}{8{c_r}^3}, \tag{1}$$

where E is the Young modulus, d the thickness of the lower plate, h the displacement of the rod and c_r the distance between the position of the crack and the point of application of the loading force. The energy release rate G, also called driving force, is a measure of the mechanical energy for an increment of crack extension, therefore the following expression for the energy release rate can be deduced:

$$G = -\frac{dU_E}{dc_r} = \frac{3Ed^3h^2}{8{c_r}^4}. \tag{2}$$

Equation (2) shows that the energy of the system decreases when the crack extension c_r increases. The fracture is accordingly controlled by the displacement of the loading rod. The loading value to reach for the front to start moving is of the order of a few tens of Newton.

The propagation of the crack front is followed optically with an optical microscope. Using a high-speed camera mounted on top of a microscope, up to 12288 consecutive images of the crack advance can be recorded with a spatial resolution of 1024×512 pixels (1 pixel $\approx 10 \times 10\ \mu m^2$) at an acquisition rate of 1000 frames per second (fps) which corresponds to an acquisition duration of 12 seconds. To provide sufficient light intensity at this small closure time (1 ms), the sample is lighted from below using optical fibers linked to an intense light source. These optical fibers contain lens elements at their end that allow the light source to focus on the crack front. The width of the image acquired by the fast camera is adjusted to the width of the PMMA band. After the opening of the sample, light rescatters in the fractured part of the sample because of the microstructures previously introduced on the PMMA plates. The open part of the sample looks opaque whereas the sintered region is still transparent. The transition between these two areas corresponds to the crack front (see Fig. 3(a)).

3. *Image Processing*

3.1. *Front Extraction*

Images of the interfacial fracture are divided in two parts: A bright zone which is the cracked and open area (microstructures along the interface scatter the light of the newly open surface) and a dark region which is still soldered and transparent. The image processing aims at extracting the shape of the transition zone between the bright and dark areas of the picture which corresponds to the crack front. Some white spots ahead of the fracture front appear on the image (see the dark part of the photo in Figure 3(a)). This is due to flaws created by the sand-blasting but located slightly inside the Plexiglas plates which scatter the light. In an attempt to remove these flaws in the fracture front images, a subtraction of a background picture taken before any opening is performed systematically. An example of the resulting image is displayed in Figure 3(b). Some flaws are still visible, however the contrast has been markedly enhanced. The gray level histogram of the image is clearly bimodal (Fig. 3(c)): The peaks represent the two different regions of the image. The picture is then clipped at the threshold level that separates these two lobes (Fig. 3(d)). The norm of the gradient of the segmented image is computed to search for the boundaries (Fig. 3(e)). Isolated clusters of pixels that might appear are removed by keeping only the largest cluster that percolates from the left to the right edge of the picture (Fig. 3(f)) to obtain the front path. More details about the last four steps of the image processing can be found in DELAPLACE *et al.* (1999). Figure 3(g) shows the extracted front superimposed on the raw image. A zoom on the preceding picture is done to show the resolution of the extracted front (cf. Fig. 3(h)). The absolute position of the front deduced from the images of the high-speed camera is accurate within 3 to 4 pixels (that is to say within 30 to 40 μm). Indeed the position of the front can move from 1 to 4 pixels if, for instance, the

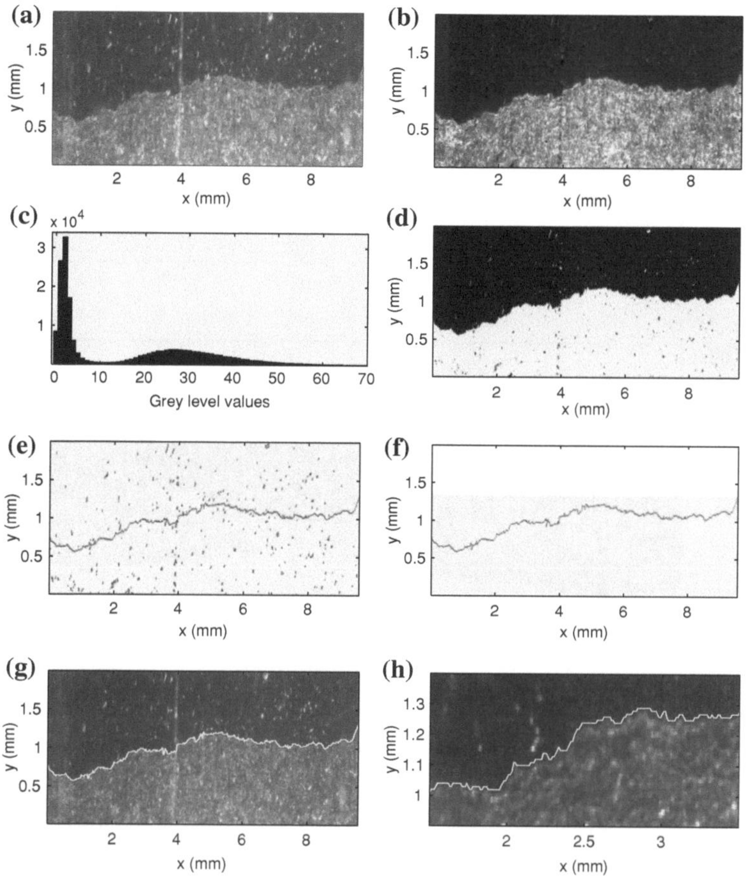

Figure 3
(a) Example of raw image taken with the high speed camera. (b) Example of the resulting picture of a subtraction between two raw images. (c) Gray level histogram of the preceding picture. (d) Image after thresholding. (e) Gradient of the clipped image. (f) Extraction of the percolating cluster from the gradient picture. (g) Extracted front (white line) superimposed on its corresponding raw image. (h) Zoom on the extracted front.

threshold level on the histogram is changed by ± 1, or if the front is defined by the maximum or minimum y position of pixels at a given x along the percolating front (the cluster that goes from one side to the other of the thresholded picture) (Delaplace *et al.*, 1999). The relative position along the front is however obtained at a much smaller precision (computed as the difference between the standard deviations of two extracted fronts with a shift in the threshold level) which is around 0.2 μm. Its irregular shape is explained by the fact that the front is pinned by local zones of high toughness during the propagation and becomes rough.

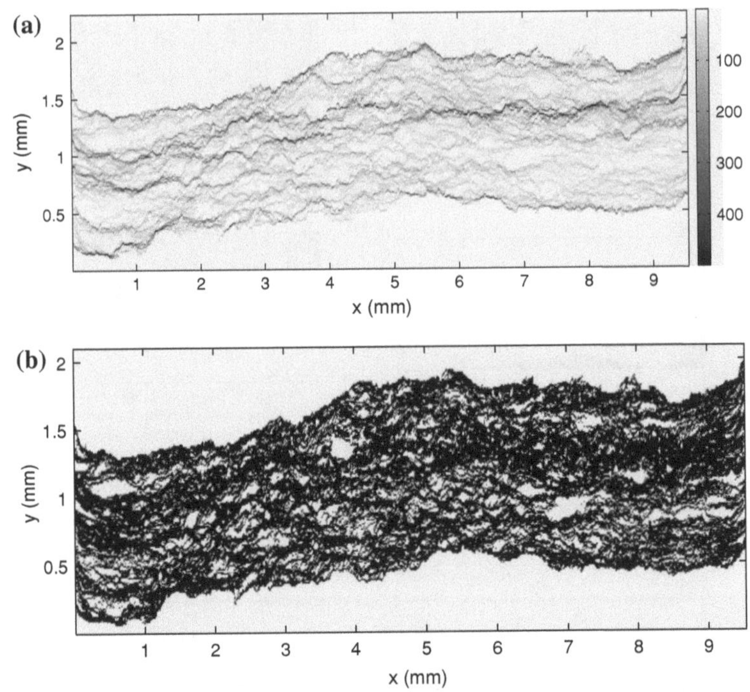

Figure 4
(a) Gray scale map of the waiting time matrix for one experiment deduced from 5908 fronts recorded at a rate of 1000 fps. The darker spots show the longer waiting times. (b) Spatial distribution of clusters (in white) for velocities 10 times larger than the average front speed (here $\langle V \rangle = 214\ \mu\text{ms}^{-1}$).

3.2. *Quake Definition*

In order to analyze the burst dynamics of the interfacial crack propagation and define rupture quakes, the fracture front lines extracted from image processing are added to form the waiting time matrix W (MÅLØY *et al.*, 2006, 2005). This matrix has the same size as the raw images and initial values equal to zero for all its elements. The value 1 is added to the matrix elements W_{ij} corresponding to pixels where a front line position (x, y) is detected. If several successive fronts are observed at the same particular position, the matrix elements W_{ij} at this position increases by one for each encountered front. This procedure was done for all frames of a given experiment to obtain the final waiting time matrix W (Fig. 4(a)). The waiting times estimate the amount of time during which the front was stuck at a precise position. The local speed of the crack front at the time when it passes a particular position (x, y) can be deduced from W by the inverse value of the corresponding matrix element W_{ij} multiplied by the ratio of the pixel linear size a (typically 10 μm) over the typical time δt between two pictures (0.001 s). The velocity matrix V is therefore defined as $V_{ij} = \frac{1}{W_{ij}} \frac{a}{\delta t}$, thus associating a local crack front speed to

Table 1

Table of parameters for the five experiments described in the article. ø is the diameter of the beads used for the sand-blasting of the sample. Nb stands for number

Experiment	ø(μm)	Duration (*s*)	$\langle V \rangle$ (μms^{-1})	*c* levels	Nb of clusters	Catalog
A	200	5.908	214	5	1990	A1
				10	2454	A2
				20	2583	A3
B	500	8.001	295	3	4756	B1
				7	6423	B2
				10	6321	B3
C	500	6.158	276	4	3736	C1
				10	4709	C2
				13	4623	C3
D	200	9.150	163	8	2290	D1
				11	2248	D2
				15	2168	D3
E	200	7.720	199	6	2995	E1
				14	3755	E2
				18	3827	E3

each pixel of the crack front in each image. It is important to mention that the image recording is performed so fast that there is basically no hole in the waiting time matrix, i.e., no element equal to zero (except below the first front, above the last one, and a few artefacts due to initial flaws in the sample).

In order to define events (or quakes), the velocity matrix V is clipped by setting its elements V_{ij} equal to 1 if $V_{ij} > c\langle V \rangle$ and 0 elsewhere, where c is a constant of the order of a few unities (MÅLØY *et al.*, 2006, 2005). Each region or cluster where the crack penetrates at a front speed larger than a few times the average front velocity is considered as an event or quake. Figure 4(b) is an example of the thresholded matrix with $c = 10$. The white clusters correspond to regions where the crack front speed is 10 times larger than the average front velocity $\langle V \rangle$. These zones are defined as quakes. The clusters connected to the first and last fronts, belonging to the upper and lower white zones in the figure, are excluded from the analysis.

3.3. *Quake Catalogs*

This data processing is applied to five experiments (named A, B, C, D and E) with different parameters. Table 1 summarizes the parameters for all experiments. Three different c values are used to threshold the velocity matrix corresponding to each experiment. We thus have fifteen data sets of events. The aim now is to build catalogs from these data sets that could be compared to real seismicity data catalogs. The most utilized quantities are: The time of occurrence, the location of hypocenter and an energy parameter of the event (seismic moment or magnitude).

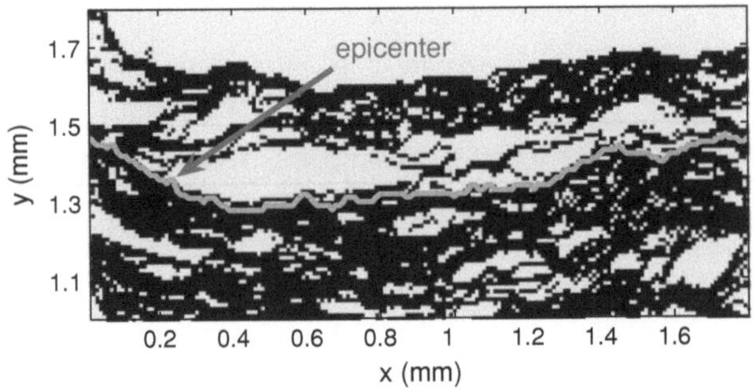

Figure 5
Determination of an epicenter for one event (white cluster) of the experimental catalog. The light gray line is the first front that enters the cluster. The arrow points to the pixel where the front touches the cluster, which defines the epicenter of the event.

For each event of each data set, the time of occurrence is found by searching for the first pixel in time that is attached to the considered cluster (Fig. 5). The time of occurrence t in ms is calculated considering the beginning of the experiment as the original time.

The epicenter of an event is given by the position (x, y) of this first pixel (Fig. 5). As the crack front propagation is constrained in a plane, the z coordinate is considered constant and equal to 0 in a first approximation. As the y axis is the direction of propagation of the crack front, a strong link exists between the time and the y coordinate of the epicenter. Hence the spatial position of the event is mainly given by the x coordinate.

The moment M of an event is given by the area of the event (which is the total number of pixels belonging to the cluster) multiplied by a characteristic opening which is assumed to be constant. The parameter M is then directly proportional to the area of the event. The three parameters t, the couple (x, y) and M, build up a quake catalog for our experiments. In total, fifteen quake catalogs named A1, A2, A3, B1, B2, B3, C1, C2, C3, D1, D2, D3, E1, E2 and E3, were created from the velocity matrix of five different experiments, each thresholded with three various level values (Table 1).

Figure 6 shows the locations of the epicenters of events superimposed on the thresholded velocity matrix for experiment A2. The diameter of the circles representing the quakes is proportional to the logarithm of the area of the clusters. Only events with $\log_{10}(M) > 1.5$ are displayed for a better visibility. The good agreement between the circles and the corresponding cluster contributes to validate the method used to build the catalogs. For a better readability, the logarithm of the moment M is denoted m in the following.

3.4. Quake Maps

As first qualitative analyses, some maps are drawn from the catalogs. Figure 7 shows the locations of epicenters for the three experimental catalogs from experiment A and for

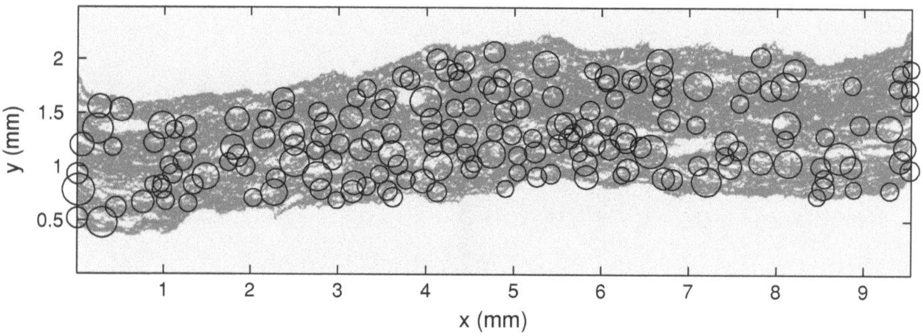

Figure 6
Epicenter localization superimposed on the thresholded velocity matrix for experiment A2. The diameters of the circles are proportional to the logarithm of the area of the clusters. Only events with $\log_{10}(M) > 1.5$ are shown for a better visibility.

one catalog from experiment B, as examples. The diameter of the circles is proportional to m. Differences can be seen between the three experimental catalogs generated from the same experiment concerning the size of the events and their localization. They are due to the removal of pixels when increasing the c level, thus dividing some large clusters into smaller ones. An effect of clustering can be observed on the maps: gaps appear in the distribution of the events.

The density of events can vary from one experiment to another. For instance, the density of events is higher for experiment B than for experiment A. This may be related to the velocity of propagation of the crack front which is larger for experiment B. The particle size used for sand-blasting the sample for experiment B is also larger than that for sample A. It can also affect the density of events by varying the toughness on the interface.

Another way of demonstrating a possible effect of clustering is to display events on a time versus x coordinate map. Figure 8 represents this spatio-temporal distribution of events for two catalogs (diameters of circles are proportional to m). The clustering effect is clearly visible on both maps: several events follow each other in time in a small zone along the x axis. This means that the advance of the crack front is very irregular and controlled by the ability of the front to go through a local region of high toughness. Indeed the front is stopped in some places because of local high toughness bounds on the interface. When it manages to penetrate the front can advance over some distance with a high velocity, which helps breaking other bounds before being stuck again.

4. *Comparison between Experimental Results and Real Seismicity Data*

4.1. *The SHLK Catalog*

To compare our experimental catalogs and real seismicity data, a subset of the Southern California catalog, the SHLK catalog (data available at http://www.data.scec.

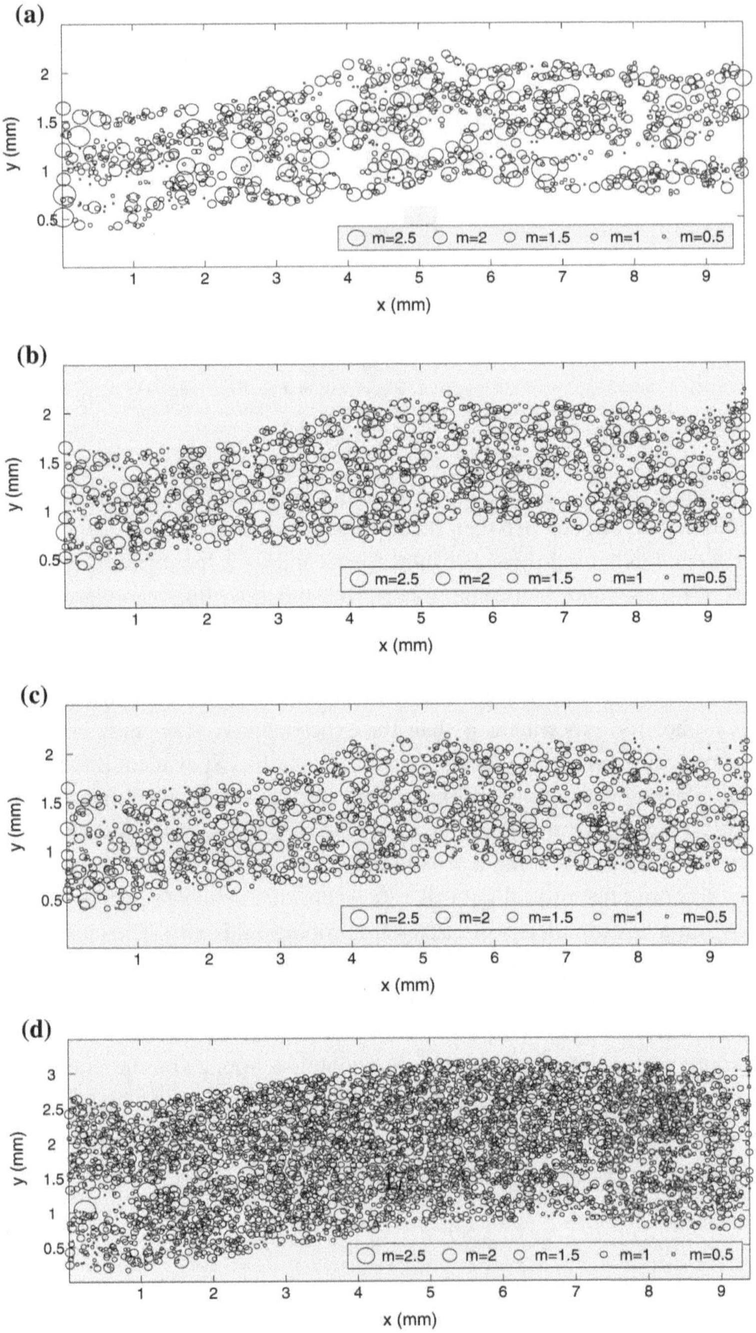
(a)
(b)
(c)
(d)
y (mm)
x (mm)
m=2.5
m=2
m=1.5
m=1
m=0.5

◄ Figure 7
Epicenters location of events for the three experimental catalogs from experiment A and the catalog B3 from experiment B. The diameters of the circles are proportional to the logarithm m of the area of the clusters. (a) Catalog A1. (b) Catalog A2. (c) Catalog A3. (d) Catalog B3.

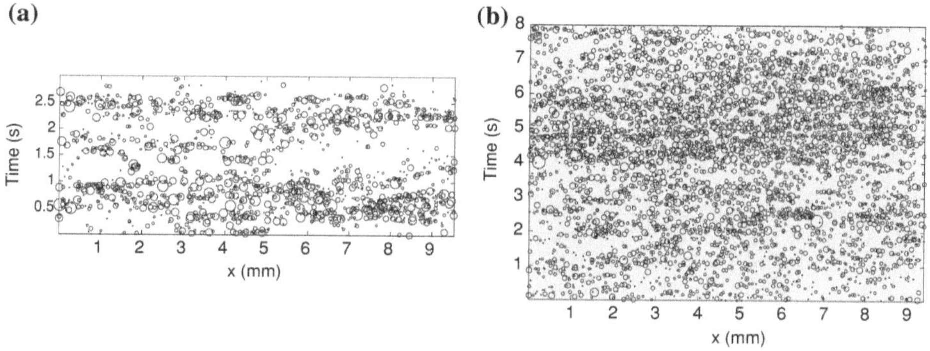

Figure 8
Spatio-temporal distribution of events for two experimental catalogs: (a) A2 and (b) B3. The diameter of the circles representing the events is proportional to the parameter m.

org/ftp/catalogs/SHLK/), is used. This catalog is built with earthquakes from a region of high seismicity. It is considered as complete for earthquakes with a magnitude larger than 2.5 (the distribution of magnitudes larger than 2.5 follow the Gutenberg-Richter relationship). The SHLK catalog is also homogeneous from January 1984 to December 2002: The seismometers in Southern California are spatially well distributed and worked continuously during this period. Davidsen and Paczuski (2005) and Davidsen *et al.* (2006) used it for their analyses and, as we wanted to compare their results to ours, we limited our own analysis to the same area (120.5°W,115.0°W) × (32.5°N,36.0°N), giving 22217 events to be considered. A detailed description of the catalog can be found in Shearer *et al.* (2005).

The exact date and time of occurrence of an earthquake are given with a precision reaching a thousandth of a second. The time of occurrence of events in the experimental catalogs is also precise up to 1 ms, although the real seismicity data catalog covers almost twenty years whereas the experiments only last a few seconds. The location of hypocenters is specified by three coordinates: latitude, longitude and depth. The configuration is clearly three-dimensional, on the contrary to the experiment in which the coordinates (x, y, z) simplify to the couple (x, y) because z is constant. Regarding the energy parameter, only the magnitude was available in the chosen catalog. The seismic moment (units in N.m) was calculated with the following relationship between seismic moment and magnitude:

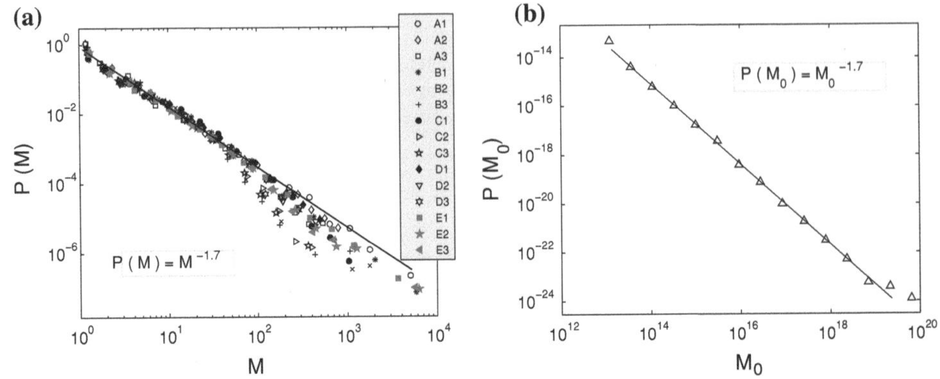

Figure 9
(a) Burst size M distributions $P(M)$ for the fifteen experimental catalogs. A fit on all the data for $0 \leq m \leq 2$ (black line) gives a slope of 1.7. A cut-off appears around $m = 2$ for large c values because of the lack of statistics for the largest clusters. (b) Seismic moment distribution $P(M_0)$ for a subset of the California seismicity catalog. The black curve shows a fit of the data with a slope equal to 1.7.

$$\log_{10}(M_0) = \frac{3}{2}M_w + 9.1, \tag{3}$$

where M_w is the so-called moment magnitude (KANAMORI, 1977) given in the catalog. The seismic moment is a quantity that combines the cracked area S with the amount of fault offset D and the shear modulus μ: $M_0 = \mu S D$. The analogy between the moment defined for the events in the experiments and the seismic moment is discussed in the last section of this paper.

4.2. Moment Distribution

Figure 9(a) shows the noncumulative quake-size distributions $P(M)$ for the fifteen experimental catalogs. A log binning was used to build the distributions which were also normalized by the total number of events for a better comparison of all the data. A robust result is found for all experiments and for the different values of c as all the data collapse. The power-law behavior of the distribution $P(M) \propto M^{-\gamma}$ proves that the burst dynamics occurs at all length scales. A fit of the data to a power-law function gives a slope equal to 1.7. A large scale cut-off appears for all the curves although at different m values, due to the lack of statistics. The exponent $\gamma = 1.7$ is comparable to exponents already found in earlier works for experiments based on the same mechanical setup over a wider range of experimental conditions (MÅLØY *et al.*, 2006). The authors in MÅLØY *et al.* (2006) used samples sand-blasted with smaller-sized beads than the ones used for the experiments described in this paper and the average front line speeds they found were also much smaller ($0.3 \mu m.s^{-1} \leq \langle v \rangle \leq 40\,\mu m.s^{-1}$). However, the fact that the value of the exponent determined for our new data is similar to the results of former experiments from MÅLØY *et al.* (2006) contributes to validate the technique.

As a first comparison between our experimental catalogs and real seismicity data, Figure 9(b) displays the noncumulative seismic moment distribution $P(M_0)$ of earthquakes taken from the SHLK catalog. The power-law behavior of $P(M_0)$ can be related to one well-known scaling law in seismology, the Gutenberg-Richter relationship (GUTENBERG and RICHTER, 1944) between seismic moment M_0 and frequency of earthquakes:

$$P(M_0) \propto M_0^{-1-\beta}, \tag{4}$$

where $P(M_0)$ is the frequency-moment distribution. The value β was found to equal 2/3 and proven to be very robust over different regions (GUTENBERG and RICHTER, 1944; SCHOLZ, 2002). Hence the exponent $\gamma = 1.7$ found for the experimental catalogs is similar to the exponent of the Gutenberg-Richter relationship (equation (4)) for earthquakes of the SHLK catalog.

4.3. Epicenter Jump Distribution

After having analyzed the size distribution of events, we focus on the spatial distances or "jumps" $\Delta r_i = \sqrt{(x_{i+1} - x_i)^2 + (y_{i+1} - y_i)^2}$ between successive events with epicenter coordinates (x_i, y_i) and (x_{i+1}, y_{i+1}). Following the work of DAVIDSEN and PACZUSKI (2005), the probability density function of the jumps $P_L(\Delta r)$ is measured for the fifteen experimental catalogs. The result can be seen in Figure 10. The straight line with a slope equal to 0.6 is displayed as a guide to the eye. A cut-off is reached around $r \approx L/2$ because of the finite spatial size of the considered areas. DAVIDSEN and PACZUSKI (2005) applied this method to characterize spatial clustering of earthquakes to the SHLK catalog. The distribution of jumps for the California catalog is superimposed to the experimental ones in the graph 10 (black triangle curve). It decays like the other curves with a trend also following a power law with an exponent $\lambda = 0.6$. Despite differences in the coordinates of hypocenters (the y coordinate in the experimental catalogs is closely related to the time and hypocenters of earthquakes are given for a 3-D configuration), a good correlation between the distributions of jumps for experimental events and real earthquakes can be observed.

4.4. Record-breaking Event Analysis

Another comparison between experimental and real data is conducted with a method developed by DAVIDSEN *et al.* (2006). They introduce the notion of record-breaking event. An event B is considered as a record with respect to an earthquake A if no event takes place in the spatial disc with radius AB centered on A during the time interval $[t_A, t_B]$ with $t_A < t_B$. Each record B is characterized by the distance $l = AB$ and the time interval (or waiting time) $T = t_B - t_A$ between the two earthquakes. Each event in the catalog has its own sequence of records (or recurrences). DAVIDSEN *et al.* (2006) tested this method on

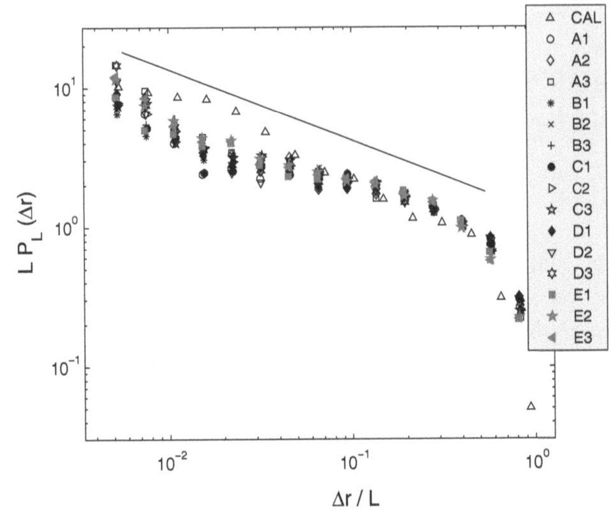

Figure 10
Distribution of jumps $P_L(\Delta r)$ for the fifteen experimental catalogs and a subset of the Southern California seismicity catalog rescaled by the longest size of the considered area L. Distances $\Delta r < 2$ km for the California catalog and $\Delta r < 40$ μm for the experimental catalogs have been discarded because of uncertainties in the locations of hypocenters. The solid black line has a slope $\lambda = 0.6$ and is shown as a guide to the eye.

the SHLK catalog. They found that the probability density functions of distances l and of recurrence times T both follow power laws spanning a few decades.

An analogy can be made here between our experiment and real seismology. Our mode I setup must to be compared to a mode II fault in friction: The y axis in the experiments is related to the time for the slipping fault (the y-coordinate value would represent a slip), while the x axis must be linked with space. The catalogs built from the experiments were processed in order to obtain a sequence of records for each cluster event but considering only the x coordinate as a position of the epicenters. Figure 11(a) shows the probability density function $P_x(l_x)$ of distances l_x of recurrent events for the different experiments. All distributions collapse and follow a power-law decay over several decades with an exponent $\delta \approx 1.1$. This value of the slope is close to the value found by DAVIDSEN *et al.* (2006) for the Southern California seismicity catalog (≈ 1.05). The probability density functions are independent of the experiments and the threshold level c proving the robustness of the process. The cut-off appearing between 4 and 5 mm is induced by the finite size of the system. The limit of linearity for the left side of the curves is due to location errors of the epicenters (distances below 40 μm, which corresponds to 4 pixels, are unreliable). The record-breaking event analysis was also applied to five experimental catalogs (A2, B1, C2, D1 and E3) considering only events with a moment parameter larger or equal to a certain value of m. Figure 11(b) shows the distributions $P_{x,\ m}(l_{x,\ m})$ resulting from this analysis. The

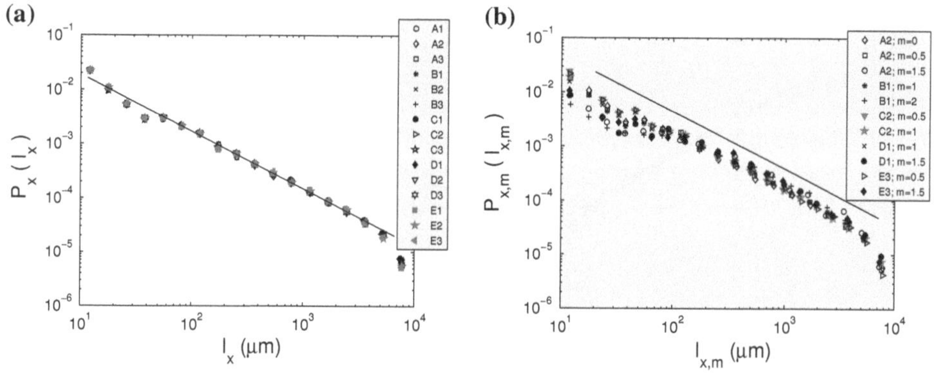

Figure 11
(a) Distribution $P_x(l_x)$ of distances l_x of recurrent events considering only the positions of epicenters on the x axis for the fifteen experimental catalogs. The straight line has a slope of 1.1. (b) Distribution $P_{x,\,m}(l_{x,\,m})$ of distances $l_{x,\,m}$ for five catalogs (A2, B1, C2, D1 and E3) varying the threshold moment m. Only events with moments greater or equal to m are considered. Data for $m = 0$ cover the whole catalog. The black line corresponds to a power law with an exponent equal to 1.1.

black straight line is a reminder of the power law with an exponent $\delta = 1.1$. The linear part of the curves for intermediate distances follows this power-law behavior. The cut-off at 4–5 mm is still observable but the peaks of the curves on the left side now vary slightly with the moment (the larger the moment threshold, the larger the minimum distance limit).

The distributions $P_x(T_x)$ for the time intervals T_x were also calculated for the record-breaking events obtained considering only the x position of the epicenters. They are shown in Figure 12(a). They obey a power law with an exponent $\alpha = 0.95$ when fitting the data for the intermediate values of recurrent times. Davidsen *et al.* (2006) found an exponent $\alpha \approx 0.9$, which is very close to our result. The variations from the general trend appearing for small T_x (between 3 and 4 ms) are due to inaccuracy concerning the time of occurrence of the events. The cut-off for large T_x indicates an upper limit of the waiting times. Figure 12(b) displays the distributions $P_{x,\,m}(T_{x,\,m})$ of recurrent times $T_{x,\,m}$ for five experimental catalogs (A2, B1, C2, D1 and E3) varying the moment threshold m. A collapse of the data is observed for intermediate $T_{x,\,m}$ values. The straight line has a slope of 0.95. Contrary to Davidsen *et al.* (2006) $P_{m,\,x}(T_{x,\,m})$ is not independent of m: The cut-off indicating a lower limit of the recurrent times varies with this threshold value. A fanning of the curves can be observed for large $T_{x,\,m}$ with the varying moment thresholds. For increasing m value, the range of the power-law regime decreases. The shift of the distributions from the power law of slope 0.95 for the experimental catalogs is probably due to the incompleteness of these catalogs. The lack of enormous events is closely related to the front propagation velocity or to the local variations of the toughness on the interface.

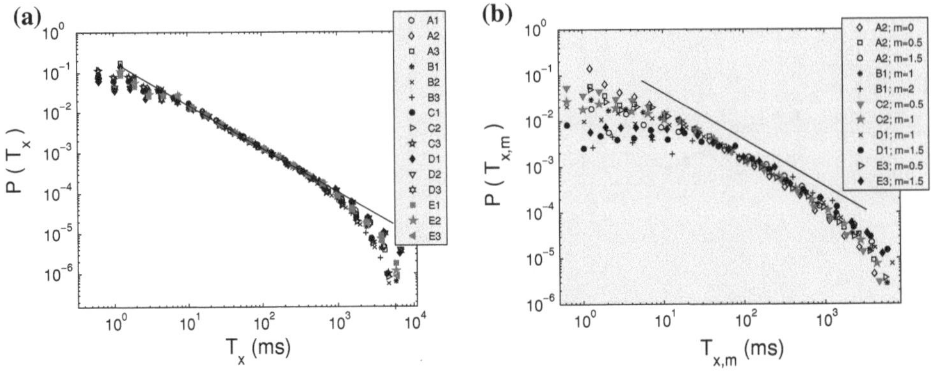

Figure 12

(a) Distribution $P_x(T_x)$ of recurrent time intervals T_x for the fifteen experimental catalogs when analyzed considering only the x-coordinate position of the events. The straight line has a slope $\approx$ 0.95. (b) Distribution $P_{x,m}(T_{x,m})$ of waiting times $T_{x,m}$ for five catalogs (A2, B1, C2, D1 and E3) varying the threshold moment m. Only events with moments larger or equal to m are taken into account. For $m = 0$ the whole catalog is involved. The black line corresponds to the power law with a slope of 0.95.

5. *Discussion*

In the previous section we compared the spatial and temporal analyses made on the experimental quake catalogs and those applied to seismicity catalogs. The distributions of size of quakes and epicenter jumps obey power laws such as the distributions of magnitude and spatial distances between epicenters of earthquakes. The record-breaking analysis of the experimental catalogs and that of the SHLK catalog also show strong analogies: The distributions of distances and waiting times between a quake and its recurrent events both follow a power-law decay. Furthermore the values of the scaling law exponents for the experimental catalogs and real seismicity data are quite close. A good data collapse can be observed between the five experiments performed on five samples with different original sand-blasting (200 μm and 500 μm bead size). No sign of a possible influence of the preparation difference can be found in the graphs. However, more experiments on samples sand-blasted with various-sized beads are needed to draw a significant conclusion on the influence of the size of disorder introduced over the crack propagation.

All these scaling properties illustrate that the dynamics of crack propagation at large scales is controlled by microstructures within the material. The good agreement between experimental and real seismicity catalogs tends to suggest that the global dynamics of rupture propagation depends on the interactions between events. These interactions are controlled by the long-range elastic coupling between heterogeneous microstructures (RAMANATHAN and FISHER, 1998; SCHMITTBUHL *et al.*, 2003a, 2003b). Indeed the modeling of SCHMITTBUHL *et al.* (2003b) for a quasistatic propagation of a crack front along a heterogeneous interface was compared successfully to experimental data of the same experiment presented in this paper. This fracture model describes the crack tip as a region of interactions between microcracks. Results from another modeling, based on an elastic contact line description (BONAMY *et al.*, 2008), were also compared with some experimental results presented here. Both modelings capture the role of the pinning and depinning of the crack front with long-range elastic coupling (RAMANATHAN and FISHER, 1998; SCHMITTBUHL *et al.*, 2003a, 2003b). These models have also been used to describe large-scale mode II ruptures (PERFETTINI *et al.*, 2001).

One main point of discussion for the comparison between the experimental catalogs and real seismicity data is the fracturing mode. Indeed an earthquake is modelled by a mode II or III shear rupture whereas the fracture experiment described in this paper is in mode I (opening). Attempts were already made to find analogies between mode I and mode II/III fracturation. GAO and RICE (1989) showed that mathematically the 2-D problem of an antiplane rupture along a frictional fault is governed by the same equations as the mode I perturbed crack-front problem. The numerical model of crack-front propagation of SCHMITTBUHL *et al.* (2003a), based on that of GAO and RICE (1989), was compared to former experiments based on the same mechanical setup described in this paper. Here the experimental data are directly compared to real seismicity data assuming an analogy between the moment for mode I and for mode II/III. The moment is directly proportional to the area since the opening is assumed to be constant (the elastic deformation gradient at the crack tip is presumed to be on a very small region, considering the size of an event). This is not the case for the seismic moment where the fault offset D varies.

As a perspective, the experiment presented here can be useful for the particular study of tensile cracks. Several authors (COX and SCHOLZ, 1988; PETIT and BARQUINS, 1988; YAMASHITA, 2000; DALGUER *et al.*, 2003) have shown through field observations, laboratory experiments or numerical simulations, that the three modes of rupture can coexist during an earhquake. They show that an array of microscopic tensile cracks initiate the rupture in the material or appear around a pre-existing macroscopically ruptured zone to form an initial damage zone. Under continued deformation increasingly more of these tensile cracks form until linkage occurs and the large-scale rupture then propagates in modes compatible with the global load possibly different from the microscopic mode I. Our experiment can be used to study the initiation and propagation of these mode I cracks at small scales that take place as a first stage during material breaking before fault growth (LEI *et al.*, 2000, 2004).

Another perspective for our experiment could be the study of properties of slow earthquakes (OBARA, 2002; ROGERS and DRAGERT, 2003; SCHWARTZ and ROKOSKY, 2007). Since our experiments involve not only dynamical events (*i.e.*, local unstable events which produce acoustic emission) but also very slow events compared to Rayleigh speed owing to the viscous rheology of the PMMA both in the bulk and along the contact plane, our experiment could be used as a paradigm for slow earthquakes. In this context it is of interest to note that we produce slow quakes without fluid presence contrasting to the first explanations for the slow earthquakes (SZELIGA *et al.*, 2004; MCCAUSLAND *et al.*, 2005;

SHELLY *et al.*, 2006; DOUBRE and PELTZER, 2007; ITO *et al.*, 2007). BRODSKY and MORI (2007) analyzed the static stress drop of slow and ordinary earthquakes, their ratio of slip to fault length. They conclude that the differences between dynamic and quasi-static rupture can explain the occurrence of ordinary earthquakes or slow ones. These authors write that the distinction between both types of earthquakes results from their ability to jump over rough patches on the fault plane, which corresponds exactly to the pinning and depinning of our experimental crack front. MELBOURNE *et al.* (2005) with inversion of GPS data for the 2003 Cascadia slow earthquake also found spatially localized non-uniform slip and evoked stress drop as a constraint for rupture mechanism.

In conclusion, the quake catalogs derived from the optical monitoring of a crack propagation experiment have strong similar scaling properties with real seismicity catalogs. This result yields a possible explanation of the rupture propagation mechanism: The rupture front is trapped by local asperities and its depinning involves local instabilities. Heterogeneities in the medium thus play a crucial part in the fracture propagation. The advantage of the experiment presented in this paper is the possibility to follow optically and continuously over an extended range of time scales, the propagation of the rupture thanks to the use of a transparent material, which can not be done for earthquake faults. Consequently analyzes of the experimental data offer good implications with which to address the question of earthquake mechanisms in complex media.

Acknowledgements

The authors would like to thank P. Srutarshi, G. Daniel and M. Bouchon for fruitful discussions regarding this project. The authors are also grateful to A. Bizzarri, S. Stanchits and S. Vinciguerra for their useful review of the manuscript. This work was supported by the programs ANR CATT MODALSIS and PICS "France/Norway".

REFERENCES

AOCHI, H. and FUKUYAMA, E. (2002), *Three-dimensional nonplanar simulation of the 1992 Landers earthquake*, J. Geophys. Res. *107*(B2), 2035.

BAK, P., CHRISTENSEN, K., DANON, L., and SCANLON, T. (2002), *Unified scaling law for earthquakes*, Phys. Rev. Lett. *88*(17), 178501.

BIZZARRI, A. and BELARDINELLI, M. (2008), *Modelling instantaneous dynamic triggering in a 3-D fault system: Application to the 2000 June South Iceland seismic sequence*, Geophys. J. Internat. *173*, 906–921.

BONAMY, D., SANTUCCI, S., and PONSON, L. (2008), *Crackling dynamics in material failure as the signature of a self-organized dynamic phase transition*, Phys. Rev. Lett. *101*, 045501.

BOUCHAUD, E. (1997), *Scaling properties of cracks*, J. Physics: Condensed Matter, *9*(21).

BOUCHAUD, E., LAPASSET, G., PLANÈS, J., and NAVEOS, S. (1993), *Statistics of branched fracture surfaces*, Phys. Rev. B *48*(5), 2917–2928.

BOUCHON, M., TOKSÖZ, N., KARABULUT, H., BOUIN, M.-P., DIETRICH, M., AKTAR, M., and EDIE, M. (2000), *Seismic imaging of the 1999 Izmit (Turkey) rupture inferred from the near-fault recordings*, Geophys. Res. Lett. *27*(18), 3013–1016.

BRODSKY, E. and MORI, J. (2007), *Creep events slip less than ordinary earthquakes*, Geophys. Res. Lett. *34*, L16309.

CLINTON, J., HAUKSSON, E., and SOLANKI, K. (2006), *An evaluation of the SCSN moment tensor solutions: robustness of the M_W magnitude scale, style of faulting and automation of the method*, Bull. Seismol. Soc. Am. *96*(5), 1689–1705.

CORRAL, A. (2004), *Long-term clustering, scaling and universality in the temporal occurrence of earthquakes*, Phys. Rev. Lett. *92*(10), 108501.

COX, S. and SCHOLZ, C. (1988), *Rupture initiation in shear fracture of rocks: An experimental study*, J. Geophys. Res. *93*(B4), 3307–3320.

DAGUIER, P., BOUCHAUD, E., and LAPASSET, G. (1995), *Roughness of a crack front pinned by microstructural obstacles*, Europhys. Lett. *30*(7), 367–372.

DALGUER, L., IRIKURA, K., and RIERA, J. (2003), *Simulation of tensile crack generation by three-dimensional dynamic shear rupture propagation during an earthquake*, J. Geophys. Res. *108*(B3), 2144–2167.

DAVIDSEN, J., GRASSBERGER, P., and PACZUSKI, M. (2006), *Earthquake recurrence as a record-breaking process*, Geophys. Res. Lett. *33*, L11304.

DAVIDSEN, J. and PACZUSKI, M. (2005), *Analysis of the spatial distribution between successive earthquakes*, Phys. Rev. Lett. *94*, 048501.

DELAPLACE, A., SCHMITTBUHL, J., and MÅLØY, K. (1999), *High resolution description of a crack front in a heterogeneous Plexiglas block*, Phys. Rev. Lett. *60*(2), 1337–1343.

DOUBRE, C. and PELTZER, G. (2007), *Fluid-controlled faulting process in the Asal Rift, Djibouti, from 8 yr of radar interferometry observations*, Geology *35*(1), 69–72.

FUKUYAMA, E., MIKUMO, T., and OLSEN, K. (2003), *Estimation of the critical slip-weakening distance: theoritical background*, Bull. Seismol. Soc. Am. *93*(4), 1835–1840.

GAO, H. and RICE, J. (1989), *A first-order perturbation analysis of crack trapping by arrays of obstacles*, ASME J. Appl. Mech. *56*, 828–836.

GUTENBERG, B. and RICHTER, C. (1944), *Frequency of earthquakes in California*, Bull. Seismol. Soc. Am. *34*, 185–188.

IDE, S. and TAKEO, M. (1997), *Determination of constitutive relations of fault slip based on seismic wave analysis*, J. Geophys. Res. *102*(B12), 27379–27391.

ITO, Y., OBARA, K., SHIOMI, K., SEKINE, S., and HIROSE, H. (2007), *Slow earthquakes coincident with episodic tremors and slow slip events*, Science *315*, 503–506.

KAGAN, Y. (1999), *Universality of the seismic moment-frequency relation*, Pure Appl. Geophys. *155*(2), 641–655.

KAGAN, Y. (2002), *Aftershock zone scaling*, Bull. Seismol. Soc. Am. *92*, 537–573.

KANAMORI, H. (1977), *The energy release in great earthquakes*, J. Geophys. Res. *82*(B20), 2981–2988.

KANAMORI, H. and ANDERSON, D. (1975), *Theoritical basis of some empirical relations in seismology*, Bull. Seismol. Soc. Am. *65*(5), 1073–1095.

KIKUCHI, M. and KANAMORI, H. (1991), *Inversion of complex body waves*, Bull. Seismol. Soc. Am. *81*(6), 2335–2350.

LAWN, B., *Fracture of Brittle Solids* (Cambridge Univ. Press 1993) second edition.

LEI, X., KUSUNOSE, K., RAO, M., NISHIZAWA, O., and SATOH, T. (2000), *Quasi-static fault growth and cracking in homogeneous brittle rock under triaxial compression using acoustic emission monitoring*, J. Geophys. Res. *105*(B3), 6127–6139.

LEI, X., MASUDA, K., NISHIZAWA, O., JOUNIAUX, L., LIU, L., MA, W., SATOH, T., and KUSUNOSE, K. (2004), *Detailed analysis of acoustic emission activity during catastrophic fracture of faults in rock*, J. Struct. Geol. *26*, 247–258.

MAI, P. and BEROZA, G. (2000), *Source scaling properties from finite-fault-rupture models*, Bullet. Seismol. Soc. Am. *90*(3), 604–615.

MAI, P. and BEROZA, G. (2002), *A spatial random field model to characterize complexity in earthquake slip*, J. Geophys. Res. *107*(B11), 2308.

MÅLØY, K., HANSEN, A., HINRICHSEN, E., and ROUX, S. (1992), *Experimental measurements of the roughness of brittle cracks*, Phys. Rev. Lett. *68*(2), 213–215.

MÅLØY, K., SANTUCCI, S., SCHMITTBUHL, J., and TOUSSAINT, R. (2006), *Local waiting time fluctuations along a randomly pinned crack front*, Phys. Rev. Lett. *96*(4).

MÅLØY, K., TOUSSAINT, R., and SCHMITTBUHL, J. (2005), *Dynamics and structure of interfacial crack fronts, Proc. 11th Internat. Congress on Fracture*, page 6.

MANDELBROT, B., PASSOJA, D., and PAULLAY, A. (1984), *Fractal character of fracture surfaces of metal*, Nature *308*, 721–722.

MARSAN, D., BEAN, C., STEACY, S., and MCCLOSKEY, J. (2000), *Observation of diffusion processes in earthquake populations and implications for the predictability of seismicity systems*, J. Geophys. Res. *105*(B12), 28081–28094.

MCCAUSLAND, W., MALONE, S., and JOHNSON, D. (2005), *Temporal and spatial occurrence of deep non-volcanic tremor: from Washington to northern California*, Geophys. Res. Lett. *32*, L24311.

MELBOURNE, T., SZELIGA, W., MILLER, M., and SANTILLAN, V. (2005), *Extent and duration of the 2003 Cascadia slow earthquake*, Geophys. Res. Lett. *32*, L04301.

OBARA, K. (2002), *Nonvolcanic deep tremor associated with subduction in southwest Japan*, Science *296*, 1679–1681.

PERFETTINI, H., SCHMITTBUHL, J., and VILOTTE, J.-P. (2001), *Slip correlations on a creeping fault*, Geophys. Res. Lett. *28*(10), 2137–2140.

PETIT, J.-P. and BARQUINS, M. (1988), *Can natural faults propagate under mode ii conditions?* Tectonics *7*(6), 1243–1256.

RAMANATHAN, S. and FISHER, D. (1998), *Onset of propagation of planar cracks in heterogeneous media*, Phys. Rev. B *58*(10), 6026–6046.

REID, H., *The California earthquake of April 18, 1906*. In *Report of the State Earthquake Investigation Committee*, vol. II (Carnegie Institution of Washington 1910).

ROGERS, G. and DRAGERT, H. (2003), *Episodic tremor and slip on The Cascadia subduction zone: The chatter of silent slip*, Science *300*, 1942–1943.

RUBIN, A. and AMPUERO, J.-P. (2007), *Aftershock asymmetry on a bimaterial interface*, J. Geophys. Res. *112*, B05307.

SANTUCCI, S., MÅLØY, K., TOUSSAINT, R., and SCHMITTBUHL, J., *Self-affine scaling during interfacial crack front propagation*. In *Proc. Dynamics of Complex Interconnected Systems: Networks and Bioprocesses*, NATO ASI Geilo (Springer 2006).

SANTUCCI, S., MATHIESEN, J., MÅLØY, K., HANSEN, A., SCHMITTBUHL, J., VANEL, L., DELAPLACE, A., HAAVIG BAKKE, J., and RAY, P. (2007), *Statistics of fracture surfaces*, Phys. Rev. E *75*(1), 016104.

SCHMITTBUHL, J., DELAPLACE, A., and MÅLØY, K., *Propagation of an interfacial crack front in an heterogeneous medium: Experimental observations*. In *Proc. NATO Advanced Study Institute on Physical Aspects of Fracture*, pages 353–369 (Kluwer Academic Publishers 2001).

SCHMITTBUHL, J., DELAPLACE, A., MÅLØY, K., PERFETTINI, H., and VILOTTE, J. (2003a), *Slow crack propagation and slip correlations*, Pure Appl. Geophys. *160*(5–6), 961–976.

SCHMITTBUHL, J., GENTIER, S., and ROUX, S. (1993), *Field measurements of the roughness of fault surfaces*, Geophys. Res. Lett. *20*(8), 639–641.

SCHMITTBUHL, J., HANSEN, A., and BATROUNI, G. (2003b), *Roughness of interfacial crack fronts: stress-weighted percolation in the damage zone*, Phys. Rev. Lett. *90*(4).

SCHMITTBUHL, J. and MÅLØY, K. (1997), *Direct observation of a self-affine crack propagation*, Phys. Rev. Lett. *78*(20), 3888–3891.

SCHMITTBUHL, J., SCHMITT, F., and SCHOLZ, C. (1995), *Scaling invariance of crack surfaces*, J. Geophys. Res. *100*(B4), 5953–5973.

SCHOLZ, C., *The Mechanics of Earthquakes and Faulting* (Cambridge Univ. Press 2002) second edition.

SCHWARTZ, S. and ROKOSKY, J. (2007), *Slow slip events and seismic tremor at circum-Pacific subduction zones*, Rev. Geophys. *45*, RG3004.

SHEARER, P., HAUKSSON, E., and LIN, G. (2005), *Southern California hypocenter relocation with waveform cross-correlation, Part 2: Results using source-specific station terms and cluster analysis*, Bull. Seismol. Soc. Am. *95*(3), 904–915.

SHELLY, D., BEROZA, G., IDE, S., and NAKAMULA, S. (2006), *Low-frequency earthquakes in Shikoku, Japan, and their relationship to episodic tremor and slip*, Nature, *442*, 188–191.

SZELIGA, W., MELBOURNE, T., MILLER, M., and SANTILLAN, V. (2004), *Southern Cascadia episodic slow earthquakes*, Geophys. Res. Lett. *31*, L16602.

TINTI, E., SPUDICH, P., and COCCO, M. (2005), *Earthquake fracture energy inferred from kinematic rupture models on extended faults*, J. Geophys. Res. *110*, B12303.

WALD, D., HELMBERGER, D., and HEATON, T. (1991), *Rupture model of the 1989 Loma Prieta earthquake from the inversion of strong-motion and broadband teleseismic data*, Bull. Seismol. Soc. Am. *81*(5), 1540–1572.

WELLS, D. and COPPERSMITH, K. (1994), *New empirical relationships among magnitude, rupture length, rupture width, rupture area and surface displacement*, Bull. Seismol. Soc. Am. *84*(4), 974–1002.

YAMASHITA, T. (2000), *Generation of microcracks by dynamic shear rupture and its effects on rupture growth and elastic wave radiation*, Geophys. J. Internat. *143*, 395–406.

ZIV, A., RUBIN, A., and KILB, D. (2003), *Spatiotemporal analyzes of earthquake productivity and size distribution: observations and simulations*, Bull. Seismol. Soc. Am. *93*(5), 2069–2081.

(Received April 1, 2008, revised September 16, 2008, accepted January 8, 2009)
Published Online First: May 20, 2009

Pure appl. geophys. 166 (2009) 801–822
0033–4553/09/050801–22
DOI 10.1007/s00024-009-0476-3

Pure and Applied Geophysics

Fracture Toughness and Fracture Roughness Interrelationship in Thermally treated Westerly Granite

M. H. B. Nasseri,[1] B. S. A. Tatone,[1] G. Grasselli,[1] and R. P. Young[1]

Abstract—This paper presents an experimental work aimed at assessing the correlation between fracture toughness (K_{IC}) and fracture roughness for a series of Westerly granite specimens thermally treated up to 850°C. Mode I fracture toughness as a function of thermal treatment is determined using Cracked Chevron Notched Brazilian Disc specimens. The degree of roughness of the resultant fracture surfaces is analyzed with the aid of a high accuracy, high precision stereo-topometric measurement system. Roughness and toughness values display a negative correlation as a function of temperature. Fracture toughness decreases with increasing temperature due to the gradual opening of grain-grain boundaries in response to thermal stresses. Mode I fractures preferentially follow these weakened grain-grain boundaries, which in addition to the thermal expansion of individual grains, result in rougher failure profiles with increasing temperature. At low temperature, a distinct anisotropy in roughness was observed in all fracture surfaces with higher roughness values perpendicular to the direction of fracture propagation. However, higher treatment temperatures resulted in the homogenization of fracture roughness in all directions. These results confirm the important link among petrofabric analysis, fracture toughness, and fracture roughness in response to thermal treatment.

Key words: Fracture toughness, fracture roughness, thermal stress, Westerly granite.

1. *Introduction*

Fracture toughness (K_{IC}), one of the basic material parameters in fracture mechanics, describes a material's resistance to crack propagation. It is considered an intrinsic material property that has application in a number of rock engineering related fields including: stability analysis, hydraulic fracturing, rock fragmentation by blasting, and earthquake seismology (Atkinson and Meredith, 1987; Whittaker *et al.*, 1992). Fracture roughness, on the other hand, describes the relief of a fracture surface. It has been shown to have an important impact on the shear strength, frictional properties, and transport properties of fractures and, therefore, has been of interest to researchers in

[1] Lassonde Institute, Department of Civil Engineering, University of Toronto, Toronto, ON, M5T 1A4, Canada. E-mail: Nasseri@ecf.utoronto.ca

M. H. B. Nasseri and B. S. A. Tatone contributed equally to this work.

geomechanical fields, ranging from oil, gas, and geothermal reservoirs to rock slope and dam foundation stability. Although the role of fracture toughness and fracture roughness in rock engineering problems has been extensively studied, few attempts have been made to characterize fracture surfaces and correlate the toughness with roughness (Brandt and Prokopski, 1993; Lange *et al.*, 1993; Issa *et al.*, 1993; Vasconcelos *et al.*, 2008). These studies show a good correlation between the fracture roughness and fracture toughness (K_{IC}). In a recent study aimed at investigating the fracture toughness anisotropy and its relation to the fracture roughness of the failed surfaces in Barre and Stanstead granite, it was found that both parameters are strongly dependent on the intrinsic microstructural features of the rock (Nasseri *et al.*, 2007a; 2009).

Thermally induced cracking resulting from the differential thermal expansion of the constituent mineral grains in polycrystalline rocks is a process that has long been known to affect elastic, strength, and transport properties (Heuze, 1983; David *et al.*, 1999; Takarli and Prince-Agbodjan, 2008; Chaki *et al.*, 2008). It has been identified as the cause of reduced strength and enhanced permeability of rock near magmatic intrusions (e.g., Engvik *et al.*, 2005), geothermal reservoirs (e.g., Pearson *et al.*, 1983; Ghassemi *et al.*, 2007), and underground nuclear waste disposal sites (e.g., Hudson *et al.*, 2001; Rutqvist *et al.*, 2005). Furthermore, exposure of fault planes to increased temperature has been found to reduce the coefficient of friction and lead to the stabilization of fault slip (Blanpied *et al.*, 1998).

By performing controlled thermal cracking experiments, the influence of microcrack density on K_{IC} has been quantitatively investigated by several researchers (Atkinson, 1984; Meredith and Atkinson, 1985; Balme *et al.*, 2004; Funatsu *et al.*, 2004; Nasseri *et al.*, 2007). Results of these studies are consistent in showing that thermal cracking significantly reduces the mechanical strength, elastic wave velocities, and fracture toughness in rock, including granite. Based on the analysis of mineral-mineral contact types along the test fracture path in thermally-treated granite specimens, it has been shown that fracture propagation affects a larger volume of rock due to the presence of pre-existing thermal damage mainly around grain-grain boundaries (Fredrich and Wong, 1986; Nasseri *et al.*, 2007). However, to the authors' knowledge there has been no systematic experimental approach to correlate fracture toughness and fracture roughness, and its sensitivity to propagation direction, as a function of temperature in granitic rock.

The aim of this study is to examine the correlation between fracture toughness and fracture roughness as a function of temperature in Westerly granite specimens. The effect of thermal treatment on mode I fracture toughness is examined by testing Cracked Chevron Notched Brazilian discs treated to various temperatures up to 850°C. The roughness of the resultant fracture surfaces is then quantitatively analyzed according to the methodology proposed by Grasselli *et al.* (2002a) to determine the effect of thermal treatment on average roughness as well as the anisotropy in roughness with respect to the direction of fracture propagation.

2. Experimental Setup and Procedure

2.1. Fracture Toughness Measurement as a Function of Temperature

To investigate fracture toughness as a function of temperature, a series of 20 Chevron Cracked Notched Brazilian Disc (CCNBD) samples of Westerly granite were thermally treated and subsequently failed in a standard Mode I tensile test at room temperature. Westerly granite was selected for the study as it is even textured, isotropic, and has been well studied in the past; thus, complexities related to pre-existing rock fabric anisotropy were avoided. Modal analysis gave the following mineralogical composition in volumetric fraction: quartz 27%, microcline 36%, plagioclase 30%, phyllosilicates 6% and others 1%. The mean grain size is 0.75 mm with a total porosity of < 1% (ATKINSON, 1984; MEREDITH and ATKINSON, 1985).

The 20 CCNBD samples were divided into five sets of four samples each. Different amounts of thermal damage were generated by heat treating four sets to 250, 450, 650, and 850°C, respectively, while leaving a fifth set, referred to as RT (room temperature) specimens, untreated. The heat-treatment process involved slowly heating the specimens to the desired temperature and cooling them back to room temperature. A heating/cooling rate of approximately 1–2°C/min was maintained at all times, which was sufficiently slow to avoid cracking due to thermal shock. After cooling to room temperature, compressional (P) wave velocities were measured radially and along the Z axis at azimuthal intervals of 30° around the circumference of each specimen. Subsequently, specimens were notched with a diamond saw according the ISRM (1995) suggested procedure, and tested using a MTS servo-control loading machine and TestStar software which allowed displacement, force and time base to be recorded during testing.

The evaluation of fracture toughness followed the ISRM suggested method (ISRM, 1995) for CCNBD specimens. As described in detail by the ISRM standard method protocol (1995) and by IQBAL and MOHANTY (2007), the fracture toughness K_{IC} can be evaluated as:

$$K_{IC} = \frac{P_{\max}}{B\sqrt{R}} Y^*_{\min}, \tag{1}$$

where $P_{\max}$ is the maximum load at failure, B is the disc's thickness, R is the radius of the disc, and $Y^*_{\min}$ is a critical dimensionless stress intensity value for the specimen, which is determined by the dimensionless parameters α_0, α_1 and α_B defined in Figure 1, and calculated according to the formula:

$$Y^*_{\min} = \mu e^{\Upsilon \alpha_1}, \tag{2}$$

where μ and Υ are constants determined from Table 2 of the ISRM standard method protocol (1995) based on the values of α_0 and α_B. A summary of the experimental parameters for the Westerly granite specimens tested in this study is provided in Table 1.

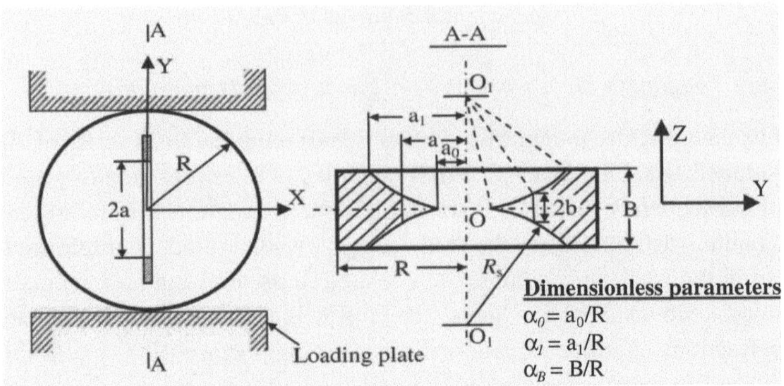

Figure 1

Geometry of the CCNBD and related parameters (after ISRM, 1995), R = Radius of disc; B = Thickness of disc; D = Diameter of disc; R_S = Radius of saw; a = Length of crack; a_0 = Initial half length of chevron notch; a_1 = Final half length of chevron notch.

Table 1

Geometrical dimensions of Westerly granite CCNBD specimens

Descriptions	Values (mm)	Dimensionless expression
Diameter D (mm)	75	
Thickness B (mm)	30	$\alpha_B = B/R = \sim 0.78$
Initial crack length a_o (mm)	6–9	$\alpha_0 = a_o/R = 0.2$–0.25
Final crack length a_1 (mm)	23.5	$\alpha_1 = a_1/R = 0.62$
Saw diameter D_s (mm)	50.0	$\alpha_S = D_s/D = 0.66$
Y^*_{min} (dimensionless)	0.78	

2.2. *Microstructural Investigation as a Function of Temperature*

To explore the effect of thermal stress on the opening of grain-grain boundaries in Westerly granite, thin sections perpendicular to the fracture propagation plane were made from the failed samples treated to different temperatures. Microstructural mapping of the grain-grain boundary type of microcrack, and its density calculation as a function of temperature were performed on areas away from the fracture plane to avoid mapping microcracks induced by the propagation of the test fracture. The analysis of the thin section images and the calculation of grain-grain boundary cracking density were performed using the program "*Intercept*" developed by LAUNEAU and ROBIN (1996). This software allows the rapid determination of microcrack density by scanning images along sets of parallel lines, and counting the number of boundaries intercepted by those lines. The total number of intercepts is then converted into an optical crack density, expressed in (mm/mm^2) which corresponds to the cumulated microcrack lengths (a_i) from area (A) using:

$$\rho_{\text{Optical}} = \frac{\sum a_i}{A}. \tag{3}$$

2.3. *Fracture Roughness Measurement Techniques and Procedure*

Quantitative description of rough surfaces in geomaterials continues to be a challenging task. For this study, a 3-D stereo-topometric measurement system, the Advanced Topometric Sensor (ATOS) II manufactured by GOM mbH, was adopted to digitize the fracture surfaces of tested CCNBD specimens treated to different temperatures. The ATOS II system consists of a measurement head containing a central projector unit and two CCD cameras, and a high-performance Linux PC to pilot the system (Fig. 2). To digitize a surface, various white-light fringe patterns are projected onto the surface and the two CCD cameras automatically capture images of these patterns, which become distorted due to the relief of the surface. From the left and right images of the distorted patterns, the software computes precise three-dimensional coordinates for each pixel based on the principle of triangulation. The resolution of the resulting point cloud can be varied by changing the lenses of the CCD cameras and projector. With the lens set-up chosen for this study, measurement points were obtained on a 250 μm *xy* grid with a computed mesh deviation of less than 10 μm. This mesh deviation is a measure of the average deviation between redundant data (i.e., overlapping scans) and, thus, serves as an estimate of the intensity of the average measurement noise in the 3-D data.

For the current research, the three-dimensional point clouds defining the fracture surfaces were reconstructed using the default triangulation algorithm (Delaunay triangulation) included in the ATOS software and exported in *STL* format for subsequent roughness analysis. This approach discretized the fracture surfaces into contiguous triangles with vertices defined by neighboring points of the point cloud and orientations defined by the vector normal to each triangle.

Following the creation of *STL* files, the roughness was analyzed following the method proposed by GRASSELLI (2006) and GRASSELLI *et al.* (2002a). This method characterizes roughness based on the apparent inclination of the individual triangles making up the surface. After establishing a best-fit plane through the entire surface and selecting one analysis direction of interest, the orientation of each triangle making up the surface can be uniquely identified by its dip, θ, and azimuth, α. The dip is defined as the maximum angle between a best-fit plane through the entire rough surface and the individual triangles, while the azimuth is defined as the angle measured clockwise between the selected analysis direction and the projection of the true dip vector, ***d,*** onto the best-fit plane (Fig. 3).

Given the dip and azimuth, it is possible to define the apparent inclination of each triangle facing the analysis direction. This apparent inclination is termed the apparent dip angle, θ^*. The apparent dip angle is obtained by projecting the true dip vector, ***d,*** onto a

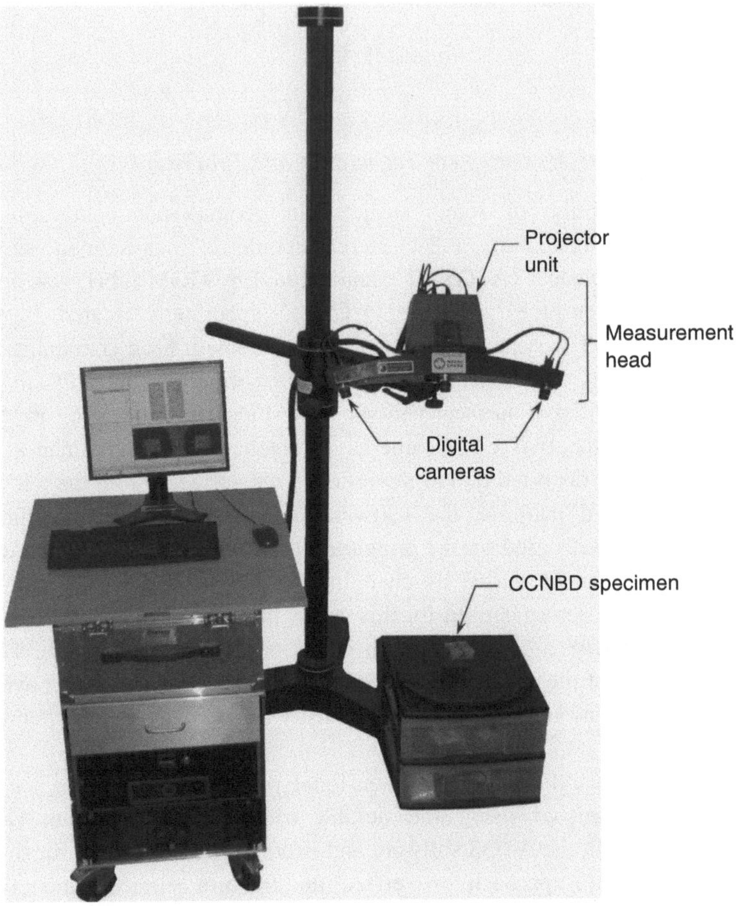

Figure 2
Schematic photograph of the three-dimensional stereo-topometric measurement system utilized for this study, the Advanced Topometric Sensor (ATOS) II manufactured by GOM mbH.

vertical plane oriented along the analysis direction, ***t*** (Fig. 3). Mathematically this occurs where:

$$\tan\theta^* = -\tan\theta\cdot\cos\alpha. \tag{4}$$

Based on the apparent dip angle of each triangle making up the surface, it is possible to distinguish the fraction of the surface area (normalized to the total surface area), $A_{\theta*}$, that is more steeply inclined than any angular threshold value, θ^*_{cr}. By considering several values of θ^*_{cr} between 0° to 90° (the upper and lower bound values), it is possible to characterize the cumulative distribution of the normalized area, $A_{\theta*}$ as a function of the angular threshold, θ^*_{cr}.

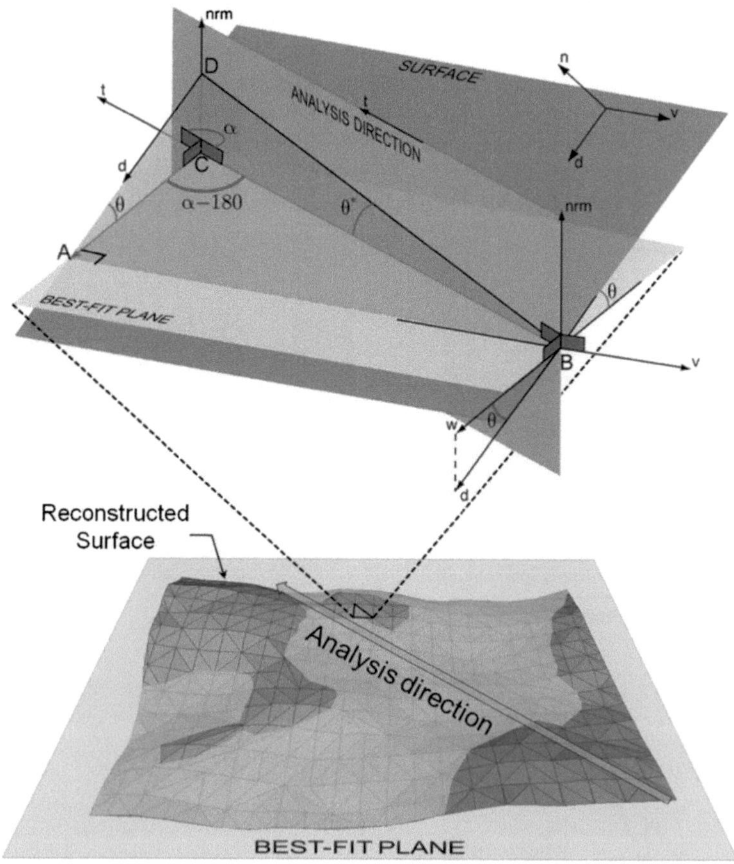

Figure 3
Schematic diagram illustrating the geometric definition of azimuth, dip and apparent dip in relation to the selected analysis direction (modified after GRASSELLI, 2006; GRASSELLI *et al.*, 2002b).

The relationship between the $A_{\theta*}$ and the threshold apparent dip angle can be expressed by the following equation:

$$A_{\theta^*} = A_0 \left(\frac{\theta^*_{\max} - \theta^*}{\theta^*_{\max}} \right)^C, \tag{5}$$

where A_0 is the percentage of the fracture calculated for a threshold angle of 0 degrees, $\theta^*_{\max}$ is the maximum apparent dip angle of the surface in the chosen analysis direction, and C is a "roughness" parameter, calculated via a least-squares regression, that characterizes the shape of distribution of $A_{\theta*}$ as a function of θ^*_{cr} (GRASSELLI *et al.*, 2002a). The resulting ratio of $\theta^*_{\max}/C$ provides a measure of roughness in the selected analysis direction. In order to fully characterize the surface roughness and visualize any roughness anisotropy, the parameters A_0, C and θ^*_{max} are calculated in several possible directions and the resulting values of θ^*_{max}/C obtained in each direction are plotted in a polar diagram (Grasselli *et al.*, 2002b). Figure 4 illustrates an example of the 3-D reconstruction and resulting polar diagram describing the directional distribution of roughness of the four fracture surfaces of a CCNBD specimen.

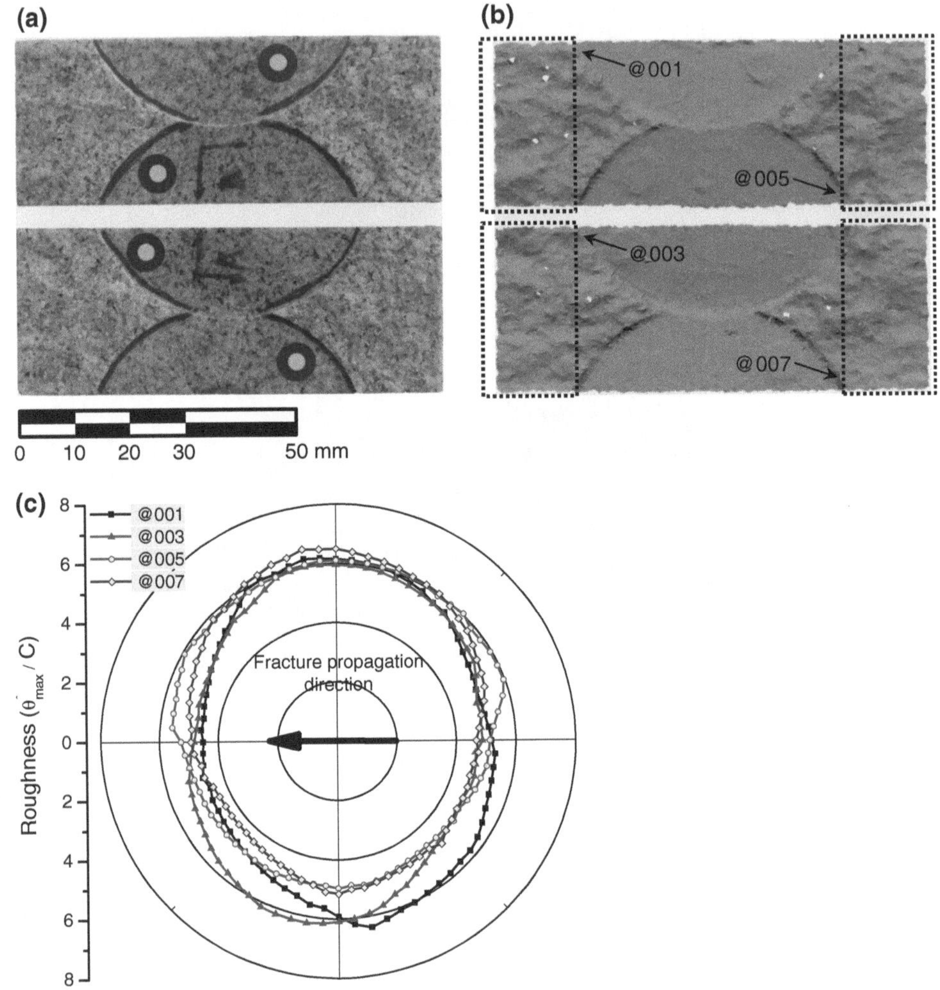

Figure 4

(a) Photograph, (b) three-dimensional surface reconstruction, and (c) polar distribution of roughness values (θ^*_{max}/C) for the fracture surfaces created during cracked chevron notched Brazilian tests of room temperature Westerly granite. Please note that in (c) the roughness values obtained from all four surfaces have been plotted such that the direction of fracture propagation is consistent. It should also be noted that the surfaces denoted as @001 and @003 correspond to each other, as do the surfaces denoted as @005 and @007 (i.e., they form the walls of the same fracture). The same plotting and surface naming convention was adopted for all results presented in this paper.

3. Experimental Results

3.1. Fracture Toughness, Vp, and Grain-grain Boundary Crack Density

Fracture toughness testing of untreated specimens (RT specimens) and those thermally treated to 250, 450, 650 and 850°C displayed a trend of decreasing K_{IC} with increasing temperature. Meanwhile, as a result of variation in the thermal coefficient of expansion among different minerals constituting the rock (i.e., quartz, feldspar and biotite), specimens treated at increasingly higher temperatures experienced more grain-grain boundary opening and thus, increased microcrack density (ρ_{GB}). The gradual opening of grain-grain boundaries with temperature is depicted in the SEM images presented in Figure 5 and the increased mapped microcrack density relative to room temperature specimens is shown in Figure 6. While values of K_{IC} decreased from 1.43 MPa.m$^{1/2}$ at room temperature to 0.22 MPa.m$^{1/2}$ at 850°C (Table 2), grain-grain boundary crack density increased from 1.4 mm/mm^2 to 6.6 mm/mm^2 (Fig. 6). Considering the P-wave velocities measured at various directions around the circumference of

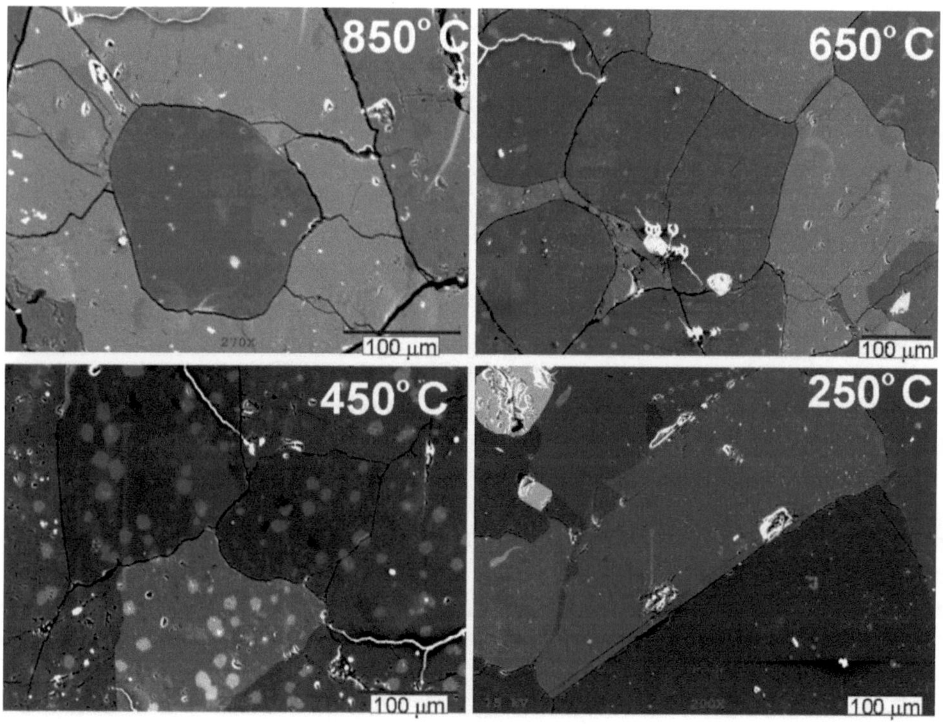

Figure 5
SEM images of thermally treated Westerly granite specimens illustrating the opening of grain-grain boundaries with increasing temperature.

Table 2

Variation of Vp, K_{IC}, and fracture roughness (θ^*_{max}/C) of CCNBD specimens of Westerly granite treated to various temperatures (mean ± standard deviation)

Temp. of thermal treatment (°C)	*Vp* (m/s)	K_{IC} (MPa .m$^{1/2}$)	Average roughness value (θ^*_{max}/C)			Roughness anisotropy ratio*
			a.	*b.*	*c.*	
RT	4488 ± 65	1.43 ± 0.05	5.33 ± 0.57	4.88 ± 0.26	5.85 ± 0.52	1.20
250	3725 ± 50	1.35 ± 0.05	5.06 ± 0.54	4.56 ± 0.24	5.77 ± 0.48	1.27
450	3040 ± 70	0.98 ± 0.07	5.24 ± 0.55	4.61 ± 0.35	5.78 ± 0.50	1.25
650	1560 ± 30	0.43 ± 0.08	6.36 ± 0.38	6.23 ± 0.23	6.51 ± 0.39	1.04
850	985 ± 40	0.22 ± 0.02	6.79 ± 0.47	6.90 ± 0.62	7.17 ± 0.26	1.04

a. Average roughness in all directions
b. Average roughness in the direction parallel to the direction of fracture propagation
c. Average roughness in the direction perpendicular to the direction of fracture propagation
* Calculated as the ratio of *c.* to *b.*

the samples, Figure 7 illustrates that *Vp* decreased uniformly in all directions with increasing temperature, which confirms the isotropic nature of the samples and of the thermally induced damage. The resulting mean P-wave velocity decreased from 4488 m/s at room temperature to 985 m/s at 850°C (Table 2).

To summarize, the thermal treatment process resulted in an 85% decrease in K_{IC}, a nearly 500% increase in ρ_{GB}, and an 80% decrease in *Vp*. Plotting of K_{IC}, *Vp*, and ρ_{GB} as a function of temperature (Fig. 8) clearly shows the decreasing positive correlation of K_{IC} and *Vp* and the negative correlation of K_{IC} and ρ_{GB}. Considering Figure 8, it is interesting to note that the largest drop in K_{IC} and *Vp* occurred between 450 and 650°C. This observation is likely the result of the transition of alpha quartz to beta quartz at 573°C, which is discussed further in Section 4.

Considering the SEM images prepared from the thin sections made perpendicular to fracture propagation plane at the tip of the notches (Fig. 9), the evolution of the test fracture path with temperature is evident. Considering the RT specimen, the test crack cuts through various minerals in its path, splitting the minerals, whereas the fracture path for 850°C specimens proves to be quite selective following the gaps between the grains (open grain-grain boundaries) induced by thermal stresses. Observing the images from specimens treated to intermediate temperatures, the gradual evolution of the fracture propagation path from traversing grains to following grain-grain boundaries is clearly demonstrated.

3.2. Fracture Roughness

The fracture surfaces produced during CCNBD fracture toughness testing were digitized and analyzed with the aim of estimating the magnitude and directional distribution of surface roughness as a function of the temperature of thermal treatment. In total, the fracture surfaces of five CCNBD specimens (one for each treatment

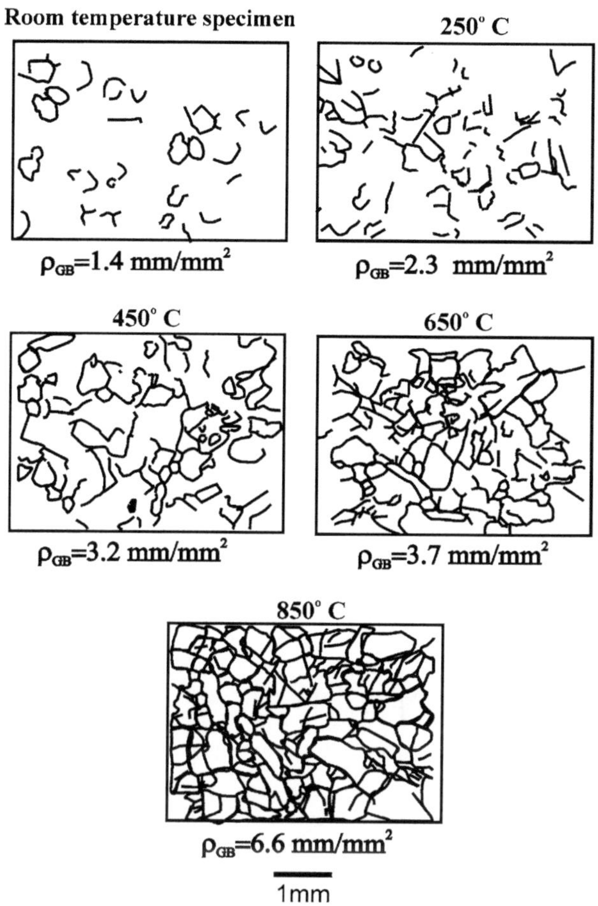

Figure 6
Variation of grain-grain boundary cracking density as a function of temperature in thermally-treated Westerly granite.

temperature) were analyzed. For each specimen, only the areas of the four fracture surfaces corresponding to unstable crack propagation were considered (Fig. 4(b)). The parameters A_0, C, and $\theta^*_{\max}$ were calculated at 5 degree increments all around the average plane of each surface. The resultant values of $\theta^*_{\max}/C$ for each direction are plotted in Figure 4(c) for the RT specimen and Figure 10 for the thermally treated specimens. To facilitate comparison of the roughness distributions for each thermal treatment temperature, the average roughness of the four surfaces of each CCNBD specimen was calculated and plotted in the same polar diagram (Fig. 11(a)). For treatment temperatures up to 450°C, the distribution of roughness has an elliptical shape, with higher values perpendicular to the direction of fracture propagation (90–270 degrees) indicating there is

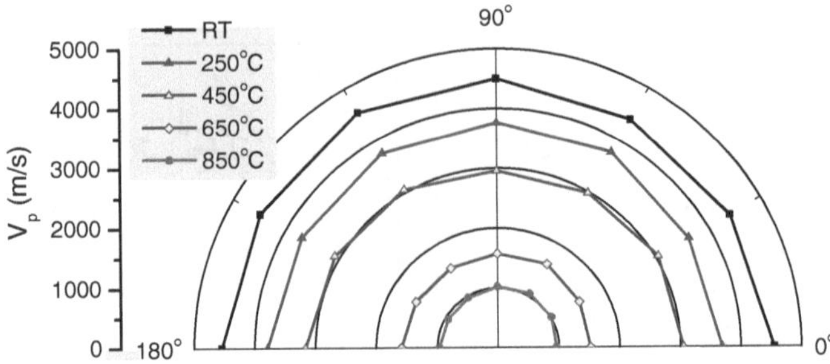

Figure 7
Variation of the P-wave velocity measured in various directions around the circumference of the Westerly granite samples as a function of the temperature of thermal treatment.

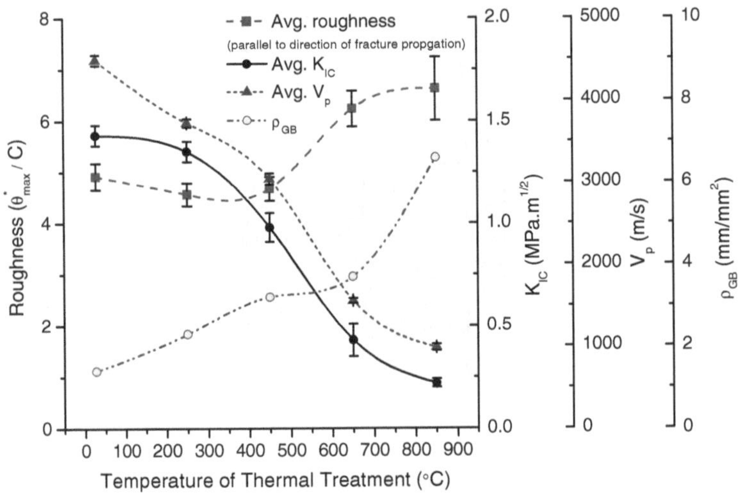

Figure 8
Variation of the average fracture roughness (θ^*_{max}/C), fracture toughness (K_{IC}), P-wave velocity (Vp), and grain boundary microcrack density (ρ_{GB}) as a function of the temperature of thermal treatment. The error bars represent +/- the standard deviation of the measurements. There are no error bars on ρ_{GB} as only one measurement was made for each treatment temperature.

a well-defined anisotropy in roughness. Beyond 450°C the polar distribution of roughness becomes nearly circular, indicating that the anisotropy in roughness nearly vanishes. To further illustrate this homogenization of roughness beyond 450°C, the average roughness values and their respective standard deviations in directions parallel and perpendicular to the direction of fracture propagation were plotted as a function of temperature (Fig. 11(b)). The plot shows that up to 450°C the average roughness values measured in each direction differ by approximately 1.0 to 1.2 roughness units (18% to 23%), while

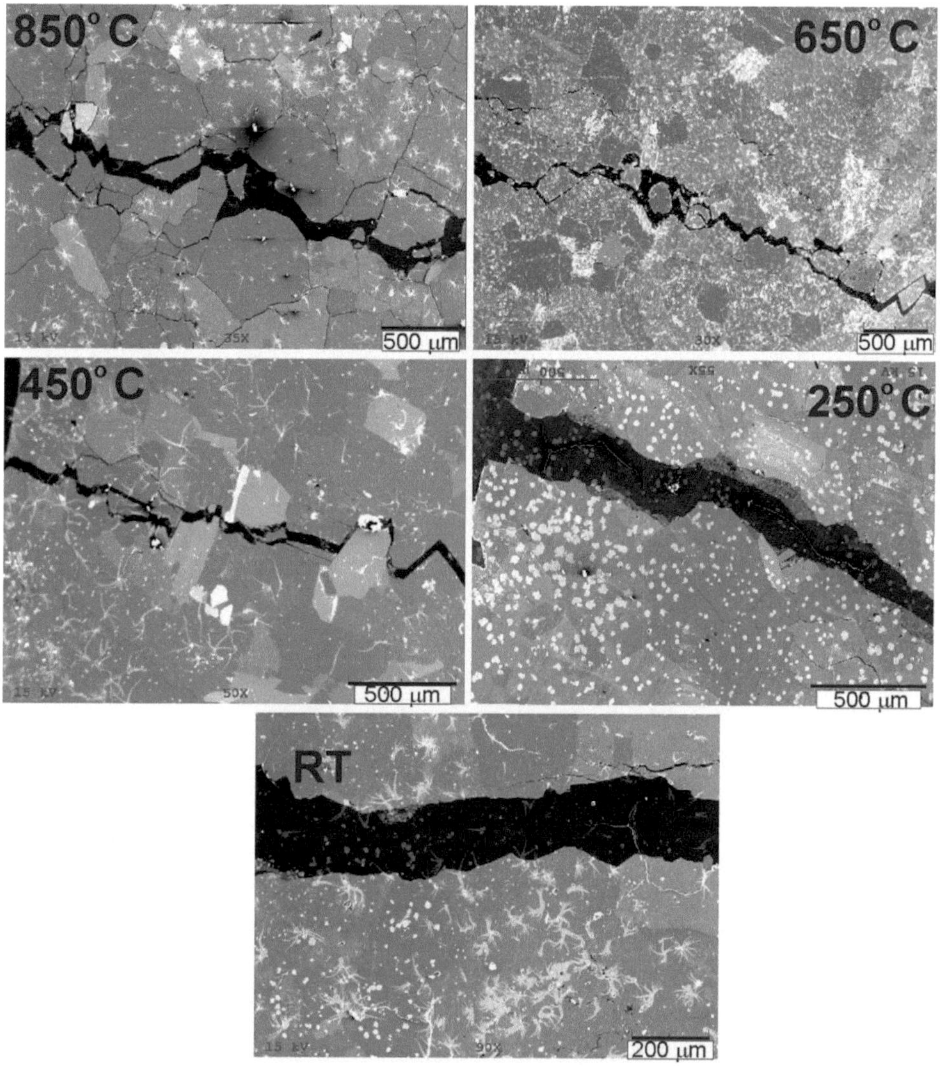

Figure 9
SEM images of thin sections taken perpendicular to direction of fracture propagation in tested CCNBD specimens of thermally-treated Westerly granite. Comparison of the fracture paths as a function of temperature shows how the fracture path transitions from cutting through grains to following grain boundaries with increasing temperature. It must be noted that the crack widths shown in the RT and 250°C images appear wider than they should as a result of the slide preparation process and, therefore, are not comparable with the crack widths of the other three specimens.

for 650°C and 850°C the values differ by only 0.28 and 0.27 roughness units (4%), respectively. Defining the "anisotropy ratio" as the roughness perpendicular to propagation divided by the roughness parallel to propagation, the surfaces measured in

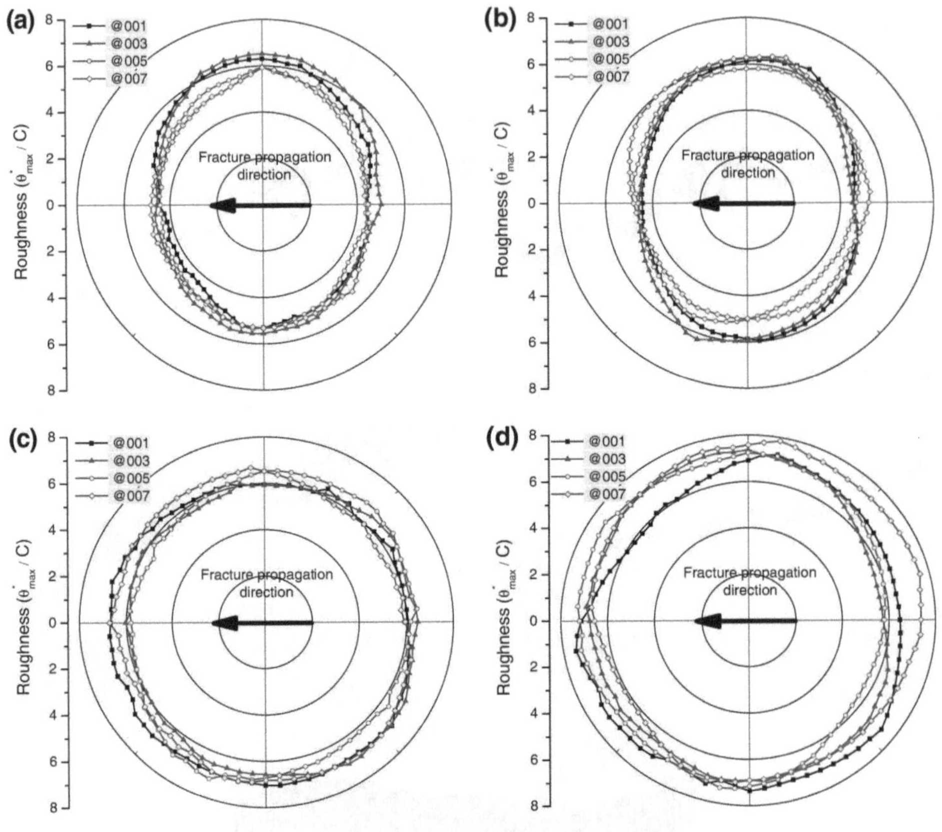

Figure 10
Polar distribution of roughness values (θ^*_{max}/C) for the fracture surfaces created during cracked chevron notched Brazilian tests of thermally-treated Westerly granite: (a) 250°C, (b) 450°C, (c) 650°C, and (d) 850°C.

this study display anisotropy ratios that range from 1.27 for 250°C to 1.04 for both 650°C and 850°C. A complete summary of the anisotropy ratios for each temperature as well as the corresponding values of roughness and their respective standard deviations are included in Table 2.

Comparison of the roughness values for specimens treated to various temperatures showed that, overall, the average roughness in all directions (Fig. 11(c)), as well as the roughness measured parallel and perpendicular to the direction of fracture propagation (Fig. 11(b)), increased with the temperature of thermal treatment. Considering Figure 11(c), the change in roughness with increasing temperature up to 450°C is minimal (less than 4% change from RT to 450°C). However, between 450°C and 650°C there is a sizeable increase in roughness as the average roughness value increases from 5.2 to 6.4, an increase of more than 20%. This increase appears to correspond to the loss of anisotropy described above. Beyond 650°C the roughness continues to increase to an average value of 6.8 at 850°C. Adding the average values of fracture roughness in the direction of propagation to Figure 8 indicates there is a negative correlation between roughness and both K_{IC} and Vp, and a positive correlation with ρ_{GB}.

4. Discussion

4.1. Correlation of Fracture Toughness and Fracture Roughness

In the present study, the measured fracture toughness and corresponding surface roughness displayed a negative correlation with increasing thermal treatment (Fig. 8). In a recent study, NASSERI *et al.* (2007a, 2009) correlated the fracture roughness with fracture toughness obtained along specific planes in two granitic rocks which were not thermally treated (see Fig. 14 in NASSERI *et al.*, 2009). These rocks were characterized by preferred petrofabric orientation and different constituent grain sizes. The roughness in both granites showed a positive correlation with measured K_{IC} as a function of preferred petrofabric orientation. The lowest values of K_{IC} and roughness were measured in a direction parallel to the preferred petrofabric orientation, while the highest values were obtained along a direction perpendicular to the preferred petrofabric orientation. It has been concluded that such a correlation between toughness and roughness within one rock type is due to the nature and type of the interaction between the propagating test crack and the pre-existing preferably oriented petrofabric elements. Crack propagation perpendicular to the latter produces a stress shielding effect that causes significant out-of-plane propagation, while propagation parallel to the pre-existing preferably oriented petrofabric elements creates stress amplification around the microcrack tips leading to much lower values of K_{IC} and a less tortuous fracture path. Considering the effect of grain sizes, it was concluded that larger grains and associated microcracks could lead to larger roughness values but not necessarily larger K_{IC} values. Another recent study by VASCONCELOS *et al.* (2008) on a suite of granitic rocks found a similar positive correlation between fracture roughness and fracture energy measured via direct tension tests. A similar increase in both roughness and fracture energy was observed when the test crack was propagated in a direction perpendicular to pre-existing, preferably oriented, petrofabric elements. The increase in fracture energy with fracture roughness was attributed to the increased energy needed to propagate a fracture along a more tortuous path.

Although the negative correlation between fracture toughness and fracture roughness observed in this study appears to contradict previous results, there are some key differences in the nature of specimens tested. In this study, Westerly granite, an even textured, isotropic granite, was selected to avoid the effects of preferred microstructural orientation on the resultant fracture toughness and roughness. Instead, the change in toughness and roughness was dictated by the amount of damage induced by thermal

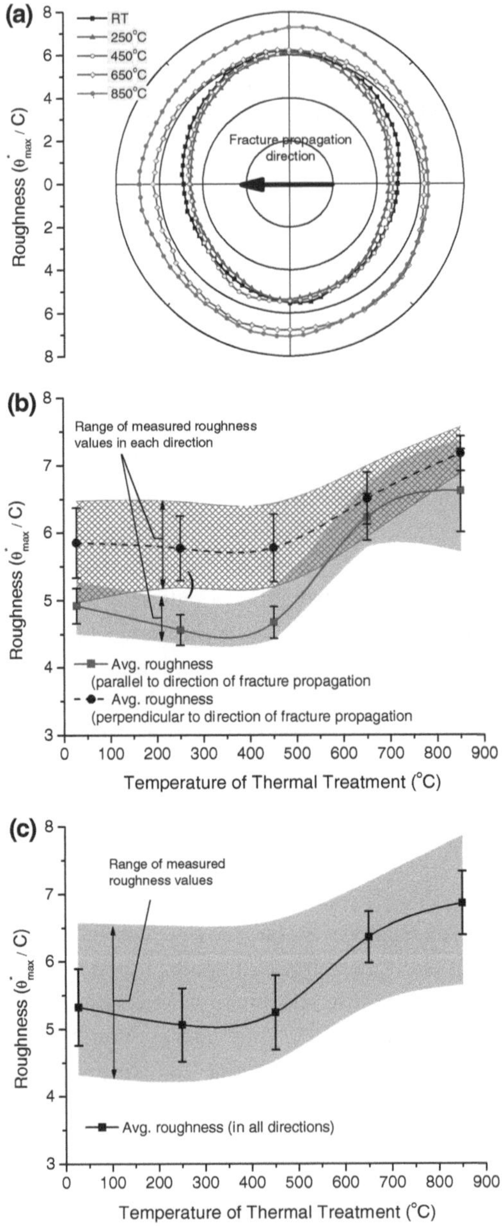

Figure 11

(a) Polar distribution of average roughness values (θ^*_{max}/C) for the fracture surfaces created during CCNBD tests of thermally-treated Westerly granite. (b) A comparison of the fracture roughness in directions parallel and perpendicular to fracture propagation as a function of the thermal treatment temperature. (c) A comparison of the average fracture roughness as a function of the thermal treatment temperature. Error bars represent ± one standard deviation.

stresses. Granitic rocks are subject to significant amounts of thermally induced damage in the form of grain-grain boundary microcracking due to differential thermal expansion of their constituent minerals (Fig. 5). Weakening and opening of grain boundaries allows the test crack to preferentially propagate along grain boundaries and thus the failed surfaces, at increasingly higher temperatures, portray the roughness of both expanded grains and grain-grain boundaries. Conversely, due to the lack of a pre-existing damaged area ahead of the propagating crack (in its process zone) in samples treated to lower temperatures (RT and 250°C), test cracks are characterized by much smoother profiles as they cut through grains rather than follow expanded grain-grain boundaries. This phenomenon can be observed when comparing the test crack's path in the RT specimens to that of the 450°C specimen, as shown in Figure 7. Since the fractures increasingly follow grain-grain boundaries with temperature, the relative increase in fracture roughness with temperature is dictated by the shape, size, and expansion of the grains which comprise the rock. It is important to note that although a finer or coarser grained granitic rock would experience increases in grain-grain boundary microcrack density when subjected to the same thermal treatment, the relative changes in fracture roughness between RT and 850°C specimens would likely differ significantly as a result of the different grain sizes.

Interestingly, the largest increase in the measured surface roughness occurred between 450°C and 650°C which corresponded with the largest decreases in both Vp, and K_{IC}. A review of previous work on thermally-treated granitic rocks found that several other physical/mechanical rock properties have been shown to undergo significant variation when the temperature of thermal treatment extends over similar temperatures. Examples of such properties include: unconfined compressive strength (NAGARAJA RAO and MURTHY, 2001), Young's modulus (DAOYING *et al.*, 2006; MOMAND-ETIENNE and HOUPERT, 1989; TAKARLI and PRINCE-AGBODJAN, 2008), tensile strength (MOMAND-ETIENNE and HOUPERT, 1989), and porosity and permeability (GLOVER *et al.*, 1995; TAKARLI and PRINCE-AGBODJAN, 2008; NASSERI *et al.*, 2007). This large variation of mechanical/physical properties is explained by the α-β quartz phase transition which is known to occur at roughly 573°C under atmospheric conditions. This phase transition increases the differential expansion between the quartz grains and other constituent minerals leading to increased microcracking of the specimen. In this study, the α-β quartz phase transition is regarded as the likely cause of the large decrease in K_{IC} and Vp, and corresponding large increase in fracture roughness observed between 450°C and 650°C.

4.2. Anisotropy in Fracture Roughness as a Function of Temperature

As presented in Section 3.2 of this paper, a distinctive anisotropy in fracture roughness was observed in the fracture surfaces of the RT, 250°C, and 450°C specimens with increased roughness perpendicular to the direction of fracture propagation. Beyond 450°C this anisotropy greatly diminished as the roughness parallel and perpendicular to the fracture propagation direction became nearly equal.

Currently, it is hypothesized that the observed anisotropy in roughness is related to the nature of energy dissipation during fracture propagation in CCNBD specimens. The initiation and propagation direction of fractures in a CCNBD test is imposed by the orientation of the applied loading. The tensile loading applied perpendicular to the plane containing the specimen notch causes a fracture front to propagate outward from the notch tip at the disc center towards its outer edge where the compressional line loading is applied (along Y direction defined in Fig. 1). The fracture must dissipate more energy in this direction compared to the direction perpendicular to crack front propagation (laterally along Z direction defined in Fig. 1). Therefore, it is thought that additional energy present in the Y direction allows the advancing crack front to cut through grains, while the lesser energy in the direction perpendicular to the propagation (Z direction) causes the fracture to preferentially follow grain-grain boundaries. The loss of the anisotropy observed between the 450°C and 650°C is again thought to be related to the α-β quartz transition. With the large increase in damage associated with the mineral phase transition, the advancing orthogonal crack front (along both Y and Z directions as shown in Fig. 1) preferentially follows the path of least resistance which is characterized by open grain-grain boundaries.

It should be emphasized that the above hypothesis is only one possible explanation for the deterioration of anisotropy in fracture roughness. Providing a conclusive explanation of the existence of the anisotropy in roughness at lower temperature continues to be a future research objective.

4.3. Impacts/application

The observations of decreased mode I fracture toughness and *Vp*, and increased grain-grain boundary microcrack density and fracture roughness with increasing thermal stress could have many important impacts on the field of rock mechanics. Considering a rock mass, the mechanical and hydraulic behavior is heavily dependent on the nature of the fractures that divide the rock into blocks and less so by the intact rock itself. Since the conditions for mode I fracture genesis and the resulting roughness of these fractures in Westerly granite have been shown to depend on the response of the intact rock to thermal stresses, it can be surmised that the strength and permeability of granitic rock masses may be significantly altered by exposing them to various thermal stresses (heating, cooling, or both).

The creation of new fractures in a rock mass would enhance its permeability and decrease its strength while the corresponding fracture roughness would have a secondary impact as it is well known that roughness controls both the shear strength and permeability of rock fractures. Examples of engineering situations that would benefit from a better understanding of fracture development resulting from thermal stress in granitic rock include:

- the optimization of thermal rock fragmentation processes to augment mechanical excavation techniques in strong rock where blasting is not desired;
- the analysis of long-term rock mass stability and permeability surrounding nuclear waste repositories in crystalline rock environments in response to the heat generated by waste canisters; and
- the prediction of the development of new fractures in engineered geothermal reservoirs in which cool water is injected and warm water extracted for energy production purposes.

Furthermore, an improved understanding of the effect of thermal stresses on the genesis of rock fractures may lead to a better understanding of natural geologic processes including: the emplacement of dykes and sills, the movement of hydrothermal fluids in the subsurface and crustal fault slip behavior.

The presence of anisotropy in roughness and its loss due to thermal treatment may also have several important impacts on the engineering behavior of rock masses. At low temperatures the anisotropy in fracture roughness may potentially lead to the creation of a preferential direction for shear displacements (anisotropic shear strength) and fluid flow (anisotropic permeability), while rocks exposed to higher temperature may display isotropic strength and transport properties.

5. *Conclusions*

In this study an attempt is made to correlate the fracture roughness of tested thermally treated Westerly granite up to 850°C with its respective fracture toughness measured under mode I. There exists a negative correlation between the former and the latter as thermal stresses target the grain-grain boundaries, which in turn, provides a path of less resistance for the test fracture to propagate under mode I. Grain-grain boundary crack density and *Vp* were found to have opposing correlation as a function of temperature as well. A distinctive anisotropy in fracture roughness was observed in the fracture surfaces of the RT, 250°C, and 450°C specimens with increased roughness perpendicular to the direction of fracture propagation. Beyond 450°C this anisotropy greatly diminished as the roughness along both directions within the fracture propagation plane became nearly equal. Reduction in the roughness anisotropy as a function of temperature is proposed to be related to the openness of grain-grain boundaries to which both the test crack propagation front and lateral front responded equally after α-β quartz phase transition is achieved.

It is important to note that, although increasing roughness of discontinuities in rocks is typically thought to increase their shear strength and decrease permeability, the intact strength of the asperities of fracture walls of the thermally damaged specimens maybe significantly reduced due to the presence of extensive grain-grain boundary microcracks. Therefore, despite the increased roughness, the shear strength of thermally treated granite may be significantly decreased relative to undamaged rock, thus, enhancing the fracture slippage potential. Overall, an improved understanding of the effect of temperature on rock fracture genesis and related properties of granitic rocks could allow for improved engineering projects and a better understanding of complex geologic phenomena.

Acknowledgements

This work has been funded by NSERC/Discovery Grant No. 341275 and NSERC/RTI Grant No. 345516. This project was also partly supported through a NSERC discovery grant held by R.P. Young. The authors would like to thank Matt Balme and Georg Dresen for their constructive comments which greatly enhance this paper.

References

Atkinson, B.K. and Meredith, P.G., *Experimental fracture mechanics data for rocks and minerals*, In *Fracture Mechanics of Rocks* (ed. B. K. Atkinson) (Academic Press, London 1987) pp. 477–525.

Atkinson, B.K. (1984), *Subcritical crack growth in geological materials,* J. Geophys. Res. *89*, 4077–114.

Balme, M.R., Rocchi, V., Jones, C., Sammonds, P.R., Meredith, P.G., and Boon, S. (2004) *Fracture toughness measurements on igneous rocks using a high-pressure, high-temperature rock fracture mechanics cell.* J. Volcanol. Geotherm. Res. *132*(2-3), 159–172.

Blanpied, M.L., Marone, C.J., Lockner, D.A., Byerlee, J.D., and King, D.P. (1998) *Quantitative measure of the variation in fault rheology due to fluid-rock interactions*, J. Geophys. Res. *103*(B5), 9691–9712.

Brandt, A.M. and Prokopski, G. (1993), *On the fractal dimension of fracture surfaces of concrete elements*, J. Mater. Sci. *28*(17), 4762–4766.

Chaki, S., Takarli, M., and Agbodjan, W.P. (2008), *Influence of thermal damage on physical properties of a granite rock: Porosity, permeability and ultrasonic wave evolutions*, Constr. Build. Mater. *22*(7), 1456–1461.

Daoying, X., Yuezhan, T., Ting, L., and Songlin, X. (2006), *Experimental investigations of thermal damage for rocks*, Key Eng. Mat. *324-325 II*, 1213–1216.

David, C., Menéndez, B., and Darot, M. (1999), *Influence of stress-induced and thermal cracking on physical properties and microstructure of La Peyratte granite*, Int. J. Rock Mech. and Min. Sci. *36*(4), 433–448.

Engvik, A.K., Bertram, A., Kalthoff, J.F., Stockhert, B., Austrheim, H., and Elvevold, S. (2005), *Magma-driven hydraulic fracturing and infiltration of fluids into the damaged host rock, an example from Dronning Maud Land, Antarctica*, J. Struct. Geol. *27*(5), 839–854.

Fredrich, J.T. and Wong, T. (1986), *Micromechanics of thermally induced cracking in three crustal rocks*, J. Geophys. Res. *91*(B12), 12,743–12,764.

Funatsu, T., Seto, M., Shimada, H., Matsui, K., and Kuruppu, M. (2004), *Combined effects of increasing temperature and confining pressure on the fracture toughness of clay bearing rocks*, Int. J. Rock. Mech. and Min. Sci. *41*(6), 927–938.

Ghassemi, A., Tarasovs, S., and Cheng, A.H.D. (2007), *A 3-D study of the effects of thermomechanical loads on fracture slip in enhanced geothermal reservoirs*, Int. J. Rock Mech. Min. Sci. *44*(8), 1132–1148.

GLOVER, P.W.J., BAUD, P., DAROT, M., MEREDITH, P.G., BONN, S.A., LERAVALEC, M., ZOUSSI, S., and REUSCHLE, T. (1995), *Alpha /beta phase transition in quartz monitored using acoustic emissions*, Geophys. J. Int. *120*(3), 775–782.

GRASSELLI, G. (2006), *Manuel Rocha medal recipient shear strength of rock joints based on quantified surface description*, Rock Mech. Rock Eng. *39*(4), 295–314.

GRASSELLI, G., WIRTH, J., and EGGER, P. (2002a), *Functional parameters for quantifying the surface anisotropy of rock discontinuities.* In Eurock 2002 – Proc. ISRM Internat. Symp. on *Rock Engineering for Mountainous Regions*, Funchal (Madeira Island), Portugal, November 25-28, 2002 (eds. Wittke, W.), pp. 722–733.

GRASSELLI, G., WIRTH, J., and EGGER, P. (2002b), *Quantitative three-dimensional description of a rough surface and parameter evolution with shearing*, Int. J. Rock. Mech. Min. Sci. *39*(6), 789–800.

HEUZE, F.E. (1983), *High-temperature mechanical, physical and thermal properties of granitic rocks— A review*, Int. Rock Mech. Min. Sci. Geomech. Abstr. *20*(1), 3–10.

HOMAND-ETIENNE, F. and HOUPERT, R. (1989), *Thermally-induced microcracking granites: characterization and analysis*, Int. Rock Mech. Min. Sci. & Geomech. Abstr. *26*(2), 125–134.

HUDSON, J.A., STEPHANSSON, O., JING, L., ANDERSSON, J., and TSANG, C.-F. (2001), *Coupled T-H-M issues relating to radioactive waste repository design and performance*, Int. J. Rock Mech. and Min. Sci. *38*(1), 143–161.

IQBAL, M.J. and MOHANTY, B. (2007), *Experimental calibration of ISRM suggested fracture toughness measurement techniques in selected brittle rocks*, Rock Mech. Rock Eng. *40*(5), 453-475.

ISRM (1995), *Suggested method for determining mode I fracture toughness using cracked chevron notched Brazilian disc (CCNBD) specimens (Fowell, R.J. - Coordinator)*, Int. J. Rock. Mech. Min. Sci. *32*(1), 57–64.

ISSA, M.A., ISLAM, M.S., and CHUDNOVSKY, A. (2003), *Fractal dimension—a measure of fracture roughness and toughness of concrete*, Eng. Fract. Mech. *70*(1), 125–137.

LANGE, D.A., JENNINGS, H.M., and SHAH, S.P. (1993), *Relationship between fracture surface roughness and fracture behavior of cement paste and mortar*, J. Am. Ceram. Soc. *76*(3), 589–597.

LAUNEAU, P. and ROBIN, P.-Y.F. (1996), *Fabric* analysis using the intercept method, Tectonophysics *267*(1–4), 91–119.

MANDELBROT, B.B., PASSOJA, D.E., and PAULLAY, A.J. (1984), *Fractal character of fracture surfaces of metals*, Nature *308*, 721–722.

MEREDITH, P.G. and ATKINSON, B.K. (1985), *Fracture toughness and subcritical crack growth during high-temperature tensile deformation of Westerly granite and Black gabbro*, Phys. Earth Planet. Inter. *39*(1), 33–51.

NAGARAJA RAO, G.M., and MURTHY, C.R.L. (2001), *Dual role of microcracks; toughening and degradation*, Can. Geotech. J. *38*(2), 427–440.

NASSERI, M.H.B., GRASSELLI, G., and MOHANTY, B. (2009), *Fracture toughness and fracture roughness interrelationship in anisotropic granitic rocks*, Rock Mech. Rock Eng. (Submitted).

NASSERI, M.H.B., GRASSELLI, G., MOHANTY, B., WIRTH, J., and BRAUN, M. (2007a), *Experimental relationship between fracture toughness and fracture roughness in anisotropic granitic rocks*, In: Rock Mechanics, meeting society's demands and challenges; Proc. 1st Canada-U.S. Rock Mechanics Symposium, Vancouver, May 27-31, 2007 (eds. Eberhardt, E., Stead, D. and Morrison, T.), pp. 617–624.

NASSERI, M.H.B., SCHUBNEL, A., and YOUNG, R.P. (2007), *Coupled evolutions of fracture toughness and elastic wave velocities at high crack density in thermally treated Westerly Granite*, Int. J. Rock Mech. Min. Sci. *44*(4), 601–616.

PEARSON, C.F., FEHLER, M.C., and ALBRIGHT, J.N. (1983), *Changes in compressional and shear-wave velocities and dynamic moduli during operation of a hot dry rock geothermal system*, J. Geophys. Res. *88*(B4), 3468–3475.

RUTQVIST, J., TSANG, C.-., TSANG, Y., BARR, D., DATTA, R., GENS, A., OLIVELLA, S., and MILLARD, A. (2005), *Coupled thermal-hydrological-mechanical analyses of the Yucca Mountain Drift Scale Test - Comparison of field measurements to predictions of four different numerical models*, Int. J. Rock Mech. Min. Sci. *42*(5-6 SPEC. ISS.), 680–697.

TAKARLI, M., and PRINCE-AGBODJAN, W. (2008), *Temperature effects on physical properties and mechanical behavior of granite: Experimental investigation of material damage*, J. ASTM Int. *5*(3), 13 pgs.

Vasconcelos, G., Lourencoo, P.B., and Costa, M.F.M. (2008), *Mode I fracture surface of granite: Measurements and correlations with mechanical properties*, J. Mater. Civ. Eng. (3), 245–254.

Whittaker, B.N., Singh, R.N., and Sun, G., *Rock fracture mechanics: principles, design, and applications* (Elsevier, Amsterdam 1992).

(Received September 19, 2007, revised July 17, 2008, accepted August 14, 2008)
Published Online First: May 7, 2009

Pure appl. geophys. 166 (2009) 823–841
0033–4553/09/050823–19
DOI 10.1007/s00024-009-0479-0

Pure and Applied Geophysics

Acoustic Emissions Monitoring during Inelastic Deformation of Porous Sandstone: Comparison of Three Modes of Deformation

JEROME FORTIN,[1] SERGEI STANCHITS,[2] GEORG DRESEN,[2] and YVES GUEGUEN[1]

Abstract—In some reservoirs, large deformations can occur during oil or gas production because of the effective stress change. For very porous rocks, these production operations can be sufficient to cause inelastic deformation and irreversible damage. Rock formations can undergo deformation by different mechanisms, including dilatancy or pore collapse. In the laboratory, it has been shown that the inelastic deformation and failure mode of porous rocks are pressure sensitive. Indeed, when subjected to an overall compressive loading, a porous rock may fail by shear localization, compaction localization, or by cataclastic compaction. Acoustic emission (AE) records provide important information to understand the failure mode of rocks: the spatial evolution of damage as well as the source mechanisms can be followed using this technique. In this paper, we present three different laboratory axisymmetric compression experiments, performed on Bleurswiller sandstone, which enable us to compare the acoustic emission signature of these three modes of deformation. Our data show that compaction localization and cataclastic compaction are characterized by similar acoustic signatures (in terms of AE sources characteristics and evolution of AE number), in comparison to the acoustic signature from shear localization. This implies similar micromechanisms involved during compaction bands formation and cataclastic compaction.

Key words: Porous sandstone, acoustic emissions, source mechanisms, shear localization, compaction bands, cataclastic compaction.

1. Introduction

When subjected to an overall compressive loading, a rock can fail by shear localization, compaction localization or cataclastic compaction (WONG *et al.*, 1997, 2004; KLEIN *et al.*, 2001; FORTIN *et al.*, 2005). The laboratory investigations provide important physical insights into tectonic processes in relation to faulting, in sandstone formation, as well as reservoir and geotechnical problems. Reservoir compaction is one of the most important processes in the oil and gas industry. Hydrocarbon production decreases the pore pressure and hence increases the effective stress. For very porous rock, the changes in effective stresses can be sufficient to induce inelastic deformation, which can i) lead to seafloor

[1] Laboratoire de Géologie, Ecole Normale Supérieure, CNRS, UMR 8538, 24 Rue Lhomond, 75005 Paris, France. E-mail: fortin@geologie.ens.fr
[2] GeoForshungsZentrum Potsdam, Telegrafenberg, D-14473 Potsdam, Germany.
E-mail: stanch@gfz-potsdam.de

subsidence (for example, the extraction of oil and gas in the Ekofisk field in the North Sea has produced a subsidence of more than 4 meters); ii) cause casing failures; or iii) can decrease significantly the permeability of the reservoir (NAGEL, 2001; OSTERMEIER, 2001).

In porous sandstone, a transition in failure mode from brittle faulting to cataclastic compaction occurs as effective pressure increases. Under a relatively low effective confining pressure the differential stress attains a peak before it undergoes strain softening; the failure of sample occurs by shear localization (WONG *et al.*, 1997). Cataclastic compaction is commonly observed as a response to purely hydrostatic loading: Under pressure, the pore space is initially reduced by elastic deformation and the rock becomes stiffer. However, as the hydrostatic loading increases, the rock reaches a critical point, where it suddenly becomes increasingly more compliant, showing a dramatic increase in compaction (ZHANG *et al.*, 1990; DAVID *et al.*, 1994; MENÉNDEZ *et al.*, 1996; WU *et al.*, 2000). In the transitional regime from brittle faulting to cataclastic compaction, different laboratory studies conducted on porous sandstones show that localization can occur purely by compaction, without shear (formation of compaction bands) (KLEIN *et al.*, 2001; BAUD *et al.*, 2004; FORTIN *et al.*, 2005, 2006). Compaction bands are zones of material which extend subperpendicular to the main compressive stress. From microstructural observations a compaction band appears as a crushed zone of reduced porosity. If the microstructure of the rock outside the band remains largely undeformed, inside the band grains are crushed and the fragments fill up the collapsed pore space (BAUD *et al.*, 2004; FORTIN *et al.*, 2006). In addition, significant reduction of permeability has been reported while a sample undergoes compaction localization, which demonstrates that a compaction band acts as a barrier for fluid flow (VAJDOVA *et al.*, 2004).

The three modes of failure, — shear localization, compaction localization, cataclastic compaction — are very different on the macroscopic scale, however they involve similar micromechanical processes. Indeed, grain-scale microcracking is commonly observed (MÉNENDEZ *et al.*, 1996) and as a consequence these three modes of deformation lead to radiation of acoustic emissions. Acoustic emission (AE) records provide important information to understand the failure mode of rocks: (i) AE activity can be recorded using one piezoelectric transducer, and the number of AE can be related to the inelastic deformation (SCHOLZ 1968; SANO *et al.*, 1981), (ii) Using multiple transducers attached to the surface of a specimen under stress, the spatial origin of the damage can be located. For example, different authors have used this technique in axisymmetric compression experiments to investigate shear-fracture nucleation (e.g., LEI, 1992; LOCKNER, 1993; ZANG *et al.*, 1998) and recently to analyze the development of discrete compaction bands (FORTIN *et al.*, 2006; STANCHITS *et al.*, this issue). (iii) Finally AE waveforms provide information on the source mechanisms taking place during the deformation of the rock, allowing a better understanding of the micromechanisms (STANCHITS *et al.*, 2001).

The aim of this study is to focus on the acoustic emission signature of the three modes of failure commonly observed in porous sandstone. In a recent paper, FORTIN *et al.* (2006) used acoustic emission analysis to investigate the formation of compaction bands in Bleurswiller sandstone. Two more experiments have been conducted on this sandstone:

A purely hydrostatic one, and a triaxial compression test at an effective confining pressure of 10 MPa, where shear localization is observed. We also used the data from a triaxial experiment at an effective confining pressure of 70 MPa, published in FORTIN *et al.* (2006), where compaction bands are observed. These three experiments enable us to compare the acoustic emission signature for the three different modes of failure: shear localization, compaction localization and cataclastic compaction.

2. *Experimental Procedures*

The Bleurswiller sandstone has a nominal porosity of 25% ± 0.5%. Grain sizes range from 80 μm to 150 μm with a mean value of 110 μm. Cylindrical specimens of 100 mm in length and 50 mm in diameter were prepared from the same block as that studied by FORTIN *et al.* (2005, 2006, 2007) and TEMBE *et al.* (2008). As in previous studies, the samples were cored perpendicular to the bedding. Figure 1a) gives a picture of an intact specimen. The modal composition of this sandstone is 50% quartz, 30% feldspars, and 20% oxydes-micas (FORTIN *et al.*, 2006). Fig. 1b) shows a SEM (scanning electron microscopy) of an intact specimen, for comparison: Porosity appears in black and quartz grains appear darker than the feldspar grains.

The experiments were performed at the GeoForschungsZentrum Potsdam. We used a servo-controlled MTS loading frame with load capacity of 4600kN (Fig. 2a). All experiments were carried out at room temperature. Following the methodology of FORTIN *et al.* (2006), the samples were saturated with distilled water and deformed under fully

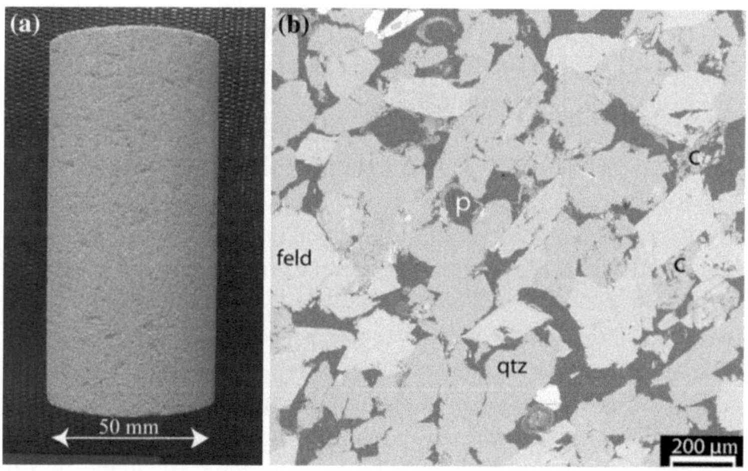

Figure 1
a) Picture of a non-deformed sample of Bleurswiller sandstone. b) SEM micrograph (backscattered) of intact sandstone. Epoxy-filled pores (p) appear in black, porosity is about 25%. Quartz, feldspar, and clay are denoted by qtz, feld, and c, respectively.

Figure 2
Experimental setup. a) MTS loading frame with 200 MPa pressure vessel. b) Cylindrical specimen encapsulated in rubber jacket with P sensors glued directly on the sample surface.

drained conditions at a constant pore pressure of 10 MPa. The pore pressure was maintained constant, and the variation of pore volume was recorded using a volumometer. From this, the evolution of connected sample porosity was deduced. For the hydrostatic experiment (VO#5), confining pressure was increased up to 120 MPa with a pressurization ramp of 1 MPa/s, then confining pressure was increased by steps of 5 MPa. At the end of each step, confining pressure was kept constant in order to reach equilibrium in pore volume variation. The two triaxial experiments (VO#3 and VO#4) were carried out at a constant strain rate of 10^{-5} s^{-1}.

Then, the deformed samples were unloaded and retrieved from the pressure vessel, the jackets were carefully removed, and samples were impregnated with epoxy for microstructural analysis.

AE activity and elastic wave velocity changes were monitored by twelve *P*-wave piezoelectric sensors, either embedded in the pistons or glued to the sample surface and sealed in a neoprene jacket using two-component epoxy (Fig. 2b). *P*-wave sensors were produced from PZT piezoceramic discs with 5 mm diameter and 2 mm thickness. Transducer signals were amplified by 40 dB using Physical Acoustic Corporation preamplifiers. Full-waveform AE data and the ultrasonic signals for P-wave velocity measurements were stored in a 12-channel transient recording system (*DaxBox, Prökel*, Germany) with an amplitude resolution of 16 bit at 10 MHz sampling rate. For periodic elastic wave speed measurements, six *P* sensors were used as senders applying 100 V pulses every $\sim$30 seconds during the loading.

Ultrasonic transmissions and AE waveforms were discriminated automatically after the experiments. Onset time of *P* wave and AE were picked automatically. Hypocenter locations were estimated, using a downhill simplex algorithm considering time-dependent changes of the anisotropic velocity field. AE events with "adjusted amplitude" exceeding 300 mV were selected. Adjusted amplitude is calculated from the maximum amplitude on each channel and for a distance of 10 mm from the AE hypocenter assuming geometrical spreading of elastic waves (LOCKNER *et al.*, 1992; ZANG *et al.*, 1998). AE hypocenter location errors are estimated to be ± 1 mm (STANCHITS *et al.*, 2006). First-motion amplitudes were picked automatically and first-motion polarities were used to discriminate AE sources types in tensile, shear, and collapse events (ZANG *et al.*, 1998).

3. Results

3.1. Mechanical Data and Failure Mode

In this section, data from two new experiments (VO#4 and VO#5) are presented. We also used data from a triaxial experiment (V0#3) performed at an effective confining pressure of 70 MPa, published in FORTIN *et al.* (2006).

We use, in this study, the convention that compressive stresses and compactive strains are positive. The maximum and minimum (compressive) stresses are denoted by σ_1 and σ_3, respectively. The pore pressure is denoted by P_p, and the difference between the confining pressure ($P_c = \sigma_3 = \sigma_2$) and the pore pressure is referred to as the "effective confining pressure". The effective mean stress $(\sigma_1 + 2\sigma_3)/3 - P_p$ will be denoted *P* and the differential stress $\sigma_1 - \sigma_3$ by *Q*.

From previous studies performed on Bleurswiller sandstone (FORTIN *et al.*, 2005; TEMBE *et al.*, 2008), the brittle faulting regime is observed for effective confining pressure lower than 30 MPa, whereas discrete compaction bands are observed for effective confining pressure higher than 60 MPa. For effective confining pressure in the range of [30–50 MPa], samples of Bleurswiller sandstone develop a hybrid failure involving both discrete compaction bands and shear bands.

Figure 3 summarizes data for the three samples deformed i) under hydrostatic pressure (VO#5), ii) under triaxial loading at an effective confining pressure of 10 MPa (VO#4), iii) and at an effective confining pressure of 70 MPa (VO#3).

Figure 3a) illustrates the hydrostatic compaction behavior. The first part of the loading is representative of elastic deformation. However at the stress state P*, the sample shows a dramatic reduction in porosity (cataclastic compaction). This inflection point P* occurs at an effective pressure of 125 MPa. This value is different from the previous results presented in FORTIN *et al.* (2005), who found 135 MPa. This difference can be partially explained by the methodology: in this study we determined P* using mechanical and AE data, whereas in FORTIN *et al.* (2005) the authors mostly relied on mechanical data.

The evolution of cumulative AE number registered during this experiment is given in Figure 4a. The critical stress state P* coincides with a surge in AE activity in agreement with previous studies (ZHANG *et al.*, 1990; BAUD *et al.*, 2004). Due to the large amount of AE during this experiment (more than 200,000), the *DaxBox* memory was overloaded when the effective pressure reached 140 MPa, and we were not able to record the AE events during an interval of pressure increasing from 140 to 145 MPa (Fig. 4a).

The sample deformed at an effective confining pressure of 10 MPa is representative of the brittle faulting regime (Figs. 3b, 4b). For reference, the hydrostatic loading is plotted as the dashed line (Fig. 3b). The differential stress attains a peak, beyond which strain softening is observed (Fig. 4b). The porosity initially decreases, but at a stress state C' it reverses to an increase indicating dilatation of the pore space (Fig. 3b). The AE data in this experiment follow a trend that has been reported in earlier studies (WONG *et al.*, 1997; ZANG *et al.*, 1998; LEI *et al.*, 2004; SCHUBNEL *et al.*, 2007): the critical stress state C' is associated with a surge in AE events (Fig. 4b) and the rupture is characterized by a surge in AE rate at an axial strain of $\sim 1\%$ (Fig. 4d). However the number of AE recorded during this experiment is only 10,000; one order of magnitude less than the AE number recorded during the experiment VO#5.

For the sample deformed at an effective confining pressure of 70 MPa, the failure mode is associated with an appreciable porosity reduction (Fig. 3c). In the first part of the loading, the triaxial data (solid curve) coincides with the hydrostat (dashed curve). However at stresses higher than C* the deviatoric stresses contribute significantly to the compactive strain (Fig. 3c). This phenomenon of inelastic yield by "shear-enhanced compaction" (WONG *et al.*, 1997) is attributed to the inception of grain crushing and pore collapse in the sandstone (MENÉNDEZ *et al.*, 1996). During this second stage, strain hardening is observed (Fig. 4c). The number of AE events during this loading is given in

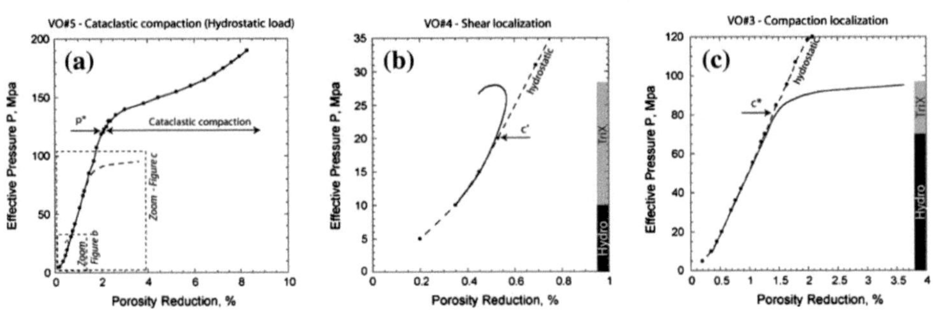

Figure 3

Effective mean pressure versus porosity reduction. a) Hydrostatic loading (for reference, the two triaxial experiments are shown as the dashed lines). b) and c) Triaxial compression experiments performed at effective confining pressure of 10 MPa and 70 MPa, respectively (for reference, the hydrostatic loading is shown as the dashed line). The critical stress state P*, C' and C* are indicated by arrows.

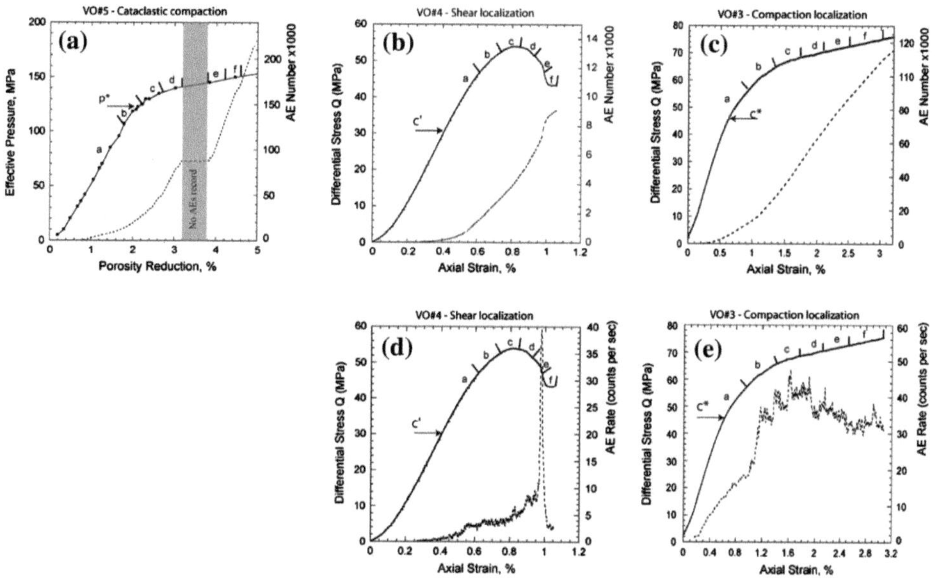

Figure 4

a) Hydrostatic experiment: Effective pressure (solid curve) and cumulative AE (dotted line) as functions of porosity reduction. Triaxial experiments: b) and c) differential stress (solid curve) and accumulative AE (dotted line) versus axial strain; d) and e) differential stress (solid curve) and AE rate (dotted line) versus axial strain. Effective confining pressure was 10 MPa in Figures b) and d) and 70 MPa in Figures c) and e). Six intervals are defined for each experiment (a to f) corresponding to the AE hypocenter distributions shown in Figure 5. The critical stress state P*, C' and C* are indicated by arrows.

Figure 4c. The critical stress state C* is also associated with a surge in AE events as reported in Wong *et al.* (1997), Wu *et al.* (2000), and Baud *et al.* (2006). Cumulative AE events increase with strain, due to the progressive increase of damage (Fig. 4c). As pointed out by Baud *et al.* (2004) important data to track compaction band is the AE rate. Baud *et al.* (2004) and more recently Stanchits *et al.* (this issue) correlated AE rate surges with the formation of discrete compaction bands. In our experiment AE rate surges can be observed (Fig. 4e). These surges are not as clear as those observed in experiments on Bentheim sandstone (Baud *et al.* 2004; Stanchits *et al.*, this issue), but can be compared to the observations from experiments on Diemelstadt sandstone (Baud *et al.*, 2004). At an axial strain of 3.16% the number of AE events is ~118,000; a number comparable to the AE records performed during the experiment VO#5.

3.2. Acoustic Emissions Hypocenters

The three modes of failure are very different on the macroscopic scale, nonetheless they involve the same micromechanical processes (microcracking at grain scale; see Section 3.4). As a consequence these three modes of deformation lead to radiation of acoustic emissions.

Cataclastic compaction. AE locations from sample deformed under hydrostatic loading are shown in Figure 5 (top) and can be compared to the stress-strain curve plotted in Figure 4a. Stress increments *a-f* in Figure 4a correspond to the AE hypocenter distributions *a-f* shown in Figure 5 (top). Each dot represents one acoustic event. Whereas the loading is isotropic, the location of the AE indicates that the processes of grain crushing and pore collapse are not distributed homogeneously. In stage *a* of Figure 5 (top), the AE location shows that the events are mostly concentrated near the ends of the specimen. This is expected as stress concentrations due to end effects are important. However, in stage *b* AE events form clusters distributed through the specimen volume and AE concentrations near the ends due to end effects tend to disappear. The clusters formed in stage *b* grow during stages *c* and *d* and form compacted zones mostly distributed in the middle part of the specimen. In latest stages *e* and *f*, the compaction seems to be more homogeneous.

Shear localization. AE hypocenter locations registered during deformation of sample VO#4 in the brittle regime are given in Figure 5 (middle) and can be compared to the stress-strain curve plotted in Figure 4b. In stage *a*, AE activity appears near the ends of the sample as initial loading begins, probably due to end cap friction. During stages *b* and *c* around 3000 AE events are located and demonstrate the early stages of shear localization in the upper part of the specimen. During stage *d*, the stress starts to drop, and the nucleation zone prior to failure extends. Within the next stages *e* and *f*, unstable failure propagates through the upper part of the sample.

The AE hypocenter locations show that the failure propagates in the last stages of the loading (stages *e* and *f*). This result is consistent with the previous experiments on sandstone performed at AE rate control by LOCKNER *et al.* (1992). The initiation of shear localization appears, in stages *c* and *d*, in the upper left of the specimen. The initiation might have been controlled by some pre-existing heterogeneities, as seen by LEI *et al.* (2004) in granite or SCHUBNEL *et al.* (2007) in sandstone.

Compaction localization. AE locations registered during deformation of sample VO#3 at an effective confining pressure of 70 MPa are shown in Figure 5 (bottom) and can be compared to the stress-strain curve plotted in Figure 4c. Stress increments *a-f* in Figure 4c correspond to the AE hypocenter distributions *a-f* given in Figure 5 (bottom). In the first stage of the loading, the spatial distribution of the AE is similar to the location obtained during the first stage in VO#5 (the specimen deformed under hydrostatic loading): Events are located at the ends of the specimen and at the same time AE clusters form through the specimen volume. However, in this experiment, the migration of AE clusters is not random: In stage *b* the clusters propagate mostly normal to the compression direction. In stages *c* to *f* several discrete compaction bands formed normal to the compression direction can be observed. The overall AE activity is also more intense: 13 times more events were recorded during compaction localization than during shear localization. Two reasons explain such behavior in the overall AE activity: i) For the sample deformed by shear localization, rupture appears at an axial strain of $\sim 1\%$

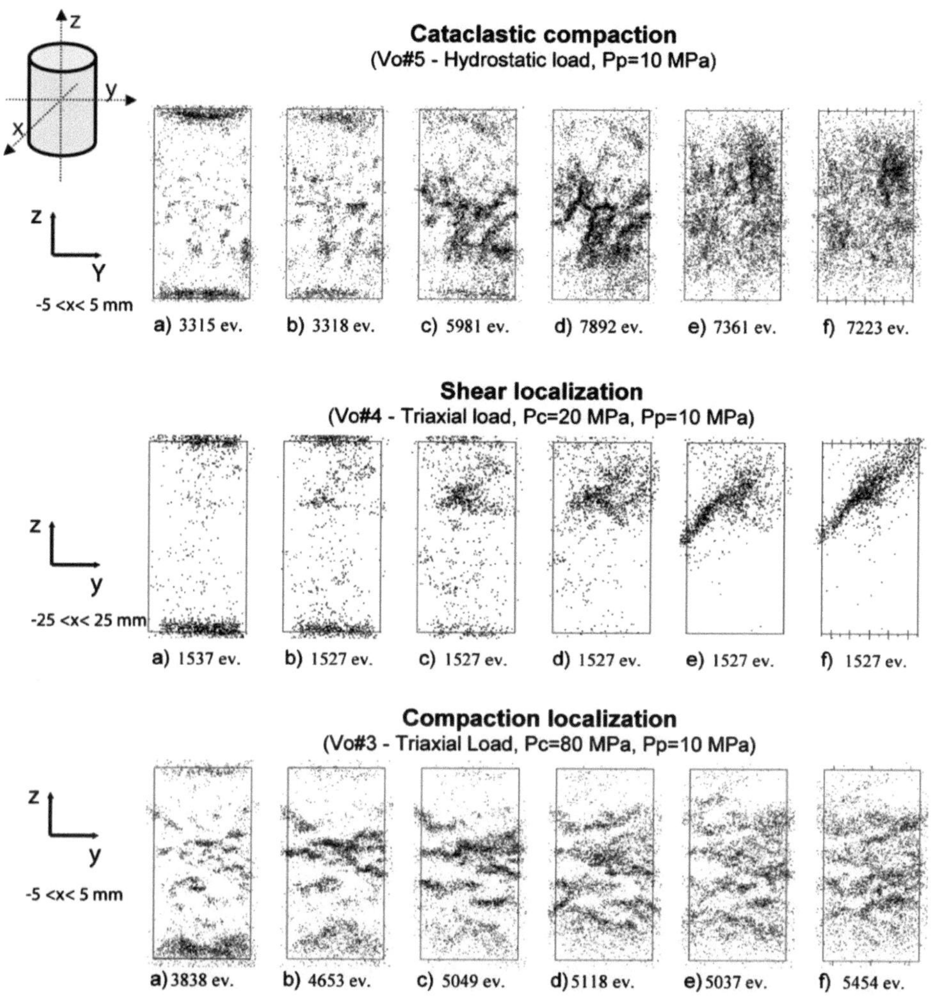

Figure 5

AE hypocenter distribution. The intervals (a to f) are defined in Figure 4. Top plot is a projection of all the events located between -5 mm $< x < 5$ mm, on the (z, y) plane recorded during the hydrostatic experiment. Middle plot is a projection of all the events on the (z, y) plane recorded during the triaxial experiment at 10 MPa effective confining pressure. Bottom plot is a projection of all the events located between -5 mm $< x < 5$ mm, on the (z,y) plane recorded during the triaxial experiment at 70 MPa effective confining pressure.

(Fig. 4d), however for the sample deformed by compaction localization, no rupture of the sample is observed and damage increases with progressive loading, ii) moreover, the damage volume induced by shear localization is less in comparison to the damage volume induced by compaction localization (Fig. 5).

In a recent study, Townend *et al.* (2008) used AE to track the spatial and temporal evolution of a single, discrete, compaction band across the full diameter of a core sample

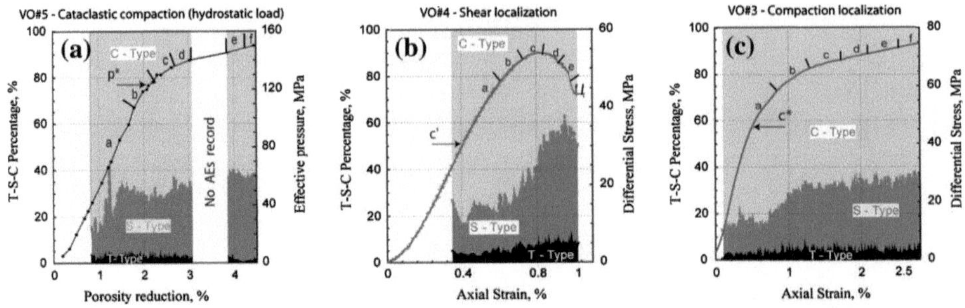

Figure 6

Top: Color code indicates AE-source type distribution during loading. C-, S-, and T-type events possibly represent signals radiated from pore collapse, shear and tensile cracks, respectively. The percentages of each part are indicated on the left vertical scale in the three figures: For example in Figure a) T-type, S-type, and C-type represent, respectively, ~5%, ~28% and ~67% of the source mechanism. For reference, the evolution of effective pressure (right vertical scale) is plotted versus porosity reduction in Figure a). The evolution of differential stress versus axial strain (right vertical scale) is plotted in Figures b) and c).

of Diemelstadt sandstone. In contrast to our observations, Townend *et al.* (2008) observed that compaction band growth in Diemelstadt sandstone originated at the sample boundary before propagating across the sample. Similar observations are also reported in Stanchits *et al.* (this issue), from experiments done on notched samples. The nucleation of compaction band will be discussed in Section 4.1.

3.3. AE Source Characteristics

During the experiments, the full-waveform AE data are recorded which allows us to conduct *P*-wave first-motion studies (see Zang *et al.*, 1996; Lei *et al.*, 1992; Stanchits *et al.*, 2001). In our experiment local AE source mechanisms were separated into pure tensile mechanisms (T), shear-type events (S) and implosion type events (C) representing tensile cracks, shear cracks and pore collapse, respectively.

The results are summarized in Figure 6a for the specimen deformed under a hydrostatic loading, in Figure 6b for the specimen deformed in the brittle regime, and in Figure 6c for the specimen deformed at an effective confining pressure of 70 MPa. In the three experiments, the source type distribution reveals i) a dominance of pore collapse (C-type), ii) a small number of tensile cracks (T-type). During cataclastic compaction, S-type and C-type represent, respectively, ~28% and ~67% of the source mechanisms. For the specimen deformed in the brittle regime, the amount of shear-type events (S-type) increases from ~20% at the beginning of the loading to ~50% of all the events in the postfailure region where the sample strength drops. At the same time, the amount of pore collapse events decreases from ~80% to ~40%. For the experiment performed at 70 MPa effective confining pressure, the amount of shear-type events increases from

~12% to ~26% during shear-enhanced compaction whereas the amount of pore collapse events decreases from ~84 % to ~70%.

The distribution of the events during cataclastic compaction and during the formation of compaction bands seems to be similar, indicating that the micromechanisms are similar. However, the development of a shear band is associated with a different acoustic signature. These results will be discussed in Section 4.2.

3.4. *Microstructural Observation*

Detailed microstructural analysis was performed on the three deformed specimens using scanning electron microscopy (SEM). To prepare SEM sections, samples 25 × 40 mm in size were cut parallel to the long specimen axis. Sections were impregnated with epoxy and subsequently polished and gold coated.

Cataclastic compaction. Figures 7a are SEM micrographs of the deformed sample and illustrate the extensive grain crushing that took place during deformation (compare Fig. 7a and Fig. 1b). The spatial distribution of damage is quite homogeneously in agreement with the last stages of the AE locations (Fig. 5, top). Grain fragments fill up the existing pores leading to a large decrease in porosity. The grain crushing process was probably triggered by extensile cracking initiated at the contact region where stress concentration had been induced by grain-grain contacts, in a manner analogous to the Hertzian fracture model proposed by Zhang *et al.* (1990).

Shear localization. Figure 7b illustrates the damage evolution in the specimen that failed by dilatant failure and shear localization. The shear localization developed in the upper part of the sample along a planar zone, which is inclined at an angle of ~45° to σ_1 (Fig. 5, center and Fig. 7b). The width of the shear band is ~0.3 mm, which is ~3 time the average of mean grain diameter (0.11 mm). The central part of the localization shows an intense grain crushing. Inside the band, mica minerals accommodate the deformation, indicating displacement of neighboring quartz or feldspar grains. Outside the band, damage is very small. These observations are in agreement with those reported by Menéndez *et al.* (1996) in Berea sandstone and by Bésuelle *et al.* (2003) in Rothbach sandstone.

Compaction localization. In Figures 7c we present SEM micrographs corresponding to the specimen deformed at 70 MPa effective confining pressure. In these pictures, a compaction band appears as a zone of intensive cracking. The orientation of this zone is subhorizontal (perpendicular to σ_1), in agreement with the AE locations (Fig. 5, bottom). Grain crushing is evident; the fragments fill the pore space leading to a decrease in the overall porosity. The thickness is about ~0.6 mm which represents ~5 mean grain diameters. A compaction band in Bleurswiller sandstone appears thicker than those observed in Diemelstadt sandstone or Bentheim sandstone, where the thickness of a

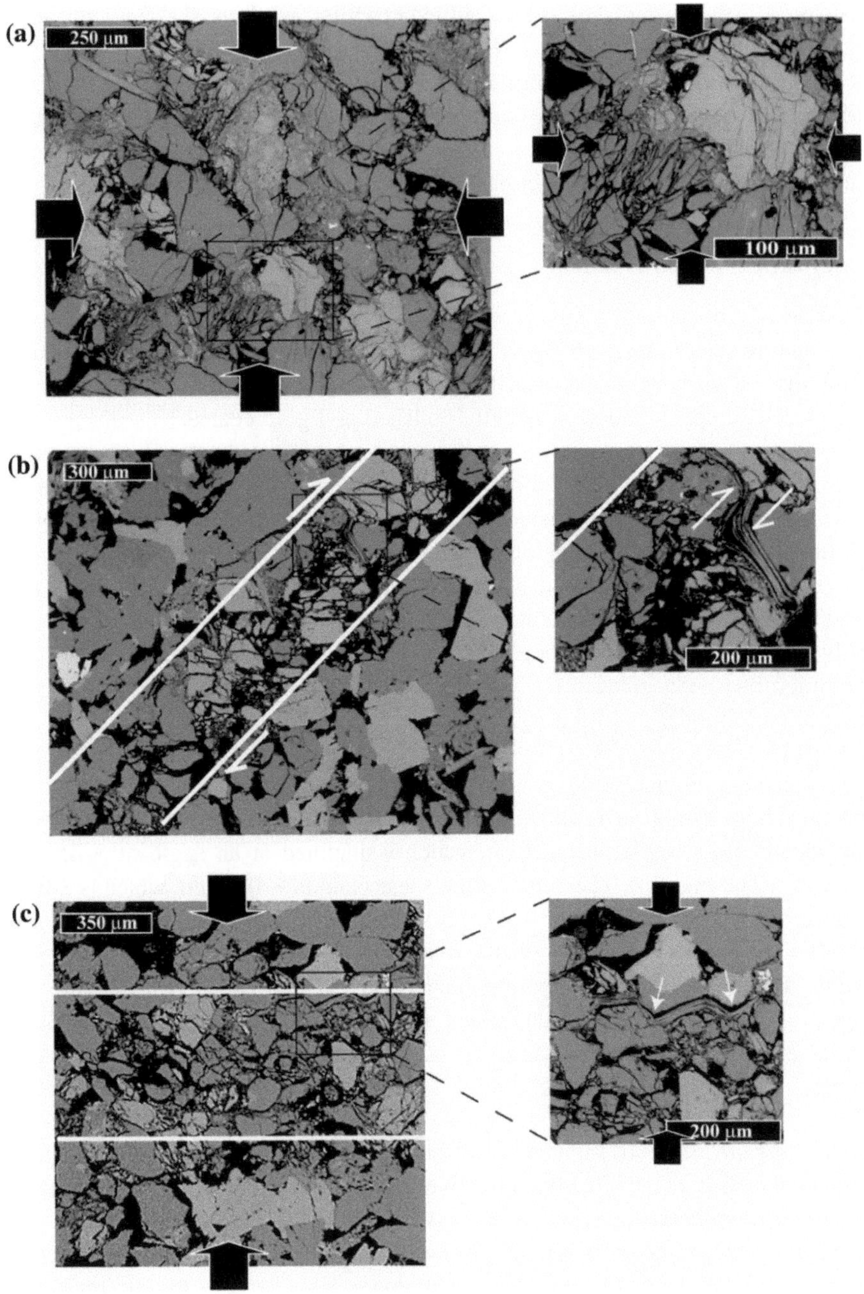
(a)
250 μm
100 μm
(b)
300 μm
200 μm
(c)
350 μm
200 μm

◀ Figure 7
SEM micrographs (backscattered) of deformed Bleurswiller sandstone loaded a) under hydrostatic pressure (cataclastic compaction), b) in triaxial compression at an effective confining pressure of 10 MPa (Shear localization), c) in triaxial compression at an effective pressure of 70 MPa (compaction band). For b) and c) maximum compressive stress is vertical. Epoxy-filled pores appear in black. Grain crushing is pervasive in the three deformed specimens, which led to radiation of acoustic emissions.

compaction band represents ~2-3 grains (Tembe *et al.*, 2008). Outside the band, the microstructure of the rocks remains largely undeformed with a porosity of ~25%. Mica minerals are present, and accommodate the deformation, showing the displacement of neighboring quartz or feldspar grains.

4. *Discussion*

4.1. Nucleation and Propagation of Damaged Zones

From Figure 5, we observed that AE events occur during the early stages of loading in the three experiments. Following Lockner (1992), Zang *et al.* (1998) and Lei *et al.* (2000) the behavior of the AE during the early stages is interpreted as indicating the presence of significant weak zones or zones of stress concentration which acted as nucleation sites. This interpretation is also confirmed by medical X-ray scanner tomography done by Fortin *et al.* (2005) on an intact specimen: This technique reveals zones of several mm^2 of high porosity, randomly located in the sample. Using X-ray CT scan, a more precise technique, Louis *et al.* (2006) report in Rothbach sandstone, porosity variation reaching 7% across the sample. Pre-existing microcracks and high porosity regions appear to be the most important factors controlling the local strength of the rock, and thus controlling the nucleation sites.

For the specimen deformed at an effective confining pressure of 10 MPa (Fig. 5, center), the early AE events are relatively diffuse, however in stage *c*, local heterogeneity appears to be sufficient in magnitude to determine where failure will initiate.

For the specimen deformed at an effective confining pressure of 70 MPa, the early AE events form clusters (stages *a* and *b* in Figure 5, bottom). Local high porosity regions may determine the location of theses clusters. Subsequently, these heterogeneities act as zones of stress concentration and initiate compaction bands. This process is similar to the effect of a notch, which induces stress concentration and facilitates the initiation of a compaction band (see Stanchits *et al.*, this issue). In notched specimens only one compaction band is initiated. In contrast, in Bleurswiller sandstone, local heterogeneities can induce several bands. However, in our specimen, when compaction bands propagate, they can merge (stages *c* and *d* on Figure 5, bottom). As a consequence, compaction bands in Bleurswiller sandstone are not perfectly plane, but are an association of parts of pure compaction localization formed perpendicular to the main compressive axis connected with small shear bands.

For the specimen deformed under hydrostatic loading, the local heterogeneities appear to be sufficient to induce AE clusters (stage *b* on Figure 5, top). However, in this experiment the orientation of the compacted zones seems to be random (stages *c* and *d* on Figure 5, bottom). With increasing pressure, the compacted zones merge, and in the last stages the AE event distribution seems to be more diffuse. Local heterogeneities may control the early stages of the deformation during the formation of compaction bands and during cataclastic compaction. To understand further the role of local heterogeneities in the process of Bleurswiller sandstone deformation, it would be useful to perform X-ray CT scan (where the resolution is in the order of $\sim$50 μm), because with this nondestructive method it would possible to acquire images of the same sample before and after failure (LOUIS *et al.*, 2006).

4.2. Micromechanics of Failure

The ratio of shear-type cracking (S-type) to pore collapse event (C-type) is related to the stress boundary conditions and porosity of the sample. Tensile cracking appears to be negligible. The ratio $\frac{S\text{–type}}{C\text{–type}}$ during cataclastic compaction is $\sim$0.41; a similar ratio of $\sim$0.37 is obtained during formation of compaction bands (Fig. 6). During the formation of shear localization the ratio increases to $\sim$1.25. Thus during the three experiments, AE source types indicate dominance of pore collapse and shear cracking. However, shear cracks are seldom seen in the microstructure (Fig. 7). What causes this source type distribution?

First, it is important to keep in mind that the AE records represent a small fraction of the total damage. For example, during the deformation of the specimen under hydrostatic loading, about 220,000 AE events were typically detected. For a grain size of $\sim$110 μm, this gives approximately one AE event for every $\sim$1,000 grains, or a mean distance $\langle l \rangle$ between two AE locations of $\langle l \rangle \approx$ *10 d*, where *d* is the mean grain diameter. During experiment VO#4, where shear localization was observed, around 9, 000 AE events were recorded. Assuming that the damage volume could be roughly estimated as the area of the shear localization (27.9 cm^2) multiplied by the thickness of the band (0.3 mm), this gives approximately one AE event for every $\sim$150 grains or a mean distance $\langle l \rangle$ between two AE locations of $\langle l \rangle \approx$ *5 d*, similar to the result found in the hydrostatic experiment. In his paper LOCKNER *et al.* (1992) recorded, using a different threshold level, $\sim$5,000 AE events during shear localization in sandstone, which is approximately one event for every $\sim$3,000 grains. Thus cracking events detected by AE represent a small fraction of the total microcrack damage occurring in the rock. However, the excellent agreement between AE locations and observed fault or compaction bands found by LOCKNER and BYERLEE (1977), LEI *et al.* (2004), FORTIN *et al.* (2006) and STANCHITS *et al.* (this issue) support the idea that microcracks detected by AE are representative of the overall microcrack population.

Figure 8 gives a conceptual model of a tensile event (Fig. 8a), a shear type event (8b) and a pore collapse event (8c). A pore collapse event is a mixed mode: a pore collapse event is a consequence of re-arrangement of the grain-packing (Fig. 8d). In order to collapse the pore, grains and parts of the cement need to be broken (8c, 8d); the breakage

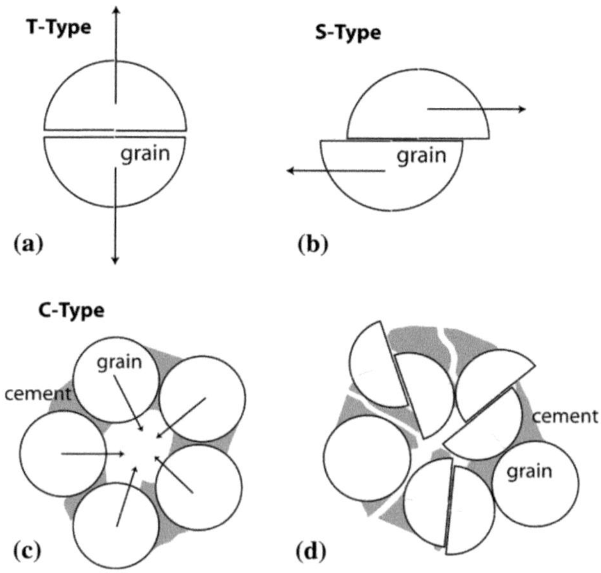

Figure 8
Conceptual model of a tensile event (a), a shear-type event (b) and a pore collapse event (c). A pore collapse event is a mixed mode: a pore collapse event is a consequence of a packed re-arrangement (d). Note that breakage of some grains or/and cement is needed to collapse a pore and radiate a C-type event.

of grains and then the collapse of the pore lead to a collapse event (C-type) (STANCHITS *et al.*, 2001).

During the formation of shear localization (Fig. 6b) our AE source type analysis indicates a ratio of shear-type cracking to pore collapse events of $\sim$1.25, in agreement with the study of ZANG *et al.* (2000). However, during the formation of compaction bands and cataclastic compaction our AE focal analysis indicates a ratio of $\sim$0.4. From Figure 8c, we would expect that these two kinds of inelastic deformation would be associated with pore collapse events and shear-type sources (see above). Two explanations can be given: i) Pore collapse induced grain cracking (Fig. 8d), thus with loading these cracks can be activated (frictional shearing) and radiate shear-type events; ii) from AE locations, we suggested in Section 4.2 that the interaction between two compaction bands can lead to small shear bands when they merge. Thus, these small shear bands can also lead to shear-type events.

4.3. *Yield Surface*

Data for the brittle strength and compactive yield stress are compiled in Figure 9. The equivalent von Mises shear stress for a compressive axisymmetric stress state are defined by $\tau = (\sigma_1 - \sigma_3)/\sqrt{3} = Q/\sqrt{3}$. The stress path of the three experiments presented in this study (VO#3; VO#4, VO#5) are drawn as grey dashed-lines. Data of FORTIN *et al.*

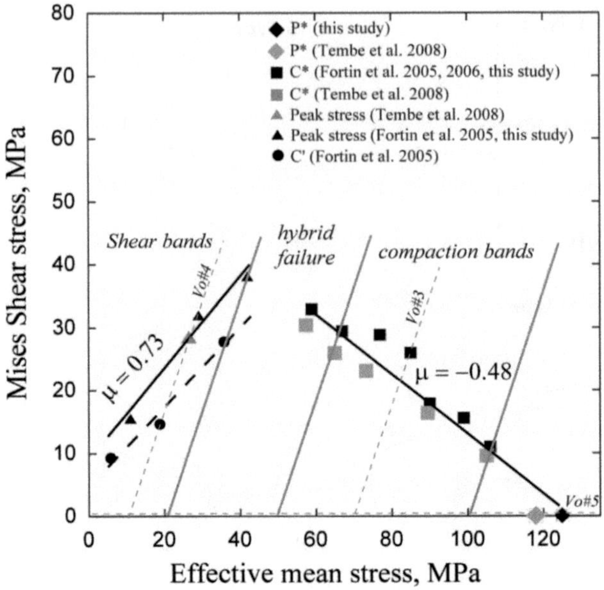

Figure 9
Peak stress, onset of dilatancy (stress state C') and onset of shear-enhanced compaction (C*) in the Von Misès shear stress versus effective mean stress plane. Note that elastic domain can be delimited by two straight lines. μ can be inferred from the slope of the lines and denotes the friction coefficient.

(2005, 2006, 2007) and TEMBE *et al.* (2008) are also included. Duplicate experiments indicate that experiments performed on cylindrical specimens of 38.1 mm in length and 18.4 mm in diameter (grey symbols on Fig. 9, TEMBE *et al.*, 2008) lead to peak stresses and compactive yield stresses lower than the values inferred from experiments performed on bigger specimens (black symbols): in FORTIN *et al.* (2005) and (2007) the length and diameter of the specimens are 80 mm and 40 mm, respectively; and in this study and in FORTIN *et al.* (2006) the length and diameter of the specimens are 100 mm and 50 mm, respectively.

The shape of the yield envelope controls the predictions of bifurcation analyses (RUDNICKI and RICE, 1975; ISSEN and RUDNICKI, 2000; BÉSUELLE, 2001, RUDNICKI, 2004). One of the important parameter is the pressure-sensitivity parameter μ, which can be inferred from the slope of the yield stress data as indicated in Figure 9. In the brittle regime the elastic domain can be bounded by a line, with a positive slope $\mu = 0.73$. This value is inferred from all the data plotted on Figure 9. In the compactive regime, yield stresses (C*) usually map an ellipse (WONG *et al.*, 1997; BAUD *et al.*, 2006), however our data show that compactive yield stresses in Bleurswiller sandstone rather map a straight line, with a negative slope $\mu = -0.48$, a value inferred from all the data plotted on Figure 9. The physical explanation for the geometry of the yield envelope in the compactive domain remains unclear: It can be due to the presence of heterogeneity in the rock; it can also be the consequence of a nonnegligible part of mica minerals in the mineralogical composition of the rock.

5. *Conclusions*

In this paper, three modes of deformation were investigated during the compaction of porous rock. Under hydrostatic pressure, a large mechanical decrease of porosity is observed at a critical pressure, where pore collapse and grain crushing occur. Under axissymmetric compression, the mechanical response of the sample deformed at 10 MPa effective confining pressure is characterized by brittle faulting. However, at 70 MPa effective confining pressure, compaction bands can be observed.

From acoustic emission records, we observed that;

i) The number of acoustic events during shear localization is around one order of magnitude less than the number of events recorded during the formation of compaction bands or during cataclastic compaction. However, the mean distance between two AE locations is, in both experiments, in the same range (between 5 and 10 *d*, where d is the mean grain diameter). In fact, the occurrence of AE activity is associated with the total damage volume, which is more important during cataclastic compaction or during the formation of compaction bands than during shear localization.

ii) AE locations allow us to characterize the spatial evolution of the damage in the rock during the three modes of deformation. Nucleation seems to be strongly affected by material heterogeneity (high local porosity). The AE locations indicate that the early stages of deformation during the formation of compaction bands and during cataclastic compaction are similar.

iii) First-motion studies show that the micromechanisms taking place during the formation of compaction bands and cataclastic compaction are similar (predominance of pore collapse events), whereas the shear localization is characterized by a different acoustic emission signature (predominance of shear events).

Acknowledgments

We are grateful to Patrick Baud and the anonymous reviewer for their careful and critical comments. The authors would particularly like to thank Stefan Gehrmann (GeoForschnugsZentrum, Potsdam) for aid in preparing the thin sections, and Nathanael Findling (Ecole normale supérieure, Paris) for aid in SEM analysis. This work was partially supported by the French-German grant “Procope”.

References

BAUD, P., KLEIN, E., and WONG, T.-F. (2004), *Compaction localization in porous sandstones: Spatial evolution of damage and acoustic emission activity*, J. Struct. Geol. *26*, 603– 624.

BAUD, P., VAJDOVA, V., and WONG, T.-F. (2006), *Shear-enhanced compaction and strain localization: Inelastic deformation and constitutive modeling of four porous sandstones*, J. Geophys. Res. *111*, B12401, doi:10.1029/2005JB004101.

BÉSUELLE, P. (2001), *Compacting and dilating shear bands in porous rocks: Theoretical and experimental conditions*, J. Geophys. Res. *106*, 13435–13442.

BÉSUELLE, P., BAUD, P., and WONG, T.-F. (2003), *Failure mode and spatial distribution of damage in Rothbach sandstone in the brittle-ductile transition*, Pure Appl. Geophys. *160*, 851–868.

DAVID, C., WONG, T.-F., ZHU, W., and ZHANG, J. (1994), *Laboratory measurement of compaction induced permeability change in porous rocks: Implication for the generation and maintenance of pore pressure excess in the crust*, Pure Appl. Geophys. *143*, 425–456.

FORTIN, J., GUÉGUEN, Y., and SCHUBNEL, A. (2007), *Effect of pore collapse and grain crushing on ultrasonic velocities and Vp/Vs*, J. Geophys. Res. *112*, doi:10.129/2005JB004005.

FORTIN, J., STANCHITS, S., DRESEN, G., and GUÉGUEN, Y. (2006), *Acoustic emission and velocities associated with the formation of compaction bands*, J. Geophys. Res. *111*, B10203, doi:10.1029/2005JB003854.

FORTIN, J., SCHUBNEL, A., and GUÉGUEN, Y. (2005), *Elastic wave velocities and permeability evolution during compaction of Bleurswiller sandstone*, Int. J. Rock. Mech. Min. Sci. Geomech. *42*, 2005, 873–889.

GRUESCHOW, E. (2005a), *Yield cap constitutive models for predicting compaction localization in high porosity sandstone*, Ph.D. Thesis, Northwestern Univ., Evanston, Ill.

GRUESCHOW, E. and RUDNICKI, J. W. (2005b), *Elliptic yield cap constitutive modeling for high porosity sandstone*, Int. J. Solids Struct. *42*, 4574– 4587.

ISSEN, K. A. and RUDNICKI, R. J. (2000), *Conditions for compaction bands in porous rock*, J. Geophys. Res. *105*, 21529–21536.

KLEIN, E., BAUD, P., REUSCHLE, T., and WONG, T.-F. (2001), *Mechanical behavior and failure mode of Bentheim sandstone under triaxial compression*, Phys. Chem. Earth, Part A *26*, 21– 25.

LEI, X, NISHIZAWA, O., KOSE, K., and SATOH, T. (1992), *Fractal structure of the hypocenter distributions and focal mechanism solutions of acoustic emission in two granites of different grain sizes*, J. Phys. Earth. *40*, 617–634.

LEI, X., NISHIZAWA, O., KUSUNOSE, K., CHO, A., and SATOH, T. (2000), *On the compressive failure of shale samples containing quartz-healed joints using rapid AE monitoring: the role of asperities*, Tectonophysics *328*, 329–340.

LEI, X.-L., MASUDA, K., NISHIZAWA, O., JOURNIAUX, L., LIU, L, MA, W., SATOH, T., and KUSUNOSE, K. (2004), *Detailed analysis of acoustic emission activity during catastrophic fracture of faults in rock*, J. Struct. Geol. *26*, 247–258.

LOCKNER, D. A. and BYERLEE, J. D., *Acoustic emission and fault formation in rocks*, In *Proc 1st Conf. on Acoustic Emission/Microseismic Activity in Geol. Structures and Materials* (eds. H.R., Hardy and F.W., Leighton), (Trans-Tech. Publications, Clausthal-Zellerfeld 1977), pp. 99–107.

LOCKNER, D. A. (1993), *The role of acoustic emission in the study of rock fracture*, Int. J. Rock Mech. Min. Sci. Geomech. *30*, 883–899.

LOCKNER, D.A., BYERLEE, J. D., KUKSENKO, V., PONOMAREV, A., and SIDORIN, A., *Observation of quasistatic fault growth from acoustic emissions*. In *Fault Mechanics and Transport Properties of Rocks* (eds. B. Evans and T.-f. Wong), (Academic Press 1992) pp. 3–31.

LOUIS, L., WONG, T-F, BAUD, P., and TEMBE, S. (2006), *Imaging strain localization by X-ray computed tomography: Discrete compaction bands in Diemelstadt sandstone*, J. Struc. Geol. *28*, 5, 762–775.

MÉNENDEZ, B., ZHU, W., and WONG, T.-F. (1996), *Micromechanics of brittle faulting and cataclastic flow in Berea sandstone*, J. Struct. Geol. *18*, 1–16.

NAGEL, N. B. (2001), *Compaction and subsidence issues within the petroleum industry: From Wilmington to Ekofisk and beyond*, Phys. Chem. Earth *26*, 3–14.

OLSSON, W. A. and HOLCOMB, D. J. (2000), *Compaction localization in porous rock*, Geophys. Res. Lett. *27*, 3537–3540.

OSTERMEIER, R. M. (2001), *Compaction effects on porosity and permeability: Deepwater Gulf of Mexico turbidites*, J. Pet. Technol. 68–74.

RUDNICKI, J. W. and RICE, J. R. (1975), *Conditions for the localization of deformation in pressure sensitive dilatant materials*, J. Mech. Phys. Solids *23*, 371–394.

RUDNICKI, J. W. (2004), *Shear and compaction band formation on an elliptic yield cap*, J. Geophys. Res. *109*, B03402, doi:10.1029/2003JB002633.

SANO, O., ITO, I., and TERADA, M. (1981), *Influence of strain rate on dilatancy and strength of oshima granite under uniaxial compression*, J. Geophys. Res. *86* (B10), 9299–9311.

SCHOLZ, C. H. (1968), *Microfracturing and the inelastic deformation of rock in compression*, J. of Geophys. Res. *73*, 1471–1432.

SCHUBNEL, A., THOMPSON, B., FORTIN, J., GUÉGUEN, Y., and YOUNG, R. P. (2007), *Pore pressure induced rupture and aftershocks on intact and fractured sandstone in the laboratory*, Geophys. Res. Lett. *34*, doi:10.1029/2007GL031076.

STANCHITS, S., ZANG, A., and DRESEN, G. (2001), *Focal mechanisms of acoustic emission events during fault propagation and friction sliding*, EOS Trans. AGU *82* (47), Fall Meet. Suppl., Abstract T51A-0847.

STANCHITS, S., DRESEN, G., and VINCIGUERRA, S. (2006), *Ultrasonic velocities, acoustic emission characteristics and crack damage of basalt and granite*, Pure Applied Geophys. *163*, 5–6, 975–994.

STANCHITS, S., FORTIN, J., GUÉGUEN, Y., and DRESEN, G. (2009), *Initiation and propagation of compaction bands in dry and wet Bentheim sandstone*, this issue.

TEMBE, S., BAUD, P., and WONG, T.-F. (2008), *Stress conditions for the propagation of discrete compaction bands in porous sandstone*, accepted in J. Geophys. Res.

TOWNEND, E., THOMPSON, B., BENSON, P. M., MEREDITH, P. G., BAUD, P., and YOUNG, R. P. (2008), *Imaging compaction band propagation in Diemelstadt sandstone using acoustic emission locations*, Geophys. Res. Lett. *35*, L15301, doi:10.1029/2008GL034723.

VAJDOVA, V., BAUD, P., and WONG, T.-F. (2004), *Permeability evolution during localized deformation in Bentheim sandstone*, J. Geophys. Res. *109*, B10406, doi:10.1029/2003JB002942.

WU, X. Y., BAUD, P., and WONG, T.-F. (2000), *Micromechanics of compressive failure and spatial evolution of anisotropic damage in Darley Dale sandstone*, Int. J. Rock Mech. Min. Sci. *37*, 143–160.

WONG, T.-F., DAVID, C., and ZHU, W. (1997), *The transition from brittle faulting to cataclastic flow in porous sandstone: Mechanical deformation*, J. Geophys. Res. *102*, 3009–3025.

WONG, T.-F., DAVID, C., and MENÉNDEZ, B., *Mechanical compaction*. In *Mechanics of Fluid-Saturated Rocks*, Int. Geophys. Ser., vol. 89 (eds. Y., Guéguen and M., Bouteca), (Elsevier, New York 2004), pp. 55–114.

ZANG, A., WAGNER, F. C., and DRESEN, G. (1996), *Acoustic emission, microstructure, and damage model of dry and wet sandstone stressed to failure*, J. Geophys. Res. *101*, 17507–17521.

ZANG, A., WAGNER, C. F, STANCHITS, S., DRESEN, G., ANDRESEN, R., and HAIDEKKER, M. A. (1998), *Source analysis of acoustic emissions in Aue granite cores under symmetric and asymmetric compressive loads*, Geophys. J. Int. *135*, 1113–1130.

ZANG, A., WAGNER, C. F, STANCHITS, S., JANSSEN, C., and DRESEN, G. (2000), *Fracture process zone in granite*, J. Geophys. Res. *105*, 651–661.

ZHANG, J., WONG, T.-F., and DAVIS, D. M. (1990), *Micromechanics of pressure-induced grain crushing in porous rocks*, J. Geophys. Res. *95*, 341–352.

Received June 6, 2008, revised January 27, 2009, accepted February 13, 2009)
Published Online First: May 12, 2009

Pure appl. geophys. 166 (2009) 843–868
0033–4553/09/050843–26
DOI 10.1007/s00024-009-0478-1

Pure and Applied Geophysics

Initiation and Propagation of Compaction Bands in Dry and Wet Bentheim Sandstone

Sergei Stanchits,[1] Jerome Fortin,[2] Yves Gueguen,[2] and Georg Dresen[1]

Abstract—We investigated initiation and propagation of compaction bands (CB) in six wet and four dry Bentheim sandstone samples deformed in axial compression tests with strain rates ranging from 3.2×10^{-8} s^{-1} to 3.2×10^{-4} s^{-1}. Circumferential notches with 0.8-mm width and 5-mm depth served to initiate CB at mid-sample length. Wet samples were saturated with distilled water and deformed at 195 MPa confining pressure and 10 MPa pore pressure. Dry samples were deformed at 185 MPa confining pressure. Twelve P-wave sensors, eight S-wave sensors and two pairs of orthogonally oriented strain-gages were glued to the sample surface to monitor acoustic emission (AE), velocities and local strain during the loading process. Nucleation of compaction bands is indicated by AE clusters close to the notch tips. With progressive loading, AE activity increased and AE hypocenters indicated propagation of a single CB normal to the sample axis. CB propagation from the sample periphery towards the centre was monitored. Microstructural analysis of deformed samples shows excellent agreement between location of AE clusters and CBs. In both dry and wet samples the lateral propagation of CBs was about 100 times faster than axial shortening rates. At the slowest displacement rate, AE activity during band propagation was reduced and CB nucleation in wet samples occurred at 20% lower stresses. This may indicate an increasing contribution of stress corrosion processes to the formation of the compaction bands. In dry and wet samples inelastic compaction energy per area ranged between 16 and 80 kJ m^{-2}. This is in good agreement with previous estimates from laboratory and field studies.

Key words: Compaction bands, acoustic emission, fracture, rock.

1. *Introduction*

Localized compaction bands in high porosity rocks are narrow zones with significantly reduced local porosity. Mollema and Antonellini (1996) described the existence of CBs in the Navajo sandstone formation of southern Utah (USA). Sternof *et al.* (2005) observed CB arrays in Aztec sandstone exposed in the southeastern part of Nevada (USA). Inside the CBs, porosity is only a few percent and significantly reduced with respect to the initial 20–25% porosity of the sandstone.

[1] GeoForschungsZentrum Potsdam, Department 3.2, Telegrafenberg D420, 14473 Potsdam, Germany. E-mail: stanch@gfz-potsdam.de
[2] Laboratoire de Geologie, Ecole Normale Superieure, CNRS, UMR 8538, 24 rue Lhomond, 75005 Paris, France. E-mail: fortin@geologie.ens.fr

Compaction bands may form permeability barriers affecting fluid circulation and extraction of oil and gas from reservoirs, and may have important implications for the management of both groundwater and hydrocarbon resources (TAYLOR and POLLARD, 2000; STERNLOF *et al.*, 2004 and 2006). Recent laboratory studies have demonstrated that compaction localization is an important failure mode in different sandstones with porosities ranging from 13 to 28% (OLSSON and HOLCOMB, 2000; DIGIOVANNI *et al.*, 2000; KLEIN *et al.*, 2001; WONG *et al.*, 2001; BAUD *et al.*, 2004, 2006, FORTIN *et al.*, 2005). FORTIN *et al.* (2006) analyzed the AE hypocenter distribution in order to investigate CB propagation in Bleurswiller sandstone. They found a close spatial correspondence of AE clusters and CB observed in the deformed samples.

Some field observations suggest development of compaction bands induced by local stress concentrations. For example, MOLLEMA and ANTONELLINI (1996), observed formation of compaction bands in the compressional quadrant close to the tip of a shear band. In laboratory tests on sandstone specimens, nucleation of CBs was found to occur at boreholes or notches. Haimson and co-workers (HAIMSON 2001, 2003; KLAETSCH and HAIMSON, 2002; HAIMSON and KOVACHICH, 2003; HAIMSON and LEE, 2004) in true-triaxial tests of porous sandstones observed the formation of slot-like borehole breakouts that are very similar to compaction bands.

VAJDOVA and WONG (2003) investigated formation of CBs in Bentheim sandstone samples with a circumferential V-shaped notch. They observed propagation of CBs from the notch tip in sequential increments. CB formation was accompanied by bursts of AE activity and reduction of axial force applied to the sample. TEMBE *et al.* (2006) report CB initiation in sandstones with porosities ranging from 21% to 23% subjected to confining pressures between 150 MPa and 350 MPa. These authors found a close correspondence of experimental results with a spring network model presented by KATSMAN and AHARONOV (2006), which simulates local compactive failure by shortening the bond length. The model of KATSMAN and AHARONOV (2006) shows that discrete compaction bands could be nucleated at the notch tip and propagate incrementally through the entire sample cross section, similar to experimental observations. In the spring-network model the shortening of the bonds is also accompanied by a local stress drop.

The energy required to develop compaction bands was estimated to range from 6 to 43 kJ m^{-2} (VAJDOVA and WONG, 2003; TEMBE *et al.*, 2006). This is more than 1000 times higher than the specific surface energy (of the order of 1 J m^{-2}) of tensile fractures of minerals (ATKINSON and MEREDITH, 1987), yet comparable to shear fracture energies for rocks under triaxial compression (WONG, 1982; LOCKNER *et al.*, 1991). TEMBE *et al.* (2006) suggested that micromechanisms involved in the formation of compaction and shear bands are similar (intense microcracking and grain crushing), leading to dissipation of significant amounts of inelastic energy. The energy estimates based on laboratory tests are in a good agreement with field observations (about 40 kJ m^{-2}, STERNLOF *et al.*, 2005), and with energies predicted from mechanical models (RUDNICKI and STERNLOF, 2005; RUDNICKI, 2007). RUDNICKI (2007) recently presented an anti-crack-dislocation model of CB propagation. The model predicts the band thickness (or midpoint closure) to increase with the square root of the band half-length, assuming that energy release rate of CB propagation remains constant. This theoretical prediction is in good agreement with the results of field observations (MOLLEMA and ANTONELLINI, 1996; STERNLOF *et al.*, 2005).

VAJDOVA and WONG (2003) identified three stages of CB formation in notched Bentheim sandstone: (a) Nucleation of a compaction band, (b) extension of a diffuse compaction band, and (c) development of an array of compaction bands. AE activity suggested that growth of CBs occurred in increments and is roughly 100 times faster than axial shortening of the samples. Brittle faulting of low-porosity rock has been thoroughly studied in the laboratory during the last two decades using AE hypocenter localization (LOCKNER, 1993; LEI *et al.*, 1992, 2000; ZANG *et al.*, 1998, 2000; THOMPSON *et al.*, 2006; STANCHITS *et al.*, 2006). It has been demonstrated that AE data provide important constraints for the analysis of the onset of shear localization. In a recent paper FORTIN *et al.* (2006) successfully applied AE hypocenter analysis and showed a close spatial agreement between the position of individual CBs and the corresponding AE hypocenter clusters. In this study we use a similar technique to monitor geometrical and AE characteristics and energy dissipation during CB propagation in dry and wet Bentheim sandstone.

2. *Experimental Techniques*

2.1. *Sample Material and Testing Procedure*

Experiments were performed on cylindrical samples of 50 mm diameter and 105 mm length cored from Bentheim sandstone (Gildehaus quarry, Germany). Bentheim sandstone (BS) is a Lower Cretaceous, homogeneous, yellow sandstone, containing 95% quartz, 3% feldspar, and 2% kaolinite (KLEIN and REUSCHLE, 2003). Porosity of BS is about 22%, grain size varies from 18 μm to 500 μm and the mean grain size is about 300 μm (REINICKE *et al.*, 2008). Specimens were prepared with a circumferential notch at mid-sample length. The notch is 0.8 mm wide with a depth of 5 mm (Fig. 1). A teflon O-ring of about 0.7 mm thickness was used to fill the notch to prevent rupture of the Neoprene jacket used to insulate samples from the oil confining medium. Experiments were performed on dry and wet Bentheim samples. Dry samples were kept in an oven at 50°C temperature under vacuum ($\sim 10^{-2}$ bar) for at least 12 hours. After jacketing and installation in the pressure vessel the samples were subjected to a vacuum of $\sim 10^{-2}$ bar for about 12 hours before loading. The vacuum was maintained during the experiment. To prepare wet samples, specimens were first saturated with distilled water for more than 12 hours. Tests were carried out at drained conditions with a constant pore fluid (water) pressure $P_P = 10$ MPa using a QUIZIX pore pressure pump.

For all tests, effective confining pressure was maintained constant at $P_{\text{eff}} = (P_C - P_P) =$ 185 MPa, where P_C and P_P are confining pressure and pore pressure, respectively. Experiments were performed at constant displacement rate in a servo-hydraulic loading frame from Material Testing Systems (MTS) with a load capacity of 4600 KN. Strain

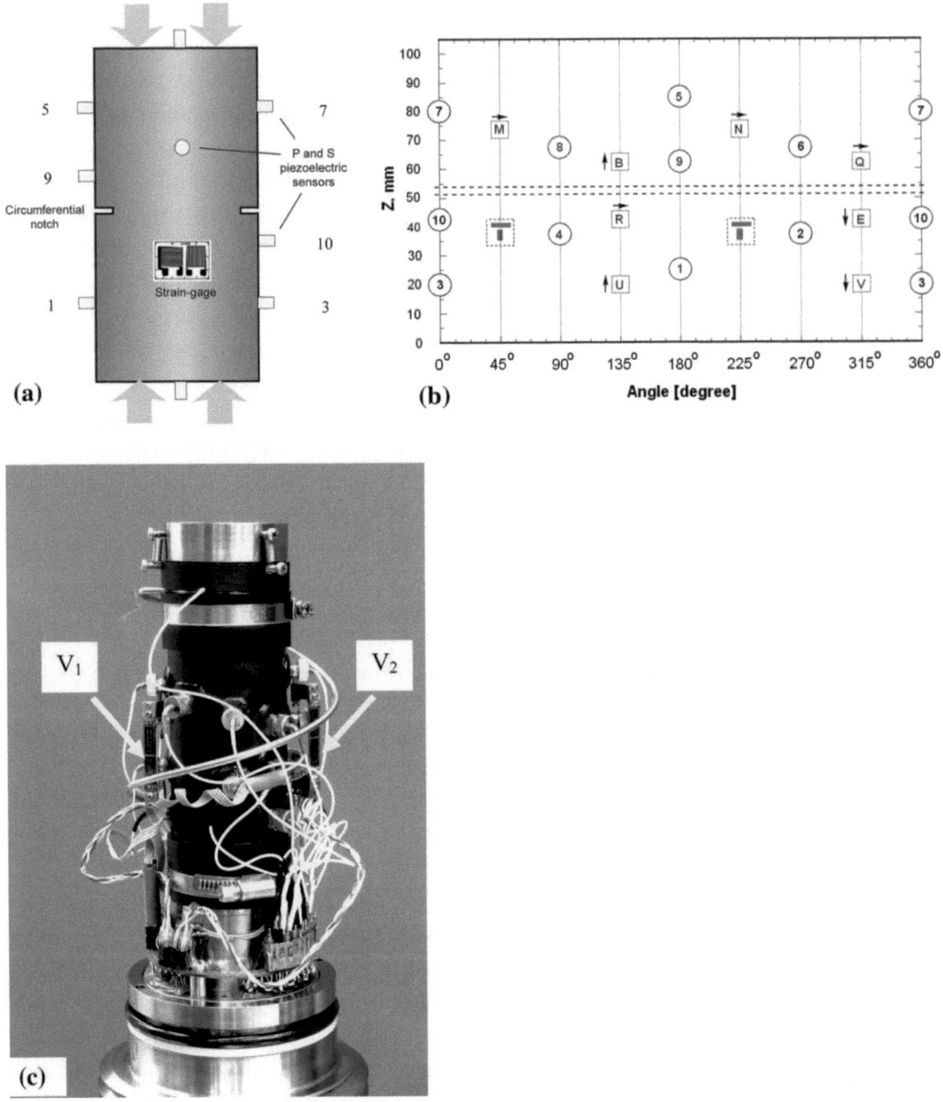

Figure 1

(a) Sketch of notched sample; (b) projected sample surface showing position of *P*-piezoelectric sensors (circles), *S* sensors (squares) and strain-gages (dashed-squares). Arrows indicate direction of *S*-sensors polarization; (c) photo of sample assembly. Vertical extensometers, measuring shortening between upper and lower halves of the sample are marked as V_1 and V_2.

rates were varied between $3.2 \times 10^{-8}\ s^{-1}$ and $3.2 \times 10^{-4}\ s^{-1}$. Axial load was measured with an external load cell with an accuracy of 1 kN corrected for seal friction of the loading piston.

2.2. *Strain Measurements and Acoustic Emission (AE) Monitoring*

Axial strain ε_{ax} was measured using a linear variable displacement transducer (LVDT) mounted at the end of the loading piston. Data were corrected for the effective stiffness (793 kN/mm) of the loading frame. In addition, two pairs of orthogonally oriented strain gages were glued directly to the specimen surface in axial and circumferential direction (Fig. 1b). Volumetric strain Δ_S was calculated using: $\Delta_S = \varepsilon_1 + 2\varepsilon_3$, where ε_1 and ε_3 are axial and circumferential strain, respectively. For wet specimens the change in pore fluid volume and connected porosity Δ_W was monitored using a volumometer with accuracy of about 0.001 cm^3. Compressive stresses, compactive strains and porosity reduction are considered positive. The maximum and minimum compressive stresses are denoted by σ_1 and σ_3, and the differential stress by $Q = (\sigma_1 - \sigma_3)$.

To monitor AE and ultrasonic velocities, twelve *P*-wave and eight polarized *S*-wave sensors were glued directly to the surface of the rock and sealed in the jacket with two-component epoxy (Fig. 1c). Piezoelectric *P*- and *S*-wave sensors were fabricated from PZT piezoceramic discs with 5 mm diameter and 2 mm thickness and rectangular piezoceramic plates 5 × 5 × 1 mm, respectively. The thickness and diameter-related resonant frequencies of the *P*-wave sensors are about 1 MHz and 400 KHz, respectively. Transducer signals were amplified by 40 dB using Physical Acoustic Corporation (PAC) preamplifiers equipped with 100 kHz high-pass filters. To measure *P*- and *S*- wave velocities in different sample directions and monitor velocity changes during the loading, some sensors were used periodically for elastic wave emission applying 100 V electrical pulses. Intervals between velocity measurements varied with sample loading rate from 5 to 1000 seconds. Full-waveform AE and ultrasonic signals were stored in a 12- channel transient recording system (DAXBox PRÖKEL, Germany) with an amplitude resolution of 16 bit at 10 MHz sampling rate (STANCHITS *et al.*, 2006). Ultrasonic transmissions and AE waveforms were discriminated automatically after the experiments. During the experiments, all waveforms were recorded in transient memory with no dead time between consecutive signals. During periods of high AE activity signals were recorded continuously.

The AE hypocenters location procedure includes automatic picking of *P*-wave onset time based on the Akaike information criterion (LEONARD and KENNETT, 1999) and minimization of travel-time residuals using the downhill simplex algorithm (NELDER and MEAD, 1965) considering time-dependent changes of velocities (STANCHITS *et al.*, 2006). The recording system allows locating successfully up to 1000 AE signals per second. We estimate the AE hypocenter location accuracy to be < 2 mm.

3. *Experimental Results*

3.1. *Mechanical Data*

In total, ten experiments were performed on notched samples. Six samples were deformed wet and four samples were dry. Based on the analysis of mechanical and AE

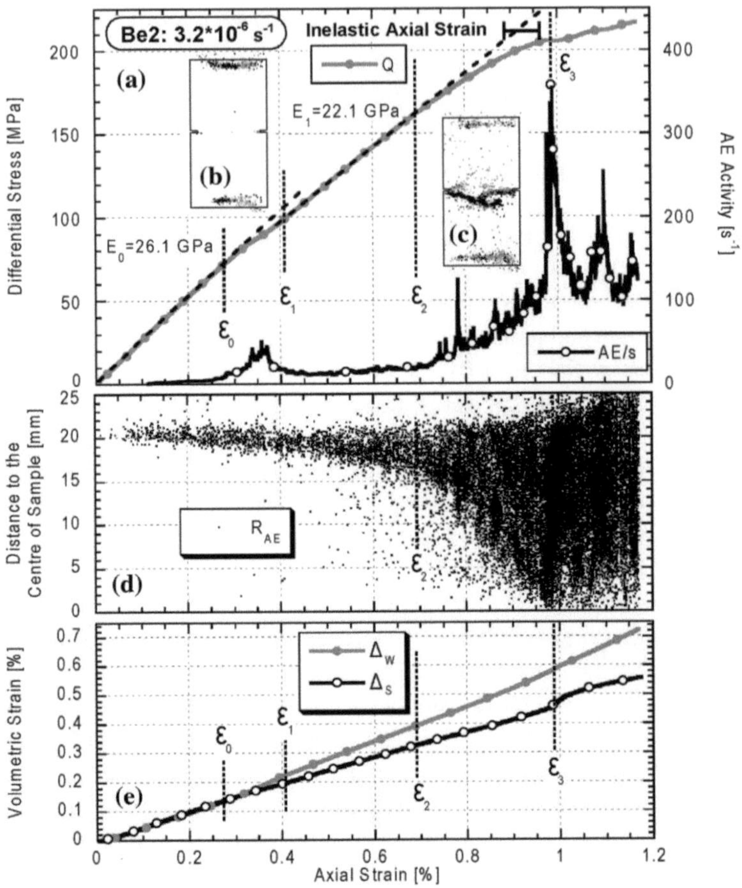

Figure 2

(a) Stress-strain curve and acoustic emission activity registered during deformation of wet sample Be2 with intermediate strain rate 3.2×10^{-6} s^{-1}; (b) AE hypocenters registered during interval of axial deformation ($\varepsilon_0 - \varepsilon_1$); (c) AE hypocenters registered during CB propagation ($\varepsilon_2 - \varepsilon_3$); (d) distance between AE hypocenters in the notched area (32.5 < Z < 72.5 mm) and the sample axis ($r = \sqrt{X^2 + Y^2}$;) (e) volumetric strains measured from a change in the pore volume (Δ_W) and by strain gages (Δ_S) versus axial strain.

data, we separate several deformation stages (Fig. 2). The initial part of the loading curve ($\varepsilon_{ax} < \varepsilon_0$) is almost linear (Fig. 2a), and the volumetric strain monitored by strain gages is close to the strain estimated from the change in the pore fluid volume (Fig. 2e). In all experiments we observed the appearance of diffuse AE clusters close to the sample top and bottom. About 13% of AEs registered during this stage were located in the notched area, whereas 87% were located at the top and bottom part of the sample (Fig. 2b). These clusters most likely result from stress concentration due to end-cap friction, beginning at a stress level of 40%–60% of the stress required for CBs initiation at the notch tip. Similar concentration of AE events near the ends of the specimens was observed for other sandstone and loading conditions in FORTIN *et al.* (this issue). With progressive loading the Young's modulus decreases (Fig. 2a). At axial strain $\varepsilon_{ax} > \varepsilon_1$, the volumetric strains Δ_W and Δ_S diverge (Fig. 2e). The change in pore fluid volume Δ_W is a representative measure for the entire specimen, as opposed Δ_S from the strain gages, which represent local measurements of areas of about 10 mm^2 below the gage.

CB nucleation is characterized by the onset of AE activity near the notch tip, and the nonlinear shape of the loading curve (Fig. 2a). At CB nucleation about 72% of AE hypocenters are located at the notched area of the sample, indicating to the propagation of CB through the entire crosssection (Fig. 2c). For each AE event we calculated the distance (R_{AE}) of AE hypocenter (X, Y) to the center of sample as $R_{AE} = \sqrt{X^2 + Y^2}$. Propagation of the AE front starting from the notch tip (Fig. 2d) towards the specimen center coincides with an increase of AE activity (Fig. 2a). Propagation of the CB front is not uniform, but involves emission of episodic AE bursts (Figs. 2a, 2d), probably indicating individual increments of CB propagation.

Young's moduli from all experiments from different loading stages are given in Table 1. We separate elastic Δ_{El}, ε_{ax_El} and inelastic Δ_{Inel}, $\varepsilon_{ax\_Inel.}$ volumetric and axial strain, respectively, assuming that $\Delta_{Tot} = \Delta_{El} + \Delta_{Inel}$ and $\varepsilon_{ax\_Inel.} = \varepsilon_{ax} - \varepsilon_{ax\_El}$ (SCHOLZ, 1968; Fig. 3b). The cumulative AE number registered in the central part of the sample

Table 1

Summary of experimental results. E_0 and E_1 —Young's modules registered during two elastic stages of loading; ε_1 and ε_2 —axial strain at the moments of CB initiation and completion, respectively; Q_1 and Q_2 — differential stress at the moments of CB initiation and completion, respectively; h_{CB} — compactive shortening across the band determined as the slope of curves Figure 14; E_{AE_Eff} and W_{CB} — cumulative effective AE energy and dissipated inelastic mechanical energy, respectively, at the moment of CB completion; last three columns — energy release densities determined by different techniques: G_{CB} —from mechanical data according to equation (5); G_{CB_AC} — according to anti-crack model of RUDNICKI and STERNLOF (2005) by equation (6); G_{KI} — using K_I according to TEMBE et al. (2006) by equation (8)

Wet Samples

Sample	Strain Rate	E_0, GPa	E_1, GPa	ε_1,%	ε_2,%	Q_1, MPa	Q_2, MPa	h_{CB}, mm	W_{CB}, J	E_{AE_Eff}, V^2 s	G_{CB}, kJ/m^2	G_{CB_AC}, kJ/m^2	G_{KI}, kJ/m^2
Be4	3.2×10^{-8}	24.7	22.5	0.59	0.76	132	156	0.073	30.5	0.86	24.3	23.3	16.0
Be11	3.2×10^{-7}	24.8	20.7	0.66	0.90	148	179	0.107	25.0	0.76	19.9	35.6	20.0
Be2	3.2×10^{-6}	26.1	22.3	0.75	0.99	173	205	0.148	62.1	2.09	49.4	52.9	26.1
Be10	3.2×10^{-5}	27.1	17.5	0.79	0.99	183	210	0.094	27.2	3.06	21.7	34.6	28.1
Be17	1×10^{-5}	26.6	18.7	0.77	0.97	171	197	0.076	20.3	2.08	16.2	31.9	24.9
Be3	3.2×10^{-4}	26.2	19.8	0.81	0.98	176	198	0.089	34.2	4.53	27.2	32.2	27.2

Dry Samples

Sample	Strain Rate	E_0, GPa	E_1, GPa	ε_1, %	ε_2, %	Q_1, MPa	Q_2, MPa	h_{CB}, mm	W_{CB}, J	E_{AE_Eff}, V^2 s	G_{CB}, kJ/m^2	G_{CB_AC}, kJ/m^2	G_{KI}, kJ/m^2
Be15	3.2×10^{-8}	26.6	22.8	0.71	1.0	168	201	0.228	100	4.56	79.6	82.6	24.1
Be14	3.2×10^{-6}	23.9	21.1	0.66	0.92	147	179	0.111	63.1	8.66	50.3	38.8	20.4
Be6	3.2×10^{-5}	25.8	22.6	0.67	0.84	149	176	0.075	39.8	13.9	31.7	23.7	19.7
Be19	3.2×10^{-4}	25.8	22.0	0.62	0.92	134	181	0.104	38.9	15.0	31.0	63.5	16.0

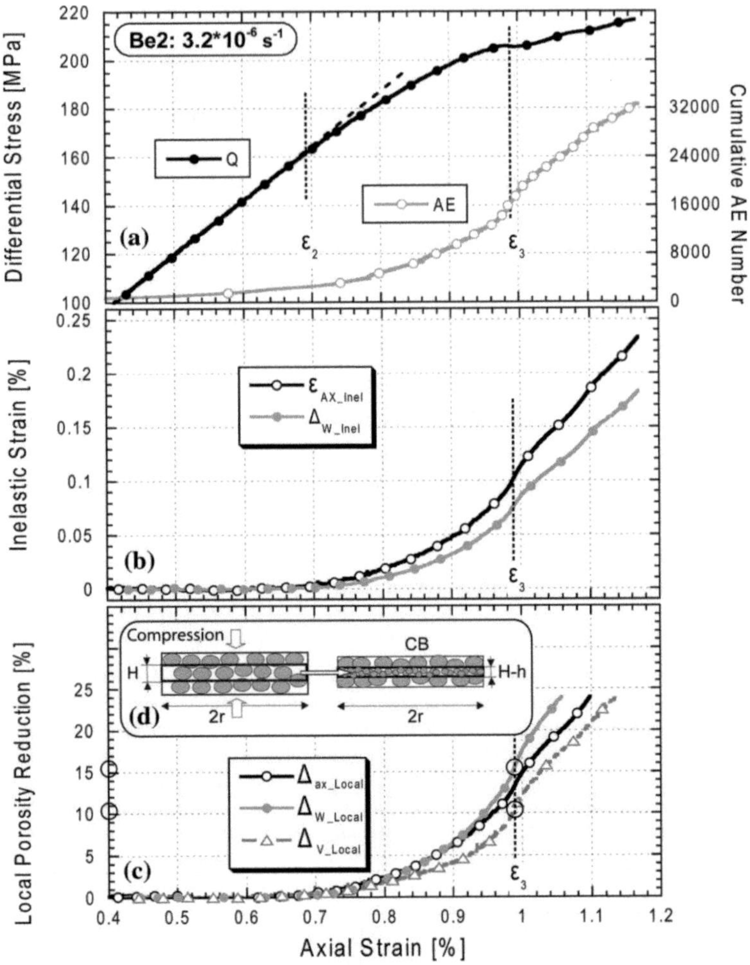

Figure 3

(a) Loading curve and cumulative AE number registered during deformation of wet sample Be2 at intermediate strain rate 3.2×10^{-6} s^{-1} versus axial strain; (b) inelastic axial and volumetric strains; (c) Local porosity reduction estimated by three independent measuring techniques, assuming that deformation of notched sample is localized in the notched area (radius $r = 20$ mm and height $H_{CB} = 0.8$ mm); (d) synoptic picture indicating CB formation adopted from KATSMAN and AHARONOV (2006).

corresponds closely to the increase of the inelastic axial strain (Figs. 3a, b). Inelastic axial strain ε_{ax_Inel} monitored by axial shortening of the sample and inelastic volumetric strain Δ_{W_Inel} estimated independently from a change in the pore fluid volume are very similar during CB propagation stage $\varepsilon_2 < \varepsilon_{ax} < \varepsilon_3$. Assuming that sample deformation is localized in the notched area having radius $r = 20$ mm and height $H_{CB} = 0.8$ mm (equal to the notch width), we can roughly estimate local inelastic porosity reduction from ε_{ax_Inel} and Δ_{W_Inel} (Fig. 3c). For comparison we also included results, obtained from

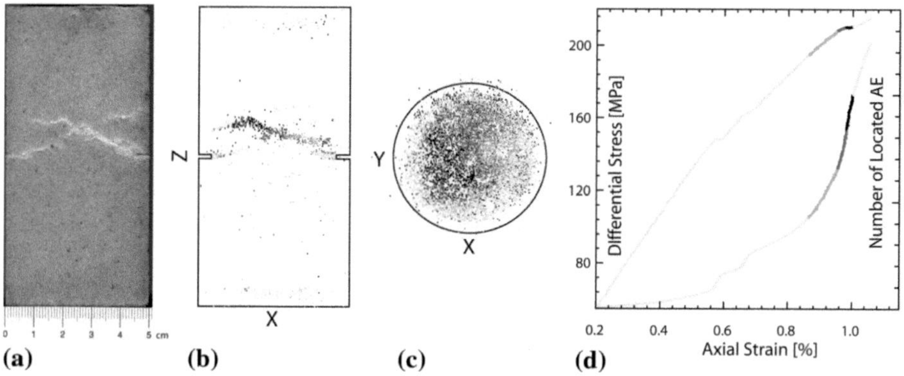

Figure 4

(a) Photo of deformed sample Be10 (strain rate 3.2×10^{-5} s^{-1}, wet conditions). Light areas in the photo (Fig. 4a) correspond to highly damaged areas of sandstone, or CBs; (b) hypocenter locations of acoustic emission events in (X–Z) plane between Y = −1.5 mm and Y = +1.5 mm, selected with AE adjusted amplitude higher than 2 Volts (we determined AdjAmp according to ZANG *et al.* (1998), as amplitude adjusted for geometrical spreading of AE); (c) hypocenter locations of acoustic emission events in (X–Y) plane between Z = 32.5 mm and Z = 72.5 mm; (d) loading history and cumulative AE number versus axial strain. Gray color code corresponds to the time sequences of AE events appearance: lighter color — earlier events, darker color — closest to the completion of CB events. To highlight stage of CB propagation, AEs registered after CB completion are marked by light color.

vertical extensometers, Δ_{V_Local} (marked as V_1 and V_2 in Fig. 1c), directly measuring shortening between upper and lower parts of the sample. From these measurements we conclude that during CB propagation radial strain is negligibly small (Fig. 3c). At CB completion, (ε_3 in Fig. 3c), local porosity reduction Δ_{Inel_Local} inside the CB obtained by three independent techniques ranges from 10.4%–15.5%.

3.2. Analysis of AE Hypocenters and CB Propagation

In all experiments discrete compaction bands nucleated at the notch tip and propagated towards the center of the specimen covering the entire crosssection. Loading was stopped after completion of compaction band propagation. After the experiment, the samples were cut in two halves parallel to the long sample axis and the maximum compressive stress direction. Locations of AE hypocenters and trace of CB on cut through the sample are in good agreement indicating high location accuracy (Figs. 4a, 4b). This suggests that the uncertainty in hypocenter location for high amplitude AEs is possibly < 1 mm, as it has been observed in our previous study (FORTIN *et al.*, 2006).

Analysis of AE hypocenter locations allows monitoring of CB nucleation and propagation (Figs. 5, 6). Usually CB propagation towards the center of the specimens follows a corrugated path, in agreement with microstructural observations (see section 3.3, Figs. 10, 11). In general, CB propagation involves coalescence of individual defect clusters, finally resulting in complex through-going and curved bands slightly

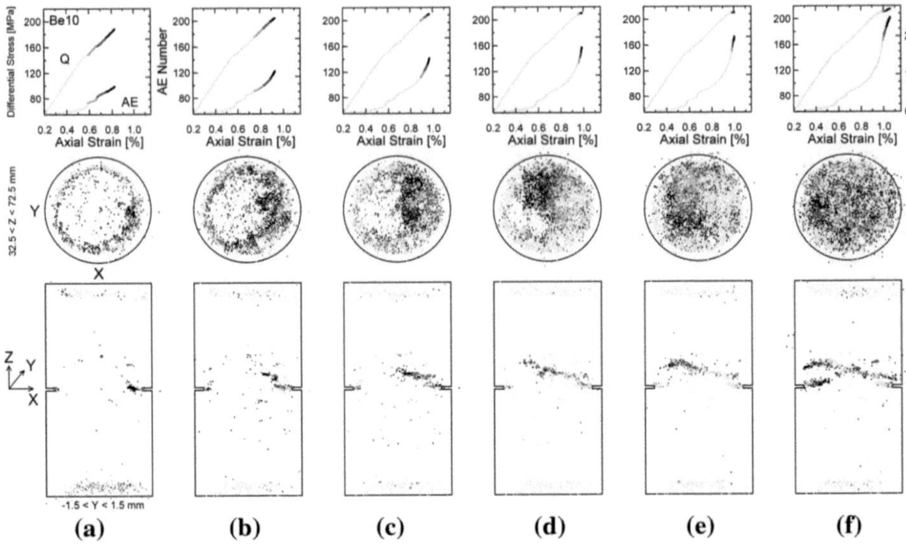

Figure 5

Compaction band initiation and propagation in sample Be10 during loading at intermediate strain rate (3.2×10^{-5} s^{-1}) at wet conditions. In the lower and middle rows two projections of cumulative hypocenter distributions are shown, divided by 6 time sequences (a)–(f). For the middle row, demonstrating X–Y projection, we selected AE events from the central part of the sample at 20 mm distance to the notch (32.5 < Z < 72.5 mm). For the lower row we selected AE events located in the same (Z–X) plane as shown in Figure 4b (−1.5 < Y <+ 1.5 mm). Upper row plots show dependence of differential stress and cumulative AE number versus axial strain. Gray color code is the same for all three rows, demonstrating time sequences of AE events appearance for each snap-shot (a)–(f): lighter color — earlier events, darker color — most recent events. In total in experiment Be10 we have successfully located about 25000 AE events.

inclined to the plane of the circumferential notch (Figs. 5, 6). Completion of compaction band propagation is associated with a small stress drop in the loading curve (Fig. 5e, 6e), similar mechanical characteristics were observed by VAJDOVA and WONG (2003) and TEMBE *et al.* (2006). After compaction band completion we can see some branching and a reduced AE activity (Figs. 5f, 6f).

AE hypocenter locations, onset of inelastic deformation, P-wave velocities, and abrupt increase in AE activity all indicate CB nucleation and reveal the evolution of individual bands (Fig. 7, 8). To define CB nucleation we investigated the spatial distribution of AE, calculating the correlation integral (HIRATA *et al.*, 1987):

$$C(R) = \frac{2}{N(N-1)} N_{(r<R)}, \tag{1}$$

where $N_{(r<R)}$ is the number of AE hypocenter pairs separated by a distance smaller than R. $C(R)$ is a measure for the degree of clustering and localization varying between 0 and 1. For $R \to \infty$, $C(R) = 1$ and for $R = 0$, $C(R) = 0$. Correlation coefficient

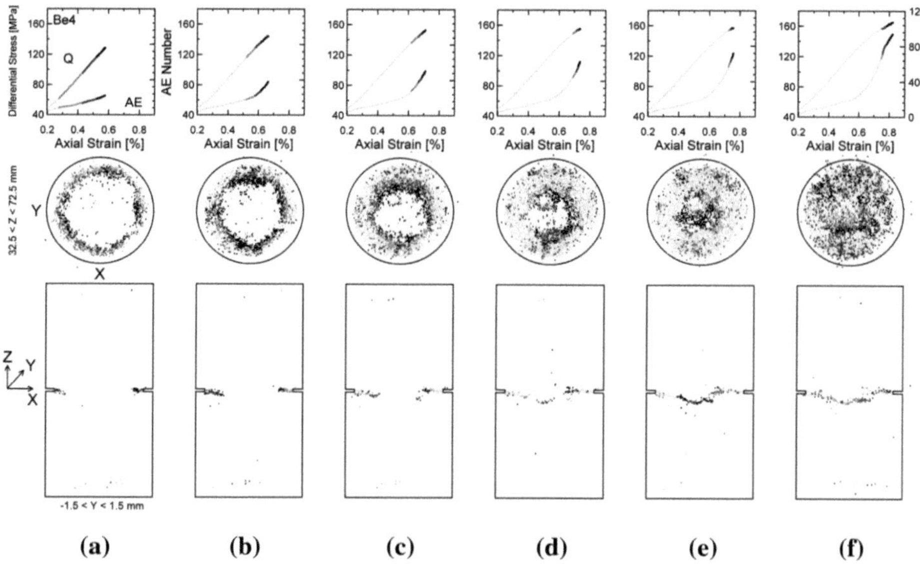

Figure 6
Time sequences of compaction band initiation and propagation in sample Be4 — slowest strain rate ($3.2 \times 10^{-8}\ s^{-1}$, wet conditions). Figure captions are the same as in Figure 5. In total in experiment Be4 we have successfully located about 10000 AE events.

(*C* value) is calculated by (1) for $R = 1$ cm (ZANG *et al.*, 1996). Drastic increase of *C* value indicates CB nucleation coinciding with the increase in AE activity (Figs. 7c, 7d and 8c, 8d).

P-wave velocities help identifying completion of the CB (Fig. 7e, 8e). As shown by FORTIN *et al.* (2006) elastic wave velocities decrease during formation of compaction bands. This is explained by the nucleation of new cracks involved in the process of compaction localization. In our experimental setup, *P*-wave velocities along ray paths outside and across the CB were recorded. Close to the moment of CB completion *P*-wave velocities along ray paths outside and across the CB start to significantly diverge. The velocity of ultrasonic signals crossing single CB decreases by almost 1% (Figs. 7e, 8e). In addition, AE magnitudes and AE activity (Figs. 7c, 8c) decrease after the CB reached the specimen centre.

The cross-section area covered by AE is estimated from (X–Y) projection of AE hypocenter distribution of events with adjusted amplitude more than 3 V that exceed a critical density varying between samples in the range 2.5–13.5 AE events/mm^2. Adjusted amplitude is calculated for a distance of 10 mm from the AE hypocenter assuming geometrical spreading of elastic waves (LOCKNER *et al.*, 1992; ZANG *et al.*, 1998). The cross-section area defined in this way shows an increase with axial strain (Figs. 7b, 8b), that corresponds closely to the increase in inelastic axial strain.

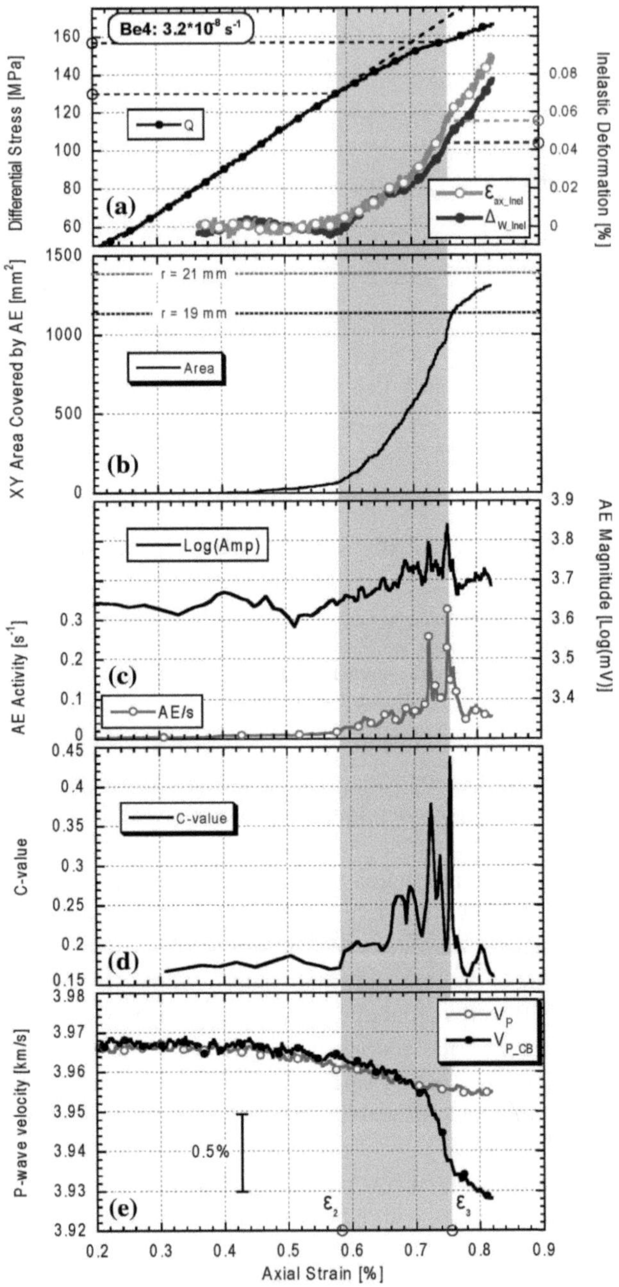

Be4: 3.2*10^-8 s^-1
Differential Stress [MPa]
Inelastic Deformation [%]
Q
ε_ax_Inel
Δ_W_Inel
(a)
XY Area Covered by AE [mm²]
r = 21 mm
r = 19 mm
Area
(b)
Log(Amp)
AE/s
AE Activity [s⁻¹]
AE Magnitude [Log(mV)]
(c)
C-value
(d)
P-wave velocity [km/s]
V_P
V_P_CB
0.5%
ε2
ε3
(e)
Axial Strain [%]

◄ Figure 7
Parameters indicating initiation and completion of compaction bands in experiment Be4 (slowest strain rate, wet conditions) versus strain rate. (a) Differential stress (Q), inelastic axial strain (opened circles) and inelastic volumetric strain (filled circles) measured by pore pressure pump volumometer; (b) CB area calculated from AE hypocenter distribution analysis; (c) AE activity (opened circles) and logarithm of adjusted AE amplitude (solid line); (d) correlation coefficient calculated by equation (1) for $R = 1$ cm; (e) P-wave velocity measured outside of notched area (opened circles) and crossing notched area along ultrasonic transmission trace (9)–(10) (solid circles).

3.3. *Microstructural Analysis of CB*

After the tests, samples were unloaded and impregnated with blue epoxy. Optical thin sections (30 μm thickness) were prepared from the notched part of the samples. We analyzed wet samples deformed at the fastest (Be3) and the slowest (Be4) loading rates by optical microscopy at GFZ Potsdam and by scanning electron microscopy (SEM) at ENS Paris. The photo mosaics of Figure 9a reveal that starting from the notch the CBs are zones of intense grain damage and reduced porosity, following an angular path through the specimen. Inside the bands, grains are crushed and fragments fill up the collapsed pore space (Figs. 9, 10). Regardless of the applied loading rate, the reduced porosity inside the bands is in the range of 10–15%, as determined by SEM picture analysis. Outside the band, the rock matrix remains mostly undeformed with a porosity of about 24%.

From SEM pictures, we determined the damage zones (Figs. 11a, 12a). The comparison with the AE hypocenter locations reveals close spatial agreement (Fig. 11). Analysis of SEM micrographs for the slowest loading rate (Fig. 11) gives a total CB length, $L_{CB} = 47.6$ mm and the damaged area, $S_{CB} \sim 37.8$ mm^2. For the fastest loading rate (Fig. 12), CB length and damage area are $L_{CB} = 55.6$ mm and $S_{CB} = 51.7$ mm, respectively. However, we found slightly higher mean width of CB in the fastest experiment Be3 ($H_{CB} = 0.92$ mm) compared to the slowest experiment Be4 ($H_{CB} = 0.8$ mm). Note that in both cases, the thicknesses of the compaction bands are close to the thickness of the notch (0.8 mm).

4. *Discussion*

4.1. *Rate Dependence of Stress Required to Initiate CB and Propagation Velocity*

Differential stress Q, and axial strain ε_{ax}, at the nucleation and completion of the CBs increases slightly for samples deformed in the presence of a pore fluid, but remains about constant for dry samples (Figs. 13a, 13b). We refer to completion of a CB once the entire sample diameter is covered with located AE hypocenters. In wet samples, the stress needed to initiate CB is strain rate sensitive, and increases with increasing strain rate. Similar results were reported for the stress needed to initiate shear localization in creep experiments performed on Darley Dale sandstone by Baud

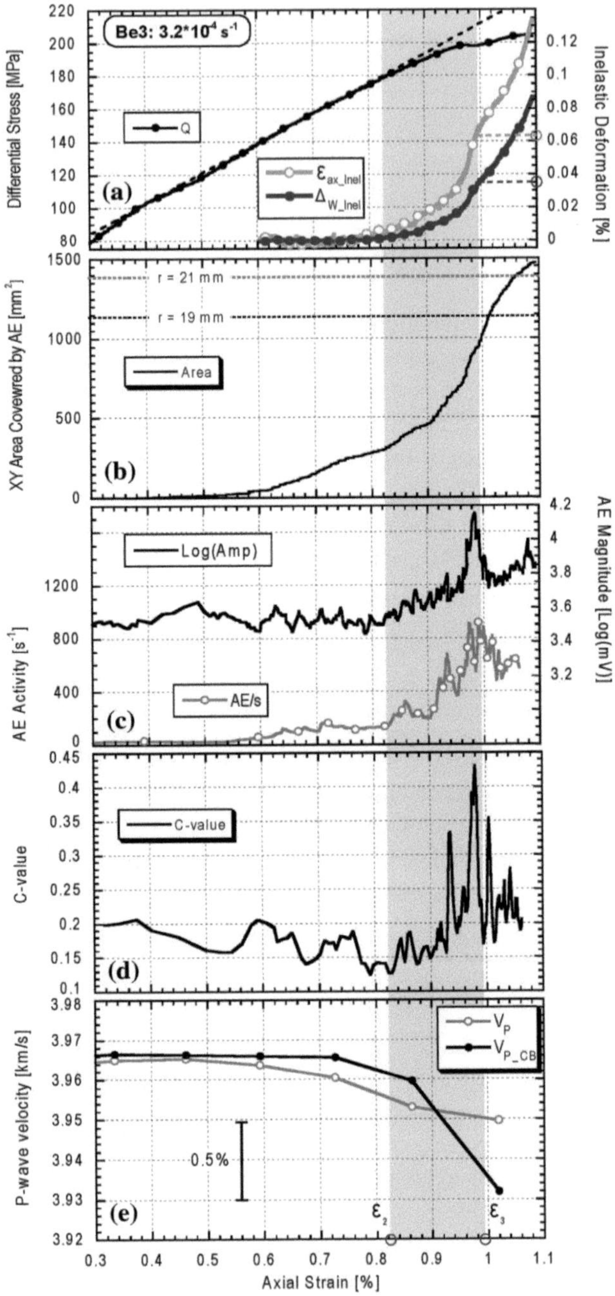

Figure 8
Parameters indicating initiation and completion of compaction bands in experiment Be3 (fastest strain rate, wet conditions) versus strain rate. Figure captions are the same as for Figure 7.

(a)

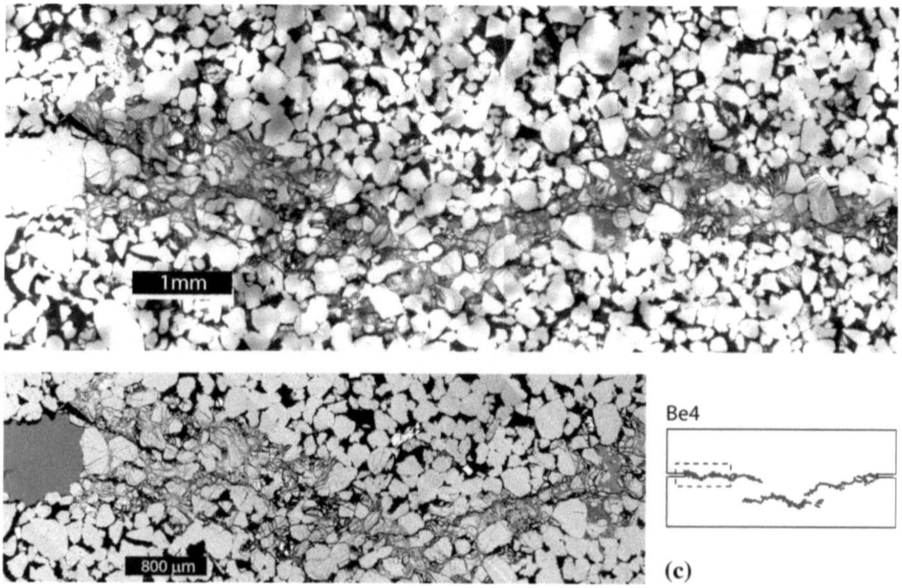

(b) (c)

Figure 9
(a) Optical and (b) SEM microphotos (backscattered electron image) of sample Be4 thin section (slowest strain rate 3.2×10^{-8} s^{-1}, wet conditions); (c) dash rectangle indicates position of taken microphotos on the Be4 sample map.

and Meredith (1997). It is conceivable that the loading rate dependence of CB formation of wet samples may be related to stress corrosion processes contributing significantly to sample compaction at the lowest rates (ATKINSON and MEREDITH, 1981). This explanation is confirmed by the results obtained for dry samples. In dry samples, no strain rate dependence for the stress at the onset of CB is observed (Fig. 13c). The stress corrosion process is time-dependent and should depend on the stress field because the chemical activity of the pore fluid is affected by pressure. Similar influence of strain rate on the strength of dry and wet Tennessee sandstone was reported by RUTTER and MAINPRICE (1978).

Irrespective of the loading rate, the total axial strain required to propagate individual CBs across the sample diameter (axial strain at completion minus axial strain at initiation) remains roughly constant between about 0.1–0.3% (Figs. 13b, 13d) corresponding to vertical displacements between 0.1–0.3 mm. The radial distance from the notch tip to the specimen center is 20 mm. This indicates that the propagation velocity of the CB tip across the sample is approximately a factor of 100 faster than the axial shortening velocity perpendicular to the band in agreement with observations of VAJDOVA and WONG (2003), and TEMBE *et al.* (2006). This ratio remains almost constant over about four orders of magnitude shortening rates for wet and dry samples (Fig. 13).

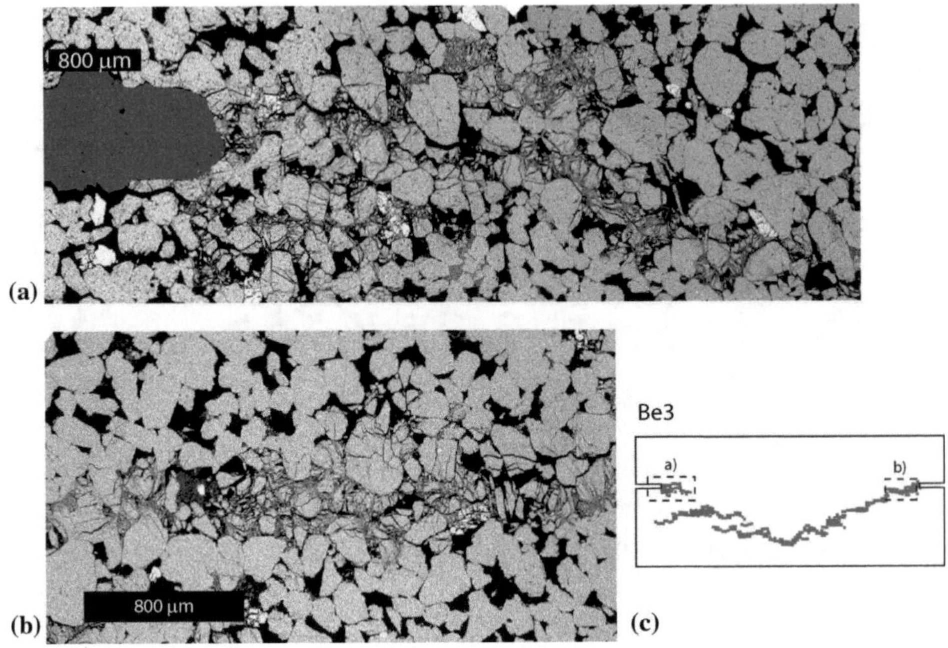

Figure 10
(a), (b) SEM microphotos (backscattered electron image) of sample Be3 thin section (fastest strain rate $3.2 \times 10^{-4}\ s^{-1}$, wet conditions). (c) Dash rectangles indicate positions of taken microphotos on the Be3 sample map.

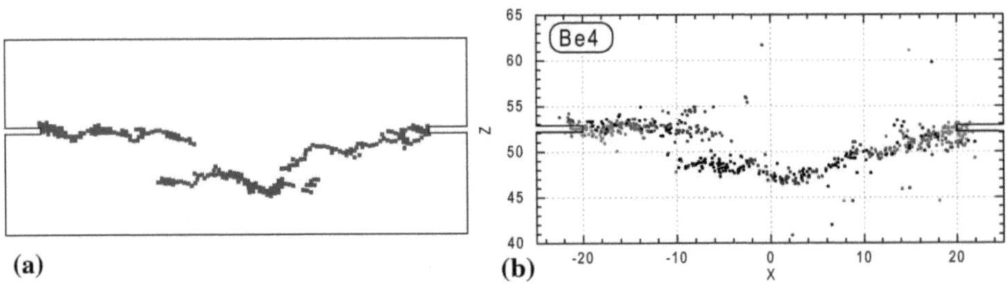

Figure 11
Comparison of damage zones obtained by SEM microstructural analysis (a) and AE hypocenter locations (b) of sample Be4 (slowest strain rate $3.2 \times 10^{-8}\ s^{-1}$, wet conditions).

4.2. Compaction Band Shortening during CB Propagation

The inelastic component of the axial strain of the sample increases during propagation of CB towards the sample center (Fig. 3b). Inelastic axial strain monitored by strain gages and inelastic volumetric strain monitored by the volumometer are almost identical within error of the measurements (Fig. 3b). This indicates that the radial component of

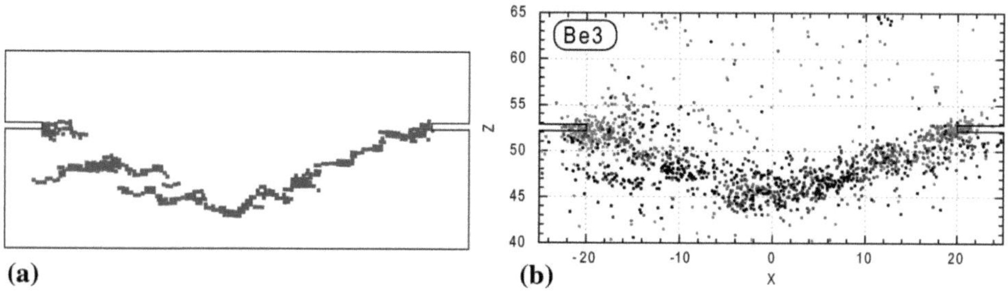

Figure 12
Comparison of damage zones obtained by SEM microstructural analysis (a) and AE hypocenter locations (b) of sample Be3 (fastest strain rate 3.2×10^{-4} s^{-1}, wet conditions).

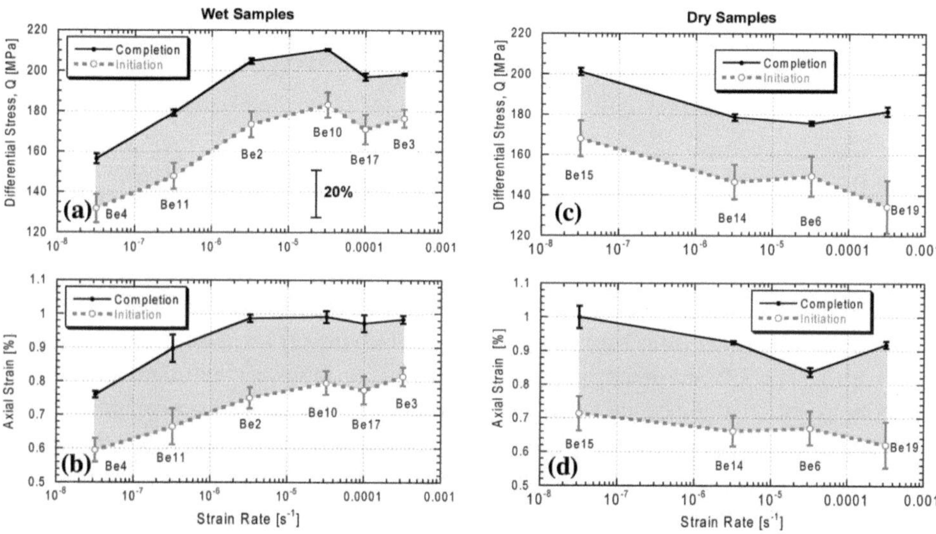

Figure 13
Differential stress (a), (c) and axial strain (b), (d) versus strain rate for wet (a), (b) and dry (c), (d) experiments. Error bars presented in the figure indicate accuracy of moments of CB initiation (gray color) and completion (black color) estimation. They are calculated for each experiment as maximal and minimal values determined by combination of different mechanical, ultrasonic and AE criteria.

the plastic strain is zero and that compaction only involves axial shortening. Consequently, the slope of the change in plastic volumetric strain and plastic shortening with change in area covered by the compaction band are measures for the compactive shortening across the band (h_{CB}):

$$h_{CB} = \frac{\partial(P_{Vol\_Inel})}{\partial(S_{CB})} \cong \frac{\partial(d_{Inel})}{\partial(S_{CB})} * S_N, \qquad (2)$$

where P_{Vol_Inel} is inelastic pore volume reduction, S_{CB} is area covered by AE cloud, d_{Inel} is inelastic displacement and $S_N = \pi r^2$ is area of the notched part of the sample

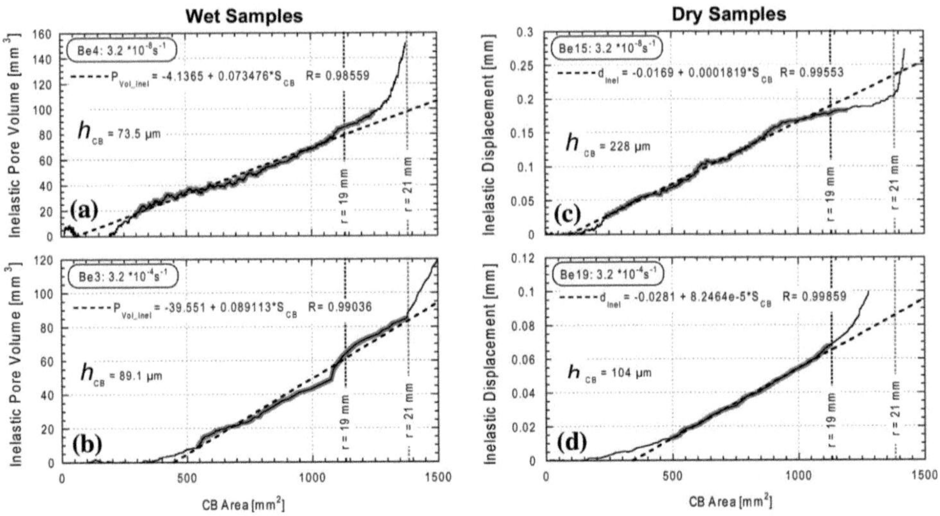

Figure 14
Inelastic pore volume reduction in wet experiments (a), (b) and inelastic displacement in dry experiments (c), (d) versus area covered by AE events. Experiments were performed at slowest (a), (c) and fastest (b), (d) loading rates.

($r = 20$ mm). The data for wet and dry samples indicate a linear relation of inelastic pore volume reduction (Figs. 14a, 14b) and inelastic displacement (Figs. 14c, d) with a change in CB area. The CB shortening h_{CB} remains relatively constant between 70–230 µm during propagation of the band (Table 1). This value is similar to the average grain size in Bentheim sandstone and significantly smaller than the thickness of the notch. From this the local axial and volumetric strain are estimated to be ~10%–20% for most samples.

A linear relation between local compaction and CB area was also suggested by KATSMAN and AHARONOV (2006) and KATSMAN *et al.* (2006) using a Spring Network Model (SNM). Our data indicate that CB propagation is a stable process that occurs by episodic increments of unstable compactive events indicated by bursts of AE activity (Figs. 2a, 2d). This is reminiscent of the run–away instabilities suggested in the Katsman – Aharonov model.

4.3. Energy Dissipation during CB Propagation

The total inelastic work per unit volume, d*w*, done on the sample by the inelastic axial deformation $d\varepsilon_{ax\_inel}$, is (EDMOND and PATERSON, 1972):

$$dw = Q(d\varepsilon_{ax\_inel}) + P_C d\Delta_{inel}, \tag{3}$$

where $d\Delta_{inel}$ is the increment of inelastic volumetric strain. The total inelastic work per unit volume is then (WONG, 1982):

$$w_{CB} = \int_{0}^{\varepsilon_{inel\_C}} Q(d\varepsilon_{ax\_inel}) + (P_C - P_P)\Delta_{Vol\_Inel}, \tag{4}$$

where Δ_{Vol_Inel} is inelastic volumetric strain, and $\varepsilon_{\text{inel\_c}}$ is the inelastic strain at completion of the compaction band.

From equation (4), the total dissipated inelastic mechanical energy W during formation of CB can be written as a function of the inelastic sample shortening h_{inel}:

$$W_{CB}(h_{inel}) = \int_{0}^{h_{inel}} F_{Dif}(dh_{inel}) + (P_C - P_P) * \Omega(h_{inel}), \tag{5}$$

where differential force $F_{Dif} = Q\pi R^2$ and $\Omega(h_{inel})$ is pore volume reduction expected in the notched part of the sample. Neglecting inelastic radial deformation of the notched part of the sample ($\Delta_{Vol\_inel} \approx \varepsilon_{ax\_inel}$), we assume that inelastic pore volume reduction is given by $\Omega(h_{inel}) \approx \pi r^2 h_{inel}$. ($R$ is radius of the sample; r is radius of notched part of the sample). Compactive shortening h_{inel} is measured directly in our experiments as described in Section 4.2. The total energy dissipated during CB formation is in the range of $W_{CB} = 20$–100 J. The results for different samples are summarized in Table 1. The dissipated energy release rate G_{CB} is in the range 16–80 kJ m^{-2} (Table 1).

For all samples the dissipated mechanical energy is linearly related to CB area as estimated from AE analysis (Fig. 15), with a slope representing G_{CB_AE}. G_{CB_AE} is in the range of 27 to 110 kJ m^{-2} irrespective of pore fluid content and loading rate (Fig. 15).

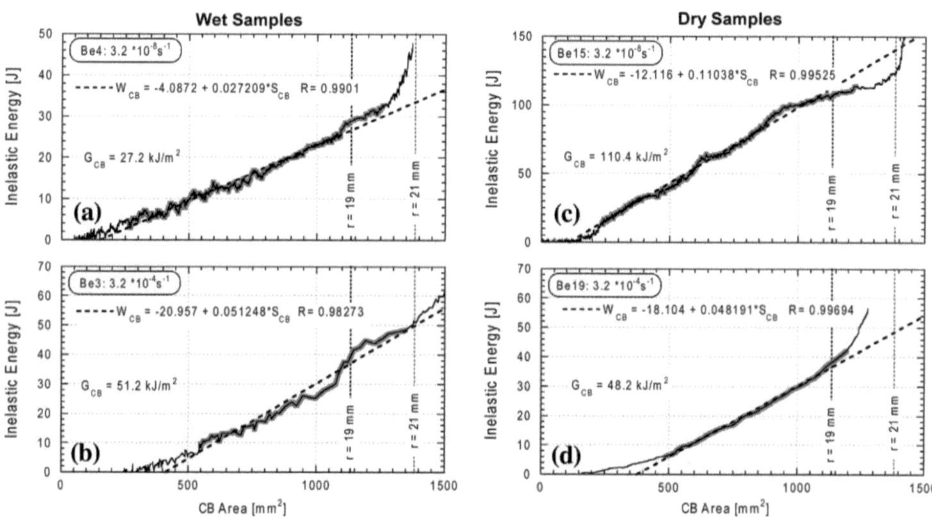

Figure 15

Dissipated inelastic energy versus area covered by AE events in wet (a), (b) and dry (c), (d) experiments performed at slowest (a), (c) and fastest (b), (d) loading rates.

RUDNICKI and STERNLOF (2005) and RUDNICKI (2007) derived an expression for the energy released during CB propagation:

$$G_{CB\_AC} = \sigma_{1*}\zeta h * \varepsilon_p, \tag{6}$$

where σ_1 is uniaxial stress far ahead of the band tip, ε_p is the local plastic compactive strain and ζh is the band thickness. Although the model does not necessarily imply a specific shape of the band or distribution of ε_p, it does assume that the thickness and inelastic strain reach asymptotic values at some lengthy distance behind the tip.

For the specimens investigated in this study, the CB thickness ζh is in the range of 0.8–0.9 mm (Section 3.3, samples Be3, Be4) and ε_p is roughly 0.1. The G_{CB_AC} values for our specimens predicted by the Rudnicki model are in the range 23–83 kJ m^{-2} in good agreement with our experimental measurements G_{CB} (Table 1, Fig. 16). In wet samples, the stress required to initiate CB is strain rate sensitive. Stress increases with rising strain rate, which may indicate a stress corrosion process (Section 5.2, Fig. 13). However, in wet and dry samples the energy release for CB propagation appears to be almost independent of the strain rate within error (Fig. 16).

TEMBE *et al.* (2006) proposed a formulation of the stress intensity factor K_I for notched cylindrical samples, deformed under uniaxial loading:

$$K_1 = (\sigma_1 - P_C)\sqrt{\pi t}\frac{(1 + 0.5\eta + 0.375\eta^2 - 0.363\eta^3 + 0.731\eta^4)}{2\eta^{3/2}}, \tag{7}$$

where $\eta = r/R$; t is the notch depth ($t = R - r$) and σ_I the axial stress. Equation (7) shows that K_I is a function of the notch geometry and the stress state. From our measurements, we estimate the axial stress σ_I at the moment of CB initiation and calculate a critical stress intensity factor K_{Ic} for CB nucleation from equation (7). From K_{Ic} the energy release rate may be calculated using the relation (LAWN, 1993):

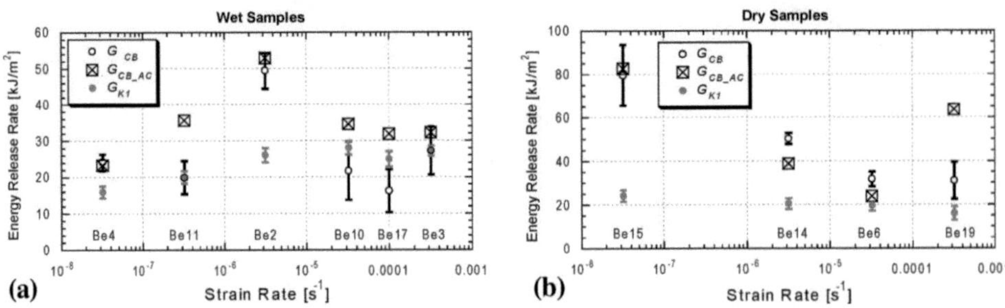

Figure 16
Strain rate dependence of normalized dissipated energy in wet (a) and dry (b) sandstones. Open circles — G_{CB} dissipated inelastic mechanical energy at the CB completion, equation (5); square symbols — G_{CB_AC} critical release rate calculated according to anti-crack model, RUDNICKI and STERNLOF, (2005), equation (6); Gray circles — G_{KI} critical release rate calculated according to equation (8).

$$G_{K1} = \frac{K_{Ic}^2}{E}(1 - \nu^2), \tag{8}$$

where E is Young's modulus, ν is Poisson ratio. Using these relations, we estimated energy release rates G_{KI} for our samples in the range 16–28 kJ m^{-2} (Table 1).

For most wet and dry specimens experimental measurements of G_{CB} and estimates predicted by the models agree within a factor of two (Fig. 16, Table 1). However, for some samples the model predictions G_{CB_AC} and G_{KI} vary by up to a factor of 5, and for all samples G_{KI} is lower than G_{CB_AC} (Table 1). Probably, the calculation of energy release rate G_{KI} using the critical stress intensity factor K_{Ic} (equation 7) underestimates G. For all samples, the experimentally measured energies and those predicted by the models are in the range 16–83 kJ m^{-2} (Table 1) — roughly similar to the results of VAJDOVA and WONG (2003) and TEMBE *et al.* (2006). Based on the field observations of STERNLOF *et al.* (2005), RUDNICKI and STERNLOF (2005) estimate G = 40 kJ m^{-2}. This value is also in the range of our results.

4.4. Energy Radiated by AE during CB Propagation

The plastic work W_{CB} done on the sample during formation of CBs is dissipated in the formation of structural defects, as heat, and as kinetic energy radiated as elastic waves or AE. The radiated AE energy E_{AE} is proportional to A^2 where A is the amplitude of the electrical signal registered by the piezoelectric sensors (LOCKNER *et al.*, 1991). Our recording system has no rearming time, so AEs could be registered continuously. This allows calculation of the cumulative electrical energy E_{AE} recorded by ten sensors as a proxy for the radiated kinetic energy (ZANG *et al*, (1996)):

$$E_{AE}(N) = \sum_{1}^{N} \sum_{i=1}^{10} \sum_{j=1}^{T} (A_{ij}^2 \Delta t), \tag{9}$$

where A_{ij} is the amplitude registered by sensor i at the sampling time t_j amplified by 40 dB; $\Delta t = 10$ ns is the sampling interval; T is the duration time of a single AE exceeding the voltage threshold (2 V) for at least one channel.

For all experiments during CB formation electrical energy E_{AE} is proportional to dissipated inelastic energy (Fig. 17). Strain rate dependence of radiated electrical energy E_{AE} at the moment of CB completion is shown in Figure 18a, the values are summarized in Table 1. The radiation efficiency ΔE_{AE} (i.e., the ratio of electrical energy E_{AE} and total inelastic energy W_{CB}, Fig. 17) is higher for dry compared to wet samples and it increases with increasing loading rate (Fig. 18b). At slow loading rates, wet and dry samples show similar ΔE_{AE}. This indicates that the observed difference could not be attributed to a difference in attenuation of wet and dry samples. We suggest that at slow loading rates and in wet samples subcritical crack growth contributes more efficiently to compaction and CB formation compared to dry

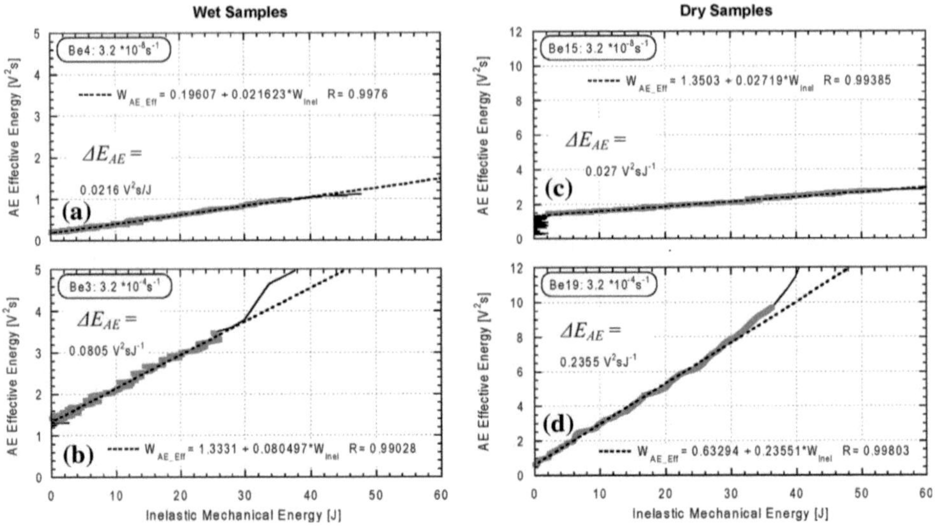

Figure 17

Effective radiated AE energy versus dissipated inelastic energy in wet (a), (b) and dry (c), (d) experiments performed at slowest (a), (c) and fastest (b), (d) loading rates.

conditions and fast loading rates. At high loading rates unstable cracking is dominant. This is in agreement with observations of BACKERS *et al.* (2005) on mode I fracturing of sandstone. They found that increasing the loading rate by four orders of magnitude resulted in a five-fold increase of AE events.

5. *Conclusions*

Nucleation of compaction bands in notched specimens is indicated by the appearance of AE clusters close to the notch tips, a drastic increase of correlation coefficient (C value), an increase in AE activity and the onset of inelastic deformation. With progressive loading, AE activity increased and AE hypocenters indicated propagation of a single CB normal to the sample axis. Completion of CB was indicated by maximum of AE amplitudes and maximum AE activity, coinciding with AE hypocenter location near the center of specimens and a decrease of P-wave velocities across the CB. Location of CB and AE hypocenter locations shows excellent spatial agreement. Microstructural analysis of deformed samples confirmed that regardless of the applied loading rate, the reduced porosity inside the band ranges from 10–15%, while outside the band, the rock matrix remains mostly undeformed with porosity of about 24%. We found slightly higher mean width of CB in the fastest experiment Be3 ($H_{CB} = 0.92$ mm) in comparison with the slowest experiment Be4 ($H_{CB} = 0.8$ mm), but both values are very close to the initial

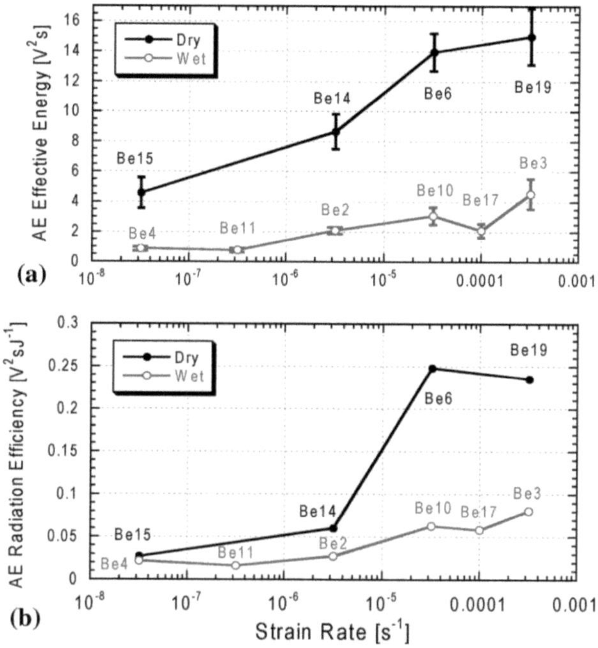

Figure 18
Effective cumulative radiated AE energy at the moments of CB completion (a) and AE radiation efficiency (b) versus strain rate. Solid symbols — dry experiments, open symbols — wet experiments. Error bars correspond to the uncertainty of CB completion moments determination shown in Figure 13.

notch width. During propagation of CB pore volume reduction is linearly related to the area covered by AE hypocenters, and compactive shortening of the CB damage zone remains constant during CB propagation. Local porosity reduction inside the CB is about 10%–15% in agreement with microstructural observation. In all samples, lateral propagation of CBs was about 100 times faster than axial shortening rates. At the slowest displacement rate, AE activity during band propagation was reduced and CB nucleation in wet conditions occurred at 20% lower stresses. In both dry and wet samples, inelastic compaction energy per area ranged between 16 and 80 kJ m^{-2}, in good agreement with previous estimates from laboratory and field studies.

Acknowledgments

We acknowledge the constructive reviews of Philip Benson and the anonymous reviewer. The authors would like to thank particularly Stefan Gehrmann (GeoForschungsZentrum, Potsdam) for aid in preparing the thin sections. This work was partially supported by the French-German grant "Procope".

References

ATKINSON, B.K. and MEREDITH, P.G (1981), *Stress corrosion of quartz: A note on the influence of chemical environment*, Tectonophysics *77*, 1–11.

ATKINSON, B.K. and MEREDITH, P.G., *Experimental fracture mechanics data for rocks and minerals*. in *Fracture Mechanics of Rock*, ed. B. K. Atkinson, pp. 477–525 (Academic Press, London. (1987)).

BACKERS, T., STANCHITS, S., and DRESEN, G. (2005), *Tensile fracture propagation and acoustic emission activity in sandstone: The effect of loading rate*, Int. J. Rock. Mech. Min. Sci. *42*, 7–8, 1094–1101.

BAUD, P. and MEREDITH, P.B. (1997), *Damage accumulation during triaxial creep of Darley Dale sandstone from pore volumometry and acoustic emission*, Int. J. Rock. Mech. Min. Sci. *34*, Art. 024.

BAUD, P., KLEIN, E., and WONG, T.-F. (2004), *Compaction localization in porous sandstones: Spatial evolution of damage and acoustic emission activity*, J. Struct. Geol. *26*, 603–624.

BAUD, P., VAJDOVA, V., and WONG, T.-F. (2006), *Shear-enhanced compaction and strain localization: Inelastic deformation and constitutive modeling of four porous sandstones*, J. Geophys. Res. 111, B12401, doi:10.1029/2005JB004101.

BORJA, R.I. and AYDIN, A., (2004), *Mathematical and geologic framework for failure modes in granular rocks*, Comput. Methods Appl. Mech. Eng. *193*, 2667–2698.

BRACE, W.F., PAULDING Jr. B.W., and SCHOLZ, C. (1966), *Dilatancy in the fracture of crystalline rocks*, J. Geophys. Res. *71*(16), 3939–3953.

COSTIN, L.S. (1983), *A microcrack model for the deformation and failure of brittle rock*, J. Geophys. Res. *88*(B11), 9485–9492.

DIGIOVANNI, A.A., FREDRICH, J.T., HOLCOMB, D.J., and OLSSON, W.A., *Micromechanics of compaction in an analogue reservoir sandstone*. In Girard, J., Liebman, M., Breeds, C., Doe, T. (eds.), Proc. *the North American Rock Mechanics Symposium*, July 31. (A.A. Balkema, Rotterdam, (2000)), pp 1153–1158.

EDMOND, J.M. and PATERSON, M.S. (1972), *Volume changes during the deformation of rocks at high pressures*. Int. J. Rock. Mech. Min. Sci. *9*, 161–182.

ESHELBY, J.D. (1957), *The determination of the elastic field of an ellipsoidal inclusion, and related problems*. Proc. Royal Society of London (A), *Mathematical and Physical Sciences 241* (1226), 376–396.

ISSEN, K.A. and RUDNICKI, J.W. (2000), *Conditions for compaction bands in porous rock*, J. Geophys. Res. *105*, 21529–21536.

FORTIN, J. SCHUBNEL, A., and GUEGUEN, Y. (2005), *Elastic wave velocities and permeability evolution during compaction of Bleurswiller sandstone*, Int. J. Rock. Mech. Min. Sci and Geomech. *42*, 873–889.

FORTIN J., STANCHITS S., DRESEN, G., and GUEGUEN, Y. (2006), *Acoustic emission and velocities associated with the formation of compaction bands in sandstone*, J. Geophys. Res. 111, B10203, doi:10.1029/2005JB003854.

FORTIN, J., STANCHITS, S., DRESEN, G., and GUEGUEN, Y. (2008), *Acoustic emissions monitoring during inelastic deformation of porous sandstone: Comparison of three modes of deformation*, Pure Appl. Geophys. this issue.

HAIMSON, B. (2001), *Fracture-like borehole breakouts in highporosity sandstone: are they caused by compaction bands?* Phys. Chem. Earth. A *26*, 15–20.

HAIMSON, B.C. (2003), *Borehole breakouts in Berea sandstone reveal a new fracture mechanism*, Pure Appl. Geophys. *160*, 813–831.

HAIMSON, B. and KOVACICH, J. (2003), *Borehole instability in highporosity berea sandstone and factors affecting dimensions and shape of fracture-like breakouts*, Eng. Geol. *69*, 219–23.

HAIMSON, B. and LEE, H. (2004), *Borehole breakouts and compaction bands in two high-porosity sandstones*, Int. J. Rock Mech. Min. Sci. *41*, 287–301.

HIRATA, T., SATOH, T., and ITO, K. (1987), *Fractal structure of spatial distribution of microcracking in rock*, Geophys. J. R. astr. Soc. *90*, 369–374.

KATSMAN, R. AHARONOV, E. and SCHER, H. (2006), *Localized compaction in rocks: Eshelby's inclusion and the Spring Network Model*, Geophys. Res. Lett. 33, L10311, doi:10.1029/2005GL025628.

KATSMAN, R. and AHARONOV, E. (2006), *A study of compaction bands originating from cracks, notches, and compacted defects*. J. Struct. Geol. *28*, 508–518.

KLAETSCH, A.R. and HAIMSON, B.C., *Porosity-dependent fracture-like breakouts in St. Peter sandstone*. In HAMMAH *et al.* (eds.), *Mining and tunneling innovation and opportunity* (University of Toronto Press, Toronto, (2002)) pp. 1365–1372.

KLEIN, E., BAUD, P., REUSCHLE, T., and WONG, T.-F. (2001), *Mechanical behaviour and failure mode of Bentheim sandstone under triaxial compression*, Phys Chem Earth (A) *26*, 21–25.

KLEIN, E. and REUSCHLÉ, T. (2003), *A model for the mechanical behaviour of bentheimsandstone in the brittle regime*, Pure Appl. Geophys. *160*, (5,6), 833–849.

LEI, X., NISHIZAWA, O., KUSUNOSE, K., and SATOH, T. (1992), *Fractal structure of the hypocenter distributions and focal mechanism solutions of acoustic emission in two granites of different grain sizes*. J. Phys. Earth. *40*, 617–634.

LEI, X., KUSUNOSE, K., RAO, M.V.M.S., NISHIZAWA, O., and SATOH, T. (2000), *Quasi-static fault growth and cracking in homogeneous brittle rock under triaxial compression using acoustic emission monitoring*. J. Geophys. Res. *105*, 6127–6139.

LEONARD, M. and KENNETT, B.L.N. (1999), *Multi-component autoregressive techniques for the analysis of seismograms*, Phys. Earth Planet. Int. *113* (1–4), 247.

LOCKNER, D.A., and BYERLEE, J.D. (1977), *Acoustic emission and creep in rock at high confining pressure and differential stress*, Bull. Seismol. Soc. Am. *67*, 247–258.

LOCKNER, D.A., BYERLEE, J.D., KUKSENKO, V., PONOMAREV, A. and SIDORIN, A. (1991), *Quasi-static fault growth and shear fracture energy in granite*, Nature *350*, 39–42.

LOCKNER, D.A., BYERLEE, J.D., KUKSENKO, V., PONOMAREV, A. and SIDORIN, A., *Observations of quasistatic fault growth from acoustic emissions*. In Fault Mechanics and Transport Properties of Rocks (Eds. B. Evans and T.-F. Wong), pp. 3–31 (Academic Press, London. (1992)).

LOCKNER, D.A. (1993), *The role of acoustic emission in the study of rock fracture*, Int. J. Rock Mech. Min. Sci. Geomech. *30*, 883–899.

MOLLEMA, P.N., and ANTONELLINI, M.A. (1996), *Compaction bands: A structural analog for anti-mode I cracks in Eolian sandstone*, Tectonophysics *267*, 209–228.

NELDER, J. and MEAD, R. (1965), *A simplex method for function minimisation*, Computer J. *7*, 308–312.

ODA, H., KOAMI, H. and SEYA, K. (1989), *Effect of occurrence rate of acoustic emissions on their statistical behaviour*. Pure Appl. Geophys. *130, 5–29.*

OLSSON, W. and HOLCOMB, D. (2000), *Compaction localization in porous rock*, Geophys. Res. Lett. *27*(21), 3537–3540.

A. REINICKE, A., STANCHITS, S., NAUMOV, D.J., LEGARTH, B.A., HUENGES, E. and DRESEN, G. (2008), *Mechanically Induced Permeability Reduction at the Fracture Proppant Interface*, Int. J. Rock Mech. Min. Sci. Geomech., submitted.

RUDNICKI, J.W. and STERNLOF, K. (2005) *Energy release model of compaction band propagation*. Geophys Res Lett. *32*, (L16303), doi:10.1029/2005GL023602.

RUDNICKI, J.W. (2007), *Models for compaction band propagation*, Geological Society, London, Special Publications 2007; v. *284*; pp. 107–125, doi:10.1144/SP284.8.

RUTTER, E.H. and MAINPRICE D.H. (1978), *The effect of water on stress relaxation of faulted and unfaulted sandstone*, Pure Appl. Geophys. *116*, 634–654.

SCHOLZ, C.H. (1968), *Microfracturing and the Inelastic Deformation of Rock in Compression*, J. Geophys. Res. *73*(4), 1417–1432.

STANCHITS, S., VINCIGUERRA, S., and DRESEN, G. (2006), *Ultrasonic Velocities, Acoustic Emission Characteristics and Crack Damage of Basalt and Granite*. Pure Appl. Geophys. *163* (5–6), 975–994, doi: 10.1007/s00024-006-0059-5.

STERNLOF, K.R., CHAPIN, J.R., POLLARD, D.D., and DURLOFSKY, L.J. (2004), *Permeability effects of deformation band arrays in sandstone*, AAPG Bull *88*, 1315–1329.

STERNLOF, K.R., RUDNICKI, J.W. and POLLARD, D.D. (2005), *Anticrack inclusion model for compaction bands in sandstone*, J. Geophys. Res. 110, B11403, doi:10.1029/2005JB003764.

STERNLOF, K.R., KARIMI-FARD, M., POLLARD, D.D., and DURLOFSKY, L.J. (2006), *Flow and transport effects of compaction bands in sandstone at scales relevant to aquifer and reservoir management*, Water Resour. Res. 42, W07425, doi:10.1029/2005WR004664.

TAYLOR, W.L., and POLLARD, D.D. (2000), *Estimation of in-situ permeability of deformation bands in porous sandstone, Valley of Fire, Nevada*, Water Resour. Res. *36*, 2595–2606.

TEMBE, S., VAJDOVA, V., WONG, T.-F., and ZHU, W. (2006), *Initiation and propagation of strain localization in circumferentially notched samples of two porous sandstones*. J. Geophys. Res. 111, B02409, doi:10.1029/2005JB003611.

THOMPSON, B.D., YOUNG, P. and LOCKNER, D. (2006), *Fracture in Westerly granite under AE feedback and constant strain rate loading: Nucleation, quasi-static propagation, and the transition to unstable fracture propagation*, Pure Appl. Geophys. *163*, 995–1019.

VAJDOVA, V. and WONG, T.-F. (2003), *Incremental propagation of discrete compaction bands: acoustic emission and microstructural observations on circumferentially notched samples of Bentheim sandstone*. Geophys. Res. Lett. 30(14), 1775, doi:10.1029/2003GL017750.

VAJDOVA, V., BAUD, P., and WONG, T.-F. (2004) *Permeability evolution during localized deformation in Bentheim sandstone*. J Geophys Res 109, B10406, doi:10.1029/2003JB002942.

WONG, T.-F. (1982), *Micromechanic of faulting in westerly granite*, Int. J. Rock Mech. Min. Sci. Geomech. Abstr. *19*, 49–64.

WONG, T.-F., BAUD, P., and KLEIN, E. (2001), *Localized failure modes in a compactant porous rock*. Geophys. Res. Lett. *28*, 2521–2524.

ZANG, A., WAGNER, F.C., and DRESEN, G. (1996), *Acoustic emission, microstructure, and damage model of dry and wet sandstone stressed to failure*, J. Geophys. Res. *101, 8, 17507–17521.*

ZANG, A., WAGNER, F.C., STANCHITS, S., DRESEN, G., ANDRESEN, R., and HAIDEKKER, M.A. (1998), *Source analysis of acoustic emissions in Aue granite cores under symmetric and asymmetric compressive loads*, Geophys. J. Internat. *135*, 1113–1130.

ZANG, A., WAGNER, F.C., STANCHITS, S., JANSSEN, C. and DRESEN, G. (2000), *Fracture process zone in granite*, J. Geophys. Res. *105*, 23651–23661.

Received June 6, 2008, revised July 30, 2008, accepted November 28, 2008)
Published Online First: May 12, 2009

Pure appl. geophys. 166 (2009) 869–898
0033–4553/09/050869–30
DOI 10.1007/s00024-009-0493-2

Pure and Applied Geophysics

Compaction and Failure in High Porosity Carbonates: Mechanical Data and Microstructural Observations

P. Baud,[1] S. Vinciguerra,[2] C. David,[3] A. Cavallo,[2] E. Walker,[1] and T. Reuschlé[1]

Abstract—We investigated systematically the micromechanics of compaction in two carbonates of porosity above 30%, Majella grainstone and Saint Maximin limestone. The composition, grain size and pore surface area of these rocks were determined. Hydrostatic compression experiments were performed under dry and wet conditions beyond the onset of grain crushing. A significant water weakening effect was observed in both rocks. A set of conventional triaxial experiments was also performed on both rocks under dry conditions at confining pressures ranging from 3 to 31 MPa. Microstructural observations were carried out on the deformed samples. The mechanical behavior of these high porosity carbonates is dominated by shear-enhanced compaction associated in most cases with strain hardening. Stress-induced cracking and grain crushing are the dominant micromechanisms of deformation in both rocks. In Majella grainstone, compactive shear bands appeared at low confinement, in qualitative agreement with the deformation bands observed in the field. At higher confining pressures, compaction localization was inhibited and homogeneous cataclastic flow developed. In Saint Maximin limestone, compaction localization was observed at all confining pressures. An increasing number of compactive shear bands at various orientations appeared with increasing strain. These new data suggest that compaction localization is important in the mechanical compaction of high porosity carbonates.

Key words: High porosity carbonates, triaxial deformation, microstructural observations, microcracking, compaction localization.

1. *Introduction*

Compactive and dilatant deformation in porous rocks is a crucial problem in fault development, geotechnical engineering and reservoir/aquifer management. Active tectonics and extraction of hydrocarbons and groundwater modify the pore pressure in a reservoir/aquifer, causing variations of the effective stress possibly leading to faulting and inelastic deformation. The ability to interpret and predict the occurrence and extent of

[1] Institut de Physique du Globe de Strasbourg (UMR 7516 CNRS, Université de Strasbourg/EOST), 5 rue René Descartes, 67084 Strasbourg Cedex, France. E-mail: patrick.baud@eost.u-strasbg.fr
[2] Istituto Nazionale di Geofisica e Vulcanologia, Sezione di Roma1, via di Vigna Murata 605, 00143 Rome, Italy. E-mail: Sergio.vinciguerra@ingv.it.
[3] Université de Cergy-Pontoise, 5 mail Gay-Lussac, Neuville-sur-Oise, 95031 Cergy-Pontoise Cedex, France.

such deformation depends on a fundamental understanding of inelastic behavior, failure mode and brittle-ductile transition in porous rocks.

Numerous studies have described mechanical compaction in porous silicate rocks (HANDIN and HAGER, 1957; HADIZADEH and RUTTER, 1983; BAUD *et al.*, 2004). At room temperature, the failure mode evolves with increasing pressure from brittle fracture to either cataclastic flow or more complex failure modes (WONG *et al.*, 1997). In the brittle regime, stress-induced dilatancy precedes the development of macroscopic fractures (BRACE, 1978). In the compactive regime, the pore space undergoes significant inelastic compaction while the rock strain hardens (EDMOND and PATERSON, 1972; WONG *et al.*, 1997; BAUD *et al.*, 2006). In porous sandstone, recent studies (WONG *et al.*, 2001; BAUD *et al.*, 2004; FORTIN *et al.*, 2005; TEMBE *et al.*, 2008) show that in many cases the rocks do not fail by cataclastic flow but by discrete compaction bands developed at the brittle-ductile transition. The different failure modes, all involving microcracking (MENÉNDEZ *et al.*, 1996; BAUD *et al.*, 2004), can be recognized in the laboratory by different acoustic emission activity at the onset of inelastic deformation.

In carbonates, the brittle-ductile transition shows some different attributes. Limestones and marbles undergo the brittle to plastic transition at room temperature for confining pressures accessible in the laboratory (ROBERTSON, 1955; PATERSON, 1958; HEARD, 1960; RUTTER, 1974; EVANS *et al.*, 1990) because calcite requires relatively low shear stresses to initiate mechanical twinning and dislocation. Many previous studies focused on the mechanical behavior of very low porosity rocks such as Carrara marble (RUTTER, 1974; FREDRICH *et al.*, 1989), or very porous chalks (TEUFEL *et al.*, 1991, RISNES *et al.*, 2005). In limestones of intermediate porosity (from 3% to 18%), a similar phenomenology of failure was observed (BAUD *et al.*, 2000a; VAJDOVA *et al.*, 2004). Dilatancy and shear localization developed under low confining pressure, while strain-hardening and shear-enhanced compaction were observed at high confining pressure. After a certain amount of strain-hardening, the samples consistently switched from compaction to dilatancy. This phenomenology of dilatant and compactant failure is similar to that of many porous sandstones. Several micromechanical models were used to capture the interplay of dilatancy and compaction in carbonate rocks (VAJDOVA *et al.*, 2004). In carbonates, unlike for sandstone, because of crystal plasticity, microcracking is not always the dominant micromechanism of deformation. In the compactive regime, a model involving plastic pore collapse (CURRAN and CARROLL, 1979) was tested to interpret the experimental data. However, this approach was unsuccessful, suggesting that macroscopic yielding is controlled by a complex interplay of cracking and crystal plasticity (VAJDOVA *et al.*, 2004). To date there is however a paucity of microstructural data to constrain more elaborate models.

In this study we have extended the scope of VAJDOVA *et al.*'s (2004) work on limestones to investigate systematically the micromechanics of compaction in two carbonates of porosity greater than 30%. We specifically focused on several related questions:

- Is the phenomenology of compaction in these very high porosity rocks similar to that of less porous limestones and silicate rocks? In Tavel and Indiana limestone, VAJDOVA *et al.* (2006) reported significant twinning and microcracking during cataclastic flow. The extent and intensity of twinning increased with increasing pressure and strain, and its coupling to microcracking can be very complicated, in particular if the pore structure is complex.
- What is the dominant micromechanism of deformation leading to macroscopic compaction in very high porosity carbonates? Unlike that for sandstone, no compaction localization has been reported in laboratory experiments on calcitic rocks. Recent field observations in porous carbonates (TONDI *et al.*, 2006) suggested however that complex failure modes involving deformation bands and compaction bands could be observed in this type of rock.
- How does the starting microstructure of porous carbonates trigger or inhibit the development of compaction localization over a wide range of pressures?
- Is the weakening effect of water comparable to what has been quantified in porous sandstones (RUTTER and MAINPRICE, 1978; BAUD *et al.*, 2000b)? Most recent mechanical data on porous carbonates were obtained under dry conditions.

In this paper we address these issues by presenting new mechanical data; detailed analysis of rock microstructure, and by comparing our results with the prediction of bifurcation analysis.

2. *Petrophysical Description of the Studied Materials*

2.1. *Source of Materials*

Two high porosity carbonates were selected in this study: Majella grainstone (MA) from the Madonna delle Mazze quarry in the Apennines (Central Italy) and Saint-Maximin limestone (SML) from the Rocamat quarry in Saint-Maximin-sur-Oise, north of Paris (France). The blocks of Majella considered in this study belong to the undeformed host rock outcropping in the quarry, which is situated on the inner part of the forelimb of the Majella anticline (TONDI *et al.*, 2006). The petrophysical description of both rocks is summarized in Table 1.

Table 1

Petrographic description of the rocks used in this study

Rock Name	Porosity (%)	Average grain radius (μm)	Origin	Composition	Specific Area (m^2/g)
Majella grainstone	30	27	Central Apennines (Italy)	100% calcite	0.15
Saint-Maximin limestone	37	70	Paris (France)	61% calcite, 39% quartz	1.4

2.2. Porosity, Composition, Grain and Pore Characterization

Starting porosity was determined for both rocks using two techniques: water saturation and gas pycnometer (Accupyc 1330). Porosity values of 30% and 31% for MA and 37% and 39% for SML were respectively found, through the two techniques.

We studied the composition and the grain size of both carbonates from petrographic thin sections. Intact samples were immersed in a bath of epoxy, ground flat and polished. The thin sections were carbon-coated and analyzed with a Field Emission Gun-Scanning Electron Microscope (FESEM) and an Electronic Probe Micro Analyser (EPMA). FESEM and EPMA allowed us to semi-quantitatively analyze the chemical composition of grains over an area of about 5 μm^2. FESEM images have been collected using the backscattered electron (BSE) mode at magnification 50 to 5000. Grain size was determined from the obtained images from the mean of the minimum and maximum Feret diameters.

The semi-quantitative chemical analysis combined with the image analysis indicate that MA is a pure calcite grainstone formed from rudist bivalve fragments (Tondi *et al.*, 2006), without a preferred orientation, cast into a poorly consolidated calcite cement (Fig.1a). The grain size is widely variable from a few microns to mm-sized with an average grain diameter of ~ 54 μm.

SML conversely shows subangular and poorly sorted grains, mainly formed by $CaCO_3$ (61%) and SiO_2 (39%) embedded into a calcite matrix and frequent fossils fragments (Figs. 1 b-c). Grain size variability is between 1–500 microns with an average grain diameter of ~ 140 microns. The silica grains are slightly larger (~ 150 μm) than the calcite grains (~ 100 μm).

Pore throat statistics were determined on intact samples from mercury injection experiments using a micrometrics Autopore IV 9500. SML has a broader spectrum of pore throat sizes than Majella (see Fig. 2). A strong peak is observed for SML at 26.4 microns. In Majella, the peak value is centered at 16.2 μm and there is no pore throat with a diameter lower than 0.2 μm. The internal pore surface area of the connected porosity was measured using the BET method (Brunauer *et al.*, 1938). The pore surface area was found to be 1.4 m^2/g for SML and 0.15 m^2/g for Majella grainstone. Since no significant crack porosity was observed in either rock, the surface area difference is likely due to the presence of smaller particles in SML, and the grain shape and pore size distributions.

2.3. Fabric Anisotropy

The variability of physical properties was used to assess the orientation of the bedding plane in both rocks, which was not obvious from visual inspection of the blocks. P-wave velocity was measured on core samples in different directions following the method described in Louis *et al.* (2004). A set of 21 independent P-wave velocity measurements

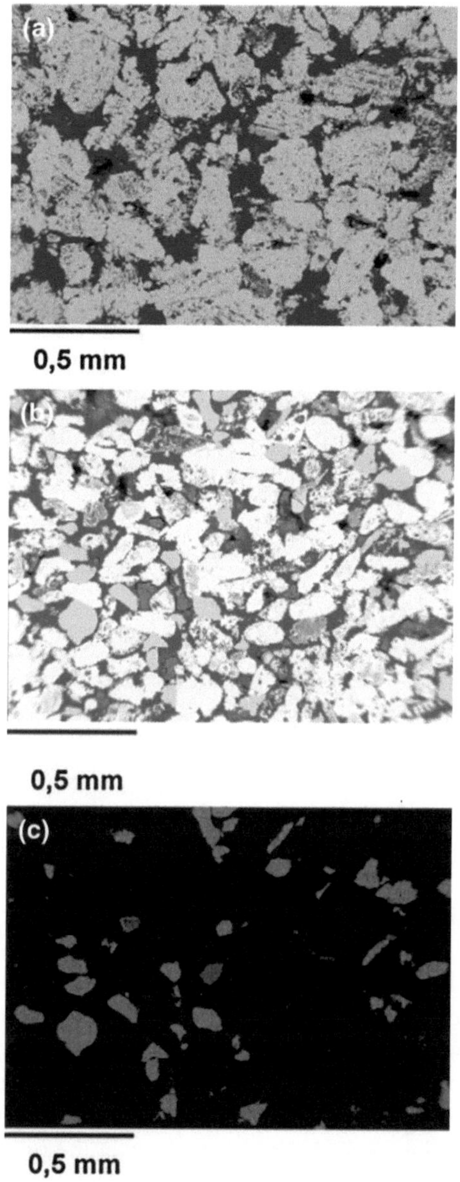

Figure 1
Micrographs of intact samples of Majella grainstone (a) and Saint Maximin limestone (b). (c) Semi-quantitative analysis of Si in SML. High Si concentrations are brighter.

was obtained on samples cored in three orthogonal directions. For each sample 8 measurements were made across diameters with an angular spacing of 22.5°. From this data set it is possible to build a second-rank velocity pseudo-tensor from which principal

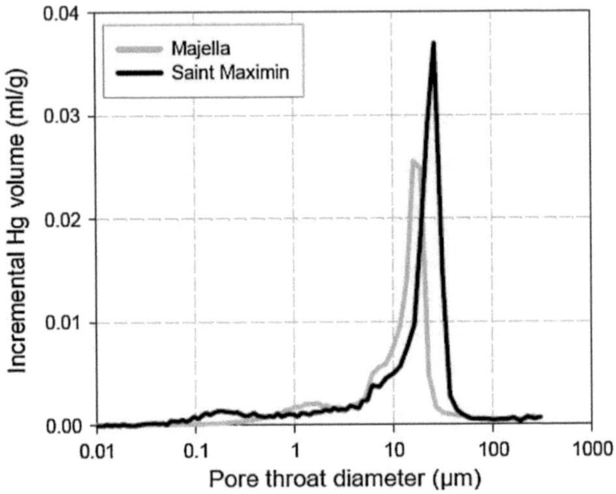

Figure 2
Pore throat statistics derived from mercury injection tests on intact samples of Majella grainstone (light grey) and Saint-Maximin limestone (dark grey).

directions can be inferred. These directions, corresponding to minimum, intermediate and maximum eigenvalues of the velocity pseudo-tensor can be compared to geological features such as bedding, strike and dip of folded structures, as well as direction and magnitude of regional stresses. These measurements were made first on dry samples, then in water saturated conditions. Figure 3 shows the evolution of P-wave velocity as a function of the measurement angle for the samples for which the largest variations were observed, for dry and wet samples of both Majella grainstone and Saint-Maximin limestone. For both rocks the velocity of the water saturated samples is systematically higher than that of the dry samples as expected. The angular velocity variations, however, are smaller in the saturated samples than the dry. This decrease in anisotropy occurs when the pore space is filled with a stiffer phase and can be interpreted as a pore shape anisotropy effect, as measured from Baud *et al.* (2005) for Bentheim sandstone and confirmed by both theoretical and microstructural studies (Louis *et al.*, 2005). Although other sources of anisotropy are possible, pore shape anisotropy might be important in both Majella and SML. Figures 3b,d show stereoplots of the orientation of the principal values for the P-wave velocities derived from the pseudo-tensor analysis on the data set corresponding to the water-saturated samples. For both rocks, the fabric is triaxial, with 3 different eigenvalues. The anisotropy factor A (Fig. 3) is calculated as the difference between maximum and minimum eigenvalues normalized to the mean velocity. For SML, the anisotropy in the saturated samples is negligible and should not have a strong influence on the mechanical behavior. However we found that A is close to 13% in dry SML.

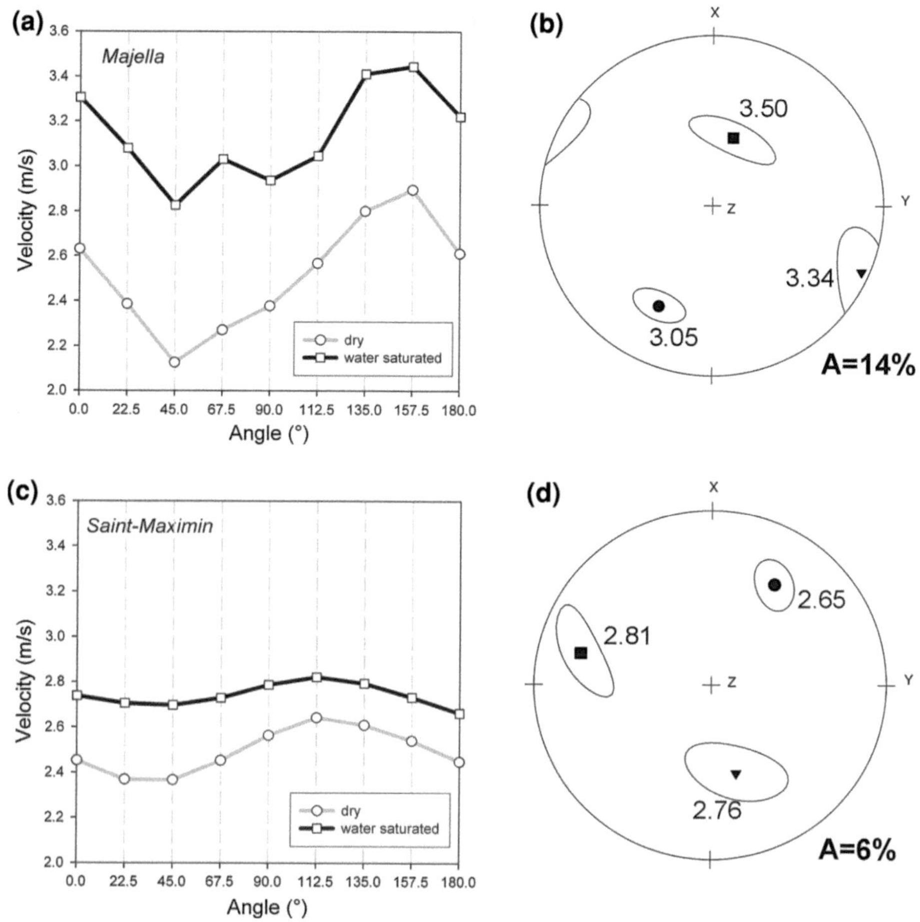

Figure 3
Evolution of P-wave velocity (a-c) across 8 different diameters in dry (circles) and water-saturated (squares) core samples for (a) Majella grainstone and (c) Saint. Maximin limestone. The directions X, Y and Z correspond to the reference frame linked to the blocks from which the core samples have been prepared. Stereoplots (b, d) with the orientation of the principal velocity values derived from the pseudo-tensor analysis with the corresponding velocity magnitude (in km/s), the 95% confidence ellipses and the anisotropy factor A for wet samples of the two rocks investigated.

2.4. Sample Orientation for Mechanical Tests

The core sample axis for the mechanical tests corresponds to the Z direction for SML. Assuming that the bedding plane incorporates two of the principal directions, the direction in which the samples were cored is oblique with respect to both the bedding plane and the bedding pole. Compared to SML, the P-wave velocity anisotropy for MA is significantly higher in both the saturated (A = 14%, Fig. 3) and the dry samples (A = 27%). The direction in which the samples for the mechanical tests were cored corresponds to the X-axis, which approaches the direction corresponding to the minimum velocity. Assuming that the minimum velocity indicates the direction perpendicular to the bedding (BAUD *et al.*, 2005; DAVID *et al.*, 2007), the direction of the core axis is almost parallel to the bedding pole for MA samples tested in the triaxial experiments. Due to the high elastic anisotropy in MA, it is expected that the behavior would be significantly influenced by anisotropic features such as bedding orientation, pore shape anisotropy or distribution of contacts between grains at a small scale.

3. Experimental Procedure

Cylindrical samples of 20-mm of diameter and 40-mm of length were cored in the Z direction (oblique with respect to the sedimentary bedding, see previous section) for SML and in the X direction (assumed to be almost perpendicular to bedding) for MA. Jacketed samples were deformed in conventional triaxial configuration at room temperature at École et Observatoire des Sciences de la Terre (EOST) in Strasbourg, see KLEIN and REUSCHLÉ (2004) for apparatus and measurements details. Most tests were performed at confining pressures ranging from 3 to 50 MPa in nominally dry conditions. A few tests were performed on wet samples saturated with deionized water. A computer-controlled stepping motor connected to a pressure transducer regulated the confining pressure with an accuracy of 0.05 MPa. The axial load was applied by a piston controlled by a second computer-controlled stepping motor. Axial displacement was measured outside the pressure vessel with a capacitive transducer with accuracy 0.2 μm mounted on the moving piston and servo-controlled at a fixed rate (corresponding to a nominal strain rate of 10^{-5}/s).

Hydrostatic experiments were performed at a fixed pore pressure of 10 MPa under fully drained conditions. The pore volume change was recorded by monitoring the displacement of the pore pressure generator with an angular encoder. The porosity change was calculated from the ratio of the pore volume change to the initial bulk volume of the sample. Due to the very porous and weak nature of the rocks investigated in this study, it was impossible to use strain gauges to measure axial and radial strain in the dry experiments. In all attempts, stress-induced pore-collapse resulted in breakage of the gauges even when the surfaced pores were filled with a thin layer of fast-cured epoxy. Volumetric strain was thus recorded in dry and wet tests by monitoring the piston displacement of the confining pressure generator with an angular encoder. This method was tested over the pressure range investigated in this study by comparing the volumetric signal (ε_{kk}) with the porosity change inferred from a comparable pore volume monitoring ($\Delta\phi$) during experiments on water-saturated samples in drained conditions. The independent regulations of pore and confining pressures lead to comparable estimations of the volumetric strain and porosity reduction: $\varepsilon_{kk} = \Delta V/V \approx \Delta\phi$. Because this method is a measurement of the whole sample volume, the volumetric strain was not affected by strain localization.

4. *Mechanical Data*

In this paper we will use the convention that the compressive stresses and compactive strains (i.e., shortening and porosity decrease) are considered positive. We will denote the maximum and minimum (compressive) principal stresses by σ_1 and σ_3, respectively. The pore pressure will be denoted by P_p, and the difference $P_c - P_p$ between the confining pressure ($P_c = \sigma_2 = \sigma_3$) and pore pressure will be referred to as the "effective pressure" P_{eff}. The effective mean stress $(\sigma_1 + 2\sigma_3)/3 - P_p$ will be denoted by P and the differential stress $\sigma_1 - \sigma_3$ by Q.

4.1. *Majella Grainstone*

Sixteen dry and four wet experiments were performed on MA. Figure 4 presents our data for hydrostatic compaction in wet (Fig. 4a) and dry (Fig. 4b) conditions. The wet hydrostat (symbols) shows two stages; (1) an approximately linear poroelastic stage at low pressure, indicative of no significant crack porosity; and (2) a second stage showing deviation from linearity, corresponding to the onset of grain crushing and pore collapse

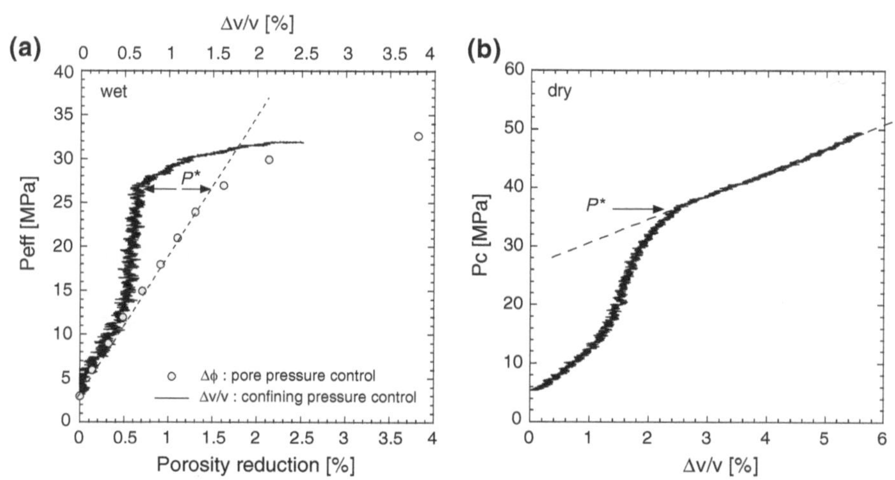

Figure 4
Representative mechanical data for hydrostatic compression experiments on Majella grainstone. (a) Effective pressure versus porosity reduction (symbols) and volumetric strain (line) for a sample deformed in wet conditions at a constant pore pressure of 10 MPa. (b) Confining pressure versus volumetric strain for a sample deformed in dry conditions. The arrow indicates the critical pressure P^* for the onset of pore collapse.

(Zhang *et al.*, 1990). The effective pressure at the transition is denoted P^*, which corresponds to an inflexion point in the hydrostat. The critical effective pressure P^* for saturated MA was observed to be 26.5 MPa. Figure 4a also shows the volumetric response recorded from the confining pressure monitoring. When the confining pressure increased during the hydrostatic test, the calibrated volumetric strain did not match the porosity reduction. However, for a weak and compressible rock such as Majella grainstone, grain crushing and pore collapse resulted in a very strong acceleration of the compaction which appeared clearly and simultaneously on both volumetric signals from pore and confining pressure monitoring. The volumetric signal from the confinement control can therefore be used to estimate P^* in dry conditions. Several dry hydrostatic tests were performed and gave comparable results. P^* appeared more evident if instead of increasing the confining pressure by small steps, as was done in the wet experiment presented in Figure 4a, a very slow continuous rate was applied. A representative example is presented in Figure 4b where the confining pressure was increased at a rate of 0.001 MPa/s. P^* was found to be 37 MPa in dry conditions.

Mechanical data for selected experiments performed at confining pressures ranging from 5 to 31 MPa are shown in Figure 5. The differential stress as a function of axial strain is shown in Figure 5a while the volumetric strain is shown as the function of the mean stress in Figure 5b. Up to 10 MPa of confining pressure, behavior typical of the brittle faulting regime, was observed with stress-strain curves reaching a peak beyond which strain softening was recorded. At confining pressure of 16 MPa and above, a more ductile behavior was observed. At the highest pressure tested (26 and 31 MPa), significant strain-hardening was observed beyond the yield point. Wong *et al.* (1997) showed that the hydrostatic and nonhydrostatic loadings are coupled in a triaxial compression experiment. In the compactive regime, they identified the yield point as the onset of shear-enhanced compaction (Curran and Carroll, 1979), the critical pressure, C^*, at which the triaxial volumetric curve manifests an accelerated compaction with respect to a hydrostatic test. They also showed that for porous sandstones this critical pressure corresponds to a dramatic increase in the acoustic emission activity. Baud *et al.* (2000a) and Vajdova *et al.* (2004) determined C^* for a series of experiments on limestone relying on strain gage volumetric data only since they recorded no significant acoustic emission activity during their tests. A similar behavior was observed here: The mean stress-volumetric strain curve was first linear. At C^*, compaction accelerated abruptly and significantly. Shear-enhanced compaction was observed in all experiments, even at low confining pressures (Fig. 5b).

4.2. Saint-Maximin Limestone

Seventeen dry and four wet experiments were performed on Saint Maximin limestone. The hydrostatic data for SML are qualitatively similar to those for Majella grainstone (Fig. 6). SML is weaker than Majella grainstone and the onset of grain crushing and pore collapse P^* was equal to 13 MPa in wet conditions. For SML the

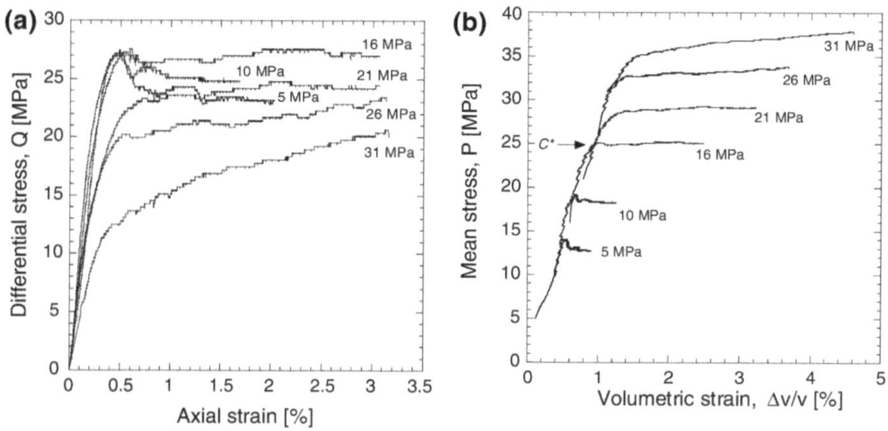

Figure 5
Representative mechanical data for selected triaxial compression experiments on dry Majella grainstone. (a) Differential stress versus axial strain for experiments at confining pressures. (b) Mean stress versus volumetric strain. The mechanical behavior of Majella grainstone is purely compactant.

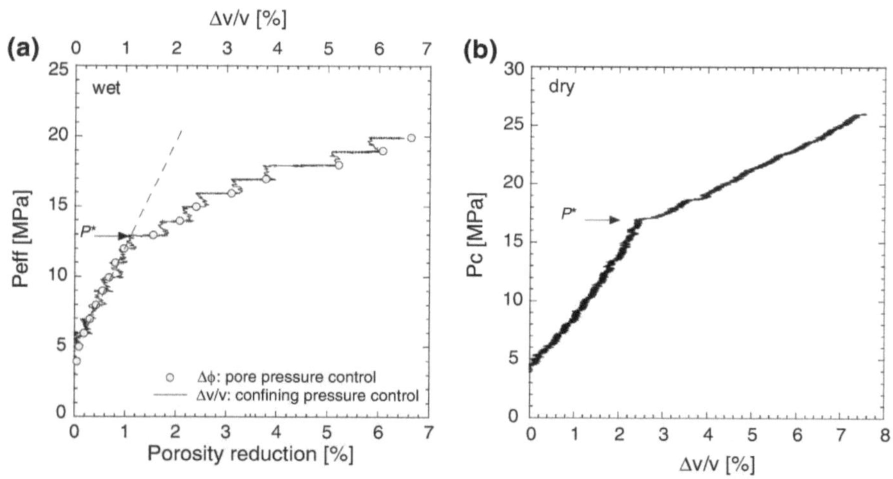

Figure 6
Mechanical data for hydrostatic compression experiments on Saint Maximin limestone. (a) Effective pressure versus porosity reduction (symbols) and volumetric strain (line) for a sample deformed in wet conditions at a constant pore pressure of 10 MPa. (b) Confining pressure versus volumetric strain for a sample deformed in dry conditions. The arrow indicates the critical pressure P^* for the onset of pore collapse.

volumetric strain inferred from the confining pressure monitoring gave results very similar to those inferred from the porosity reduction (Fig. 6a). The volumetric signal for SML represents a proxy for the hydrostat for wet and dry hydrostatic loadings. For

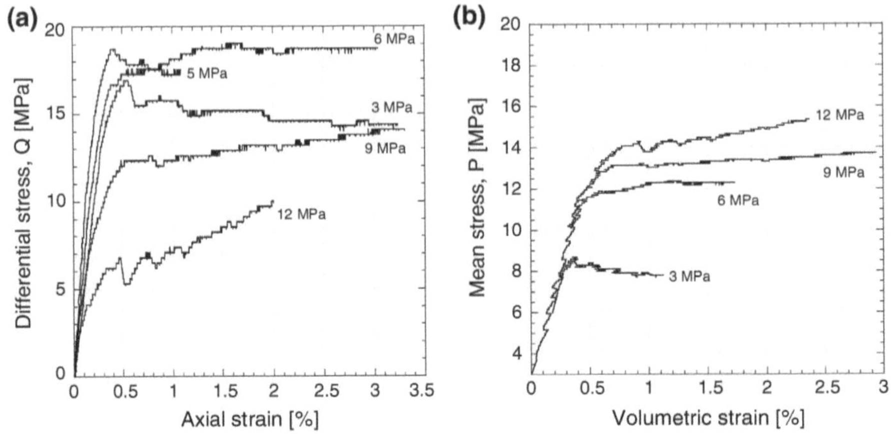

Figure 7
Representative mechanical data for selected triaxial compression experiments on dry Saint Maximin limestone. (a) Differential stress versus axial strain for experiments at confining pressures indicated next to each curve. (b) Mean stress versus volumetric strain. The mechanical behavior of Saint Maximin limestone is purely compactant.

the dry case, the same method as for Majella grainstone was used. The confining pressure was increased at 0.001 MPa/s, giving the onset of grain crushing, P^* equal to 17 MPa.

Triaxial experiments were performed on SML at confining pressures ranging from 3 to 12 MPa. Dissimilar to Majella grainstone, the response of this rock exhibited variability at all pressures. Mechanical data for representative experiments are presented in Figure 7. Three different behaviors were observed: 1) at the lowest confining pressures (3–5 MPa), the stress-strain curves reached a peak beyond which strain softening was observed (Fig. 7a), behavior typical of the brittle faulting regime. Visual inspection of samples indeed revealed the presence of macroscopic failures at relatively high angles with respect to the major principal stress. 2) At confining pressures between 6 and 12 MPa, the stress-strain curves were mostly flat beyond the yield point. In porous sandstone this behavior is typical of a transitional regime (Bésuelle, 2001; Baud *et al.*, 2004) where the failure modes usually involve high angle conjugate shear bands. 3) At the highest confining pressures (12 MPa), the stress-strain curves were punctuated by episodic stress drops while the rock strain hardened. This behavior is comparable to the observations of Baud *et al.* (2004) on Bentheim sandstone in which discrete compaction bands developed. However, in both cases, visual inspection of the samples revealed no clear evidence of strain localization. The failure modes of SML samples were therefore unclear above 5 MPa of confining pressure and a detailed microstructural analysis was performed to describe them (see section 5). Shear-enhanced compaction was observed at all pressure conditions.

Table 2

Summary of mechanical data for the samples investigated in this study

Confining pressure, MPa	Compactive yield stress C^*		Comments on failure modes
	Differential stress $Q = \sigma_1 - \sigma_3$, MPa	Mean stress $P = (\sigma_1 + 2\sigma_3)/3 - P_p$, MPa	
Majella grainstone			
5	25.5	13.5	compactive shear bands
10	24.6	18.2	compactive shear bands
16	22.9	23.7	homogeneous cataclastic flow
21	19.6	27.5	homogeneous cataclastic flow
26	15.7	31.3	homogeneous cataclastic flow
28	13.2	32.40	homogeneous cataclastic flow
31	10.5	34.5	homogeneous cataclastic flow
P^*		37	homogeneous cataclastic flow
$P^{*(1)}$	0	26.5	homogeneous cataclastic flow
Saint Maximin limestone			
3	8	15	high-angle compactive shear bands
3	8.1	15.4	high-angle compactive shear bands
5	10.	14.8	high-angle compactive shear bands
6	10.6	13.6	compactive shear bands
6	10.4	13.2	compactive shear bands
9	12.8	11.5	compactive shear bands
9	12.8	11.4	compactive shear bands
12	14.62	7.3	compactive shear bands
12	14.1	6.2	compactive shear bands
P^*	0	17	homogeneous cataclastic flow
$P^{*(1)}$	0	13	homogeneous cataclastic flow

(1) Wet hydrostatic experiment with $P_p = 10$ MPa

4.3. *Critical Pressures*

The critical pressures C^* for the onset of shear-enhanced compaction are summarized in Table 2. Because no dilatancy was observed in our experiments, for both rocks our data mapped a single failure envelop for C^* for each rock (Fig. 8). For MA, the variability between samples was relatively small and the compactive envelop is well constrained with a few experiments (Fig. 8). For SML we observed some variability between samples and we therefore duplicated all the experiments. Since we measured no significant variations of porosity in our blocks, this variability could be the result of some local variations of the silica content of our samples.

5. *Microstructural Observations*

Selected samples at different stages of deformation were unloaded and retrieved from the pressure vessel. The deformed samples were first impregnated with epoxy and then

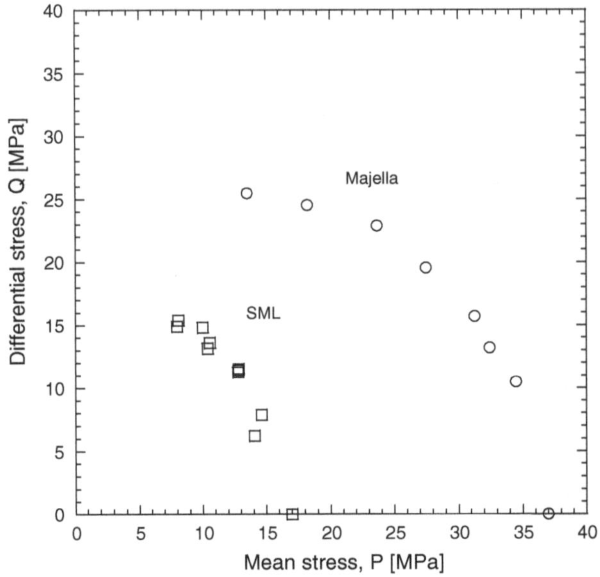

Figure 8
Yield stresses in the stress space *P*-*Q* for: (a) Majella grainstone, (b) Saint Maximin limestone.

sawed along a plane parallel to the axial direction to prepare petrographic thin sections. The thin sections of about 150 μm of thickness were sputter-coated with 0.1 μm of carbon. Back-scattered electron (BSE) images have been taken through a FESEM (see section 2).

5.1. Hydrostatic Compaction

Figure 9 presents a selection of micrographs for two samples of Majella grainstone deformed hydrostatically beyond P^* in dry (Figs. 9a-b) and wet conditions (Fig. 9c). In both wet and dry samples, grain crushing was pervasive. Damage was distributed homogeneously in the samples. The dry sample reached a peak confining pressure of 50 MPa (13 MPa beyond P^*) which corresponded to a relatively modest level of plastic strain $<4\%$. Figure 9a shows that a large proportion of the grains are undamaged and therefore the sample is in an early stage of cataclastic flow. Microcracks appeared to radiate from grain contacts in qualitative agreement with the Hertzian fracture model developed by Zhang *et al.* (1990). Twinning was observed in a few grains (Fig. 9b) but twinning activity did not appear as an important deformation mechanism at this level of compaction. More intense grain crushing was observed in the wet sample deformed to about 5% of plastic strain, leading to significant grain size reduction (Fig. 9c).

Similar observations were made in samples of SML deformed hydrostatically (Fig. 10). The microstructure of this rock is more complex than Majella grainstone, with

Figure 9
Field emission SEM back-scattered micrographs of Majella grainstone samples deformed hydrostatically beyond P^* in (a) dry and (c) wet conditions. Grain crushing is pervasively observed. Limited twinning activity is visible under transmitted light (b).

calcite and quartz grains and fossils. Figures 10a-c present micrographs of a sample deformed in dry conditions to a confining pressure of 27 MPa (10 MPa beyond P^*). Damage appeared homogeneously distributed in this sample and cracking was mostly observed in the quartz and calcite grains and less in the fossils (Figs. 10a-b). Limited

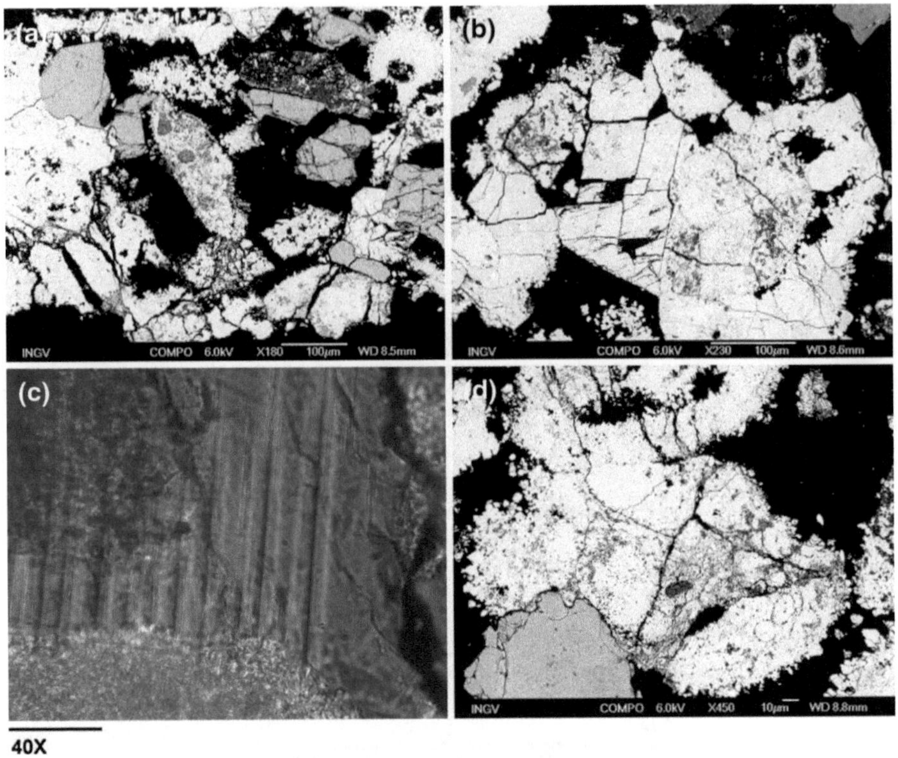

Figure 10
Micrographs of Saint Maximin samples deformed hydrostatically beyond P^* in (a, b, c) dry and (d) wet conditions. Grain crushing is observed in both silicate and carbonate grains. Limited twinning activity is visible under transmitted light (c).

twinning activity was observed in a few calcite grains (Fig. 10c). For a sample deformed in wet conditions to an effective pressure of 20 MPa (7 MPa beyond P^*), we observed a similar development of cataclastic flow. The intensity of grain crushing appeared somehow more important in this sample (Fig. 10d). As in Majella grainstone, microcracking initiated from grain contacts and did not appear to differently affect silica and calcite grains. Our observations on both rocks are qualitatively similar to what has been described for porous sandstone of different porosities by MENÉNDEZ *et al.* (1996) and WU *et al.* (2000).

Preliminary measurements revealed an increase of the internal pore surface in hydrostatically deformed samples of Majella grainstone, from 0.15 m^2/g to 0.30 m^2/g for a dry sample pressurized to 50 MPa of confinement. The increase in the pore surface due to microcracking was found directly proportional to the plastic strain and was comparable in dry and wet samples. In SML, no clear increase of the internal pore surface was observed in samples deformed hydrostatically just beyond P^*. More intense microcracking (i.e., higher plastic strain) was presumably needed to impact significantly the much larger internal pore surface in SML compared to Majella. A systematic study is underway to confirm this first series of observations.

5.2. *Shear-Enhanced Compaction*

Mechanical data on dry Majella grainstone (section 4.1) suggest that shear-enhanced compaction is followed by the two main failure modes in this rock: shear failure and cataclastic flow. Microstructural observations performed on a series of triaxially deformed samples confirm these conclusions. At low confining pressure (5–10 MPa), we observed compactive shear bands in all the samples. Figure 11a presents an example for a sample deformed at 5 MPa of confining pressure to an axial strain of 1%. The band has a thickness of about 2 to 3 grains and is oriented at a low angle (45–50°) with respect to the major principal stress. The compactive shear band appears as alternating fine

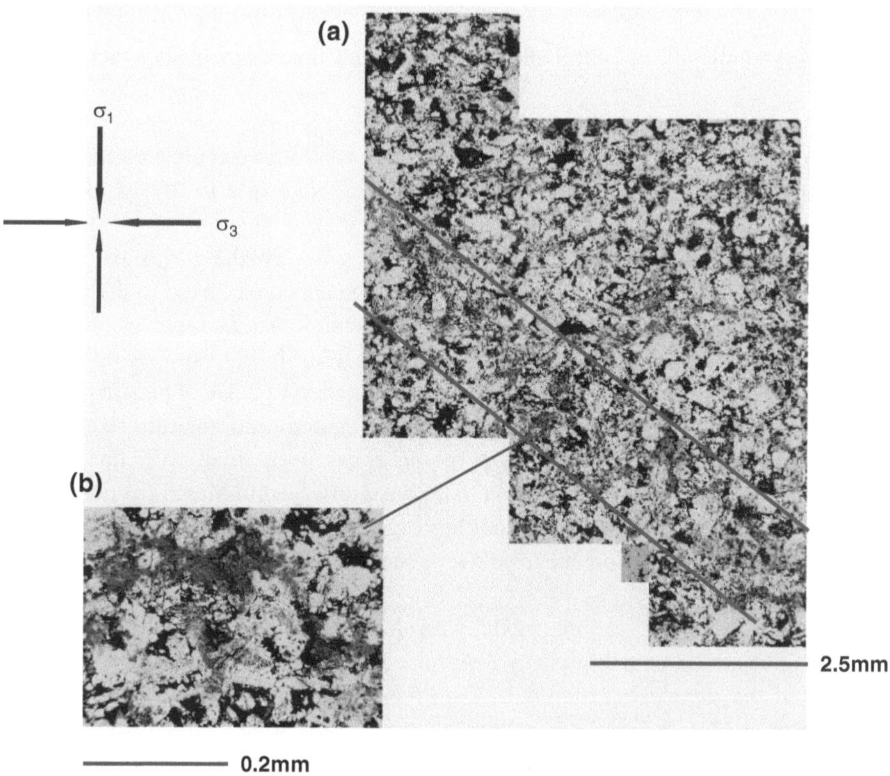

Figure 11
(a) Mosaic of micrographs showing the details of a compactive shear band in a sample of Majella grainstone deformed at 5 MPa of confining pressure. (b) extensive grain crushing inside the band. Maximum principal stress σ_1 was along the vertical direction.

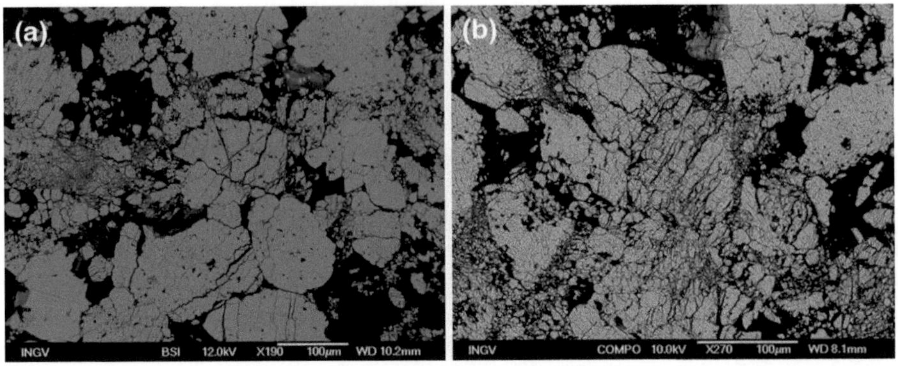

Figure 12
Micrographs of Majella samples deformed in dry conditions at (a) 16 MPa of confining pressure to 3% axial strain; and (b) at 26 MPa confining pressure to 5.8% axial strain. Maximum principal stress σ_1 was along the vertical direction.

powder layers subparallel to the band orientation and heterogeneously cracked grains (Fig. 11b).

At higher confining pressures (>10 MPa), homogeneous cataclastic flow was observed in all the deformed samples. Figures 12a-b present two typical damaged zones in two samples deformed at 16 MPa of confining pressure and to 3% of axial strain (Fig. 12a) and 26 MPa of confining pressure and 5.8% of axial strain (Fig. 12b). In both cases, the samples were deformed beyond C^* and grain crushing was the dominant deformation mechanism. The level of isotropic damage appeared linked to the amount of plastic strain accumulated in the various studied samples. An example of microcracks initiating from grain contacts is shown in Figure 12a. It corresponds to a plastic volumetric strain of about 1.5% while after an accumulation of 2.3% of plastic volumetric strain some grains are completely crushed, comminuted into micron size particles (Fig. 12b). In some areas, part of the pore space has been filled with fine material. Therefore two distinctly different failure modes were observed in Majella grainstone, i.e., shear failure and cataclastic flow, without any sign of hybrid failure modes, suggesting that the brittle ductile transition occurred over a narrow window of confining pressures in this rock.

The evolution of the failure mode with increasing pressure in Saint-Maximin limestone could not be unambiguously determined from the mechanical data and visual inspections of the samples (section 4.2). Over the confining pressure range of 3 MPa to 12 MPa (P^* is at 17 MPa), various patterns of strain localization were observed in all the samples. At pressures of 3 and 5 MPa, after only a few percent of plastic strain beyond C^*, a single band oriented at relatively high angle with respect to the major principal stress was observed. The macroscopic band started in the middle of the sample (Fig. 13) or from one end. In the example presented in Figure 13, the band orientation was about 80 degrees with respect to the major principal stress. The band has a thickness of 3 to 4

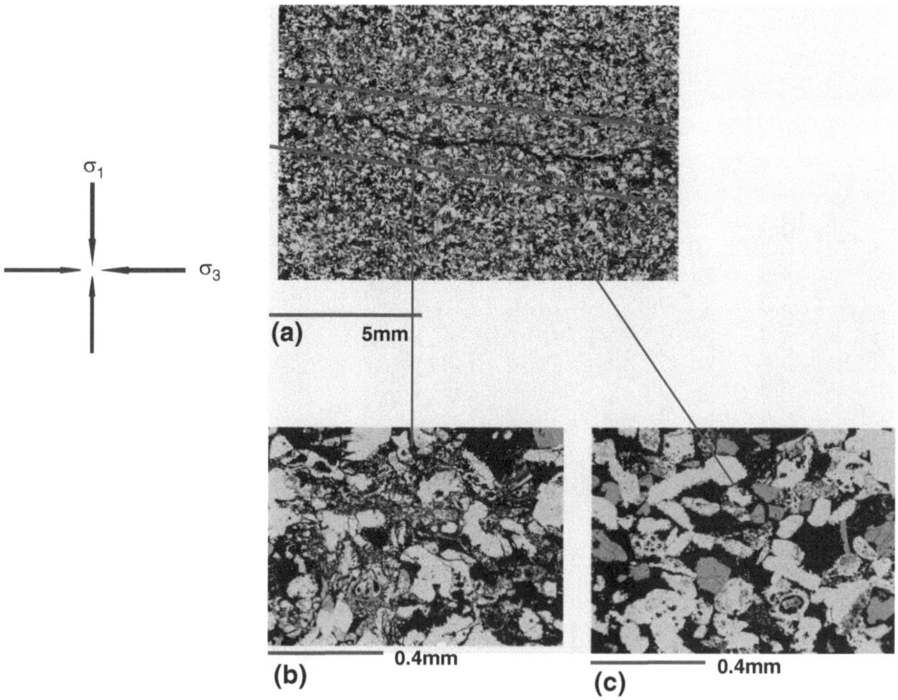

Figure 13
(a) Mosaic of micrographs showing the details of a high angle compactive shear band in a sample of Saint Maximin limestone deformed in dry conditions at 5 MPa of confining pressure. Details of (b) intense grain crushing inside the band and (c) the absence of damage outside the band. Maximum principal stress σ_1 was along the vertical direction.

grains and no significant damage was visible outside the band. A macroscopic crack developed in the central part of the band, suggesting that this structure is not purely compactant and therefore is interpreted as a compactive shear band. At higher confining pressures (9, 10 and 12 MPa), more complex and very variable spatial distribution of damage is observed. Damage appeared in all cases in a few compactive shear bands (from 2 to 4) oriented at various orientations with respect to the major principal stress. Figure 14 shows two examples for samples deformed at 9 MPa (Fig. 14a) and 10 MPa (Fig. 14b) of confining pressure. The compactive shear bands appear as dark bands if the sample is saturated with colored epoxy. Inside the bands, intense grain crushing was observed (Fig. 14d) while no visible damage appeared outside (Fig. 14c). Since for the same pressure conditions significantly different microstructures were observed in different samples, it is likely that pre-existing heterogeneities in the samples played a role in the nucleation and development of the compactive shear bands. Differing from Majella limestone, compaction localization was the dominant feature of compaction in triaxially deformed SML samples.

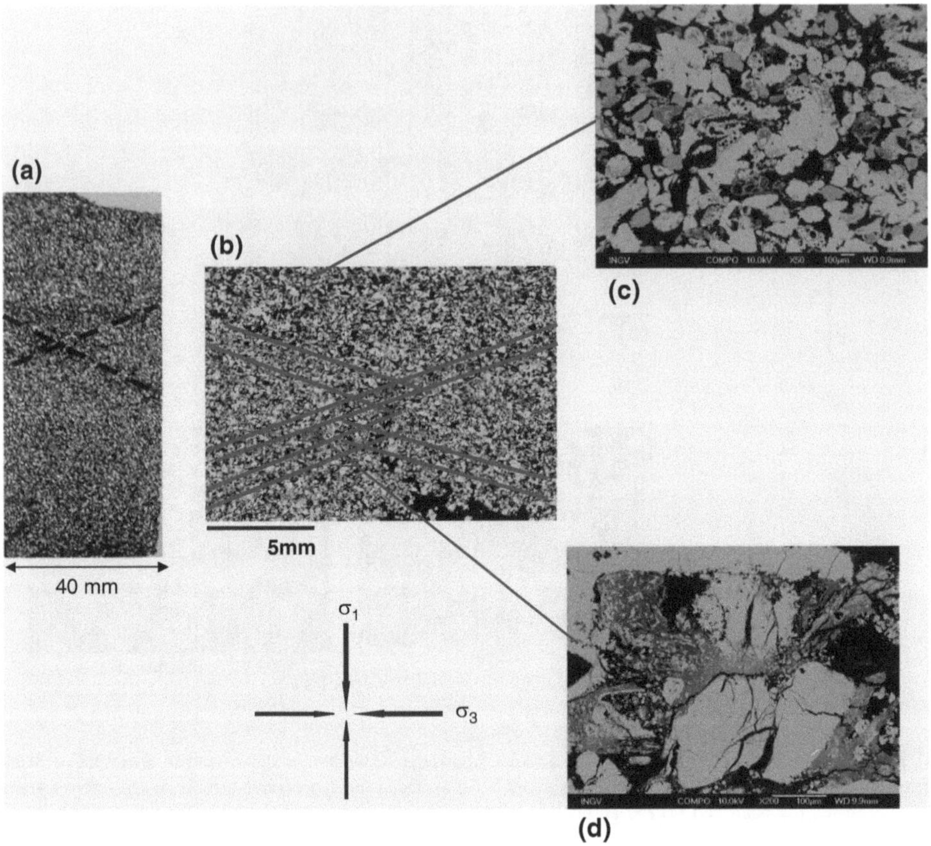

Figure 14
Optical (a) and Field Emission SEM micrographs mosaic (b) of compactive shear bands in samples of Saint Maximin limestone deformed in dry conditions at 9 and 10 MPa of confining pressure, respectively. Details showing (c) no damage outside the band; (d) intense grain crushing inside the band. Maximum principal stress σ_1 was along the vertical direction.

6. Discussion

6.1. Microstructural Control of Strength in Porous Carbonates

Microstructural observations indicate that microcracking was the dominant mechanism of deformation in both Majella grainstone and SML at both hydrostatic and triaxial conditions. In hydrostatically deformed samples, extensive cracking occurred mostly from grain contacts in qualitative agreement with the Hertzian fracture model developed by ZHANG *et al.* (1990). This model was based on observations on sandstones and predicts that for a rock of porosity ϕ and grain radius R, the critical pressure P^* follows the trend $P* \propto (\phi R)^{-3/2}$. VAJDOVA *et al.* (2004) compiled published values of P^* for carbonates and siliciclastic rocks and compared them with this model. For the same value of ϕR, they showed that the

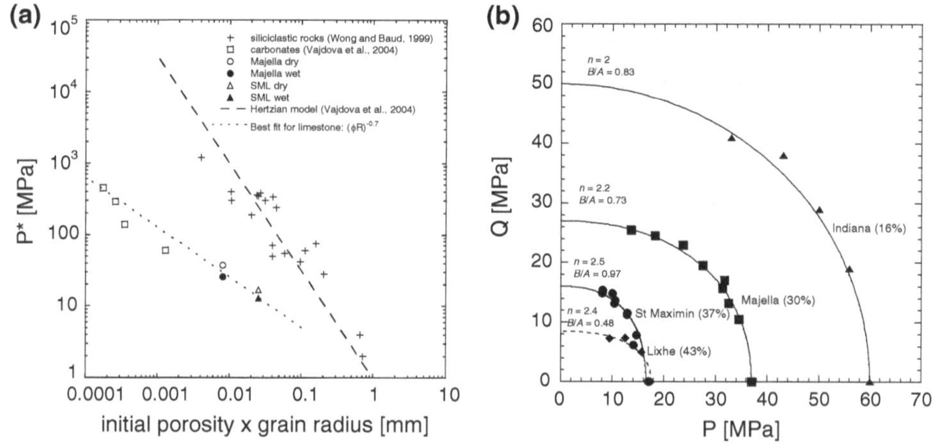

Figure 15

(a) Critical pressure P^* for the onset of grain crushing in various carbonates under hydrostatic loading as a function of the product of initial porosity and grain radius. Open symbols represent data for dry conditions and dark symbols represent data for wet conditions, for Solnhofen limestone (Baud *et al.*, 2000a), Tavel limestone, Indiana limestone (Vajdova *et al.*, 2004), Chauvigny limestone (Fabre and Gustkiewicz, 1997), Majella grainstone and Saint Maximin limestone (this study). The crosses represent data for siliciclastic rocks from Wong and Baud (1999). The dashed line is a prediction of the Hertzian fracture model. The dotted line is the best fit for the carbonate rocks. (b) Compilation of yield envelopes for carbonate rocks. Initial porosity is indicated for each rock. Data sources: Lixhe chalk: Homand and Shao (2000), Indiana limestone: Vajdova *et al.* (2004), Majella grainstone and Saint Maximin limestone: this study. The curves represent the best fit using the model of Grueschow and Rudnicki (2005).

limestones are significantly weaker than the siliciclastic rocks and that they do not follow the trend predicted by the Hertzian model. In Figure 15a, we added the critical pressure P^* for Majella and SML in dry and wet conditions to Vajdova *et al.* (2004) compilation. Our new data confirmed their conclusions and the whole set of limestone data follows a much less steep trend $P^* \propto (\phi R)^{-0.7}$. Because crystal plasticity did not appear to be a major factor leading to the observed pore collapse in our samples, the discrepancy between the Hertzian model and the measured P^* for carbonates reflects an intrinsic limit of the model in particular because it considers uniform defect structure (Vajdova *et al.*, 2004). The results shown here indicate that more data, particularly for weaker materials, will be needed to confirm this trend, and more elaborate models should be developed. Baud *et al.* (2000b) analyzed the effect of water in conventional triaxial experiments on porous sandstone. They interpreted the water-weakening effect as due to the reduction of the surface energy in the presence of water and quantified this effect using a parameter λ defined as:

$$\lambda = \left(\frac{\gamma'}{\gamma}\right) = \left(\frac{P^*_{\mathrm{wet}}}{P^*_{\mathrm{dry}}}\right)^{2/3},$$

where γ and γ' are the surface energies in dry and wet conditions, respectively. From our hydrostatic dry and wet data, we inferred $\lambda = 0.80$ and $\lambda = 0.84$ for Majella grainstone

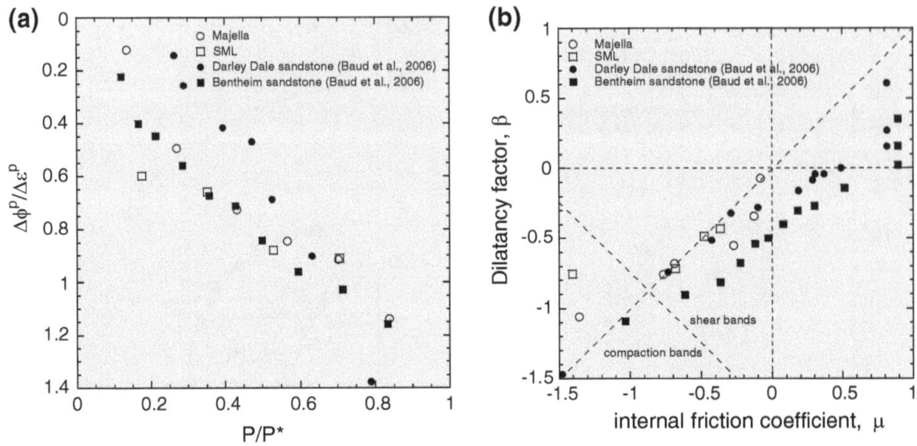

Figure 16

(a) Compaction factor at the onset of shear-enhanced compaction C^*. Open symbols represent data for Majella grainstone and Saint Maximin limestone (this study), dark symbols represent data for Bentheim and Darley Dale sandstone (Baud *et al.*, 2006). (b) Constitutive parameters for the nonassociative model of Rudnicki and Rice (1975). The dilatancy factor β is plotted as a function of the internal friction coefficient μ. The open symbols represent the data for Majella grainstone and Saint Maximin limestone (this study). For reference data for samples of Bentheim and Darley Dale sandstone (Baud *et al.*, 2006) are given as dark symbols. Two of the failure modes (compaction bands and shear bands) predicted from bifurcation analysis (Issen and Rudnicki, 2000) are separated by the dashed line $\beta + \mu = -\sqrt{3}$. The normality line $\beta = \mu$ is also represented as a dashed line.

and SML, respectively. Baud *et al.* (2000b) found values in the range [0.79; 0.97] with the lower value for Boise sandstone in the same porosity range as Majella and SML. This suggests that the weakening effect of water measured in our carbonates can also be due to the reduction of the surface energy. Previous studies reported a weakening effect of water of various magnitudes in carbonates (Boozer *et al.*, 1963; Rutter, 1974; Risnes *et al.*, 2005). Several chemical mechanisms, in particular pressure solution (Hellmann *et al.*, 1996) and stress corrosion (Hadizadeh and Law, 1991) were proposed to explain the observed reduction in strength. Pressure solution was unlikely to be a factor in our experiments performed at room temperature over short periods of time. The stress corrosion effect leads to macroscopic creep and time-dependent behavior both in extension and in compression (Atkinson and Meredith, 1987; Baud and Meredith, 1997). Our wet experiments were performed under fully servo-controlled pore fluid pressure and no significant macroscopic creep was observed. It suggests that stress corrosion only slightly impacted on our results, if any. Risnes and Flaageng (1999) and Heggheim *et al.* (2005) studied the weakening effect of pore fluid on chalks. They showed that the strength of chalks is strongly dependent on the nature of the pore fluid and increases according to: water < methanol < oil < dry. Risnes *et al.* (2005) performed a systematic study on Lixhe chalk of porosity > 40% using water-glycol mixtures. Their data for water saturated and glycol saturated cases would lead to $\lambda < 0.60$ for Lixhe chalk. However, RISNES *et al.* (2005) interpreted part of this very important weakening as the result of complex capillary effects (since dry chalk still contains water). Because the strength reduction observed in our wet samples was much less pronounced and comparable for two rocks of very different internal surfaces, it is unlikely that such capillary effects contributed significantly to the water weakening in Majella grainstone and SML.

In Figure 15b we compile our data in the stress space ($P-Q$) and compare them with VAJDOVA *et al.*'s (2004) data on Indiana limestone of 16% porosity and data on Lixhe chalk of 43% porosity from HOMAND and SHAO (2000). Figure 15b shows clearly that porosity significantly influences the strength of porous carbonates. Unlike for porous sandstone for which the compactive failure envelopes can in most cases be fitted with an elliptical cap (WONG *et al.*, 1997; BAUD *et al.*, 2006), the failure envelopes of porous carbonates display a wide variety of shapes. To quantify the differences in the shapes, we used an expression proposed by GRUESCHOW and RUDNICKI (2005):

$$\left(\frac{P}{A}\right)^n + \left(\frac{Q}{B}\right)^n = 1. \tag{1}$$

The exponent n varies from 2 to 2.5 and seems to increase with increasing porosity. The ratio B/A also shows a wide variability from 0.48 to 0.97. This variability underlines the difficulty of the constitutive modeling of porous carbonates. Attempts using a model based on plastic pore collapse (CURRAN and CARROLL, 1979; BAUD *et al.*, 2000a; VAJDOVA *et al.*, 2004) yielded in most cases limited success. VAJDOVA *et al.* (2004) suggested that a more appropriate model should consider the coupling of plasticity and microcracking. Such an approach is however unlikely to apply to our new data, considering the lack of evidence of crystal plasticity in deformed samples of Majella grainstone and SML. A more appropriate micromechanical model should take into account the geometric complexity of the rock microstructure (pore shape, grain size distribution, etc). It would therefore require more mechanical data, and particularly a very accurate description of the rock microstructure employing techniques such as mercury porosimetry, BET and further quantitative microstructural analyses using various techniques of imaging such as Computed Tomography (CT). Building such a model is beyond the scope of this study, however mechanical and microstructural data presented here could be adopted as a referring base.

6.2. Compaction Localization in Porous Carbonates

Several studies have described experimentally the brittle-ductile transition in sandstone (WONG *et al.*, 1997; BÉSUELLE, 2001; BAUD *et al.*, 2004) and carbonates (PATERSON, 1958; HEARD, 1960; EVANS *et al.*, 1990; VAJDOVA *et al.*, 2004). In porous sandstone, recent experimental work revealed a wide variety of failure modes involving compaction localization at effective pressure higher than that of brittle failure. With increasing effective pressure, a transition from dilatancy to compaction is observed while the failure evolves from dilating shear-band to high angle compactive shear bands mostly on the left-hand side of the failure envelope. Further increase in the effective pressure leads to the developments of homogeneous cataclastic flow or in many cases to the sequential growth of discrete compaction bands (Baud *et al.*, 2004; 2006). The brittle-ductile transition in sandstone is therefore not a sharp transition and not concomitant to the dilatant/compactant transition. In rocks where the pervasive development of compaction bands is observed, the terminology brittle-ductile transition is actually not really suitable since ductile deformation occurred mostly under hydrostatic conditions. In porous limestone, the data of Vajdova *et al.* (2004; 2006) suggest that the transition between the dilatant brittle regime and homogeneous catatclastic flow is sharper and relatively closer to the dilatant/compactant transition. However, more microstructural data are needed to confirm this tendency.

The behavior of Majella and SML is different from that of less porous limestones, in particular because of the absence of dilatancy. Compactive shear bands were observed at the lowest tested pressures in both rocks. Unlike for sandstone, we had no evidence of the band angle decreasing with increasing pressure. A quantitative description of the band geometries (orientation, tortuosity, thickness) would be very difficult to give because of the variability between samples and the small interval of pressures at which the bands were observed to form for these weak materials. In Majella limestone this failure mode is followed by homogeneous cataclastic flow, and the compactive behavior of this rock at high pressures appears similar to that of Indiana and Tavel limestones (Vajdova *et al.*, 2004). The Majella limestone used in this study was from the host rock in the S. Madonna delle Mazze quarry described by Tondi *et al.* (2006). A variety of structures involving deformation and compaction bands and stylolites was observed in the quarry. The deformation bands are oriented at various angles to the bedding and show both porosity reduction and shearing. In this sense, they are comparable to the compactive shear bands observed at low confinement in the laboratory experiments. The compaction bands were the first structures of this type observed in a carbonate rock (Tondi *et al.*, 2006). These field compaction bands are mostly oriented parallel to the bedding direction. Recent experiments on porous sandstones (Townend *et al.*, 2008; Louis *et al.*, 2009, this volume) indeed support the idea that in a material with significant bedding anisotropy, compaction bands tend to develop preferentially in the bedding plane. Starting from a comparable host material taken from the same formation, we observed no evidence of compaction bands in our laboratory samples, although they were cored in the most favorable orientation (i.e., perpendicular to bedding). This suggests that the field compaction bands could have appeared after the deformation bands and were the results of complex stress interactions between different structures. It is also possible that water played a major role in the evolution of the failure in this type of material. Wet triaxial experiments are currently being performed in our laboratory to study this question.

Compaction localization appeared in all triaxially deformed samples of Saint Maximin limestone and can be considered for this carbonate as the dominant feature of compaction. However, a certain number of differences were observed with respect to the patterns of compaction localization reported in porous sandstones. First, when the deformation is localized, the bands observed in SML were in most cases not orthogonal to the major principal stress and thus not labeled as compaction bands. Second, bands appeared less tortuous and thicker than the discrete compaction bands observed in Bentheim and Diemelstadt sandstones (BAUD *et al.*, 2004). The sequential growth of the bands, similarly to the sandstone, seemed strongly influenced by the presence of heterogeneities in the sample. Given the weak velocity anisotropy measured in this rock (see section 2), it is unlikely that the sample orientation with respect to the sedimentary bedding played a key role in our observations. Further analysis is needed to understand the nucleation and growth of these bands. Some variability in the quartz content and spatial distribution of quartz grains in the samples could explain the relative scattering observed in the mechanical data, as well as the variability in the failure modes. DRESEN *et al.* (1998) showed that a synthetic marble containing about 20% quartz is about 5 times stronger than without quartz. The relative weakness of SML and the fact that P^* follows the carbonate trend (Fig. 15a) suggest some complex quartz-calcite interaction in this rock. It will be difficult to improve our understanding of these hard/soft grain interactions and their impact on strain localization relying solely on more microstructural work. Recent studies using X-ray computed tomography (LOUIS *et al.*, 2006, 2007) showed that it is possible to image anisotropic damage in a nondestructive way. This technique could be used on SML to characterize the heterogeneity of this material and facilitate understanding of the onset of compaction localization. The pilot use of Digital image correlation on CT images of intact and deformed samples gave promising results on sandstone (LOUIS *et al.*, 2007). It will be the subject of future work on SML.

In a recent study, BAUD *et al.* (2006) determined the constitutive parameters for a series of porous sandstones and checked the prediction of bifurcation analysis (RUDNICKI and RICE, 1975; ISSEN and RUDNICKI, 2000) focusing on compaction localization. Because we observed compaction localization in both studied carbonates, we quantified as a pilot study the constitutive parameters for both Majella and SML. Following WONG *et al.* (1997) we inferred the plastic compaction factor from:

$$\frac{\Delta\varepsilon^p_{kk}}{\Delta\varepsilon^p_1} = \frac{(\Delta\varepsilon_{kk}/\Delta\varepsilon_1) - \phi\beta_\phi[\Delta Q/\Delta\varepsilon_1]}{1 - [\Delta Q/\Delta\varepsilon_1]/E}, \tag{2}$$

where β_ϕ is the pore compressibility, ε_l the axial strain and E Young's modulus.

In Figure 16a, the plastic compaction factor is presented as a function of the effective pressure normalized by P^*. The data of BAUD *et al.* (2006) for Darley Dale and Bentheim sandstone are given for reference. Since only compaction was observed in the experiments reported here, the plastic compaction factor for the dry experiments is positive. The evolution with pressure is very similar the one observed in the more porous Bentheim sandstone ($\phi = 23\%$). Darley Dale sandstone behaves somewhat differently, most probably because of its lower porosity (13%). Most of our measurements fall in the range $0 < \Delta\varepsilon^p_{kk}/\Delta\varepsilon^p_1 < 1$, with the implication that $0 < \Delta\varepsilon^p_{kk} < \Delta\varepsilon^p_1$. Since $d\varepsilon^p_{kk} = d\varepsilon^p_1 + 2d\varepsilon^p_3$ in a conventional triaxial experiment, our data require that inelastic yield in these carbonates be accompanied by an overall reduction in volume that involves the interplay of axial shortening ($d\varepsilon^p_1 > 0$ with our sign convention) and lateral expansion $(d\varepsilon^p_2 = d\varepsilon^p_3 < 0)$.

Conditions for strain localization in the context of the bifurcation analysis were formulated by RUDNICKI and RICE (1975) for shear localization and more recently by ISSEN and RUDNICKI (2000) for compaction localization. In this framework, these conditions were expressed in terms of a dilatancy factor β and a friction parameter μ defined for a conventional triaxial experiment as:

$$\beta = -\sqrt{3}\frac{\Delta\varepsilon^p_{kk}/\Delta\varepsilon^p_1}{\left(3 - \Delta\varepsilon^p_{kk}/\Delta\varepsilon^p_1\right)} \quad \text{and} \quad \mu = \frac{1}{\sqrt{3}}\frac{\partial Q}{\partial P}, \text{ respectively.}$$

In particular, ISSEN and RUDNICKI (2000) showed that compaction bands should develop if:

$$\mu + \beta < -\sqrt{3}. \tag{3}$$

Figure 16b illustrates that overall our new data on carbonates are relatively close to the normality line $\beta = \mu$. For a comparable value of μ, the compaction factor is significantly lower than the estimations for Bentheim sandstone in which discrete compaction bands were observed. A majority of the points for Majella grainstone and Saint Maximin limestone falls in the domain where shear bands are predicted, in agreement with our microstructural observations. WONG *et al.* (2001) showed that in this parameter range, the theory also predicts strain-softening. Although we did not calculate the hardening parameter, because both strain-softening and strain-hardening were observed in the mechanical data (Figs. 5 and 7), some discrepancies exist between our observations on carbonates and the theoretical predictions. WONG *et al.* (2001) observed more significant discrepancies in sandstone, for which the theory failed to predict the apparition of compaction bands. They attributed the observed discrepancies to the inadequacy of the constitutive model to capture the partitioning of several mechanisms, such as stress-induced microcracks, pore collapse and grain crushing. Similar conclusions may be applicable here. Work is ongoing to collect more data to confirm these preliminary results, particularly tests performed under wet conditions, within which the dilatancy factor β is better constrained.

7. Conclusion

We presented recent advances of a work in progress on two very porous carbonates, Majella grainstone and Saint Maximin limestone. New data indicate that the mechanical behavior of porous carbonates is dominated by shear-enhanced compaction associated, in most cases, with strain-hardening. Stress-induced cracking and grain crushing are the dominant micromechanisms of deformation in both rocks, with no major crystal plasticity being observed. In Majella limestone, compactive shear bands comparable to the field deformation bands were observed at low confinement. At higher pressures, compaction localization was inhibited and homogeneous cataclastic flow developed. Recent field observations suggested the development of compaction bands in the Majella formation. Further laboratory and microstructural work will be needed to explain the discrepancies between the failure mode observed in the laboratory and the structures described in the field. In the more porous Saint-Maximin limestone, compaction localization was observed in all of the triaxially deformed samples. Localized bands developed at various orientations with respect to major principal stress. These bands are interpreted as compactive shear bands. The number of bands increases with increasing strain. Although compaction localization can develop in porous carbonates, it remains unclear whether discrete compaction bands can develop in such materials. Identifying compaction localization in porous carbonates has proven more difficult than for porous sandstones, mostly because the carbonates are essentially made of calcite. To further understand the development of the localized deformation structures in porous carbonates, microstructural analyses are being combined with X-ray computed tomography to capture the 3D geometry of the bands and to better understand their formation.

Acknowledgments

We are grateful to Emanuele Tondi who kindly furnished us with a block of Majella grainstone. In Strasbourg, we benefited from the very valuable technical assistance of Jean-Daniel Bernard. In Cergy-Pontoise, part of the measurements were done by Jean-Christian Colombier, Matthieu Angeli, Lisa Casteleyn and Romain Giros. We thank Silvio Mollo for assisting with the grain size analysis. We also benefited from discussions with Teng-fong Wong, Veronika Vajdova, and Laurent Louis. We are grateful to Dave Olgaard and Alexandre Schubnel for thoughtful reviews. This research was partially supported by the CNRS/CNR bilateral agreement (Project 19138). SV was supported from FIRB-MIUR Project "Research and Development of New Technologies for Protection and Defense of Territory from Natural Risks".

References

Angeli, M., Benavente, D., Bigas, J.-P., Menéndez, B., Hébert, R., and David, C. (2008), *Modification of the porous network by salt crystallization in experimentally weathered sedimentary stones*, Mater. Struct., *41*, 1091–1108.

Atkinson, B.K. and Meredith, P.G. *The theory of subcritical crack growth with applications to minerals and rocks*. In (ed. B.K. Atkinson) *Fracture Mechanics of Rocks*, pp. 111–166, (Academic Press, London. 1987).

BAUD, P. and MEREDITH, P.G. (1997), *Damage accumulation during triaxial creep of Darley Dale sandstone from pore volumometry and acoustic emission*, Int. J. Rock Mech. Min., *34*, 3–4.

BAUD, P., SCHUBNEL, A., and WONG, T.-F. (2000a), *Dilatancy, compaction and failure mode in Solnhofen limestone*, J. Geophys. Res. *195*, 19289–19303.

BAUD, P., ZHU, W., and WONG, T.-F. (2000b), *Failure mode and weakening effect of water on sandstone*, J. Geophys. Res. *105*, 16371–16389.

BAUD, P., KLEIN, E., and WONG, T.-F. (2004), *Compaction localization in porous sandstones: Spatial evolution of damage and acoustic emission activity*, J. Struct. Geol. *26*, 603–624.

BAUD, P., LOUIS, L., DAVID, C., RAWLING, G.C., and WONG, T.-F. (2005), *Effects of bedding and foliation on mechanical anisotropy, damage evolution and failure mode*. In (eds Bruhn, D. and Burlini, L.) *High-Strain Zones: Structure and Physical Properties*. Geological Society of London Special Publications *245*, 223–249.

BAUD, P., VAJDOVA, V., and WONG, T.-F. (2006), *Shear-enhanced compaction and strain localization: Inelastic deformation and constitutive modeling of four porous sandstones*, J. Geophys. Res. *111* (B12401), doi: 10.1029/2005JB004101.

BÉSUELLE, P. (2001), *Compacting and dilating shear bands in porous rocks: Theoretical and experimental conditions*, J. Geophys. Res. *106*, 16371–16389.

BOOZER, G.D., HILLER, K.H., and SERDENGECTI, S. (1963), *Effect of pore fluids on the deformation behavior of rocks subjected to triaxial compression*, 5th Symp. on Rock Mechanics, Univ. Minnesota, 1962, 579–625.

BRACE, W.F. (1978), *Volume change during fracture and frictional sliding: A review*, Pure Appl. Geophys. *116*, 603–614.

BRUNAUR, S., EMMET, P.H., and TELLER, E. (1938), *Adsorption of gasses in multimolecular layers*, J. Amer. Chem. Soc. *60*, 309.

CURRAN, J.H. and CARROLL, M.M. (1979), *Shear stress enhancement of void compaction*, J. Geophys. Res. *84*, 1105–1112.

DAVID, C., ROBION, P., and MENÉNDEZ, B. (2007), *Anisotropy of elastic, magnetic, and microstructural properties of the Callovo-Oxfordian argillite*, Phys. Chem. Earth *32*, 145–153.

DRESEN, G., EVANS, B., and OLGAARD D.L. (1998), *Effect of quartz inclusions on plastic flow in marble*, Geophys. Res. Lett. *25*, 1245–1248.

EDMOND, J.M. and PATERSON, M.S. (1972), *Volume change during the deformation of rocks at high pressures*, Int. Rock Mech. Min. Sci. *9*, 161–182.

EVANS, B., FREDRICH, J.T., and WONG, T.-F. (1990), The brittle-ductile transition in rocks: Recent theoretical and experimental progress. In (eds. A.G. Duba, W.B. Durham, J.W. Handin, H.F. Wang), *The Brittle-Ductile Transition in Rocks (The Heard Volume)*, Geophys. Monogr. Vol. *56*, pp. 1–20.

FABRE, D. and GUSTKIEWICZ, J. (1997), *Poroelastic properties of limestones and sandstones under hydrostatic conditions*, Int. J. Rock Mech. Min. Sci. *34*, 127–134.

FORTIN, J., SCHUBNEL, A., and GUÉGUEN, Y. (2005), *Elastic wave velocities and permeability evolution during compaction of Bleurwiller sandstone*, Int. Rock Mech. Sci. *42*(7–8), 873–889.

FORTIN, J., STANCHITS, S., DRESEN, G., and GUÉGUEN, Y. (2006), *Acoustic Emission and velocities associated with the formation of compaction bands in sandstone*, J. Geophys. Res. *111*, B10203, doi:10.1029/2005JB003854.

FREDRICH, J.T., EVANS, B., and WONG, T.-F. (1989), *Micromechanics of the brittle to plastic transition in Carrara marble*, J. Geophys. Res. *95*, 10,907–10,920.

GRUESCHOW, E. and RUDNICKI, J.W. (2005), *Elliptical yield cap constitutive modeling for high porosity sandstone*, Int. J. Solids Struct. *42*, 4574–4587.

HADIZADEH, J. and LAW, R.D. (1991), *Water-weakening in quartzite deformed at various stress and strain rates*, Int. J. Rock. Mech. Min. Sci. Geomech. Abstr. *28*(5), 431–439.

HADIZADEH, J. and RUTTER, E.H. (1983), *The low temperature brittle-ductile transition in a quartzite and the occurrence of cataclastic flow in nature*, Geol. Rundsch. *72*, 493–509.

HANDIN, J. and HAGER, R.V. (1957), *Experimental deformation of sedimentary rocks under confining pressure: Tests at room temperature on dry sample*, Am. Assoc. Pet. Geol. Bull. *41*, 1–50.

HEARD, H.C. (1960), *Transition from brittle fracture to ductile flow in Solnhofen limestone as a function of temperature, confining pressure and interstitial fluid pressure*. In (eds. D.T. Griggs and J. Handin) *Rock Deformation*, Mem. Geol. Soc. Am. *79*, 193–226.

HEGGHEIM, T., MADLAND, M.V., RISNES, R., and AUSTAD, T. (2005), *A chemical-induced enhanced weakening of chalk by seawater*, J. Petrol. Sci. Eng. *46*, 171–184.

HELLMANN, R., GRATIER, J.P., and RENDERS, P. (1996), *Deformation of Chalk by pressure solution*, V.M. Goldschmidt Conf., Heidelberg.

HOMAND, S. and SHAO, J.F. (2000), *Mechanical behaviour of a porous chalk and effect of saturating fluid*, Mech. Cohes.-Frict. Mater. *5*, 583–606.

ISSEN, K.A. and RUDNICKI, J.W. (2000), *Conditions for compaction bands in porous rock*, J. Geophys. Res. *105*, 21529–21536.

KLEIN, E. and REUSCHLÉ, T. (2004), *A pore crack model for the mechanical behaviour of porous granular rocks in the brittle deformation regime*, Int. J. Rock Mech. Min. Sci. *41*, 975–986.

LOUIS, L., ROBION, P., and DAVID, C. (2004), *A single method for the inversion of anisotropic data sets with application to structural studies*, J. Struct. Geol. *26*, 2065–2072.

LOUIS, L., DAVID, C., METZ, V., ROBION, P., MENÉNDEZ, B., and KISSEL, C. (2005), *Microstructural control on the anisotropy and transport properties in undeformed sandstones*, Int. J. Rock Mech., *42*/7-8, 911–923.

LOUIS, L., WONG, T.-F., BAUD, P., and TEMBE, S. (2006), *Imaging strain localization by X-ray computed tomography: Discrete compaction bands in Diemelstadt sandstone*, J. Struct. Geol. *28*, 762–775.

LOUIS, L., WONG, T.-F., and BAUD, P. (2007), *Imaging strain localization by X-ray radiography and digital image correlation: Deformation bands in Rothbach sandstone*, J. Struct. Geol. *29*, 129–140.

LOUIS, L., BAUD, P., and WONG, T.-F. (2009) *Microstructural inhomogeneity and mechanical anisotropy associated with bedding in Rothbach sandstone*, Pure Applied Geophysics *166*, 517.

MENÉNDEZ, B., ZHU, W., and WONG, T.-F. (1996), *Micromechanics of brittle faulting and cataclastic flow in Berea sandstone*, J. Struct. Geol. *18*, 1–16.

PATERSON, M.S. (1958), *Experimental deformation and faulting in Wombeyan marble*, Geol. Soc. Am. Bull. *69*, 465–476.

RISNES, R. and FLAAGENG, O. (1999), *Mechanical properties of chalk with emphasis on chalk-fluid interactions and micromechanical aspects*, Oil and Gas Sci. Tech., *54*, 6, 751–758.

RISNES, R., MADLAND, M.V., HOLE, M., and KWABIAH, N.K. (2005), *Water weakening of chalk – Mechanical effects of water-glycol mixtures*, J. Petrol. Sci. Eng. *48*, 21–36.

ROBERTSON, E.C. (1955), *Experimental study of the strength of rocks,* Bull. Geol. Soc. Am. *66*, 1275–1314.

RUDNICKI, J.W. and RICE, J.R. (1975), *Conditions for the localization of deformation in pressure sensitive dilatant materials*, J. Mech. Phys. Sol. *23*, 371–394.

RUTTER, E.H. (1974), *The influence of temperature, strain rate and interstitial water in the experimental deformation of calcite rocks*, Tectonophysics *22*, 311–334.

RUTTER, E.H. and MAINPRICE, D.H. (1978), *The effect of water on stress relaxation on faulted and unfaulted sandstone*, Pure Appl. Geophys. 116, 634–654.

TEMBE, S., BAUD P., and T.-F. WONG (2008), *Stress conditions for the propagation of discrete compaction bands in porous sandstone*, J. Geophys. Res., *113*, B09409, doi:10.1029/2007JB005439.

TEUFEL, L.W., RHETT, D.H., and FARRELL, H.E. (1991), *Effect of reservoir depletion and pore pressure drawdown on in situ stress and deformation in the Ekofisk field*, North Sea, Proc. U.S. Rock Mech. Symp. *32*, 63–72.

TONDI, E., ANTONELLINI, M., AYDIN, A., MARCHEGIANI, L., and CELLO G. (2006), *The role of deformation bands, stylolites in fault development in carbonate grainstones of Majella mountain, Italy*, J. Struct. Geol. *28*, 376–391.

TOWNEND, E., BAUD, P., and MEREDITH, P.G. (submitted), *Permeability anisotropy during triaxial deformation of Diemelstadt sandstone*, J. Geophys. Res.

VAJDOVA, V., BAUD, P., and WONG T.-F. (2004), *Compaction, dilatancy and failure in porous carbonate rocks,* J. Geophys. Res. *109*, B10406, doi: 10.1029/2003JB002942.

VAJDOVA, V., CHEN, T.-M.N., and WONG, T.-F. (2006), *Micromechanics of the brittle-ductile transition in Tavel and Indiana limestones*, EOS Trans. AGU *87*(52), Fall Meeting Suppl., Abstract T23D-0521.

WONG, T.-F., DAVID C., and ZHU, W. (1997), *The transition from brittle faulting to cataclastic flow in porous sandstones: Mechanical deformation*, J. Geophys. Res. *102*, 3009–3025.

WONG, T.-F. and BAUD, P. (1999), *Mechanical compaction of porous sandstone*, Oil Gas Sci. Tech. *54*, 715–727.

WONG, T.-F., BAUD, P., and KLEIN, E. (2001), *Localized failure modes in a compactant porous rock*, Geophys. Res. Lett. *28*, 2521–2524.

Wu, X.-Y., Baud, P., and Wong, T.-F. (2000), *Micromechanics of compressive failure and spatial distribution of anisotropic damage in Darley Dale sandstone*, Int. J. Rock Min. Sci. *37*, 143–160.

Zhang, J., Wong, T.-F., and Davis, M.D. (1990), *Micromechanics of pressure induced grain crushing in porous rocks*, J. Geophys. Res. *95*, 341–352.

(Received August 9, 2008, revised December 8, 2008, accepted March 6, 2009)
Published Online First: May 20, 2009

Pure appl. geophys. 166 (2009) 899–925
0033–4553/09/050899–27
DOI 10.1007/s00024-009-0477-2

Pure and Applied Geophysics

Stress Transfer During Pressure Solution Compression of Rigidly Coupled Axisymmetric Asperities Pressed Against a Flat Semi-Infinite Solid

Y. Bernabé,[1] B. Evans,[1] and D. D. Fitzenz[2]

Abstract—In a previous work, we developed a numerical model of compression by pressure solution (PS) of a single axisymmetric asperity pressed against a flat semi-infinite solid. The dissolution rate at any point along the contact and at any time t was determined by (1) computing the normal stress distribution from the current shape of the asperity, and (2) solving the diffusion equation inside the fluid-saturated solid-solid interface, including local dissolution source terms corresponding to the stress field previously determined. The change in shape of the asperity during an infinitesimal time interval δt can then be calculated and the entire procedure repeated as many times as desired. Our results showed that, as the contact flattens and grows during PS, the initial elastic deformation is partially relaxed and the stress transferred from the contact center to the edge. Our goal in the present paper is to demonstrate that, among a population of asperities, stress can also be transferred from one contact to another and that the overall compaction rate can be significantly affected by this process. For this purpose we extended our previous numerical model to simulate PS of two rigidly coupled spherical asperities simultaneously pressed against a flat semi-infinite solid. We considered two end-member cases: 1) transfer of stress to a newly created, not initially present contact, 2) transfer of stress between asperities with different sizes. In both cases, stress was transferred from the most stressed asperity to the least, and, the overall PS displacement rate was reduced. Thus, formation of new contacts and heterogeneous distribution of asperity sizes, which are both expected to exist in rough fractures with self-affine aperture or in heterogeneous granular materials with variable grain-packing geometry, may significantly slow down PS creep compaction.

Key words: Pressure solution, creep.

1. *Introduction*

Pressure solution (PS) has been widely recognized for many years as an important deformation mechanism in the upper crust (e.g., Thomson, 1959; Heald, 1961; Renton and Heald, 1968; Groshong, 1972; Elliot, 1973; Durney, 1976; Rutter, 1983; Gratier, 1987) and extensive effort has been expended for the development of a macroscopic PS creep law based on experimental data and theoretical arguments (among many others, e.g., Spiers et al., 1990; Paterson, 1995; Niemeijer *et al.*, 2002; Fitzenz *et al.*, 2007b). Applications of the proposed PS creep laws include modeling fault gouge healing during

[1] Earth, Atmospheric and Planetary Sciences Department, Massachusetts Institute of Technology, Cambridge, MA 02139, U.S.A.
[2] Centro de Geofisica de Evora, Colegio Luis Verney, 59 Rua Romao Ramalho, 7002-554 Evora, Portugal.

the seismic cycle (e.g., BLANPIED *et al.*, 1992; SEGALL and RICE, 1995; FITZENZ *et al.*, 2007b), predicting porosity decrease with depth in sedimentary basins (e.g., RENARD *et al.*, 2000; REVIL, 2001), or modeling deformation rates at the (time-dependent) transition depths between the seismogenic and ductile zones (ROLANDONE *et al.*, 2004). The transition zone is presently under accrued scrutiny because its behavior during coseismic slip could help resolve the stress drop scaling debate (e.g., SHAW and WESNOUSKY, 2008, HILLERS and WESNOUSKY, 2008). Many treatments use constitutive laws constructed by analogy to high-temperature diffusion creep where the deformation is limited by dissolution or transport processes occurring along the grain boundary region (ELLIOT, 1973; RAJ, 1982; PATERSON, 1995). In generalized form, the strain rate ε' in steady-state conditions is given as:

$$\varepsilon' = \varepsilon'_0 \exp\left(-H_{\mathrm{PS}}/_{\mathrm{R}T}\right) \frac{\sigma_{\mathrm{eff}}}{\mathrm{R}T}, \tag{1}$$

where ε'_0 is a prefactor (which must be dimensionally consistent with the rest of equation (1) and is assumed to contain, among other things, the grain size dependence), R the gas constant, T the absolute temperature, H_{PS} the PS activation enthalpy (its experimental value is often used to identify the rate-limiting process) and σ_{eff} the effective normal stress at a typical solid-solid contact. Note that this relation is only a generic version of many postulated relations, which might involve many different mechanisms, each of which could involve important changes to its mathematical form. For example, in a common variant of equation (1), the linear σ_{eff} term is replaced by σ^n_{eff}, where the exponent n must be empirically determined. In the case of isostatic compaction of a granular aggregate, a similar generic equation has been proposed for the time rate of change of porosity ϕ' (NIEMEIJER *et al.*, 2002):

$$\phi' = \phi'_0 \exp\left(-H_{\mathrm{PS}}/_{\mathrm{R}T}\right) \frac{\sigma_{\mathrm{eff}}}{\mathrm{R}T} \theta(\phi), \tag{2}$$

where the porosity factor $\theta(\phi)$ accounts for the expected reduction in ϕ' with decreasing porosity (obviously compaction must stop when porosity ϕ reaches zero). However, the data from laboratory experiments may be quite complex and are often, at least partially, in conflict with the simplified constitutive laws. There are many possible reasons for such discrepancies. For example, all experiments have unavoidable uncertainties; equations like those above, despite their appealing form and simplicity, could be inadequate; the experimental data could contain significant contributions from other mechanisms, including dislocation creep or brittle fracturing; equations such as those above are often based on simple geometric assumptions that may be inappropriate owing to the presence of a very broad distribution of contact sizes and shapes or to changes in grain-packing geometry occurring during compaction. Motivated by such discrepancies, workers often reformulate the constitutive law to account for multiple mechanisms or more complex geometry. In this analysis, rather than try to match a simple constitutive law to complex data from laboratory experiments, we take an alternate approach. First, we postulate a very simple contact-scale model, but one which includes coupling of the elementary PS processes. A numerical analysis of the model system is used to generate "data", which can then be compared with macroscopic scaling laws of which equations (1) and (2) are examples.

The numerical model describes compression by pressure solution of a single axi-symmetric asperity pressed against a flat semi-infinite solid (see BERNABÉ and EVANS, 2007, hereafter referred to as BE, who followed a theoretical work by LEHNER and LEROY, 2004, hereafter referred to as LL). The dissolution rate at any radial distance r along the contact and any time t was determined by (1) computing the normal stress distribution from the current shape of the asperity, and (2) solving the diffusion equation inside the fluid-saturated contact interface, taking into account the local dissolution source terms corresponding to the stress field previously determined. The change in shape of the asperity during an infinitesimal time interval δt can then be calculated and the entire procedure repeated as many times as desired. Under these assumptions, the asperity flattens with time, the contact area grows, and stress is transferred from the contact center to the edge. If the simulated creep data are interpreted in terms of a creep law such as equation (2), the "constants" are not constant but vary with time, even when they are determined by a Bayesian inference method (FITZENZ *et al.*, 2007a). In particular, the PS activation enthalpy, H_{PS}, was observed to decrease with time, leading the latter authors to suggest the mechanism of deformation evolved from being primarily controlled by contact dissolution to being limited by diffusion of the dissolved solid along the grain interface.

In this paper we extended that single contact models to include two coupled contacts, thus allowing stress to be transferred from one asperity to the other. As intuitively anticipated, the compaction rate slowed as a result of the stress transfer. Nonetheless, despite the obvious oversimplification of the model, when realistic material constants are added to the calculation, the details of the stress transfer, the length of transient behavior, and the deformation rates of the system are considerably more complex than can be described by a simple one-mechanism constitutive law.

2. *Pressure Solution Model*

2.1. *Basic Assumptions*

It is important, at the outset, to state the basic assumptions included in our model. We essentially followed the classic approach (e.g., PATERSON, 1973, 1995), according to which pressure solution creep is understood as granular flow accommodated by mass transfer through an intergranular fluid. All other mechanisms such as dislocation creep or brittle fracturing are excluded from this definition, although they might play an important role in natural situations. Equally important, an infinitesimally thin, dynamically stable interface that remains fully saturated by the pore fluid at pressure p, but nevertheless transmits arbitrary normal and tangential tractions, must exist at each solid-solid contact.

Several possible structures have been proposed for this interface (e.g., strongly adsorbed films, RENARD and ORTOLEVA, 1997, RUTTER, 1983, WEYL, 1958; a dynamic island-channel structure, LEHNER, 1990, PATERSON, 1995, RAJ, 1982; GHOUSSOUB and LEROY, 2001; stress corrosion microcracks and fluid inclusions, DEN BROK, 1998, GRATZ, 1991; altered mineral layer, REVIL, 1999, 2001). The properties of the contact interface may be enhanced by the presence of clay particles (e.g., HICKMAN and EVANS, 1995; RENARD *et al.*, 1997). Conversely, healing may occur at grain contacts for some mineral systems and "wet" interfaces may then be inhibited (HICKMAN and EVANS, 1991; ZUBTSOV *et al.*, 2004). Here, for the purpose of numerical modeling, we only need to stipulate the interface width w and the interface diffusion coefficient D^*. For simplicity sake, we assume that w is constant, insensitive to stress and temperature and that D^* has the same temperature dependence as the diffusion coefficient D of the "free" pore fluid (RENARD *et al.*, 2000). We can therefore write $D^* = fD$ with the (stress independent) correction factor $0 < f < 1$ accounting for various retardation effects depending on the contact interface structure. We further limited the numerical model to the case in which the pore fluid was buffered in some way so the concentration in the "free" pore space (i.e., outside the contact interface) was maintained constant at the equilibrium value c_{eq} of the fluid (at pressure p and temperature T) in contact with the solid. Consequently, there is no precipitation of solid either in the contact interface or on the free surface of the pores. This assumption greatly simplifies the model but prevents investigation of the grain-size dependence of PS creep. Also, it is clear that precipitation and, therefore, advective and/or diffusive transport in the "free" pore space may be essential ingredients of PS deformation in natural conditions (e.g., HE *et al.*, 2003; LEHNER, 1995). In this paper, we do not specify the mechanism buffering the pore fluid composition (one possibility in natural conditions is fast advective transport by the equilibrated fluid), and we do not claim that such conditions are always obtained in geologic settings. Our aim is simply to investigate the complexity of behavior that may occur even with a simple model. In summary, PS creep is assumed here to be controlled by the interplay of two coupled processes: contact dissolution, a driving mechanism, and interface diffusion, a resisting one.

2.2. *Numerical Procedures*

We begin by describing the geometry of the contact system we are analyzing in this study. We consider two solidly attached hemi-spherical asperities of radii R_{C1} and R_{C2} pressed against a semi-infinite, flat solid of the same material (here, quartz). Each contact is therefore axisymmetric and local polar coordinates can be used. We will hereafter assume that the same vertical motion is imposed to both asperities simultaneously and neglect any bending or other elastic deformation between the two asperities. One way to justify this assumption qualitatively is to view the two-asperity system as one elementary cell of a dense periodic array (see Fig. 1a). In this case, the elastic deformation is almost exclusively located around the tips of the asperities and can be calculated knowing the unstressed separation between each asperity and the flat semi-infinite solid. As illustrated

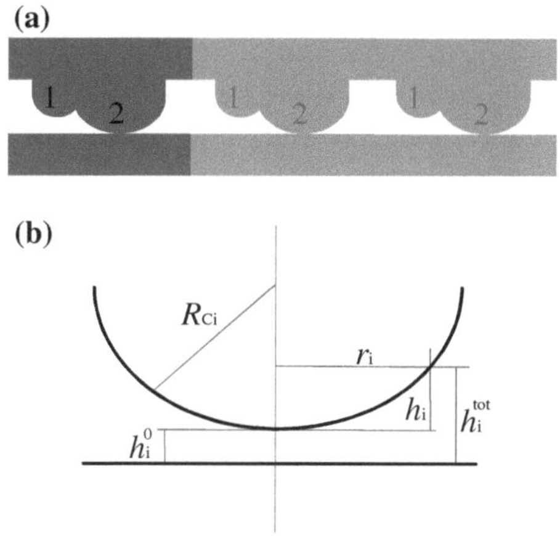

Figure 1
a) A schematic representation of the two-asperity system studied here. The rigidly coupled asperities 1 and 2 are assumed to be part of a dense infinite periodic array, therefore bending or other deformations in the solid between the asperities are negligible. b) Graphic definition of the unstressed separation $h_i^{tot} = h_i^0 + h_i$. It is important to note that, for the sake of visibility (i.e., to avoid representing the asperity surface as nearly planar), this diagram does not really show the assumed case of a radius of curvature R_{Ci} much greater than r_i in the contact region.

in Figure 1b, the total unstressed separation h_i^{tot} is a function of the local radial distance r_i and time t. It can be divided into two parts:

$$h_i^{tot}(r_i, t) = h_i^0(t) + h_i(r_i, t), \tag{3}$$

where the index $i = 1$ or 2 indicates which asperity is being considered. The unstressed separation $h_i^0(t)$ at the tip of asperity i is a nonnegative quantity that can only increase with time or remain constant, since precipitation is not allowed to occur within the contact interface. Note that, if both $h_1^0(t)$ and $h_2^0(t)$ are strictly positive, a downward rigid-body motion equal to the smallest of the two is needed to achieve unstressed contact at one of the two asperities. For convenience, we will always start a simulation with at least one of the two $h_i^0(0)$ equal to zero (i.e., one of the two asperities in unstressed contact) and the global dissolution displacement at any time t will therefore be defined as $u_{dis}(t) = \min(h_i^0(t))$. The other separation term $h_i(r_i, t)$ describes the current, unstressed shape of asperity i. For initially hemi-spherical asperities, we have:

$$h_i(r_i, 0) \approx \frac{r_i^2}{2R_{Ci}}, \tag{4}$$

where $r_i \leq a_i(0) << R_{Ci}$ (with $a_i(0)$ denoting the initial contact radius).

Thus two end-member cases can be investigated: Case 1) identical asperities ($R_{C1} = R_{C2}$) with different initial tip separations (by convention, we chose

$h_1^0(0) > h_2^0(0) = 0$), and, case 2) different asperities (by convention, $R_{C1} < R_{C2}$) with identical initial tip separations ($h_1^0(0) = h_2^0(0) = 0$). Case 1 corresponds to creation of a new contact during PS creep. Indeed, when the elastic relative displacement u_{el} initially imposed is lower than the separation gap $h_1^0(0)$, asperity 1 is not initially in contact even under load, but will eventually make contact at some later time during PS creep. Case 2 is a first step towards investigating the effect of asperity (or grain) size distribution. Obviously, more work is needed to go from this very simple binary distribution to a broad continuous one.

2.2.1 Interface diffusion equations. Under the assumptions listed in section 2.1, the mass of solid dissolved at any given time t and at any point r_i of contact i per unit time and unit area is, to a very good approximation, given by (see LL):

$$J^{(i)}(r_i,t) \approx \rho_S K_{dis}\left[\frac{\Omega_S}{kT}\sigma_{eff}^{(i)}(r_i,t) - \ln\left(c^{(i)}(r_i,t)\Big/c_{eq}\right)\right], \tag{5}$$

where ρ_S denotes the solid density, Ω_S the molecular volume of the solid, k the Boltzmann constant, K_{dis} a phenomenological rate coefficient for contact dissolution (here, K_{dis} is assumed identical to the rate constant of dissolution for a solid immersed in a fluid at pressure p and temperature T), c_{eq} the equilibrium concentration of the interstitial fluid in the "free" pore space (i.e., outside the contact interfaces) and $\sigma_{eff}^{(i)}$ the normal effective stress at the point considered. The concentration $c^{(i)}(r_i)$ of the fluid inside the interface of contact i is the solution of the quasi-static diffusion equation with source terms $J(r_i)^{(i)}/w\rho_F$, where ρ_F is the fluid density:

$$\frac{d^2c^{(i)}(r_i,t)}{dr_i^2} + \frac{1}{r_i}\frac{dc^{(i)}(r_i,t)}{dr_i} + \frac{K_{dis}\rho_S}{wD^*\rho_F}\left[\frac{\Omega_S}{kT}\sigma_{eff}^{(i)}(r_i,t) - \ln\left(c^{(i)}(r_i,t)\Big/c_{eq}\right)\right] = 0. \tag{6}$$

If the stress fields $\sigma_{eff}^{(i)}$ are known at time t, the two equations 6 (i.e., one for each value of the index i) can be numerically solved independently of each other, using the technique described in BE. Thus, assuming that quasi-static conditions hold, the thickness of solid dissolved in each contact during an infinitesimal time increment δt is:

$$\delta h_i(r_i,t) = \frac{J^{(i)}(r_i,t)\delta t}{\rho_S}. \tag{7}$$

The separation h_i^{tot} can then be updated by adding δh_i (i.e., $h_i(r_i, t+\delta t) = h_i(r_i, t) + \delta h_i(r_i, t) - \delta h_i(0, t)$ and $h_i^0(t+\delta t) = h_i^0(t) + \delta h_i(0, t)$) and the global dissolution displacement $u_{dis}(t+\delta t) = \min(h_i^0(t+\delta t))$ can be determined. One final remark is that, as explained in BE, the thickness of dissolved solid at the contact edges $\delta h_i(a_i, t)$ may be large enough to produce a small local irregularity in the asperity shape, which must be smoothed out to avoid numerical instability when solving the subsequent contact mechanics problem.

2.2.2 Contact mechanics equations. At any given time t, the asperities normally pressed against the planar body form circular contact zones of radius $a_i(t)$. The relative normal displacements $u_i(r_i, t)$ and the effective normal stress $\sigma_{eff}^{(i)}(r_i, t)$ must verify the following conditions:

$$u_i(r_i,t)=u_{el}(t)-\left[h_i^0(t)+h_i(r_i,t)-u_{dis}(t)\right] \text{ and } \sigma_{eff}^{(i)}(r_i,t)\geq 0 \text{ for } r_i\leq a_i(t), \tag{8a}$$

$$u_i(r_i,t)\geq u_{el}(t)-\left[h_i^0(t)+h_i(r_i,t)-u_{dis}(t)\right] \text{ and } \sigma_{eff}^{(i)}(r_i,t)=0 \text{ for } r_i> a_i(t), \tag{8b}$$

where $u_{el}(t)$ is the global elastic relative displacement. Note that a negative value of $u_i(0, t)$ (i.e., $h_i^{tot}(0, t) > u_{el}(t) + u_{dis}(t) = u_{tot}(t)$, or, total separation greater than total relative displacement) indicates that the corresponding asperity is not in contact and therefore $\sigma_{eff}^{(i)}(r_i, t) = 0$. To simplify the analysis we used the classic Hertzian contact mechanics assumptions (i.e., no friction, no adhesion, contact radius much smaller than the asperity size, and so forth; see JOHNSON, 1985). Accordingly, $u_i(r_i, t)$ is given for $r_i \leq a_i(t)$ by (LI and DEMPSEY, 1990):

$$u_i(r_i,t)=\frac{8(1-\nu^2)}{\pi E}\int_0^{a_i(t)}\sigma_{eff}^{(i)}(s_i,t)\frac{s_i}{\max(s_i,r_i)}\mathrm{K}\left(\frac{\min(s_i,r_i)}{\max(s_i,r_i)}\right)\mathrm{d}s_i, \tag{9}$$

where E is Young's modulus, ν Poisson's ratio, and K denotes the complete elliptical integral of the first kind. The two equations (9) (for $i = 1$ and 2) are coupled through the global equilibrium condition:

$$\int_0^{a_1(t)}2\pi s_1\sigma_{eff}^{(1)}(s_1,t)\mathrm{d}s_1+\int_0^{a_2(t)}2\pi s_2\sigma_{eff}^{(2)}(s_2,t)\mathrm{d}s_2=F_1(t)+F_2(t)=F, \tag{10}$$

where F is the (constant) total effective normal load externally applied to the asperities.

These equations can be numerically solved for $a_i(t)$, $u_{el}(t)$ and $\sigma_{eff}^{(i)}(r_i,t)$ using a generalization of the iterative method utilized by BE. The first step of this method is to choose test-values for $a_i(t)$ and rewrite equations (9) and (10) in nondimensional discretized form (distances are normalized with respect to the unknown a_i, e.g., $R_i = r_i/a_i$; note also that the time variable t is omitted in the rest of this section):

$$\frac{4}{\pi}\mathbf{B}.\zeta^{(i)}=\frac{u_{el}}{a_i}\mathbf{1}-\mathbf{H}^{(i)}, \tag{11a}$$

$$a_1^2\mathbf{A}^T.\zeta^{(1)}+a_2^2\mathbf{A}^T.\zeta^{(2)}=\frac{F(1-\nu^2)}{\pi E}=F*, \tag{11b}$$

where the matrix $\mathbf{B}$ and transposed vector $\mathbf{A}^T$ are calculated using Gauss-Chebyshev discretization scheme (note that $\mathbf{B}$ and $\mathbf{A}^T$ only depend on the number of discretization nodes; see BE and LI and DEMPSEY, 1990, for more details), $\mathbf{1}$ is a vector whose components are all equal to 1, and, the symbol **.** denotes matrix product. The components of the vectors $\zeta^{(i)}$ and $\mathbf{H}^{(i)}$ are equal to the (nondimensional) values of $2(1-\nu^2)\sigma_{eff}^{(i)}/E$ and

$(h_i^0 + h_i - u_{dis})/a_i$, respectively, at the discretization nodes along one contact radius. The next step is to decompose $\zeta^{(i)}$ as:

$$\zeta^{(i)} = \frac{u_{el}}{a_i}\zeta - \zeta_h^{(i)}, \tag{12}$$

where ζ and $\zeta_h^{(i)}$ are solutions of $4/\pi\ \mathbf{B}.\zeta = \mathbf{1}$ and $4/\pi\ \mathbf{B}.\zeta_h^{(i)} = \mathbf{H}^{(i)}$, respectively. Thus, tentative stress fields $\sigma_{eff}^{(i)}(r_i)$ in both contact regions can be determined. Finally, plugging equations (12) into equation (11b) yields the global elastic relative displacement:

$$u_{el} = \frac{F * + a_1^2\mathbf{A}^T.\zeta_h^{(1)} + a_2^2\mathbf{A}^T.\zeta_h^{(2)}}{(a_1 + a_2)\mathbf{A}^T.\zeta}. \tag{13}$$

If the test-values used for a_i are equal to the correct ones, the conditions (8ab) must be verified. Otherwise, the above computations are reiterated with new test-values until solutions for a_i are found. In the case of a single asperity the one-dimensional search strategy for the solution contact radius a is rather simple: 1) Start with an over-estimated test-value for the contact radius a^- and verify that the tentative normal stress at the contact edge $\sigma_{eff}(a^-)$ is negative, 2) find a lower test-value a^+ for which $\sigma_{eff}(a^+)$ is positive, and, 3) explores the interval $[a^+, a^-]$ by dichotomy until the value at which the change of sign occurs is found with sufficient accuracy. As illustrated in Figure 2, for two coupled asperities the solution corresponds to the intersection of two change-of-sign lines in the (a_1, a_2) plan. Unfortunately these lines are not horizontal and vertical straight lines. In particular, if one of the contacts is much smaller than the other (as happens if a new contact is being formed), one of the lines becomes very strongly bent towards the negative region (see Fig. 2). As a consequence, several sequences of horizontal, vertical and oblique one-dimensional searches in the (a_1, a_2) plan must be used.

3. *Simulations*

As in BE, we consider that asperities and planar solid all consist of quartz ($E = 95600$ MPa, $\nu = 0.08$, $\Omega_S = 3.7\ 10^{-29}$ m^3, $\rho_S = 2.65$ g cm^{-3}) and that the interstitial fluid is water ($\rho_F = 1.0$ g cm^{-3}). We assume that the thickness of the contact interfaces w is equal to 10^{-8} m and the diffusion correction factor f to 0.001 (i.e., the middle of the range explored by BE). Although we believe that these values are reasonable, the qualitative conclusions of this work are not very sensitive to the actual numerical values used. The three temperature-dependent parameters used here, K_{dis}, $D^* = fD$ and c_{eq}, have an approximate Arrhenius behavior: $K_{dis} \approx 3.39\ 10^{-4}$ Exp $(-79600/RT)\ 10^{-0.002037T}$ ms^{-1} (corresponding to an activation enthalpy of around 72 kJ mol^{-1} in the temperature range explored here), $D \approx 9\ 10^{-7}$ Exp$(-15000/RT)$ m^2 s^{-1} and $c_{eq} \approx 0.0307$ Exp$(-21200/RT)$ (see BE, LL and references therein). In both case 1 and case 2 the complete set of simulations was repeated for three values of the temperature, $T = 435$, 508 and 574 K. In case 1 we considered three values of the total

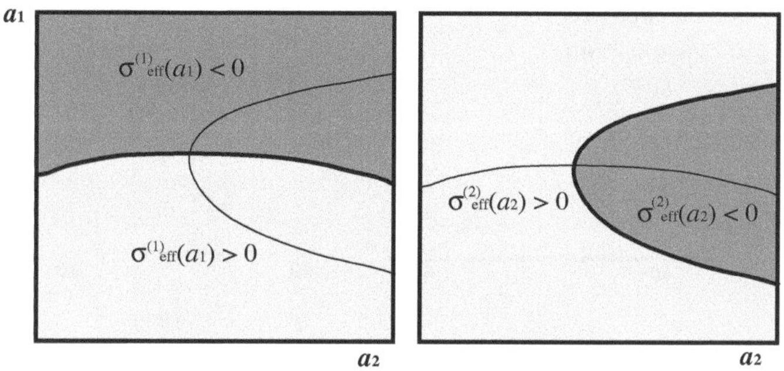

Figure 2
Maps in the (a_1, a_2) plane (where a_1 and a_2 refer to the trial values of the contact radii of the two asperities) showing the regions of positive and negative $\sigma^{(1)}_{\rm eff}(a_1)$ and $\sigma^{(2)}_{\rm eff}(a_2)$ and the change-of-sign lines that delimit them. The solution of the contact mechanics problem is the intersection of these two lines.

applied load, $F = 0.11$, 0.56 and 2.89 N and four initial separation gaps $h^0_1(0) = 0$, 0.6, 1.2 and ∞ μm. Note that the simulations for $h^0_1(0) = 0$ and ∞ were performed as single asperity simulations in which the total applied load was $F/2$ and F, respectively. In case 2, we used $F = 0.21$, 1.11 and 5.77 N, and, $[R_{\rm C1}, R_{\rm C2}] =$ [600 μm, 600 μm], [500 μm, 696 μm], [400 μm, 784 μm] and [300 μm, 862 μm]. It is important to note that these values were chosen so that the initial, total area of contact $\pi(a^2_1(0) + a^2_2(0))$ for a given load F was the same for all pairs $[R_{\rm C1}, R_{\rm C2}]$. If we define the normalized contrast in asperity radius as $\Delta_{\rm R} = (R_{\rm C2}-R_{\rm C1})/600$ μm, the above $[R_{\rm C1}, R_{\rm C2}]$ pairs correspond to $\Delta_{\rm R} = 0.00$, 0.33, 0.64 and 0.94, respectively. In all case 1 and case 2 runs, the amount of time simulated was $4{\cdot}10^7$ s or about 1.3 years (in addition to the simulation sets mentioned above, a few individual runs with different values of the parameters were also investigated).

3.1. Results: Creation of a New Contact (Case 1)

In order to visualize the evolution of the asperities and understand how they interact with each other during PS compaction, it is helpful to look at snapshots of the unstressed separations and stress fields at increasing times $t_{\rm k}$. An example corresponding to $T = 508$ K, $F = 0.56$ N and $h^0_1(0) = 0.7$ μm, is shown in Figure 3 (in these diagrams only half of each asperity is visible). In the left-side diagrams, we plotted the unstressed separation $h^{\rm tot}_{\rm i}$ versus the local radial distance $r_{\rm i}$. The thick black line visualizes the dissolution displacement, $u_{\rm dis}(t_{\rm k})$, while the thin grey lines allow comparison of $u_{\rm el}(t_{\rm k})$, the overall elastic displacement, and of $h^0_1(t_{\rm k})-h^0_2(t_{\rm k})$, the effective separation of asperity 1 and the flat solid. We see that $h^0_1(t_{\rm k})-h^0_2(t_{\rm k})$ decreases with time faster than $u_{\rm el}(t_{\rm k})$, so asperity 1 makes contact slightly before $t_{\rm k} = 1{,}750{,}000$ s. Note that these curves can be interpreted as showing the evolution of the unstressed shapes of the asperities (assuming that the flat solid does not dissolve). In the right-side diagrams we plotted the

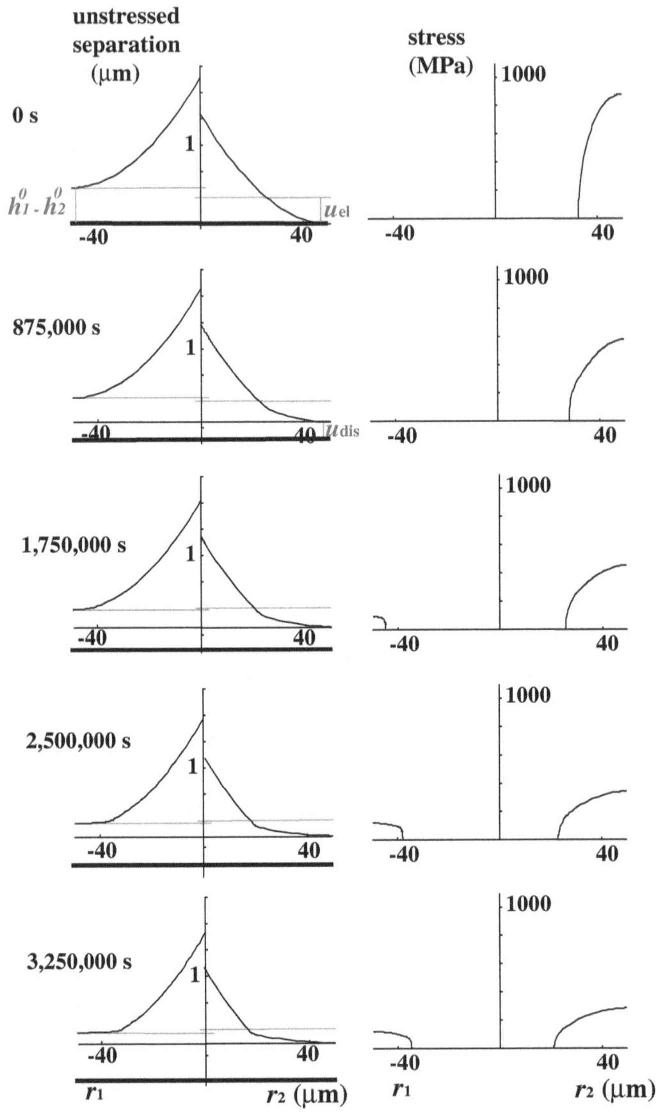

Figure 3

Snapshots of the unstressed separations and stress fields at increasing times t_k for case 1 (i.e., creation of a new contact). This example corresponds to $T = 508$ K, $F = 0.56$ N and $h_1^0(0) = 0.7$ μm. The left-side diagrams show the unstressed separation h_i^{tot} versus the local radial distance r_i. The thick black line visualizes the dissolution displacement, $u_{dis}(t_k)$, while the thin grey lines allow comparison of $u_{el}(t_k)$, the overall elastic displacement, and of $h_1^0(t_k)-h_2^0(t_k)$, the effective separation of asperity 1 and the flat solid. These parameters, u_{dis}, u_{el} and $h_1^0-h_2^0$, are indicated in light grey color. The right-side diagrams display the corresponding normal effective stress fields $\sigma_{eff}^{(i)}(r_i, t_k)$.

corresponding $\sigma_{\rm eff}^{(i)}(r_i, t_k)$, illustrating how flattening and growth of asperity 2 are accompanied by a decrease of the maximum contact stress (i.e., stress at the contact center). As the new contact at asperity 1 is formed, very rapid growth of contact 1, moderate increase of $\sigma_{\rm eff}^{(1)}$ and accelerated decrease of $\sigma_{\rm eff}^{(2)}$ occur simultaneously. Thus stress is transferred from the old contact to the new one, and, as will be shown in the next paragraph, this corresponds to a sizable reduction in dissolution displacement rate $u'_{\rm dis}$.

In Figure 4 we compare the simulated elastic and dissolution displacements for initial separation gaps $h_1^0(0) = 0$, 0.6, 1.2 and ∞ μm ($T = 508$ K and $F = 0.56$ N). The elastic decay is fastest for $h_1^0(0) = 0$ (curve "a" in Fig. 4a) and slowest for $h_1^0(0) = \infty$ (curve "d"), whereas $u'_{\rm dis}$ is respectively minimum (curve "a" in Fig. 4b) and maximum (curve "d"). As mentioned, these two end-member cases correspond to single asperities (both with $R_C = 600$ μm) submitted to external loads $F/2$ and F, respectively (in other words,

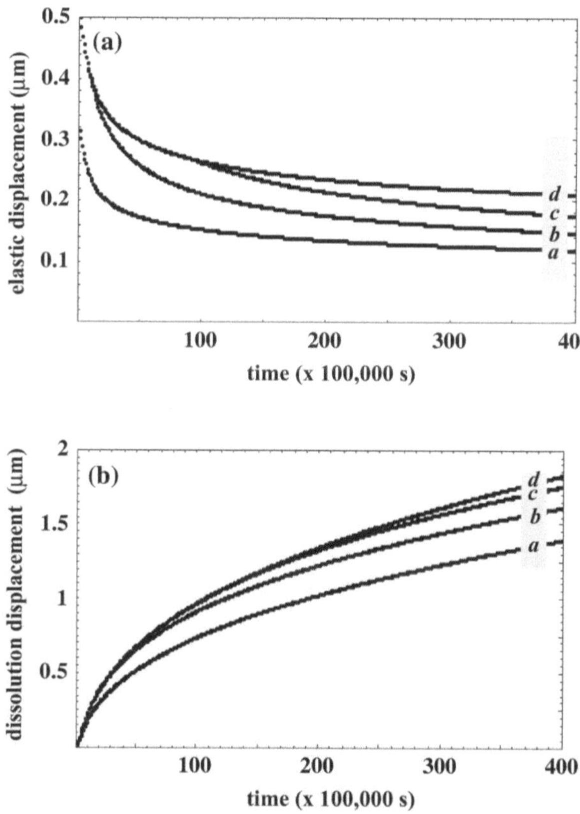

Figure 4

a) Simulated elastic relative displacement $u_{\rm el}(t)$ versus time for case 1 (i.e., the separation gaps $h_1^0(0) = 0$, 0.6, 1.2 and ∞ μm correspond to the curves labeled **a**, **b**, **c**, and **d**, respectively). The physical conditions are $T = 508$ K and $F = 0.56$ N. b) Dissolution displacement $u_{\rm dis}(t)$ versus time for the same simulations.

$F_1 = F_2 = F/2$ in the former case while $F_1 = 0$ and $F_2 = F$ in the latter). For all temperature and load conditions used in single contact simulations, we found that replacing F by $F/2$ produced a reduction of $u'_{dis}(t)$ by a factor of about 1.2 at $t = 0$ and 1.3 at $t = 1.3$ years. Note that the relationship between u'_{dis} and F is not linear. This is expected because reducing the total applied load also reduces the total area of contact (see BE and the discussion in section 4). Extrapolating this behavior to times greater than 1.3 years suggests that the curves "a" and "d" in Figure 4b continue to diverge from each other indefinitely. In the intermediate cases (i.e., for $h_1^0(0) = 0.6$ and 1.2 μm), the simulated u_{el} and u_{dis} curves are coincident with the $h_1^0(0) = \infty$ curves at early times and, then, bend down towards the $h_1^0(0) = 0$ curves (this is clearer for u_{el} than u_{dis} because a discontinuous change in displacement rate occurs for u_{el} whereas the evolution is gradual for u_{dis}). This switching process is slow at the time-scale investigated here and we could not observe its final stage even at the highest temperature considered. Therefore, we cannot be totally sure that the curves "b" and "c" asymptotically would approach their lower limit (curve "a") in very long simulations, although it seems quite reasonable to imagine such a behavior. The load transfer associated with the formation of a new contact is shown in Figure 5 (F_2 remains equal to F at early times, suddenly drops at the onset of new contact creation, and, asymptotically reaches $F/2$ at late times).

In summary, the formation of a new contact produces a reduction of the PS compaction rate. Clearly, this process will be compounded if, for each contact initially present, N new contacts are formed during creep compaction. We obtained an upper limit of the reduction in $u'_{dis}(t)$ by running single asperity simulations in which F was reduced to F/N (i.e., the N new contacts are formed right at the beginning). For $N = 10$, we found that the creep rate was reduced by a factor of about 2 at $t = 0$ and 3 at $t = 1.3$ years. For $N = 50$, the reduction factors were approximately 3 and 6 for $t = 0$ and $t = 1.3$ years, respectively. Interestingly, the long-term reduction factors tended to decrease with

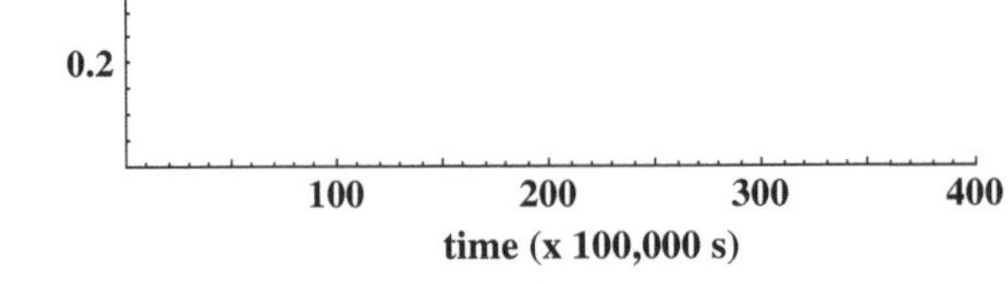

Figure 5
Normalized load, $F_2(t)/F$, supported by asperity 2 versus time for the same simulations as in Figure 4.

increasing temperature (e.g., 6.3 for $T = 435$ K, 4.5 for $T = 508$ K and 4.1 for $T = 574$ K, in the case $N = 50$). Formation of new contacts should occur when a population of asperities with a broad distribution of separations is considered (e.g., in a rough fracture with self-affine aperture). Tightening of grain packing during PS compaction of a granular aggregate also corresponds to creation of new contacts.

3.2. Results: Asperities with Different Sizes (Case 2)

The evolution of two different asperities in initial unstressed contact (i.e., $h_1^0(0) = h_2^0(0) = 0$) is illustrated in Figure 6. The example shown corresponds to $T = 508$ K, $F = 1.11$ N and $[R_{C1}, R_{C2}] = [400\ \mu m, 784\ \mu m]$ (or $\Delta_R = 0.64$). On the left-side diagrams, we plotted $h_i^{tot}(r_i,t_k) - h_2^0(t_k)$ to allow easy observation of the increase with time of $h_1^0(t_k) - h_2^0(t_k)$, that is to say, the lifting of the smallest asperity (i.e., asperity 1) away from the flat solid. The right-side diagrams show that $\sigma_{eff}^{(1)}$ initially has a higher maximum, is more concentrated (i.e., $a_1(0) < a_2(0)$), and subsequently decreases faster than $\sigma_{eff}^{(2)}$. Thus, stress is transferred from the smallest asperity to the largest one ($R_{C1} < R_{C2}$), and, as will be shown in the next paragraph (see Fig. 7), this corresponds to a reduction in displacement rate. It is important to point out right away that, although the stress magnitude is greater in contact 1 than contact 2, it corresponds to a lower load ($F_1 < F_2$) owing to concentration on a smaller area (see Fig. 8). Thus, the combination of a higher stress magnitude and a lower diffusion distance leads to a higher dissolution rate in contact 1 than in contact 2. Consequently, as surprising as it might seem, load is transferred from the asperity experiencing the lowest load to that subjected to the largest.

In Figures 7 and 8 we compare the simulated $u_{el}(t)$, $u_{dis}(t)$, $h_1^0(t) - h_2^0(t)$ and $F_1(t)/F$ for $\Delta_R = 0.00$, 0.33, 0.64 and 0.94 ($T = 508$ K and $F = 1.11$ N). It can be seen that the greater the contrast in asperity radius, the slower the elastic decay and dissolution displacement rate, the larger the lifting of asperity 1 (but note that, for the largest Δ_R, this effect appears to reach a maximum and to reverse after a certain time) period and the greater the drop in load F_1. By contrast with case 1 for which, as explained earlier, the very long time behavior can be reasonably imagined, the very long time evolution in case 2 is impossible to extrapolate from the simulated curves (the only, although impractical, solution is to run very long simulations). In particular, the transfer of load from the smallest to the largest contact observed here may continue indefinitely or may turn back at later times. The magnitude of the compaction rate reduction (defined as the ratio of $u'_{dis}(t)$ for $\Delta_R = 0$ by $u'_{dis}(t)$ for the Δ_R considered) is generally modest, increases with Δ_R and decreases with time. We found reduction factors of the initial displacement rates of 1.11, 1.22 and 1.30 for $\Delta_R = 0.33$, 0.64 and 0.94, respectively. For $t = 1.3$ years and for all Δ_R's the rate reduction factors drop to values very near 1. Finally, we observed a very small but systematic increase of the rate reduction factors with increasing T and F. The reason for the rate reduction is easily understood by noting that the effectiveness of the PS resisting process, interface diffusion, increases with contact area. Therefore, interface diffusion is most effective as a resisting process when the total area of contact is assigned

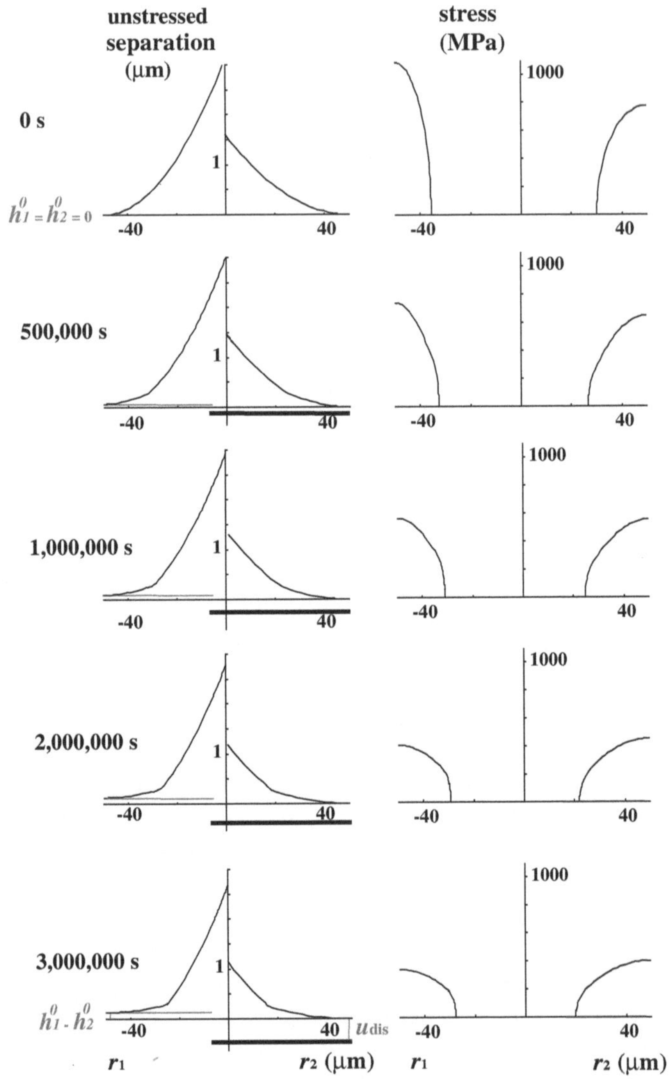

Figure 6

Snapshots of the unstressed separations and stress fields at increasing times t_k for case 2 (i.e., different asperities, both in initial contact). This example corresponds to $T = 508$ K, $F = 1.1$ N and $[R_{C1}, R_{C2}] = [400$ μm, 784 μm] (or $\Delta_R = 0.64$). The left-side diagrams show the unstressed separation h_i^{tot} versus the local radial distance r_i. The thick black line visualizes the dissolution displacement, $u_{dis}(t_k)$, while the thin grey line allows observation of the increase with time of $h_1^0(t_k)-h_2^0(t_k)$, that is to say, the lifting of the asperity 1 away from the flat solid (note that $u_{el}(t_k)$ is not shown here). These parameters, u_{dis} and $h_1^0-h_2^0$, are indicated in light grey color. The right-side diagrams display the corresponding normal effective stress fields $\sigma_{eff}^{(i)}(r_i, t_k)$.

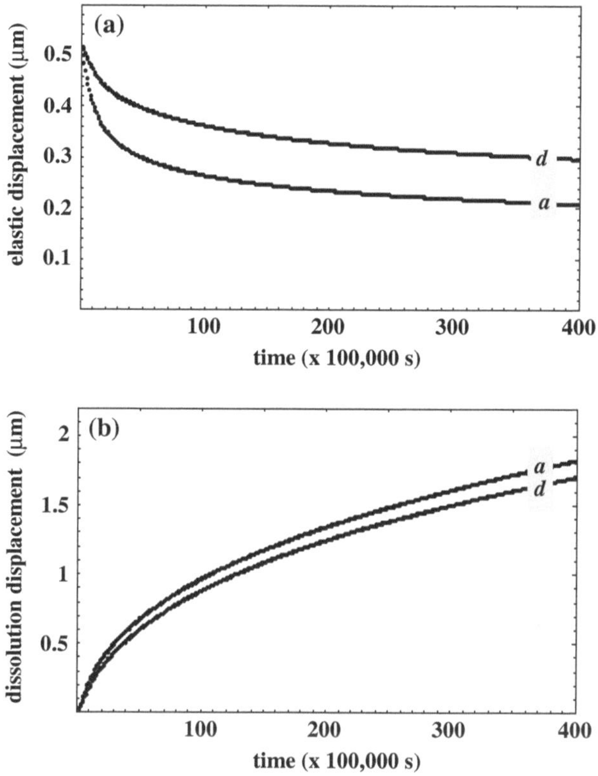

Figure 7
a) Simulated elastic relative displacement $u_{el}(t)$ versus time for case 2 (i.e., the asperity radius contrast $\Delta_R = 0$ and 0.94 correspond to the curves labeled **a** and **d**, respectively). The physical conditions are $T = 508$ K and $F = 1.1$ N. b) Dissolution displacement $u_{dis}(t)$ versus time for the same simulations.

to only one (large) contact (the other one shrinking to a point) and least effective if it is split into two equal (smaller) pieces.

As in case 1, the rate reduction process may be multiplied if a population of asperities (or grains) with a broad size distribution is considered. However, we should remember that the creep rate is reduced for non-identical asperities with respect to identical ones only when the other state parameters are kept constant (i.e., same load, same initial total area of contact and therefore same initial mean stress). As a consequence, we cannot conclude that poorly sorted granular aggregates must necessarily compact more slowly than well-sorted ones (recent experiments actually showed the opposite, A. NIEMEIJER, personal communication), unless the different aggregates were prepared in equivalent initial states (i.e., same grain packing and same total area of contact under the same effective pressure). Such a preparation would be very difficult to achieve since brittle fracturing of the mineral grains during the cold pressing stage is impossible to control

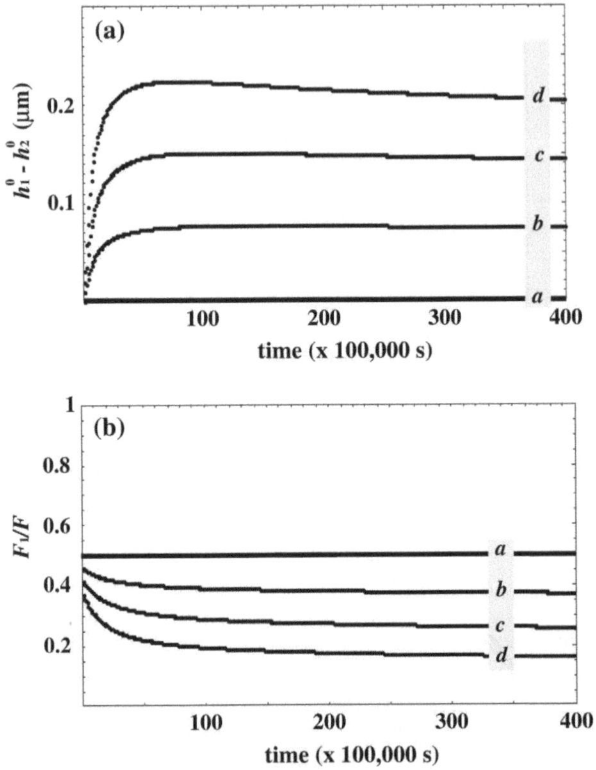

Figure 8
a) Lifting of asperity 1, $h_1^0(t)-h_2^0(t)$, versus time for the same simulations as in Figure 7. b) Normalized load, $F_1(t)/F$, supported by asperity 1 versus time. The curves labeled **a**, **b**, **c**, and **d** correspond to the asperity radius contrast $\Delta_R = 0.00$, 0.33, 0.64 and 0.94, respectively.

(analysis of the data of NIEMEIJER *et al.*, 2002, by REVIL *et al.*, 2006, suggests that the grain size distribution of cold-pressed quartz aggregates was much broader than the nominal grain size distribution used).

4. *Interpretation and Discussion*

Our next goal now is to fit the simulated $u_{\text{dis}}(t)$ and $u_{\text{tot}}(t) = u_{\text{el}}(t) + u_{\text{dis}}(t)$ to a heuristically posited creep law such as equations (1) or (2) and estimate the effective values of its parameters. Here, we decided to test the following creep law:

$$u' = u'_0 F^n \exp\left(-H_{\text{PS}}/_{RT}\right) u^{-m}, \tag{14}$$

where n is the load exponent and H_{PS} the effective activation enthalpy (the 1/RT factor in equation (1) can be ignored since it does not produce a significant discrepancy with a pure Arrhenius behavior in the temperature range considered here). We prefer to use the factor F^n rather than $\langle\sigma_{\mathrm{eff}}\rangle^n$ (where $\langle\sigma_{\mathrm{eff}}\rangle$ refers to the mean contact stress) because F is held constant during the creep simulations whereas $\langle\sigma_{\mathrm{eff}}\rangle$ changes with time. Moreover, the total load F is really a macro-scale quantity akin to the effective confining pressure in granular aggregate compaction experiments while $\langle\sigma_{\mathrm{eff}}\rangle$ is a micro-scale parameter usually not measurable directly in laboratory tests. The displacement variable u may represent either u_{dis} or u_{tot}. The power-law displacement factor u^{-m} was included in equation (14) because it arises naturally from an analytical relation derived from the LL model (equation (16) in BE), which was shown to approximate the BE results within 20%. In the case of a hemispherical asperity initially in point contact, we have $u(t) \approx a^2(t)/2R_C$ and the LL relation becomes:

$$u'(t) = \frac{\dfrac{\Omega_S}{kT}\dfrac{F}{\pi R_C u(t)}}{\dfrac{2}{K_{\mathrm{dis}}} + \dfrac{\rho_S}{\rho_F}\dfrac{R_C u(t)}{2c_{\mathrm{eq}} w D^*}}. \tag{15}$$

This ordinary differential equation takes a simple form in two extreme cases, 1) diffusion much faster than dissolution (infinite D^*) yielding $u' \propto u^{-1}$, and, 2) diffusion much slower than dissolution (infinite K_{dis}) leading to $u' \propto u^{-2}$. According to this view, m should lie between 1 and 2, and, its value gives an indication of the rate-limiting process. Pressure solution creep is dissolution-controlled for $m = 1$ while $m = 2$ corresponds to diffusion-control.

Equation (14) is easily solved, yielding:

$$u(t) = \left\{(m+1)\Gamma t + u(0)^{m+1}\right\}^{\frac{1}{m+1}}, \tag{16}$$

with

$$\Gamma = u_0' F^n \exp\left(-{}^{H_{PS}}\!/_{RT}\right). \tag{17}$$

The problem is then to find the values of u'_0, n, H_{PS} and m producing the best fits of equation (16) simultaneously to the 9 simulated displacement time-series corresponding to each value of $h_1^0(0)$ for case 1 or of Δ_R for case 2 (see the beginning of section 3). We followed the two-stage strategy devised by FITZENZ *et al.*, 2007b). In the first stage, we minimize the following cost function:

$$f_{\mathrm{cost}}^{(1)} = \sum_j \sum_i \left(u_{\mathrm{mod}}^{(j)}(t_i) - u_{\mathrm{sim}}^{(j)}(t_i)\right)^2, \tag{18}$$

where the subscripts "sim" and "mod" refer to the simulated and model (i.e., equation 16) displacement time-series, respectively, and the superscript (j) indicates the temperature and load conditions (j = 1 to 9, see the beginning of section 3). Thus, using a commercially available minimization program, we estimate the best-fitting values of m and $\Gamma^{(j)}$ ($j = 1$ to 9). In stage 2, we estimate the values of u_0', n and H_{PS} minimizing another cost function:

$$f_{cost}^{(2)} = \sum_{j} \left(\Gamma_{mod}^{(j)} - \Gamma_{est}^{(j)} \right)^2, \tag{19}$$

where the subscripts "est" and "mod" refer to the values estimated in stage 1 and modeled through equation (17), respectively. This technique is easier to implement but much less powerful than the Bayesian inference method of FITZENZ *et al.* (2007b). In particular, it does not yield the expected uncertainties on u_0', n, H_{PS} and m directly. One way to estimate the uncertainties is to add a small, zero-mean, Gaussian noise (standard deviation of 6 to 18 10^{-3} μm) to the simulated time-series $u_{sim}^{(j)}(t_i)$ and run the analysis on a number of realizations (this procedure is highly time-consuming and we could only use 6 to 10 realizations). Our analysis is also unable to resolve the time dependence of the creep parameters as thoroughly as the Bayesian inference method of FITZENZ *et al.* (2007a) modified to allow identification of changes of creep regime. However, FITZENZ *et al.* (2007a) reported difficulties in assessing which regimes are common to several simulated creep curves, in particular when the T and F conditions are wide apart. Following their discussion of these unresolved issues, we decided to simply analyze shorter time-series constructed by selecting the first N data-points of the original time-series ($N = 400$, 200 and 100 correspond to creep durations of 1.3, 0.6 and 0.3 years, respectively) and observe that the parameters estimated vary when the time-series duration is changed.

The estimated values of n, H_{PS} and m for all the datasets are given in Tables 1 to 6. Note that the results of Table 1 were obtained from the single asperity simulations described in section 3.1. Thus, Table 1 characterizes the standard situation (i.e., $h_1^0(0) = 0$ for case 1 or $\Delta_R = 0$ for case 2), to which all the other ones must be compared. One important property already pointed out by BERNABÉ and FITZENZ *et al.* (2007a) is that the creep parameters depend on time. Considering the entire set of u_{dis} data, we found that n increased from about 0.7 to 0.9, H_{PS} decreased from around 65 to 58 kJ mol^{-1} and m increased from 1.0 to 1.3 when the creep duration increased from 0.3 to 1.3 years. For u_{tot} the variations with time were smaller (decrease of n from 1.5 to 1.4,

Table 1

Values of the creep law parameters for the case 1 datasets corresponding to separation gaps $h_1^0(0) = 0$ and ∞, and case 2 dataset $\Delta_R = 0$. The values in parentheses refer to u_{tot} and the other ones to u_{dis}. The uncertainty values represent plus or minus two ensemble standard deviations

Duration (years)	n	E_{PS} (kJ mol^{-1})	m
0.3	0.76 ± 0.02 (1.44 ± 0.02)	65.2 ± 0.8 (58.5 ± 0.9)	1.05 ± 0.01 (2.01 ± 0.03)
0.6	0.84 ± 0.01	61.5 ± 0.3	1.20 ± 0.01
1.3	0.91 ± 0.01 (1.34 ± 0.01)	58.2 ± 0.3 (57.1 ± 0.1)	1.33 ± 0.01 (1.97 ± 0.01)

Table 2

Same as Table 1 for case 1 dataset, $h_1^0(0) = 0.6\ \mu m$

Duration (years)	n	E_{PS} (kJ mol^{-1})	m
0.3	0.70 ± 0.01 (1.38 ± 0.01)	67.2 ± 0.6 (58.1 ± 0.3)	1.15 ± 0.01 (2.28 ± 0.03)
0.6	0.78 ± 0.01	62.9 ± 0.3	1.32 ± 0.01
1.3	0.85 ± 0.01 (1.28 ± 0.01)	58.7 ± 0.2 (56.6 ± 0.2)	1.47 ± 0.01 (2.23 ± 0.01)

Table 3

Same as Table 1 for case 1 dataset $h_1^0(0) = 1.2\ \mu m$

Duration (years)	n	E_{PS} (kJ mol^{-1})	m
0.3	0.74 ± 0.01 (1.45 ± 0.01)	67.3 ± 0.7 (58.7 ± 0.8)	1.13 ± 0.01 (2.25 ± 0.02)
0.6	0.82 ± 0.01	63.6 ± 0.3	1.31 ± 0.01
1.3	0.88 ± 0.02 (1.36 ± 0.01)	59.9 ± 0.3 (58.2 ± 0.2)	1.49 ± 0.01 (2.28 ± 0.01)

Table 4

Same as Table 1 for case 2 dataset, $\Delta_R = 0.33$

Duration (years)	n	E_{PS} (kJ mol^{-1})	m
0.3	0.70 ± 0.01 (1.52 ± 0.01)	64.1 ± 0.5 (58.3 ± 0.3)	0.98 ± 0.01 (2.13 ± 0.02)
0.6	0.78 ± 0.01	61.0 ± 0.3	1.14 ± 0.01
1.3	0.86 ± 0.01 (1.42 ± 0.01)	57.7 ± 0.1 (57.2 ± 0.1)	1.29 ± 0.01 (2.12 ± 0.01)

Table 5

Same as Table 1 for case 2 dataset, $\Delta_R = 0.64$

Duration (years)	n	E_{PS} (kJ mol^{-1})	m
0.3	0.67 ± 0.01 (1.56 ± 0.02)	63.5 ± 0.5 (58.4 ± 1.0)	0.93 ± 0.01 (2.20 ± 0.04)
0.6	0.75 ± 0.01	60.3 ± 0.4	1.09 ± 0.01
1.3	0.83 ± 0.01 (1.45 ± 0.01)	57.2 ± 0.1 (57.3 ± 0.1)	1.25 ± 0.01 (2.17 ± 0.01)

Table 6

Same as Table 1 for case 2 dataset, $\Delta_R = 0.94$

Duration (years)	n	E_{PS} (kJ mol^{-1})	m
0.3	0.64 ± 0.01 (1.64 ± 0.02)	62.7 ± 0.4 (58.5 ± 0.8)	0.89 ± 0.01 (2.33 ± 0.04)
0.6	0.73 ± 0.01	59.8 ± 0.3	1.06 ± 0.01
1.3	0.81 ± 0.01 (1.51 ± 0.01)	56.7 ± 0.1 (57.3 ± 0.1)	1.22 ± 0.01 (2.26 ± 0.01)

decrease of H_{PS} from 58 to 57 kJ mol^{-1} and decrease of m from 2.2 to 2.1). The variations of H_{PS} and m for u_{dis} are both consistent with an evolution of the rate-limiting mechanism from contact dissolution at early times (activation enthalpy for dissolution = 72 kJ mol^{-1}

and $m = 1$) to interface diffusion at late times (activation enthalpy for diffusion = 15 kJ mol^{-1} and $m = 2$). Note that equation (15) does not contain D^* alone but the product $c_{eq}D^*$, suggesting that the effective activation enthalpy for diffusion-controlled PS may be 15 + 21 = 36 kJ mol^{-1}. In that case, our results suggest that PS may not need an overly long period of time to become diffusion-controlled. We also observe that the load exponent n for u_{dis} differs appreciably from 1 at early times. This is due to the fact that the initial radius of contact $a(0)$ varies with F (according to Hertzian contact mechanics for spherical asperities $a(0)$ must be proportional to $F^{1/3}$). As illustrated by equation (15), n is approximately equal to 1 only when $a(0)$ is a constant. If we plug $a(0) \propto F^{1/3}$ in equation (15), we find $u'_{dis}(0) \propto F^{1/3}$ when contact dissolution is rate-limiting (i.e., infinite D^*) and $u'_{dis}(0) \propto F^{-1/3}$ when interface diffusion is rate-limiting (i.e., infinite K_{dis}). In order to verify this, we calculated the initial displacement rate $u'_{dis}(0)$ for F from 0.0016 to 50 N using the diffusion correction factor $f = 0.1$ (very fast diffusion). We determined n by linear regression of $\log(u'_{dis}(0))-\log(F)$ data and found $n = 0.33$ as predicted. We also tried to perform the same calculations for $f = 10^{-6}$ (very slow diffusion) but ran into numerical problems. We could obtain reliable values of $u'_{dis}(0)$ only for small F (from 0.0016 to 0.05 N), from which we indeed estimated a negative value $n = -0.13$. For $f = 0.001$, the value of f used in the bulk of this paper, we found $n = 0.28 < 1/3$, indicating that PS creep was not perfectly dissolution-controlled in our simulations even at $t = 0$. At finite times $t > 0$, the exponent of the power law relationship between $a(0)$ and F (initially equal to 1/3) decreases with time (probably approaching 0 in the very long time limit). Consequently, this leads to an increase of the load exponent n (probably approaching 1 at late times; extrapolating our results to periods of 100–1000 years leads to much higher values of n, however this coupled model of PS is highly nonlinear and extrapolation is, therefore, extremely unreliable) as indeed observed in our simulations. Therefore, the time dependence of n for u_{dis} is also consistent with an evolution with time from dissolution-to diffusion-control of PS creep.

Another general observation is that including the elastic deformation in the analysis (i.e., replacing u_{dis} by u_{tot}) has a negligible effect on the long-term estimate of H_{PS} but leads to large increases of the exponents n and m. However, the elastic decay is a transient phenomenon and its effect on n and m must decrease with time. Since it is a transient effect, the elastic decay may be safely neglected in most geologic situations. However, porosity changes at early times after an earthquake can impact pore pressure and, therefore, both the aseismic deformation rate and the aftershocks occurrence. The elastic decay might therefore be a necessary ingredient of numerical models of postseismic fault models. In any event, our results show that the elastic decay should be taken into account when analyzing laboratory experiments. In particular, it might contribute to the high stress exponents often observed (e.g., NIEMEIJER *et al.*, 2002).

Analysis of the case 1 simulations shows that creation of a new contact tends to produce slightly greater m and H_{PS}, and lower n, for both u_{dis} and u_{tot}. However, no clear trend with $h_1^0(0)$ could be identified. This is to be expected since the time series corresponding to $h_1^0(0) = 0.6$ and 1.2 μm must lie between the $u_{dis}(t)$ or $u_{tot}(t)$ curves for

$h_1^0(0) = 0$ and ∞. These two end-member time series were calculated using the single asperity BE model with loads $F/2$ and F, respectively (see section 3.1), and, when modeled by equation (14), should only differ by the value of the prefactor u'_0. Unfortunately, u'_0 is rather difficult to study with the method described above because its estimated values are very sensitive to the variations of the other parameters. However, we did verify that u'_0 increased by factors of 1.1, 1.5 and 1.6 when $h_1^0(0)$ was increased from 0 to 0.6, 1.2 and ∞ μm. The distortions of the $u_{dis}(t)$ and $u_{tot}(t)$ curves due to the creation of a new contact and the associated changes in n, H_{PS} and m reflect the transfer of stress from the old to the new contact. They accompany an elastic transient and should, therefore, be limited in time. Only the decrease in u'_0 with respect to its initial value is expected to be permanent (corresponding to the reduction in creep rate discussed in section 3.1). In the very long time limit, the behavior should approach that corresponding to $h_1^0(0) = 0$ (identical asperities in initial contact). If a population of asperities with a broad distribution of initial separation is considered, new contacts will be continuously formed during a very long interval of time and more permanent changes of n, H_{PS} and m may be expected. The results of Tables 2 and 3 suggest that a (probably moderate) decrease of n and an increase of H_{PS} and m should occur.

For the case 2 simulations, we can identify the effect of Δ_R on n, H_{PS} and m clearly. For u_{dis}, we observe moderate (approximately linear) decreases of n, H_{PS} and m with increasing Δ_R. Moreover, the absolute values of the proportionality constants tend to decrease modestly with time. For u_{tot}, n and m display a slightly more pronounced (approximately linear) increase with increasing Δ_R while H_{PS} remains nearly constant. These variations are not consistent with acceleration or deceleration of the transition from dissolution-to diffusion-control of PS creep. They merely reflect the change in shape of the $u_{dis}(t)$ and $u_{tot}(t)$ curves caused by the transfer of load from one contact to the other. This load transfer is an elastic phenomenon and, as such, appears as a transient similar to (but probably longer than) the elastic decay observed by BE for single contacts. If a heterogeneous population of asperities is considered, load should be transferred from roughly the smallest to the largest ones but the details of this evolution (i.e., load transfer between specific asperities) must be very complicated.

A number of experimental studies of PS compression of a single indenter against a flat solid substrate have been conducted in the past. However, comparison with our simulations is difficult in many cases because sharp, rectangular or pyramidal indenters were used (TADA *et al.*, 1987; GRATIER, 1993; KARCZ *et al.*, 2008). In this configuration high stresses were produced at the edge of the contact and, therefore, PS undercutting was triggered. When smooth, hemi-spherical halite indenters were used (HICKMAN and EVANS, 1992, 1995) decelerating contact growth was observed in qualitative agreement with our simulations, nonetheless quantitative comparison is not possible owing to the difference in material (note also that surface tension, which is not included in our model, played an important role in these experiments). PS compaction experiments of quartz granular aggregates were also performed in the laboratory (e.g., among many others, GRATIER and GUIGET, 1986; SCHUTJENS, 1991; RUTTER and WANTEN, 2000; NIEMEIJER *et al.*, 2002;

CHESTER *et al.*, 2007; KARNER *et al.*, 2008). Comparison of our simulations to the results of these studies is difficult for a variety of reasons (for example, deformation in the experiments of CHESTER *et al.*, 2007, and KARNER *et al.*, 2008, is predominantly controlled by subcritical crack growth) and we will limit our discussion to the work of NIEMEIJER *et al.* (2002) who conducted isostatic hot pressing experiments on water-saturated quartz granular aggregates at temperatures T ranging from 673 to 873 K and effective pressures p_{eff} between 50 and 150 MPa. The porosity reduction was measured as a function of time. The tests lasted on the order of 200 hours. For each experiment, the final porosity was measured and the evolution of the porosity with time $\phi(t)$ was back calculated. The $\phi(t)$ curves show qualitatively similar decelerating creep behavior as our simulations. In order to interpret their data, NIEMEIJER *et al.* estimated the effective PS activation enthalpy H_{PS} and effective pressure exponent n by linear regression of selected values of $\log(\mathrm{d}\phi/\mathrm{d}t)$ with respect to $1/RT$ (i.e., Arrhenius plots) and to $\log(p_{\mathrm{eff}})$, respectively (note that we use the same symbol, n, for the effective pressure and the load exponents because they are somewhat equivalent; indeed, p_{eff} multiplied by the nominal surface area of the sample is equal to the total compressive load). One important difficulty with this method is that the estimated values of H_{PS} and n are meaningful only if the selected values of $\log(\mathrm{d}\phi/\mathrm{d}t)$ (one value for each test) correspond to identical "internal" states. NIEMEIJER *et al.* argue that the state of the compacting aggregate is uniquely determined by the ratio $\phi(t)/\phi_0$, where ϕ_0 is the back-calculated porosity at the start of the creep test (after the preliminary cold pressing stage). Using this method, they found $H_{\mathrm{PS}} = 65$, 75 and 85 kJ mol^{-1} and $n = 3.35$, 3.45 and 3.58 for $\phi(t)/\phi_0 = 0.9$, 0.8 and 0.7, respectively. These values all correspond to very early times (less than 10 hours). They are quite high compared to ours and they tend to increase with $\phi(t)/\phi_0$ or, equivalently, with time (however, the time dependence observed might be questionable because the linear regressions were run on only three points). For $\phi(t)/\phi_0 < 0.7$, values less discrepant with our results are found, i.e., $H_{\mathrm{PS}} = 62$ kJ mol^{-1} and n on the order of 2.3 (but these values may not be reliable since the linear regressions are based on only two points). FITZENZ *et al.* (2007b) re-interpreted these data using a Bayesian inference method, which has the great advantage that no assumptions about the "internal" state are needed (i.e., no data selection necessary, although removing experimental artefacts is always a good idea). They found $H_{\mathrm{PS}} = 37$ kJ mol^{-1} and $n = 2.55$ for the complete dataset (but NIEMEIJER *et al.* warned that their late-time data may contain experimental artefacts) and $H_{\mathrm{PS}} = 57.6$ kJ mol^{-1} and $n = 3.56$ for the subset corresponding to $\phi(t)/\phi_0$ between 0.5 and 0.8 (i.e., times less than 10 hours for most of the tests). The discrepancy with NIEMEIJER *et al.*'s results is clear, although there is some convergence when the restricted dataset is used. The FITZENZ *et al.*'s estimates of H_{PS} are not incompatible with ours but the values of n are much higher. These results also suggest the same kind of time-dependence as we observed in our simulations (but the late-time data may not be reliable).

In summary, it can be concluded that interpreting experimental PS creep data is a difficult and uncertain task because of its sensitivity to the method used. For NIEMEIJER *et al.*'s data, both interpretation methods yielded activation enthalpies more or less compatible with our simulations. It also appears that a possible time dependence of the creep parameters could neither be ruled out nor clearly identified. The most dependable observation is that the effective pressure exponents from NIEMEIJER *et al.*'s data are substantially higher than those from our simulations. The main question, therefore, is to find out the cause of such high values of n.

It is sometimes argued that PS creep rates should be exponentially related to p_{eff}, thus producing a high power law exponent if a limited range of p_{eff} is considered (NIEMEIJER *et al.* themselves invoke this argument). The proposed exponential dependence on p_{eff} is derived as follows: 1) because the total dissolution mass flux within a contact and the diffusion mass flux exiting it must balance each other, the relative displacement rate is proportional to an "average" concentration gradient, $u'_{\mathrm{dis}} \propto (\langle c\rangle - c_{\mathrm{eq}})/a$, where $\langle c\rangle$ is a "representative" concentration inside the contact and a the contact radius; 2) a state of equilibrium is then assumed, implying that the chemical potentials in the stressed solid and in the contact fluid are equal, leading to $\langle c\rangle = c_{\mathrm{eq}}\,\exp(p_{\mathrm{eff}}\,\Omega_S/kT)$ and, therefore, $u'_{\mathrm{dis}} \propto \exp(p_{\mathrm{eff}}\,\Omega_S/kT) - 1$. This reasoning is questionable for many reasons, the most important one being that a state of equilibrium implies a complete halt of dissolution (see equation (5) in section 2.2.1) and cannot be associated with a nonzero creep rate. Although equations (5) and (6) are linear with respect to the local stresses $\sigma_{\mathrm{eff}}(r,t)$, they do not necessarily imply a load exponent equal to unity. This is best seen in equation (15) after replacing $u(t)$ by $a^2(t)/2R_C$. It clearly allows any value of n depending on the relationship between $a(t)$ and F. Concerning the load exponent n, the present study shows two things: 1) the early-time elastic decay produces load exponents significantly greater than 1, although not as large as the values found for NIEMEIJER *et al.*'s experiments; 2) this effect is reinforced by heterogeneity in asperity size but not by creation of a new contact. Accordingly we may conclude that elastic decay (obligatorily present in a short-term compaction test) and a broad grain size distribution may have contributed to the high values of n observed by NIEMEIJER *et al.* (2002) but could not be the sole cause. Most likely, another creep mechanism such as dislocation creep or subcritical crack growth, was also active during the experiments. Microstructure observations seem to rule out microfracturing (NIEMEIJER *et al.*, 2002), but it is reasonable to think that dislocation creep, even if limited to a small region around the asperity tips, could have produced high n values. Thus we will try, in the future, to extend the BE model and simulate a single asperity simultaneously undergoing PS and power-law creep (with a high exponent).

5. *Conclusion*

The deformation of porous and fluid-saturated rocks in the upper crust of the Earth is likely to occur via complex interactions among a variety of mechanisms, including solution transfer, dissolution and precipitation, cataclastic and frictional processes, and diagenetic reactions. Even in carefully designed and executed laboratory experiments it is difficult to constrain these interactions. Most constitutive laws used to interpret the laboratory studies and to predict solution-transport deformation in nature rely upon simplifying assumptions regarding the geometry of the microstructure and the relative speed of competing mechanisms. In this work, we examined the mechanical behavior of a simple numerical model system of two asperities deforming by only two processes: dissolution of the stressed solid along the contact interface coupled with transport along an intergranular fluid phase. The degree of simplification of this model when compared to the complexities of pressure solution in natural situations is obvious. Notwithstanding the limitations of the model, it is important to note that when typical material and kinetic constants are used, the coupled model suggests that transient mechanical responses can be quite long-lived, at least from the point of view of laboratory experiments. In addition, if one inverts the numerical data to obtain material parameters to be used in the simple aggregate constitutive laws, those parameters, normally viewed as material constants, are not stationary in time. In particular, this numerical study suggests the following conclusions:

1) Our simulations show that PS creep rates are reduced by the creation of new contacts as well as by differences in asperity size (provided that the total area of contact remains unchanged). These results can reasonably be generalized in a qualitative sense to a large population of asperities. Therefore, PS creep models based on a single "representative" asperity (or grain) size and/or a fixed number of contacts may significantly overestimate PS creep rates in rough fractures with self-affine aperture or in heterogeneous granular materials with variable grain-packing geometry.
2) This study is consistent with previous results of BE and FITZENZ *et al.* (2007a), suggesting that the empirically based creep-law parameters (in particular, the activation enthalpy and the stress exponent in laws similar to eq. (1) are not constant but depend on time. The reason for this is that PS creep results from the interplay of two coupled processes, contact dissolution and interface diffusion, and that contact growth during PS produces a change in the partitioning of these two processes, namely, a transition of PS creep from dissolution-to-diffusion-control. It is important to characterize this evolution as well as possible (see FITZENZ *et al.*, 2007a) and to include it in models of long-term geological phenomena, such as fault gouge healing during the seismic cycle.
3) When interpreting the simulated creep curves, it is important to include the decay of elastic strain, which leads to a (transient) increase of the apparent load exponent. The elastic decay must indeed be present in short-term experimental data such as NIEMEIJER *et al.*'s. However, the elastic decay is probably not essential for modeling long-term field situations.
4) Finally, we saw that a broad asperity size distribution tends to enhance the above-mentioned elastic decay effect (although not enough to explain load exponents on the order of 2.5–3 as found by NIEMEIJER *et al.* (2002)). In contrast, creation of new contacts produces an attenuation of the elastic decay effect.

Acknowledgements

We are grateful to J.-P. Gratier and A. Niemeijer for reviewing this paper. This work was partially funded by the National Science Foundation under grant EAR 6914517 and the U.S. Department of Energy under grant DE-FG01-97ER14760.

References

Bernabé, Y. and Evans, B., *Numerical modeling of pressure solution deformation at axisymmetric asperities under normal load.* In *Rock Physics and Geomechanics in the Study of Reservoirs and Repositories* (eds. David, C. and Le Ravalec-Dupin, M.) (Geol. Soc., London, 2007) pp. 185–205.

Blanpied, M. L., Lockner, D. A., and Byerlee, J. D. (1992), *An earthquake mechanism based on rapid sealing of faults,* Nature *358,* 574–576.

Chester, F. M., Chester, J. S., Kronenberg, A. K., and Hajash, A. (2007), *Subcritical creep compaction of quartz sand at diagenetic conditions: Effects of water and grain size,* J. Geophys. Res. *112,* B06203, doi:06210.01029/02006JB004317.

Den Brok, S. W. J. (1998), *Effect of microcracking on pressure-solution strain rate: The Gratz grain-boundary model,* Geology *26,* 915–918.

Durney, D. W. (1976), *Pressure-solution and crystallization deformation,* Philosophical Transactions Roy. Soc. London, Series A: Mathematical Phys. Sci. *283,* 229–240.

Elliot, D. (1973), *Diffusion flow law in metamorphic rocks,* Bull. Geolog. Soc. Am. *84,* 2645–2664.

Fitzenz, D. D., Hickman, S. H., and Bernabé, Y., *Interpreting laboratory creep experiments: from single contacts to interseismic fault models (abstr.).* In *Euro-conference of Rock Physics and Geomechanics on Natural Hazards: Thermo-Hydro-Mechanical Processes in Rocks* (eds. Vinciguerra, S. and Bernabé, Y.) Erice, Italy, 2007a) pp. 120, URL: http://hdl.handle.net/2122/2632 (earth-prints.org).

Fitzenz, D. D., Jalobeanu, A., and Hickman, S. H. (2007b), *Integrating laboratory creep data with numerical fault models: A Bayesian framework,* J. Geophys. Res. *112,* B08410, doi:08410.01029/02006JB004792.

Ghoussoub, J. and Leroy, Y. M. (2001), *Solid-fluid phase transformation within grain boundaries during compaction by pressure solution,* J. Mech. Phys. Solids *49,* 2385–2430.

Gratier, J.-P. (1993), *Experimental pressure solution of halite by an indenter technique,* Geophys. Res. Lett. *20,* 1647-1650.

Gratier, J.-P. and Guiget, R. (1986), *Experimental pressure solution-deposition on quartz sand,* J. Struc. Geology *8.*

Gratier, J. P. (1987), *Pressure solution-deposition creep and associated tectonic differentiation in sedimentary rocks,* Geolog. Soc. Special Publications *29,* 25–38.

Gratz, A. J. (1991), *Solution-transfer compaction of quartzites: progress toward a rate law,* Geology *19,* 901–904.

Groshong, R. H., Jr. (1972), *Deformation by pressure solution,* Geolog. Soc. America, Northeastern Section 7th Annual Meeting, Abstracts with Programs *4,* 17–18.

He, W. W., Hajash, A., and Sparks, D. (2003), *Creep compaction of quartz aggregates: Effects of pore-fluid flow-A combined experimental and theoretical study,* Amer. J. Sci. *303,* 73–93.

Heald, M. T. (1961), *Packing and shape modifications during pressure solution,* Proc. West Virginia Academy Sci. *33,* 95.

Hickman, S. H. and Evans, B. (1991), *Experimental Pressure Solution in Halite—the effect of grain interphase boundary structure,* J. Geolog. Soc. *148,* 549–560.

Hickman, S. H. and Evans, B., *Growth of grain contacts in halite by solution-transfer; implications for diagenesis, lithification, and strength recovery.* In *Fault Mech. Transport Properties Rocks; A Festschrift in Honor W. F. Brace* (eds. Evans, B. and Wong, T.-f.) (Academic Press, London, UK, 1992) pp. 253–280.

Hickman, S. H. and Evans, B. (1995), *Kinetics of pressure solution at Halite-silica interfaces and intergranular clay films,* J. Geophys. Res. -Sol. Earth *100,* 13,113–113,132.

Hillers, G. and Wesnousky, S. G. (2008), *Scaling relations of strike-slip earthquakes with different slip-rate-dependent properties,* Bull. Seismol. Soc. Am. *98,* 1085–1101, doi: 1010.1785/0120070200.

Johnson, K. L., *Contact Mechanics* (Cambridge University Press, New York, 1985).

Karcz, Z., Aharonov, E., Ertas, D., Polizzotti, R., and Scholz, C. H. (2008), *Deformation by dissolution and plastic flow of a single crystal sodium chloride indenter: An experimental study under the confocal microscope,* J. Geophys. Res. 113, B04205, doi:04210.01029/02006JB004630.

Karner, S. L., Kronenberg, A. K., Chester, F. M., Chester, J. S., and Hajash, A. (2008), *Hydrothermal deformation of granular quartz sand,* J. Geophys. Res. *113,* B05404, doi:05410.01029/02006JB004710.

Lehner, F. K., *Thermodynamics of Rock Deformation by Pressure Solution* (Unwin Hyman, London, 1990).

Lehner, F. K. (1995), *A model for intergranular pressure solution in open systems,* Tectonophysics *245,* 153–170.

Lehner, F. K. and Leroy, Y., *Sandstone compaction by intergranular pressure solution.* In *Mech. Fluid-Saturated Rocks* (eds. Guéguen, Y. and Boutéca, M.) (Elsevier Academic Press, New York, 2004) pp. 115–168.

Li, H. and Dempsey, J. P. (1990), *Axisymmetric contact of an elastic layer underlain by a rigid base,* Int. J. Numer. Meth. Engng. *29,* 57–72.

Niemeijer, A. R., Spiers, C. J., and Bos, B. (2002), *Compaction creep of quartz sand at 400-600 degrees C: Experimental evidence for dissolution-controlled pressure solution,* Earth Planet. Sci. Lett. *195,* 261–275.

Paterson, M. S. (1973), *Nonhydrostatic thermodynamics and its geologic applications,* Res. Geophys. Space Phys. 355–389.

Paterson, M. S. (1995), *A theory for granular flow accommodated by material transfer via an intergranular fluid,* Tectonophysics *245,* 133–151.

Raj, R. (1982), *Creep in polycrystalline aggregates by matter transport through a liquid phase,* J. Geophys. Res. *87,* 4731–4739.

Renard, F., Brosse, E., and Gratier, J. P. (2000), *The different processes involved in the mechanism of pressure solution in quartz-rich rocks and their interactions,* Special Publication Int. Association Sedimentologists *29,* 67–78.

Renard, F., Ortoleva, P., and Gratier, J.-P. (1997), *Pressure solution in sandstones: Influence of clays and dependence on temperature and stress,* Tectonophysics *280,* 257–266.

Renard, F. and Ortoleva, P. J. (1997), *Water films at grain-grain contacts; Debye-Hueckel, osmotic model of stress, salinity, and mineralogy dependence,* Geochimica Cosmochimica Acta *61,* 1963–1970.

Renton, J. J. and Heald, M. T. (1968), *Experimental investigation of pressure solution,* Special Paper, Geolog. Soc. America 496.

Revil, A. (1999), *Pervasive pressure-solution transfer; a poro-viscoplastic model,* Geophys. Res. Lett. *26,* 255–258.

Revil, A. (2001), *Pervasive pressure solution transfer in a quartz sand,* J. Geophys. Res. *106,* 8665–8686.

Revil, A., Leroy, P., Ghorbani, A., Florsch, N., and Niemeijer, A. (2006), *Compaction of quartz sands by pressure solution using a Cole-Cole distribution of relaxation times,* J. Geophys. Res. *111.*

Rolandone, F., Bürgmann, R., and Nadeau, R. M. (2004), *The evolution of the seismic-aseismic transition during the earthquake cycle: Constraints from the time-dependent depth distributions of aftershocks,* Geophys. Res. Lett. *31,* L23610, doi:23610.21029/22004GL021379.

Rutter, E. H. (1983), *Pressure solution in nature, Theory and experiment,* J. Geolog. Soc. *140,* 725–740.

Rutter, E. H. and Wanten, P. H. (2000), *Experimental study of the compaction of phyllosilicate-bearing sand at elevated temperature and with controlled pore water pressure,* J. Sedimentary Research *70,* 107–116.

Schutjens, P. M. T. M. (1991), *Experimental compaction of quartz sand at low effective stress and temperature conditions,* J. Geol. Soc. London *148.*

Segall, P. and Rice, J. R. (1995), *Dilatancy, compaction, and slip instability of a fluid infiltrated fault,* J. Geophys. Res. - Sol. Earth *100,* 22,155–122,171.

Shaw, B. E. and Wesnousky, S. G. (2008), *Slip-length scaling in large earthquakes: the role of deep-penetrating slip below the seismogenic layer,* Bull. Seismol. Soc. Am. *98,* 1633–1641, doi: 1610.1785/0120070191.

Spiers, C. J., Schutjens, P. M. T. M., Brzesowsky, R. H., Peach, C. J., Liezenberg, J. L., and Zwart, H. J. (1990), *Experimental determination of constitutive parameters governing creep of rocksalt by pressure solution,* Geolog. Soc. Special Publications *54,* 215–227.

Tada, R., Maliva, R., and Siever, R. (1987), *A new mechanism for pressure solution in porous quartzose sandstone,* Geochim. Cosmochim. Acta *51,* 2295–2301.

THOMSON, A. F. (1959), *Pressure solution and porosity,* Special Publication - Soc. Economic Paleontologists Mineralogists *7,* 92-110.

WEYL, P. K. (1958), *Mechanism of stylolite formation,* Geolog. Soc. Am. Bull. *69,* 1660.

ZUBTSOV, S., RENARD, F., GRATIER, J.-P., GUIGUET, R., DYSTHE, D. K., and TRASKINE, V. (2004), *Experimental pressure solution compaction of synthetic halite/calcite aggregates,* Tectonophysics *385,* 45–57.

(Received September 17, 2008, revised January 7, 2009, accepted February 27, 2009)
Published Online First: May 7, 2009

Pure appl. geophys. 166 (2009) 927–948
0033–4553/09/050927–22
DOI 10.1007/s00024-009-0485-2

Pure and Applied Geophysics

Common Evolution of Mechanical and Transport Properties in Thermally Cracked Westerly Granite at Elevated Hydrostatic Pressure

M. H. B. Nasseri,[1] A. Schubnel,[2] P. M. Benson,[1,3] and R. P. Young[1]

Abstract—Increasing the damage and crack porosity in crustal rocks can result in significant changes to various key physical properties, including mechanical strength, elastic and mechanical anisotropy, and the enhancement of transport properties. Using a Non-Interactive Crack Effective Medium (NIC) theory as a fundamental tool, we show that elastic wave dispersion can be inverted to evaluate crack density as a function of temperature and is compared with optically determined crack density. Further, we show how the existence of embedded microcrack fabrics in rocks also significantly influences the fracture toughness (K_{IC}) of rocks as measured via a suite of tensile failure experiments (chevron cracked notch Brazilian disk). Finally, we include fluid flow in our analysis via the Guéguen and Dienes crack porosity-permeability model. Using the crack density and aspect ratio recovered from the elastic-wave velocity inversion, we successfully compare permeability evolution with pressure with the laboratory measurements of permeability.

Key words: Effective medium theories, transport properties, fracture toughness, thermal cracks, hydrostatic stress.

1. Introduction

The mechanical and transport properties of crustal rocks are profoundly influenced by cracks and pore spaces. The existence of embedded microcrack fabrics in rocks significantly influences the elastic moduli and inferred permeability (Heard and Page, 1982), fracture toughness (K_{IC}) (Nasseri *et al.*, 2005, 2006 and 2008), elastic wave velocities (Reuschle *et al.*, 2006; Schubnel and Guéguen 2003; Schubnel *et al.* 2006), and the permeability of rocks (Guéguen and Schubnel 2003; Benson *et al.*, 2006).

[1] Lassonde Institute, Department of Civil Engineering, University of Toronto, Toronto, ON, Canada. E-mail: Nasseri@ecf.utoronto.ca

[2] Laboratoire de Géologie, Ecole Normale Supérieure de Paris, CNRS UMR8538, 24 rue Lhomond, 75005 Paris, France.

[3] Rock and Ice Physics Laboratory, Department of Earth Sciences, University College London, Gower Street, London WC1E 6BT, U.K.

The need for understanding of a coupled hydro-thermo-mechanical effect on fracture toughness, elastic properties and permeability in rocks is increasingly in demand. Seismic and velocity field data and its relation to complex, coupled phenomena is important and has a role in many applied engineering fields, such as enhanced oil production methods, hydro-fracturing, and the improved understanding of the stability of underground waste repositories (RUTQVIST *et al.*, 2005; ENGVIK *et al.*, 2005). By performing controlled thermal cracking experiments, we investigate quantitatively the influence of thermal damage on fracture toughness, elastic wave velocity, and permeability. Until now, few experimental studies have been performed on the effect of thermal cracking on the mechanical strength, K_{IC} and permeability of granitic rocks together. A small number of studies have successfully measured an *in-situ* K_{IC} data, which shows a strong dependence upon other rock physical parameters, such as elastic-wave velocities and permeability (MEREDITH and ATKINSON, 1985; DAROT *et al.*, 1992; BALME *et al.*, 2004; FUNATSU *et al.*, 2004). Conversely, however, measurements of heat-treated specimens after cooling (ATKINSON, 1984) have suggested that embedded microcrack fabrics in rocks significantly influence both elastic-wave properties and permeability (SIMMONS and BRACE, 1965; KERN, 1978, 1997). NASSERI *et al.* (2007) studied the coupled evolutions of fracture toughness and elastic wave velocities at high crack density in thermally treated Westerly granite and used a noninteractive crack theory to predict K_{IC} relatively well at high crack density and therefore will not focus on the said subject in the present study. Recently, BENSON *et al.* (2006) used inverted elastic-wave velocities inversion and permeability models of GUÉGUEN and DIENES (1988) to predict the evolution of permeability with increasing pressure, from elastic-wave velocity variations.

Despite the fact that fractures generally represent only small amounts of porosity, the theoretical evolution of elastic properties with damage can generally be predicted using models developed in the framework of Effective Medium Theory (EMT), (e.g., O'CONNELL and BUDIANSKY, 1974; CHENG and TOKSÖZ, 1979; KACHANOV, 1994; SAYERS and KACHANOV, 1995; LE RAVALEC and GUEGUEN, 1996; MAVKO *et al.*, 1999). These models are used in order to predict the material properties as a function of a single non-dimensional damage parameter, the crack density and/or the porosity. Although elastic-wave velocities and the fracture toughness are both known to decrease with microcracking and permeability to increase with the latter, few attempts have been made until now to link all the three properties and compare quantitatively their coupled evolution. The present study presents the following, experimental and modeling study on samples of thermally stressed (up to 850°C) Westerly granite: i) the experimental evolution of fracture toughness and elastic-wave velocities under a dry unconfined state for a series of chevron notched Brazilian discs, (ISRM, 1995); ii) the experimental evolution of permeability as a function of effective pressure up to 35 MPa, contemporaneously with the evolution of elastic-wave velocities up to 75 MPa of effective pressure; and (iii) the theoretical evolution of permeability with increasing pressure, from elastic-wave velocity variations.

2. *Experimental Methods and Procedures*

2.1. *Sample Selection*

Westerly granite was chosen in this study due to its evenly textured, isotropic initial microstructure (BRACE *et al.*, 1968). In addition, the physical and mechanical properties of Westerly granite are well known, with numerous studies performed on this rock type over the years at elevated temperature and pressures (e.g., HADLEY, 1976; HEUZE, 1983; BAURE and JOHNSON, 1987; THOMPSON *et al.*, 2006). Compositionally, Westerly granite is composed of 27% quartz, 36% microcline, 30% plagioclase, 6% phyllosilicates and 1% accessory minerals. The mean grain size is 0.75 mm with a total porosity of less than 1% (MEREDITH and ATKINSON, 1985). Two types of samples were prepared. Firstly, standard chevron notched Brazilian disc (CCNBD) of 75 mm diameter and 30 mm thickness for K_{IC} measurement and, secondly, right cylindrical cylinders of 38 mm diameter and 78 mm length for permeability and elastic-wave velocity measurement.

2.2. *Experimental Methods*

Isotropic crack damage was generated in sixteen CCNBD samples and four cylindrical samples of Westerly granite (isotropic in nature) by thermally stressing the samples to temperatures of 250°C, 450°C, 650°C and 850°C at a rate of 1.5°C/min using a standard laboratory furnace. After cooling at the same rate, azimuthal elastic-wave velocities (Vp) were measured at an interval of 30° circumferentially around each CCNBD specimen in order to confirm and measure the level of isotropic crack damage in each sample. Four samples were prepared at each temperature in the case of the CCNBD samples, with a single example at each temperature for the cylindrical samples. Notches were machined within the disc specimens only after this step, according to the standard ISRM (1995) procedure (Fig. 1). Four additional CCNBD and one cylindrical sample were not subjected to heating to provide an 'unfractured' room temperature (RT) example.

Fracture toughness was determined using the standard ISRM (1995) technique (Fig. 1) utilizing a stiff, servo-controlled loading machine to record displacement and force as each sample was loaded until failure at a constant rate of 0.01 mm/sec. Fracture toughness is given by:

$$K_{IC} = \frac{P_{\max}}{B\sqrt{R}} Y^*_{\min}, \qquad (1)$$

where $Y^*_{\min} = \mu . e^{Y.\alpha 1}$; $Y^*_{\min}$ is a critical dimensionless stress intensity value for the specimen, determined by the specimen geometry dimensions α_0, α_1 and α_B (Fig. 1); μ and Y are constants determined by α_0, α_B (ISRM 1995, Table 2); $P_{\max}$ is the maximum load at failure; B is the disc's thickness; and R is the radius of the disc.

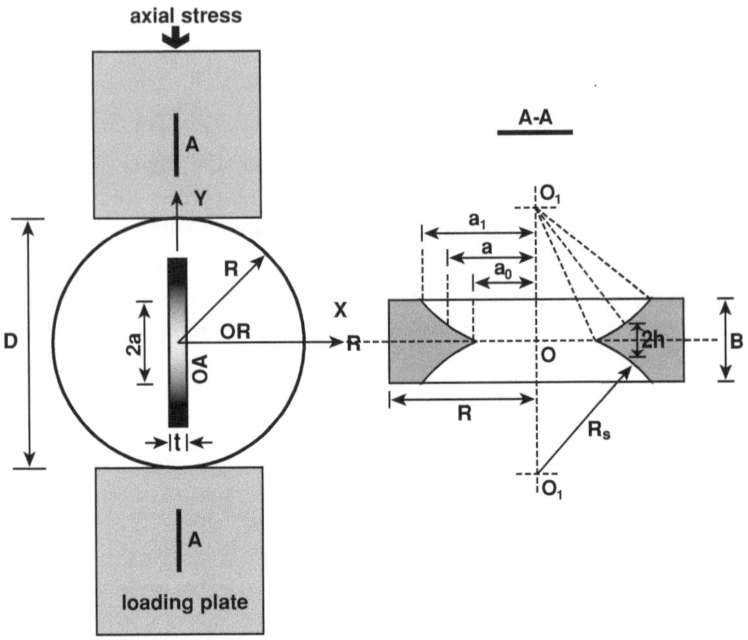

Figure 1
Geometry of the CCNBD and related parameters (after ISRM, 1995), R = Radius of disc; B = Thickness of disc; D = Diameter of disc; R_S = Radius of saw; a = Length of crack; a_0 = Initial half length of chevron notch; a_1 = Final half length of chevron notch.

Permeability was measured as a function of effective pressure ($P_{eff} = P_c - P_p$, where P_c is the confining pressure, and P_p is the pore pressure) up to 35 MPa. For cylindrical specimens thermally stressed to 450°C, 650°C and 850°C, the steady-state flow method was used, whilst for the RT and 250°C samples, the pulse decay method of Brace *et al.*, (1968) was employed, due to the much lower sample permeability. Calibration of the two methods was made using the 450°C specimen tested at 2.5 MPa of effective stress. At 2.5 MPa of effective stress for 450°C heat-treated specimen steady-state flow method was applied to measure permeability value and the result was compared to the permeability value obtained for a similar situation using the pulse decay method. This approach helps to calculate the water dead volume in the system which is an essential input parameter in the pulse decay method. Permeability was measured using a hydrostatic (instrumented Hoek-type cell made by Ergo Tech Ltd.) pressure vessel equipped with platens allowing fluid flow through the cylindrical sample (Fig. 2). Upstream and downstream pressures were maintained at approximately 5.5 to 4.5 MPa respectively, using a high precision servo-controlled microvolumetric syringe pump (Quizix QX20 k) capable of measuring cumulative fluid volume. Using this setup allows permeability to be calculated via the application of Darcy's law (e.g., Guéguen and

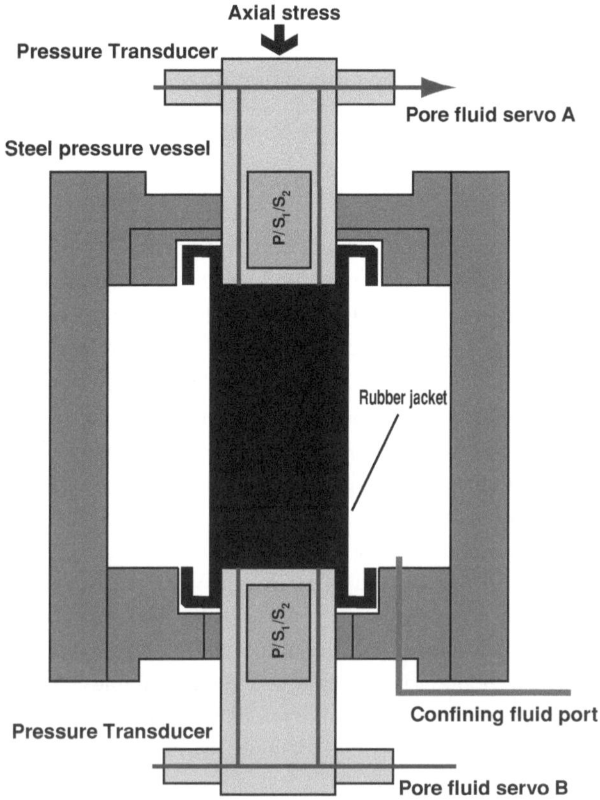

Figure 2
Hoek cell for measurement of contemporaneous permeability, elastic-wave velocity (V_p, Vs_1, Vs_2) and porosity change as a function of effective pressure to 100 MPa.

Palciauskas, 1994). In addition, contemporaneous elastic-wave velocities (V_p, V_s) are calculated axially using transducers embedded into the platens (Fig. 2). Elastic-wave velocities were measured up to effective pressures reaching 75 MPa. Finally, porosity change was determined at each step change in effective pressure by measuring the volume of pore volume expelled via the pore volumometers for all the heat-treated specimens. Permeability measurements were not continued beyond 35 MPa of effective stress and the K value was significantly reduced and application of the pulse decay method would take considerably longer and in practicality is not of great interest.

2.3. *Determination of Optical Microcrack Density, (ρ_{Optical})*

Thin sections (35 mm × 22 mm) perpendicular to the fracture propagation plane were made from the failed CCNBD specimens. Microstructural mapping, including intragranular and grain boundary microcrack mapping, and length measurements were conducted using digital images taken from the thin section. In order to be certain that this analysis was not affected by the crack propagation plane, these analyses were conducted at some distance from the thoroughgoing fracture (approximately 15 mm away from the area where test crack interacts with rock matrix). For intragranular crack measurement length HADLEY's (1976) definition of crack was applied where the crack's orientation does not deviate more than 20 degrees on any rolling 5 mm section of its length. A set cut off limit of 0.05 mm was selected for the crack length mapping from the optical images to ensure repeatability. Grain boundary microcrack lengths measurement involved unfolding of the curved boundaries into straight lines where the latter vary between 0.07 to 3.6 mm as the temperature for the specimens varied from room temperature to 850°C. For specimens treated at 650 and 850°C, where most of the grain boundary contacts were cracked, maximum grain boundary length could not exceed the perimeter of the maximum grain size. In addition to thin section analysis, scanning electron micrographs (backscatter) were also prepared, allowing the average widths of intragranular and grain boundary microcracks (50 measurements for each) to be directly measured as a function of thermal stressing temperature (NASSERI *et al.*, 2007).

The analysis of the thin section and SEM images were performed using the program "Intercept" developed by LAUNEAU and ROBIN (1996). This software allows the rapid determination of microcrack density by scanning images along sets of parallel lines, and counting the number of boundaries intercepted by those lines. The total number of intercepts is then converted into an optical crack density, expressed in (mm/mm^2) which corresponds to the accumulated microcrack lengths (a_j) from area (A) using:

$$\left(\rho_{optical} = \frac{\sum a_j}{A}\right). \tag{2}$$

3. Experimental Results

3.1. Microstructural Analysis for RT and Heat-treated Specimens

Figure 3 shows the evolution of both grain boundary (ρ_{GB}) and intragranular microcrack (ρ_{IG}) densities with temperatures reaching 850°C determined through optical microscopy. The initial ratio of ρ_{IG}/ρ_{GB} (RT specimens) decreased from approximately 2.85 to approximately 1.73 and 1.25 for 250°C and 450°C specimens respectively due to the rapid increase in ρ_{GB}. This ratio attains a value of $\sim$1.54 for 650°C specimens, reflecting the fact that intragranular density increased more than grain boundary crack density. Whereas this ratio reaches its lowest value of $\sim$0.95 for 850°C specimens, reflecting a further increase in ρ_{GB} especially observed due to the thermal splitting of individual quartz and feldspar minerals at 850°C. A similar trend of observation has been reported by FREDRICH and WONG (1986) on thermal cracking of Westerly granite.

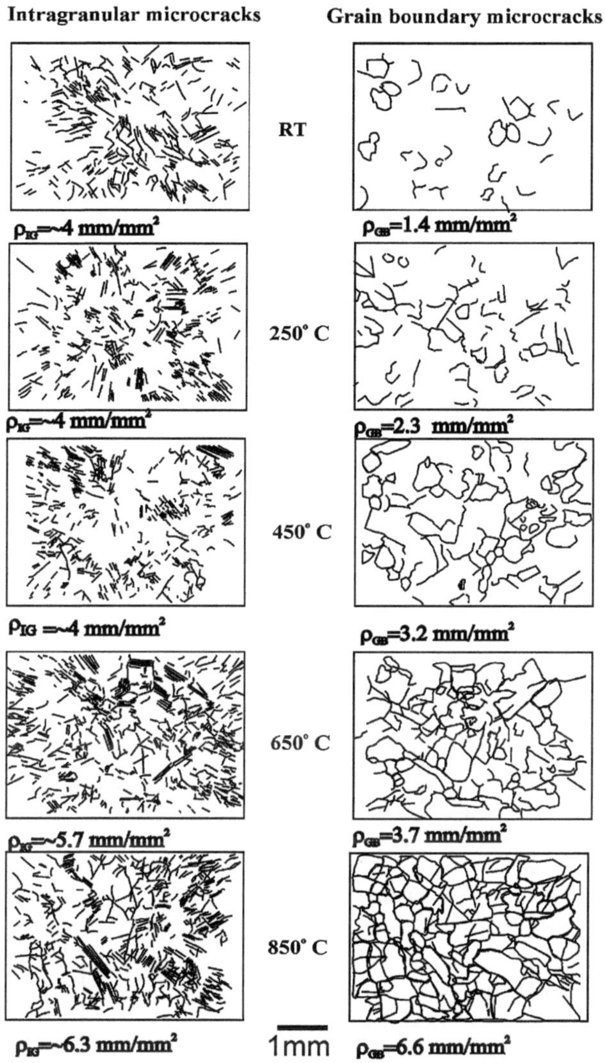

Figure 3
Variation of intragranular and grain boundary cracking density as a function of temperature in thermally-treated Westerly granite obtained using an optical method.

3.2. Fracture Toughness and P-waves Velocity Evolution with Temperature

Figure 4a shows the fracture opening displacement as a function of thermal stress level. There is a clear dependence on the damage level and both peak stress intensity factor K_I (normalized to the peak fracture toughness of 1.48 MPa.m$^{0.5}$, measured in the RT sample) and maximum fracture displacement. A 60% reduction of the fracture toughness (from 0.98 to 0.43 MPa.m$^{0.5}$) is observed for the heat-treated specimens ranging from 450 to 650°C. Figure 4b displays the coupled evolution of K_{IC} as a function of temperature treatment as compared to normalized P-wave velocities, as a proxy to damage level (normalized to RT specimen values). There is a clear negative correlation of K_{IC} and normalized V_p with thermal stressing temperature. However, the evolution of K_{IC} and V_p appears well coupled. Therefore, it is likely that V_p is a good geophysical indicator of the deterioration of K_{IC}. The maximum P-wave velocity was measured in RT specimens (4.55 km/s), which rapidly decreased to 0.94 km/s at 850°C. This corresponds to the 80% decrease in V_p, which compares favorably with the 85% decrease in K_{IC}. The biggest drop in velocity occurs between 450°C and 650°C as a result of the $\alpha-\beta$ quartz phase transition. In general, the peak stress becomes ill-defined beyond 650°C and is characterized with lower stress drops and increased post failure plasticity. These observations are in close agreement with the previous results obtained on Westerly granite (HEUZE, 1983; MEREDITH and ATKINSON, 1985).

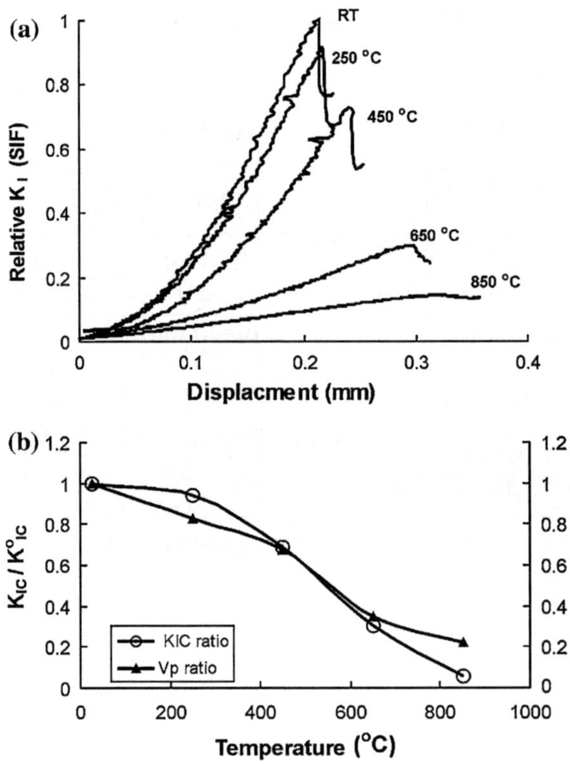

Figure 4

(a) Evolution of relative stress intensity factor *KI*, versus displacement, and as a function of temperature. (b) A comparison of the fracture toughness (K_{IC}) and the mean P- wave velocity as a function of temperature treatment. Both K_{IC} and the V_p were normalized to that of the untreated specimens (after NASSERI *et al.*, 2007).

4. *Experimental Results: Elevated Isostatic Pressure*

4.1. *Permeability and Seismic Wave Velocities at Elevated Hydrostatic Pressure to 35 MPa*

Permeability was measured with an increasing effective pressure in a Hoek type triaxial cell equipped with an elastic-wave measurement system (shown in Fig. 2). Figure 5 shows that an approximately exponential decrease of permeability (k) occurs with increasing effective pressure (where $P_{eff} = P_C - P_P$), together with appreciable hysteresis during depressurization. The phenomenon of 'crack lips sticking' is thought to cause irreversible crack closure; this effect is more pronounced for the RT and heat-treated specimens to 450°C as compared to the 650°C and 850°C specimens. This is best illustrated by reference to the change in elastic-wave velocity (both V_P and V_S) shown in Figure 6. During pressurization, elastic-wave velocities increased as a response to crack closure, which, in turn, cause a concomitant reduction in permeability. During subsequent depressurization, the viscoelastic property of the microcracks dominates and therefore the complete restoration of permeability and elastic-wave velocities does not occur. The permeability decreases by an order of magnitude (from 4.13E-19 m^2 to 4.9E-20 m^2) as effective pressure increases from 2.5 MPa to 35 MPa (RT specimen). However, this hysteresis (for RT specimen) is not as obvious when measured via the P-wave velocity, which returned to within 5% of its initial values. This trend is also seen for S-wave data

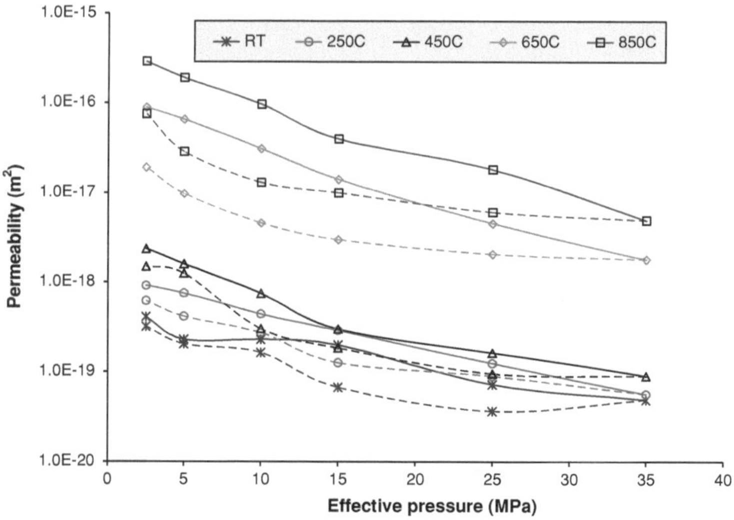

Figure 5

Variation of permeability (k in log values) with effective pressure for thermally-treated Westerly granite; the solid lines show the path during the pressurization cycle and the dashed lines show the path for depressurization cycles for different specimens.

(Fig. 6b), which returned to within 4% more than their initial values. Whereas such a hysteresis in terms of V_p and V_s as a function of effective stresses for heat-treated samples, especially the second group, is quite distinguishable during depressurization.

For the specimens thermally stressed to 250°C and 450°C, a similar trend was observed. However, there is considerable overlap with the RT data, and it is clear that the

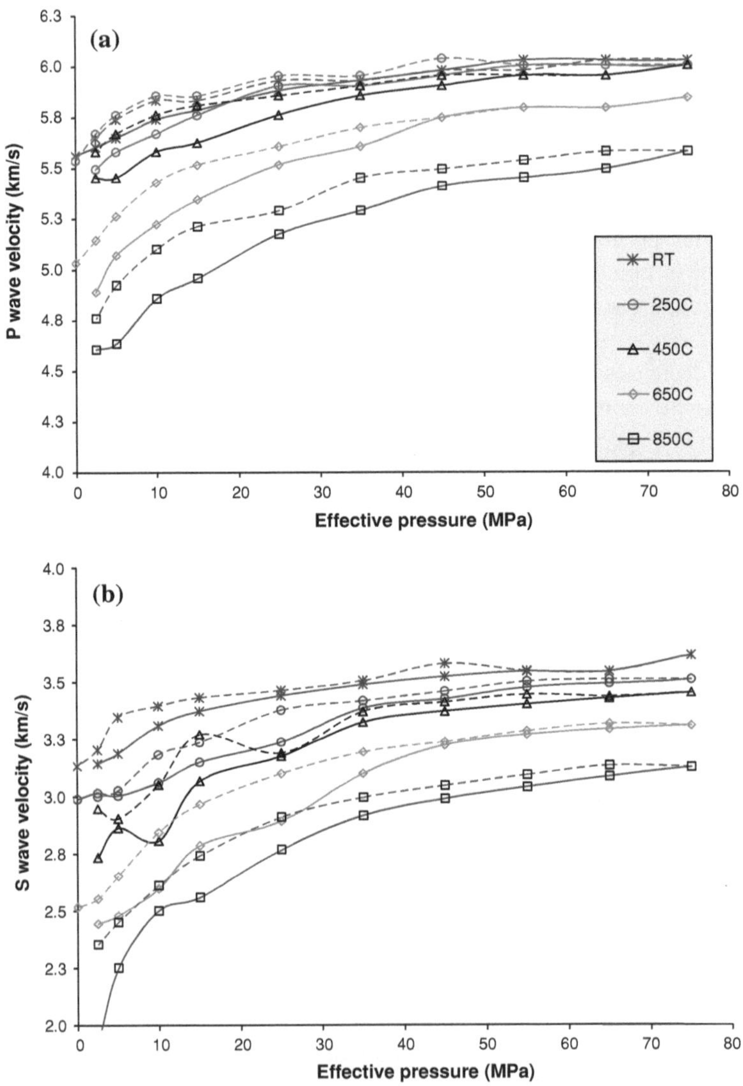

Figure 6
(a) Variation of measured V_p and (b) V_s with effective pressure for RT and thermally-treated specimens; the solid lines show the path during the pressurization cycle and the dashed lines show the path for depressurization cycles for different specimens.

enhancement of permeability does not appreciably occur until a thermal stressing temperature threshold, at between 450°C and 650°C, is crossed. 250°C and 450°C samples show a reduction in permeability with increasing effective pressure, decreasing respectively from 9.18E-19 m^2 and 2.36E-18 m^2 to 5.69E-20 m^2 and 9.05E-20 m^2. The P-wave velocity experiences an increase of approximately $\sim$7–8% during pressurization for both specimens. S-wave velocity for these two specimens increases about 19% for both specimens (Fig. 6). On depressurization, elastic-wave velocities returned to within 4–5% above their initial values measured at 2.5 MPa of effective stress. Variation of permeability for 650°C and 850°C specimens as a function of effective pressure up to 35 MPa shows a reduction of one to two orders of magnitude respectively. V_p and V_S both increase $\sim$13% for 650°C specimen where as this increase for 850°C is about 13% and 35% respectively as function of effective stress rising up to 35 MPa. On depressurization, both elastic-wave velocities recover to within approximately 4–5% more than their initial values, i.e., measured at 2.5 MPa of effective stress before pressurization.

5. Theoretical Analysis and Comparisons to Laboratory Data

The effective mechanical and physical (elastic and transport) properties of an initially isotropic cracked medium depend on a number of key parameters combining linear fracture mechanics and statistical physics. In the present method, the following key parameters are defined:

- Fluid bulk modulus (K_f) and matrix elastic properties with Young's modulus E_o and Poisson ratio ν_o.
- Crack geometry (penny shaped cracks) with an average aspect ratio defined by: $\zeta = \langle w/2c \rangle$.
- Crack density ρ, defined by: $\rho = 1/V \sum_0^N c_i^3$, where c_i is the radius of the ith crack, N being the total number of cracks embedded in the representative elementary volume V.
- Percolation factor f.

5.1. Elastic Properties

For an isotropic matrix containing a random distribution of cracks, the effective Young's modulus of a rock E^* is a linear function of the crack density which can be written in the form (first perturbation order):

$$\frac{E_0}{E^*} = 1 + H\rho. \tag{3}$$

A similar expression can be written for the effective shear modulus (μ^*). In equation (3), Young's modulus E_0 is that of the crack-free matrix, and H is a positive scaling parameter which is dependent upon the matrix and fluid properties, the geometry of the

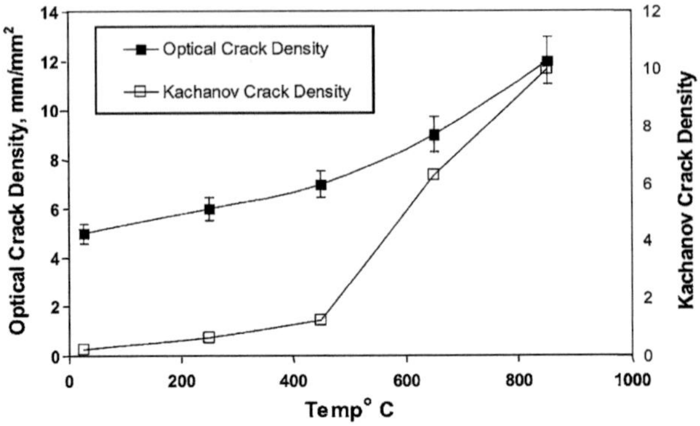

Figure 7
Crack densities determined optically and by P-wave velocity inversion as a function of temperature for the Westerly granite.

cracks and the interactions between them. The scalar H has been calculated by many authors for a wide variety of crack geometries and fluid properties. In this work we use an appropriate and widely used scalar for non-interactive penny shaped cracks (Kachanov, 1994; Schubnel and Guéguen, 2003; Schubnel *et al.*, 2006). One of the most straightforward methods is Kachanov's (1994) non-interactive effective medium theory, as it neglects stress interactions between cracks, and can therefore represent a valid approximation for low crack densities (up to ~0.5). However, it is important to point out that above 0.5, the calculation remains physical (the elastic moduli fall asymptotically to zero) such that the non-interactive theory remains a useful physical tool to quantitatively evaluate damage. Because we neglect stress interactions between cracks, calculations using the non-interactive approximation overestimate the actual crack density within the samples and in the following, we evaluate only an upper bound for crack density. More realistic values of crack density could be obtained using alternative effective medium theories taking into account stress interactions such as the Extended Differential Self Consistent scheme (Le Ravalec and Guéguen 1996). Comparing both theories at high crack densities would probably demonstrate that non-interactive theory continues to give physically interpretable results up to considerably higher crack densities than ~0.5. However, this last observation is beyond the scope of this paper.

In the present work we use the case of a random crack distribution (Kachanov, 1994; Schubnel and Guéguen, 2003). When neglecting crack interactions, effective elastic moduli of a cracked solid can be calculated exactly and rigorously in a unique manner that depends solely upon the average crack orientation and distribution. For certain distributions, where stress interactions are compensating geometrically such as a random (isotropic) or aligned crack distributions, it was shown that the non-interactive approximation is the most effective scheme when compared to other effective media theories (Kachanov, 1994; Sayers and Kachanov, 1995). Moreover, in our case such a model is pertinent because cracks are elastically opening and closing due to isostatic

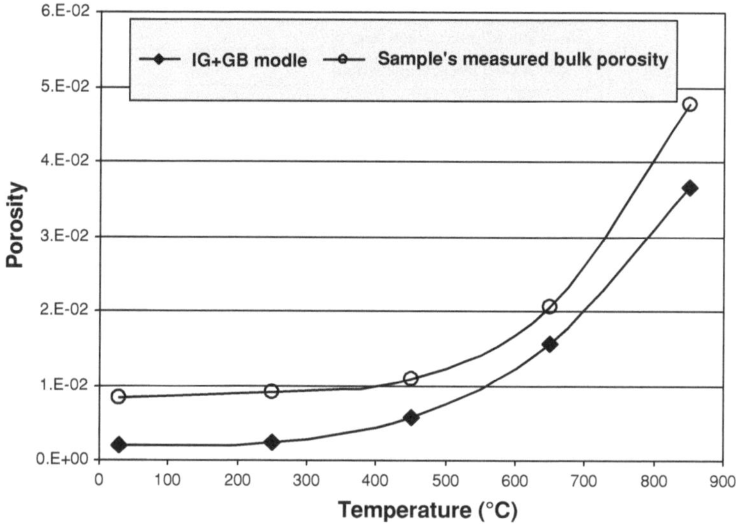

Figure 8
Comparison between sample's measured bulk porosity with microcrack porosity deduced form microstructural parameters (after NASSERI *et al.*, 2007). IG = intragranular, GB = grain boundary microcracks.

pressure solely, whereas stress interactions are most important in the case of crack propagation only. In such a scheme, the evolution of elastic-wave velocities can be used to quantify uniquely both the crack density and aspect ratios. As all of these methods and concepts have been well discussed in previous publications (e.g., BENSON *et al.*, 2006; SCHUBNEL *et al.*, 2006), the precise details are not reproduced here.

5.2. Dimensionless Microcrack Density, Inversion Method (KACHANOV)

Figure 7 plots the dimensionless crack density ρ as a function of temperature. As temperature increases, K_{IC} decrease from 1.4 to 0.2 MPa.m$^{0.5}$, and ρ increases from 0.26 to approximately 10. The optically determined crack density ρ_{optical}, which is not a dimensionless parameter, is also plotted for comparison. Although ρ_{optical} and ρ_{kachanov} are correlated and show a very similar trend, it is hard to compare two parameters of different dimensions. In order to palliate to any misinterpretation this comparison could lead us to, microcrack aspect ratios ζ measured optically (using high resolution SEM images) were calculated by simply using $\zeta = \langle \mathrm{w} \rangle / \langle 2\mathrm{c} \rangle$ where w is the crack width and c the crack radius. Remembering the fact that for penny-shaped cracks with constant aspect ratio distribution, the total microcrack porosity is also equal to $\phi = \pi \rho w/c$, then the aspect ratio can be expressed as a function of the porosity and the crack density, i.e., $\zeta = \phi/2\pi\rho$. Figure 8 shows the close comparison between the microcrack porosity as deduced from the microstructural parameter (shown in detail in NASSERI *et al.*, 2007) with the porosity measured on the bulk samples as a function of temperature. Figure 9 shows

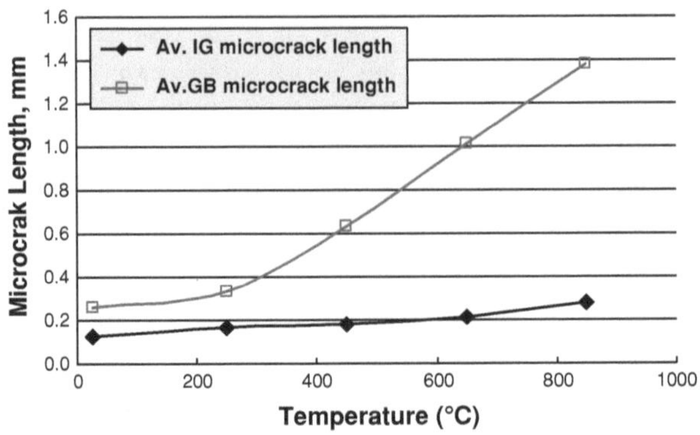

Figure 9
Evolution of intragranular and grain boundary crack length as a function of temperature at zero effective stress (measured from thin sections prepared for each heat-treated specimen).

the evolution of thermal crack length (intragranular and grain boundary) measured from a thin section prepared at various temperatures. Needless to emphasize that it is not physically possible to monitor such an evolution at various effective stresses and the responses of crack length to applied pressures.

5.3. *Simultaneous Inversion for Crack Porosity and Concomitant Permeability*

In order to link the effect of microcrack aspect ratio, density, and aperture to crack connectivity (and therefore permeability), the Guéguen and Dienes (1989) crack permeability model may be employed. Any crack represents a void space of non-zero size, with the aperture being defined as the maximum width. If the cracks intersect, this will produce a crack permeability based on the fluid movement through the network of linked crack elements. Guéguen and Dienes (1989) quantify the effect of penny-shaped cracks on the permeability of an otherwise impermeable host matrix, and derived the following relation:

$$k_{\text{cracks}} = \frac{1}{15} f_{\text{cracks}} w^2 \zeta \rho, \tag{4}$$

where, ρ is the crack density, w and ζ are the crack aperture and average aspect ratio (ratio of crack aperture to crack length), respectively, and f_{cracks} is the percolation factor (Guéguen and Dienes, 1989; Schubnel and Guéguen, 2003), approximated via:

$$f_{\text{cracks}} \approx \frac{4}{9}\left(\frac{\pi^2}{4}\rho - \frac{1}{3}\right)^2,$$

and valid for $1/3 < \rho\pi^2/4 < 1$. In essence, f_{cracks} describes the probability of two cracks intersecting ($\rho\pi^2/4$), in which case the volume element created by the intersecting cracks is discounted in the method (an excluded volume). In all cases, $0 \leq f_{\text{cracks}} \leq 1$. Here and

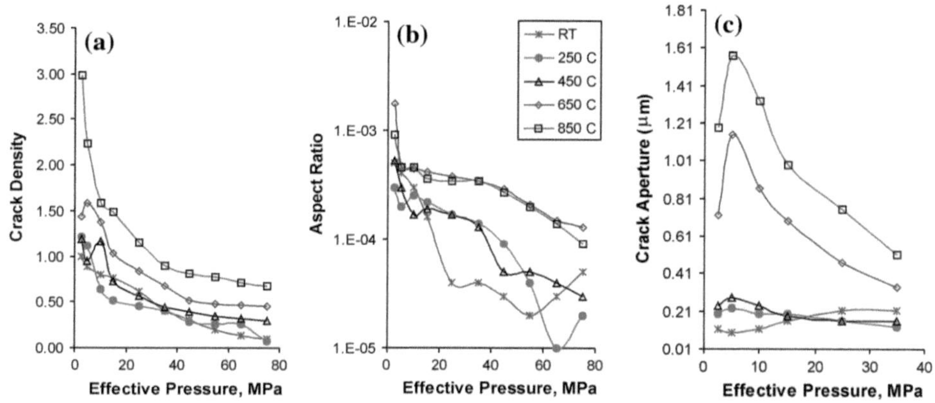

Figure 10
(a) Variation of crack density obtained from inversion of V_p (b) aspect ratio and (c) crack aperture as a function of effective pressure measured in instrumented Hoek-type cell for RT and treated specimens.

for the sake of simplicity, we assume crack network to be always fully connected so that $f_{cracks} = 1$ in all the following. As we shall see later, even a simple model can provide useful insight in helping estimate which factors dominate the bulk permeability and the permeability changes.

Figure 10 shows the variation of crack density, aspect ratio and crack aperture as a function of effective pressure for heat-treated and non-treated Westerly granite obtained from the inversion methodology explained above. The evolution of crack aspect ratio is shown up to 35 MPa of effective pressure. All the mentioned properties respond to applied pressure with a decrease in magnitude. Figure 11 shows the comparison between the experimental and modeled values for V_p, V_s, porosity (up to 75 MPa) and permeability (up to 35 MPa) as a function of effective pressure. A detailed description of the said figures is included in the next section.

6. *Discussion and Interpretation*

Two groups of samples based on intensity of induced thermal damages are distinguished in this study. The first group of samples, treated up to 450°C, is characterized with a comparable range of data in terms of porosity, crack density, elastic-wave velocities (including the nature of depressurization responses), fracture toughness and permeability (including the nature of depressurization responses) as a function of temperature. These properties are substantially different from the same properties shown by the second group of samples, which have been treated at 650° and 850°C (Figs. 3 to 6). The modeled or predicted properties as well follow the similar overall trend with respect to the range of thermal treatment acquired in this study (Figs. 10 and 11). A similar trend

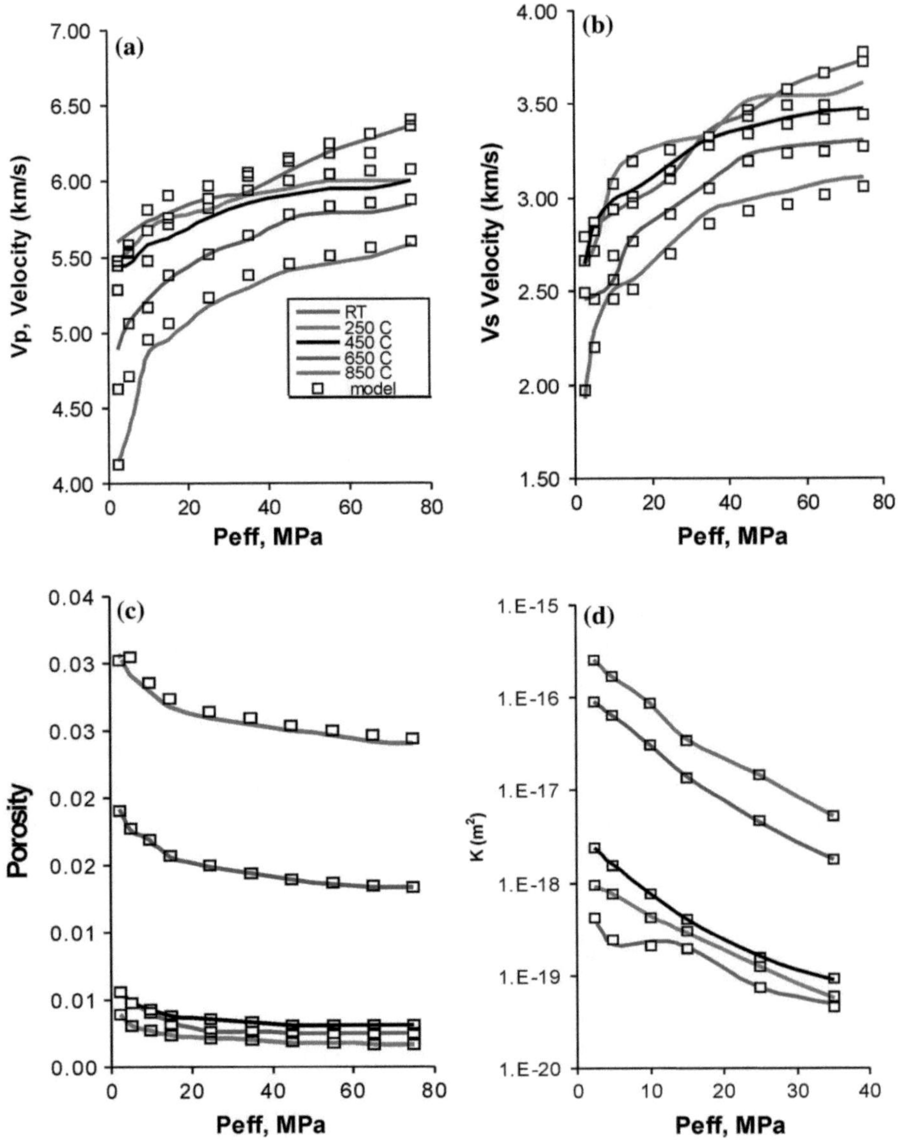

Figure 11

(a)Variation of experimentally determined and modeled V_p (b) V_S, (c) porosity and (d) permeability (k) as a function of effective pressure for RT and treated Westerly granite. The solid lines show the experimental values and square symbols show the modeled values.

of observations was noted in correlating between the fracture toughness and fracture roughness as a function of temperature for Westerly granite, (NASSERI *et al.*, Fig. 8, this issue). Such a division based on distinctly different mechanical, physical and transport properties is directly related to the extend of thermal damages caused by α–β quartz phase transition in granitic rocks, (DAROT *et al.* 1992).

The stress intensity factor generated at the tip of propagating test crack (unconfined, dry specimens) drops 25% for the 450°C specimen in comparison to the RT specimen, whereas the stress intensity factor drops about 75% and 80% for the 650°C and 850°C specimens, respectively. Increased crack density above α–β quartz phase transition for the 650°C specimen and further enlargement of grain-grain boundary apertures at the 850°C specimens is responsible for lower resistance to cracking at the onset of test crack propagation. It has been concluded that during CCNBD testing, a macroscopic effective medium approach can grasp the reduction of K_{IC} with crack density and might well explain the strong correlation observed experimentally between P-wave velocity and K_{IC} reduction (NASSERI *et al.*, 2007).

Figures 12a and 12b show the evolution of porosity and P-wave velocities as a function of effective pressure for RT and thermally treated specimens. In this graph each measured value for porosity and V_p (Φ or V_p) is divided by its corresponding values (Φ_o or V_{po}) measured at 2.5 MPa of effective pressure. Porosity in RT and treated specimens at 250°C and 450°C responds quickly to effective pressure increases, decreasing 55%-45% respectively up to 45 MPa of effective pressure, followed with a more or less constant decrease up to 75 MPa of effective pressure. This trend for 650°C and 850°C specimens follows a decrease of 25% and 15%, respectively as effective pressure increases up to 75 MPa. P-wave velocity, on the other hand, increases linearly about 10% as a function of effective pressure for RT, 250°C and 450°C specimens whereas 650°C and 850°C specimens show a maximum increase of 20% and 35%, respectively. It is shown that in the saturated specimens in the range of RT to 450°C, induced thermal cracks which are characterized with lower crack density, smaller crack aperture and lower crack aspect ratio (Fig. 11), behave differently to the sets of samples treated at 650°C and 850°C when subjected to various levels of effective pressure. In the first group of specimens, thermal crack density being less ($\sim$1) and characterized with much lower aspect ratio (ζ, = $\sim$5.00E-4) comply with the applied stress quickly, and as a result porosity drops instantaneously up to a threshold effective pressure in the range of 45 MPa. Whereas for the second range of specimens, characterized with high initial thermal crack density ($\sim$2–3), much higher crack aspect ratio (ζ = $\sim$2.00E-3) and bigger crack aperture (0.5 μm) do not comply with applied pressure, immediately resulting in comparatively little porosity reduction (15%–25%) as a function of effective pressure. The variation of V_p with effective pressure for the said two ranges of heat-treated rocks follows a similar trend. The first group of rocks show no large improvement in V_p (Fig. 12b) with applied pressure during pressurization due to their lower porosity and lack of interconnected saturated pores, where as the second group of rocks show an increase (20%-35%) in V_p with effective pressure increasing to 75 MPa. During depressurization, crack opening is delayed, suggesting irreversible closure of the cracks (Figs. 6a and 6b). Opening of cracks results in a larger V_p and V_s for the second group of rocks in comparison to first

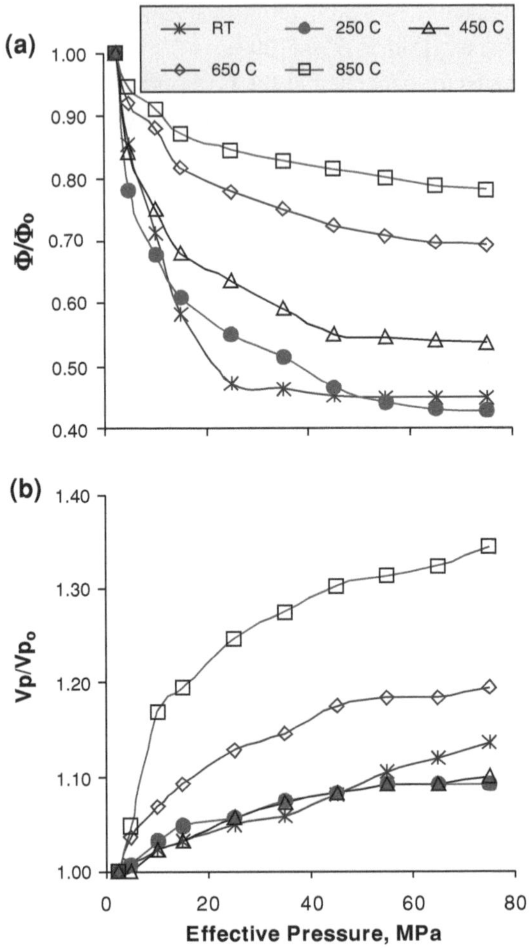

Figure 12
(a) Evolution of porosity and (b) P-wave velocities as a function of effective pressure for RT and thermally-treated specimens. In these graphs each measured value for porosity and P-wave velocity (Φ or V_p) is divided by its corresponding lowest value (Φ_o or V_{po}) measured at 2.5 MPa of effective pressure.

group of rocks. Similarly the hysteresis between the closing and opening sequence is more pronounced for the second group of rocks than for the first group.

Figure 10 compares the experimental V_p, V_s, porosity and permeability with the modeled data during the pressurization sequence. Using the crack density and aspect ratio recovered from the elastic-wave velocity inversion together with the model of Guéguen and Dienes and initial porosity data; the recovered permeability variation can be mapped (Fig. 11d) with pressure. Increase of elastic-wave velocities due to pressure is large and controlled by crack density for specimens RT 250°C and 450°C ($\sim$10%). For specimen 650°C and 850°C the increase is larger because the crack density is indeed larger and also because cracks are fully open, as seen by the permeability inversion which shows that for these samples, the crack aperture jumps almost one order of magnitude. In this case, the variation of crack density is larger during pressurization, thus a larger variation of V_p. In fact, the crack density ρ is an indirect measure of the average crack length × number of cracks. For a given specimen, the variation of ρ with pressure can only be interpreted as an average diminution of crack length (because the number of cracks remains constant) due to the increase of crack surface contact points. However, the increase of crack length with temperature treatment is harder to infer using wave velocities because both the average crack length and number increase. In this case, only our optical analysis can give bounds only for zero effective pressure (Fig. 9). However, in all cases, the permeability order of magnitude is controlled by the aperture and the jump between 450°C and 650°C samples is interpreted in a jump of crack average half aperture. However, the RT sample shows a decrease in aspect ratio and crack density, but a slight increase in crack aperture with effective pressure (Fig. 10c), which is counter-intuitive and probably shows the limitation of our microstructural inversion that neglects the important problem of crack connectivity. However, this could certainly be interpreted by introducing a percolation factor f nonequal to 1, which shows that the latter probably plays an important role at low effective pressures. The same observation can be performed in the case of the 650°C and 850°C samples, for which the crack aperture increases at low effective pressure while the aspect ratio and crack density decreases.

The excellent fit is a confirmation that the model, based on a network of cracks, is appropriate for this type of rock, which is consistent with previous experience with this type of hybrid experimental-modeling approach (Benson *et al.*, 2006). While this strategy does not predict permeability from base properties, the fact that the variation of permeability with pressure is driven almost solely by crack aspect ratio, the crack density and crack aperture thus allows important controls to be placed on how damage in rock masses (natural or induced) controls and affects transport properties. This has important implications in many areas of applied science and engineering, for example geothermal reservoir production and mining.

7. *Conclusions*

In this study a series of 20 heat-treated CCNBD samples of Westerly granite were tested in order to evaluate the effect of thermal damage on fracture toughness (K_{IC}). Four heat-treated and one untreated cylindrical samples were tested in a Hoek type of cell, designed to measure contemporaneous permeability (up to 35 MPa of effective pressure), elastic-wave velocity (V_p, V_s) and porosity change as a function of effective pressure to 75 MPa. The decrease in P-wave velocity matched well with that of fracture toughness in samples heat-treated up to 850°C. Crack densities ρ obtained from wave-velocity inversion showed similar trend to that determined optically. Thermal cracking was found to induce a significant decrease in the mechanical and dynamic elastic properties, and to increase the permeability of Westerly granite of three orders of magnitude. Seismic wave velocity measurements are a good geophysical indicator of the damage and hence the deterioration of K_{IC} and the enhancement of permeability with microcracking, and could be used as a nondestructive technique to monitor the evolution of these parameters in the laboratory and field. Permeability was estimated based on a method involving the GUÉGUEN and DIENES (1989) crack permeability model, and prediction of the evolution of permeability with increasing pressure from elastic-wave velocity variations is in good agreement with the experimental data for the pressurization cycle. These constitute part of an ongoing research project endeavoring to link mechanical properties (fracture toughness, elastic constants) and physical properties (permeability, elastic-wave velocities) to quantitative microstructural parameters (crack density, aspect ratio).

Acknowledgements

This work was supported by a Canadian Foundation for Innovation (CFI) award to RPY, a Marie-Curie International Fellowship within the 6th European Community Framework programme to PMB (Contract MOIF-CT-2005-020167) and an International Scientifique Collaboration Project (PICS) by the French National Scientific Research Agency (CNRS). The authors would like to thank D. Collins and Michael Heap (University College London) and an anonymous reviewer for the constructive remarks which helped to enhance this paper greatly.

REFERENCES

ATKINSON, B.K. and MEREDITH, P.G., *Experimental fracture mechanics data for rocks and minerals*. In *Fracture Mechanics of Rocks* (ed. B. K. Atkinson) (Academic Press, London 1987) pp. 477–525.

ATKINSON, B.K. (1984), *Subcritical crack growth in geological materials,* J. Geophys. Res. *89*, 4077-114.

BALME, M.R., ROCCHI, V., JONES, C., SAMMONDS, P.R., MEREDITH, P.G., and BOON, S. (2004), *Fracture toughness measurements on igneous rocks using a high-pressure, high-temperature rock fracture mechanics cell,* J. Volcanol. Geotherm. Res. *132*(2–3), 159–172.

BAURE, S.J. and JOHNSON, B. (1979), *Effects of slow uniform heating on the Westerly and Charcoal granites.* Proc. 20th Symp. *On Rock Mechanics*, Austin, Texas, 7–18. ASCE, New York.

BENSON, P. MEREDITH, P., and SCHUBNEL, A. (2006), *Examining the role of void space fabric in permeability evolution with from statistics and percolation,* J Geophys Res. (in press).

BRACE, W.F., WALSH, J.B., and FRANGOS, T. (1968), *Permeability of granite under pressure*, J. Geophys. Res. *6*, 2225–2236.

BRACE, W. and MARTIN, A. (1968), *A test of law of effective stress for crystalline rocks of low porosity*, Int. J, Rock Mech, Min. Sci. *5*, 416–426.

CHENG, C.H. and TOKSOZ, M.N. (1979), *Inversion of seismic velocities for the pore aspect ratio spectrum of a rock*, J. Geophys. Res. *84*, 7533–7543.

DAROT, M., GUÉGUEN, Y., and BARATIN, M.-L. (1992), *Permeability of thermally cracked granite,* Geophys. Res. Le. *19*(9), 869–872.

ENGVIK, A.K., BERTRAM, A., KALTHOFF, J.F., STOCKHERT, B., AUSTRHEIM, H. and ELVEVOLD, S. (2005), *Magma-driven hydraulic fracturing and infiltration of into the damaged host rock, and example from Drowning Maud land, Antarctica,* J. Structural Geol. *27*, 839–854.

FREDRICH, J.T. and WONG, T.-F. (1986), *Micromechanics of thermally induced cracking in three crustal rocks*, J Geophy Res. *91*, B12 12743–12764.

FUNATSU, T., SETO, M., SHIMADA, H., MATSUI, K., and KURUPPU, M. (2004), *Combined effects of increasing temperature and confining pressure on the fracture toughness of clay-bearing rocks*., Int. J. R. Mech. Min. Sci. *41*, 927–938.

GUEGUEN, Y., and DIENES, J. (1988), *Transport properties of rocks from statistics and percolation*. J. Fath. Geol. *21*, 1, 131.

GUÉGUEN, Y. and PALCIAUSKAS, V. *Introduction to the Physics of Rocks*, 292 pp. (Princeton University Press, New Jersey 1994).

GUÉGUEN, Y., and SCHUBNEL, A. (2003), *Elastic-wave velocities and permeability in cracked rocks*, Tectonophysics *370*, 163–176.

HADLEY, K. (1976), *Comparison of calculated and observed crack densities and seismic velocities in Westerly granite*, J. Geophys. Res. *87*, 9340–9348.

HEARD, H.C. and PAGE, L., (1982), *Elastic moduli, thermal expansion and inferred permeability of two granites to 350°C and 55 Megapascals*, J. Geophys. Res., *87*, 9349–9348.

HEUZE, F.E. (1983), *High-temperature mechanical, physical and thermal properties of granitic rocks-A review*, Int. J. Rock. Mech. Min. Sci. and Geomech. Abstr. *28* (1), 3–10.

ISRM (1995), *Suggested method for determining Mode I fracture toughness using cracked chevron notched Brazilian disc (CCNBD) specimens*, Int. J. Rock Mech. Min. Sci. Geomech. Abstr. *32*, 57–64.

KACHANOV, M. (1994), *Elastic solids with many cracks and related problems*, Adv. Appl. Mech. *30*, 259–445..

KERN, H. (1978), *The effect of high temperature and high confining pressure on compressional wave velocities in quartz bearing and quartz free igneous and metamorphic rocks*, Tectonophysics *44*, 185–203.

KERN, H., LIU, B., and POPP, T. (1997), *Relationship between anisotropy of P- and S- wave velocities and anisotropy of attenuation in serpentinite and amphibolite*, J. Geophys. Res. *102*, 3051–3065.

LAUNEAU, P. and ROBIN, P.-YF. (1996), Fabric analysis using the intercept method., Tectonophysics *267*(1–4), 91–119.

LE RAVALEC, M. and GUÉGUEN, Y. (1996), *High and low frequency elastic moduli for a saturated porous/cracked rock– Differential self-consistent and poroelastic theories*, Geophysics *61*, 1080–1094.

MAVKO, G. MUKERJI, T., and DVORKIN, J. *The Rock Physics Handbook* (Cambridge University Press, 1999) 329 pp.

MEREDITH, P.G. and ATKINSON, B.K. (1985), *Fracture toughness and subcritical crack growth during high-temperature tensile deformation of Westerly granite and Black gabbro*. Phys. Earth and Plan. Int. *39*, 33–51.

NASSERI, M.H.B., MOHANTY, B., and ROBIN, P.-YF. (2005), *Characterization of microstructures and fracture toughness in five granitic rocks*, Int. J. Rock Mech. Min. Sci. *42*, 450–560.

NASSERI, M.H.B., MOHANTY, B., and YOUNG, R.P. (2006), *Fracture toughness measurements and acoustic emission activity in brittle rocks*, Pure Appl. Geophys. *163*, 917–945.

NASSERI, M.H.B., SCHUBNEL, A., and YOUNG, R.P. (2007), *Coupled evolutions of fracture toughness and elastic-wave velocities at high crack density in thermally treated Westerly Granite*, Int. J. Rock. Mech. and Min. Sci. *44*(4), 601–616.

NASSERI, M.H.B. and MOHANTY, B. (2008), *Fracture toughness anisotropy in granitic rocks*. J. Rock Mech. and Min. Sci. *45*, 167–193.

NASSERI, M.H.B., TATONE, B.S.A., GRASSELI, G. and YOUNG, R.P. (2008), Fracture toughness and fracture roughness interrelationship in thermally treated Westerly granite (submitted to the same volume of special issue, Pure Appl. Geophys).

O'Connell, R., and BUDIANSKY, B. (1974), Seismic velocities in dry and saturated rocks, J. Geophys. Res. *79*, 5412–5426.

REUSCHLE., T., GBAGUIDI HAORE, S., and DAROT, M. (2006), *The effect of heating on microstructural evolution of La Peyratte granite deduced from acoustic velocity measurements*, Earth Planet. Sci. Lett. *243*, 692–700.

RUTQVIST, J., BARR, D., DATTA, R., GENS, A., MILLARD, S., OLIVELLA, S., TSANG, C.-F and TSANG, Y. (2005), *Coupled thermal-hydrological-mechanical analyses of the Yucca mountain shaft scale test-comparison of field measurements to predictions of four different numerical models*, Int. J. Rock Mech. Min. Sci. *42*, 680–697.

SAYERS, C.M. and KACHANOV, M. (1995), *Microcrack induced elastic wave anisotropy of brittle rocks*. J. Geophys Res. *100*, 4149–56.

SCHUBNEL, A, and GUEGUEN, Y. (2003), *Anisotropy and dispersion in cracked rocks*. J Geophys. Res. 108 art 2101.

SCHUBNEL, A., and BENSON, P.M., THOMPSON, B.D., HAZZARD, J.F., and YOUNG, R.P. (2006), *Quantifying damage, saturation and anisotropy in cracked rocks by inverting elastic wave velocities*, Pure Appl. Geophys. *163*, 947–973.

SIMMONS, G., and BRACE, W.F. (1965), Comparison of static and dynamic measurements of compressibility of rocks, J. Geophys. Res. *70*, 5649–5656.

THOMPSON, B.D., YOUNG, R.P., and LOCKNER, D.A. (2006), *Fracture in Westerly granite under AE feedback and constant strain rate loading: nucleation, quasi-static propagation, and the transition to unstable fracture propagation*, Pure Appl. Geophys. *163*, 995–1019.

(Received June 19, 2008, revised January 27, 2009, accepted February 13, 2009)
Published Online First: June 4, 2009

Pure appl. geophys. 166 (2009) 949–968
0033–4553/09/050949–20
DOI 10.1007/s00024-009-0480-7

Pure and Applied Geophysics

Improved Interpretation of Nuclear Magnetic Resonance T_1 and T_2 Distributions for Permeability Prediction: Simulation of Diffusion Coupling for a Fractal Cluster of Pores

H. Pape[1,2] and C. Clauser[1]

Abstract—NMR relaxometry is a powerful tool for inferring porosity and permeability data. In practice, measured magnetization decay curves are inverted for relaxation time distributions. Subsequently, one presumes a linear relationship between the pore radius distribution and the T_1 and T_2 distribution, for longitudinal and transverse magnetization, respectively. The fundamental equations used are based on a pore model, in which pores are assumed to be isolated from each other with respect to the NMR process and have smooth walls. The present study is based on a geometrical pore space model with connected pores and structured pore walls. The physical processes of surface relaxation, irreversible dephasing of magnetic spins and diffusive proton exchange between pores, are described by a system of differential equations. The solution yields a set of exponential functions representing the relaxation time distribution. We describe the difference between the distributions obtained for diffusion coupling and for isolated pores. With diffusion coupling on, the spectral width of the T_1 distribution is strongly reduced, which indicates that the influence of large and small radii according to the T_1-pore radius relationship is mixed to some extent. For a fractal pore space structure, where large pores are surrounded by adjacent minor pores, the T_1 distribution does not resolve these substructures. Nevertheless, permeability values calculated from the logarithmic mean relaxation time $T_{1,LM}$ are quite the same for diffusion coupling and for isolated pores. The T_2 distribution for diffusion coupling is little constricted and gives a better resolution of the pore wall structures than the corresponding T_1 distribution. The permeability values from T_2 distributions agree with the values from longitudinal magnetization, provided that we use a corrected relaxation time $T_{2,corr}$, accounting for the dependence of the surface relaxivity ρ_2 on pore radius. The study shows that radius distributions calculated from a T_1 and from a T_2 distribution differ from one another and both present an altered image of the true pore radius distribution. In practice, this has no serious influence on estimating permeability of medium- to high-permeability sandstones with the currently applied methods. The presented methodology of calculating the NMR response of pore space models with diffusion coupling may facilitate understanding porosity-permeability relationships of different rock types such as carbonate rocks with micro-porosity.

Key words: NMR, relaxation, diffusion, pore radius distribution, permeability, fractals.

1. *Introduction*

When considering physical processes in porous media, models with the highest degree of simplification are constructed. For example, pores of uniform pore size can be

[1] Applied Geophysics and Geothermal Energy E.ON Research Center of Energy, RWTH Aachen University, Germany.

[2] now at: Eichendorffstr. 10, D-30952 Ronnenberg, Germany. E-mail: h.pape@geophysik.rwth-aachen.de

assumed. However, natural rocks are better described by models with a pore size distribution. The capillary bundle model, for instance, is such a simple model for the interpretation of mercury intrusion measurements, neglecting the constrictions of pores and the network structure. Similarly, the relationship between pore radius distribution and distribution of NMR relaxation times is described by a linear equation valid for isolated pores. In this study, we simulate relaxation time distributions based on a fractal model with connected pores.

1.1. The Role of Diffusion Coupling between Pores of Different Size for the Interpretation of NMR Data

NMR magnetization relaxation measurements provide a measure of porosity and an estimate of permeability of reservoir rocks. Besides bulk porosity, different contributions, such as free water, capillary-bound water and clay-bound water, can be calculated from NMR data. The interpretation is based on the relationship between the relaxation time T of the exponential magnetization decay curve and the specific surface S/V of the pores in a porous medium.

$$\frac{1}{T_{1,2}} = \rho_{1,2}\frac{S}{V}, \tag{1}$$

where S and V are pore surface and volume, $\rho_{1,2}$ is surface relaxivity, and $T_{1,2}$ the relaxation time with index 1 and 2 for longitudinal and transverse magnetization, respectively.

S/V is inversely proportional to the pore radius r where the factor depends on the pore shape:

$$\frac{S}{V} = \frac{a}{r} \text{ with a} = 2 \text{ for cylindrical capillaries and a} = 3 \text{ for spheres.} \tag{2}$$

Then, permeability k can be calculated from porosity ϕ and the Kozeny-Carman equation (Kozeny, 1927; Carman, 1937, 1948, 1956).

$$k = \frac{\phi}{8T_{\text{hydr}}}r^2, \tag{3}$$

where T_{hydr} is the hydraulic tortuosity. In another study (Pape *et al.*, this issue), we substituted T_{hydr} with $1/\phi^{1.2}$.

However, natural rocks such as sandstones are characterized by a large range of pore sizes. This is expressed by the pore radius distribution curve where each radius class is represented by its volume. Accordingly, the decay curve is the integral of exponential terms characteristic for each of these classes. Therefore, a T_1 distribution can be calculated from the measured decay curve by Laplace transformation. According to equation (1), the T_1 distribution is identical to the pore radius distribution, if pores of different size are not connected. For calculating permeability, the logarithmic mean $T_{1,\text{LM}}$

or the logarithmic mean r_{LM} can be used (PAPE *et al.*, this issue). In practice, the transverse magnetization, which yields the T_2 distribution instead of the T_1 distribution, is usually measured rather than the longitudinal magnetization. This complicates the interpretation in so far as the surface relaxivity ρ_2 is not a constant, but decreases with pore radius. In the case of Raethian sandstone from the Allermoehe borehole near Hamburg (PAPE *et al.*, this issue), ρ_2 can be described by an empirical relationship:

$$\rho_2 = \rho_{2,\mathrm{eff}} \left(\frac{9.53}{3\rho_{2,\mathrm{eff}} T_2} \right)^{0.3} = \rho_{2,\mathrm{eff}} \left(\frac{9.53}{r} \right)^{0.3/0.7}, \quad \text{for } r < 9.53, \text{or } T_2 < 9.53/\left(3\rho_{2,\mathrm{eff}}\right), \tag{4}$$

where $\rho_{2,\mathrm{eff}}$ is the relaxivity for pores of radius 9.53 μm.

Up to this point we have summarized the general practice of interpretation of NMR relaxation data. This is based on the assumptions of isolated pores and the fast diffusion limit. The first assumption may be questioned because a permeable porous medium requires connected pores for sustaining fluid flow and all studies of the pore system of sandstones based on various methods show a network structure. On the other hand, this structure can be described by an arrangement of larger voids connected by constricted throats or "bottle necks". The interpretation of pore radius distributions from mercury intrusion measurements, e.g., is based on this concept. In the case of NMR relaxometry, the pore throats reduce the interchange of diffusing water molecules more or less effectively. According to another study (PAPE *et al.*, this issue), this effect will be more pronounced for T_2 than for T_1 measurements, because relaxation of transverse magnetization is influenced by an interlayer of strong field gradients along the pore walls where the magnetic spins are irreversibly dephased. In this case, diffusion coupling is measured only on those water molecules which traverse the pore throat through a small channel surrounded by the interlayer. In order to study how diffusion between pores affects the T_1 and the T_2 distributions we performed forward simulations on a geometrical pore space model with a fixed number of pore radius classes. We obtained as many relaxation times and corresponding coefficients as there are classes. In this way, we calculated directly the exponential terms, the sum of which constitutes the decay curve and which represent the accurate spectral lines of the relaxation time distribution. Conversely, inverse Laplace transformation or inversion of a decay curve would yield a similar but not the same result, if the distribution of the exponents is determined beforehand and the corresponding coefficients are calculated.

Our study was inspired by thorough investigations of the topic by ARNS *et al.* (2005) based on digitized core images of various sandstones. They calculated petrophysical properties directly from 3-D images of the pore system and simulated NMR signals by a random walk method. The acquired decay curves of transverse magnetization were interpreted by Laplace transformation to calculate the corresponding T_2 distribution. In order to study the effect of diffusion coupling, the pore system of the 3-D image was partitioned into pore bodies. Then, simulations with "diffusion on" and with "diffusion off" between the pore bodies were compared with respect to the interpreted T_2 distributions. The general effect of diffusion coupling was a "sharpening" of the distribution curve, which means that the "diffusion on" curve was compressed from both sides compared to that corresponding to the "diffusion off" run.

1.2. A Fractal Pore Space Model as a Base for Simulation of NMR Data

The relationship between surface area and pore volume according equation (2) is valid for smooth pore surfaces. Previously, we established a fractal pore space model for describing electrical and hydraulic properties of porous rocks (Pape *et al.*, 1982, 1984, 1987a, b, 1999, 2000, 2005). It was also used for deriving the pore geometry of sandstone from pulsed field gradient NMR (Pape *et al.*, 2006). From surface area measurements of sandstone with different resolution power, Pape *et al.* (1967a) calculated a mean value of 2.36 for the fractal dimension D. According to Mandelbrot (1977), the fractal dimension of a self-similar pore model is given by

$$D = \frac{\log N}{\log \frac{1}{v}}, \tag{5}$$

where N is the number of substructures and v is the ratio of the size parameters of the substructure and of the larger structures. Figure 1 presents a cluster of pores which was used for the simulations of NMR response. This model consists of connected pores on the larger scale which allows the study of diffusion coupling. Moreover the model is fractal, as the pores are surrounded by self-similar substructures. The cluster consists of 13 classes of pores with different radii; each class represented by one pore in the figure. According to the multipliers, written at the sides of the subclusters and substructures, the total number of pores is 341. The geometrical parameters of the model are presented by Table 1. For the substructures of classes 2 and 3, we find $N = 16$ and radius ratio $1/v = 3.2$ which yields $D = 2.38$ from equation (5).

1.3. Analytical Calculation of NMR Data for the Fractal Model

The function of magnetization is $P(t)$, normalized to 1 for $t = 0$. For model with n pore classes $P(t)$ is the sum of the individual magnetization functions $P_i(t)$ of each pore class.

$$P(t) = \sum_{i=1}^{n} P_i. \tag{6}$$

If the pores are isolated from pores of different size, P_i are exponential functions (Kenyon, 1992).

$$P_i = V_i \exp\left(-\frac{t}{T_i}\right), \tag{7}$$

where V_i denotes the relative volume of the pore class i and $\sum_{i=1}^{n} V_i = 1$.

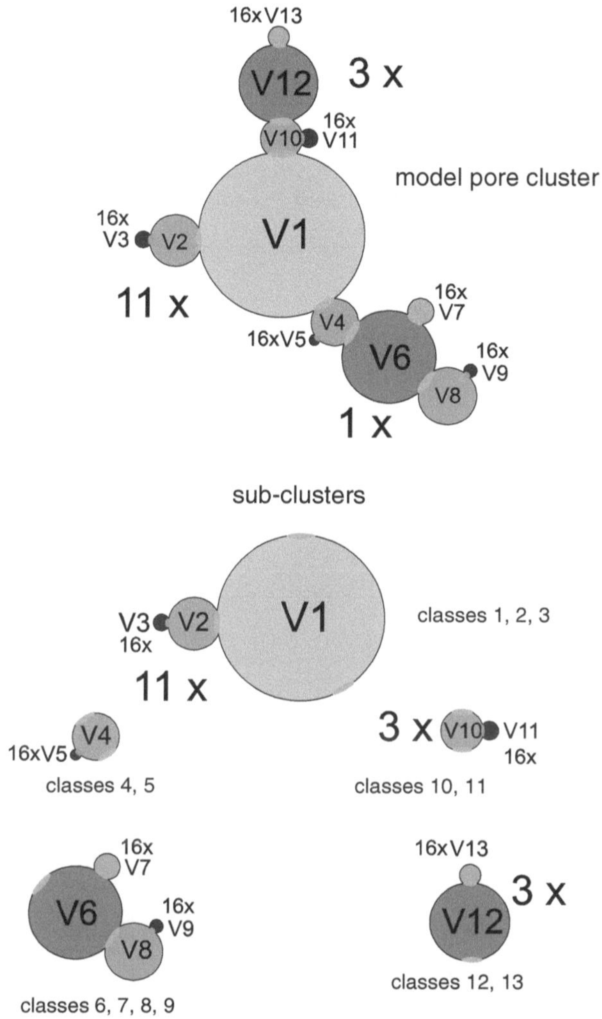

Figure 1
Schematic model of a cluster of connected pores. Each pore class consists of pores with the same radius and is represented by one pore with the volume V_i of the respective pore class. The number of substructures connected to the central pore is explained in the figure. The total cluster is subdivided into 5 sub-clusters for the study of relationships of individual relaxation time distributions due to diffusion coupling.

The relaxation time $T_{1,i}$ is calculated from equations (1) and (2) based on the surface relaxivity ρ_1 and geometrical parameters of the pore. In the case of $T_{2,i}$, the surface relaxivity is ρ_2 which depends on the pore radius according to equation (4).

Table 1

Geometrical parameters of the pore cluster model and the relaxation times for isolated pores, calculated with surface relaxivities of Table 2

Pore class	Number of pores	Relative volume of class = coefficient b	Pore radius [μm]	Radius of cross section area of pore connection			Isolated pores	
				Connection with larger pores	Longitudinal magnetization [μm]	Transverse magnetization [μm]	T_1 [ms]	T_2 [ms]
1	1	0.422	31.62	—	—	—	2635	753
6	1	0.075	17.78	—	—	—	1482	331
12	3	0.135	15.00	—	—	—	1250	260
8	1	0.018	11.00	class 6	2.7500	1.7500	917	167
2	11	0.147	10.00	class 1	2.5000	1.5000	833	145
4	1	0.010	9.00	classes1, 6	2.2500	1.2500	750	125
10	3	0.021	8.00	classes 1, 12	2.0000	1.0000	667	106
7	16	0.027	5.00	class 6	1.2500	0.7500	417	54
13	48	0.041	4.00	class 12	1.0000	0.5000	333	39
11	48	0.027	3.50	class 10	0.8750	0.3750	292	32
3	176	0.072	3.13	class 2	0.7800	0.2800	261	28
9	16	0.003	2.50	class 8	0.6250	0.2250	208	20
5	16	0.002	2.00	class 4	0.5000	0.1000	167	15

1.4. Simulation of NMR-relaxation Using a Corresponding Electrical Circuit

In the case of connected pores, surface relaxation is coupled with diffusion of protons. Both processes can be described by differential equations which have the same form as electrical circuit equations. The circuit analogy consists of capacitors and resistors, where a capacitor represents an elastic member such as a spring and a resistor stands for a friction member like a dashpot.

The original equations for the time derivatives of the amount of activated protons p and for the concentration c of activated protons are:

$$\frac{dp}{dt} = -\left(\rho_{1,2}\frac{S}{V}\right)p \quad \text{and} \quad V\frac{dc}{dt} = -(\rho_{1,2}S)c. \tag{8}$$

The analogous electrical differential equations describe decharging of a capacitor through a resistor connected to the earth :

$$\frac{dQ}{dt} = -R_a^{-1}\frac{Q}{C} \quad \text{or} \quad I = -R_a^{-1}U \quad \text{and} \quad C\frac{dU}{dt} = -R_a^{-1}U, \tag{9}$$

with the charge Q, current I, capacity C, resistance R_a and voltage U.

For simplicity, the diffusion coupling between two pores with the indices 1 and 2 is described by Fick‘s first law:

$$\frac{dp_1}{dt} = -\frac{dp_2}{dt} = -\left(\frac{D_{\text{diff}}A_{\text{cross}}}{l_{\text{diff}}}\right)(c_1 - c_2) \text{ and } \frac{V_1c_1}{dt} = -\frac{V_2c_2}{dt} = -\left(\frac{D_{\text{diff}}A_{\text{cross}}}{l_{\text{diff}}}\right)(c_1 - c_2), \tag{10}$$

where D_{diff} is the diffusion coefficient, l_{diff} is a length, and A_{cross} is the cross-sectional area.

The corresponding electrical equations for two capacitors connected by a resistor are of the same form:

$$\frac{dQ_1}{dt} = -\frac{dQ_2}{dt} = -R_b^{-1}(U_1 - U_2) \quad \text{and} \quad \frac{C_1dU_1}{dt} = -\frac{C_2dU_2}{dt} = -R_b^{-1}(U_1 - U_2), \tag{11}$$

with the resistance R_b.

Figure 2 shows the electrical circuit corresponding to the pore cluster of Figure 1.

Surface relaxation of a single pore can be represented by an RC-unit (Fig. 3) where the electrical charge Q is equivalent to the amount of activated protons, i.e., the measurable magnetization. The capacitance C is equivalent to the pore volume V and the resistance R is equivalent to

$$R_{\text{NMR}} = (\rho_{1,2}S)^{-1}. \tag{12}$$

We obtain the relaxation time $T_{1,2}$:

$$T_{1,2} = RC = (\rho_{1,2})^{-1}\frac{V}{S}, \tag{13}$$

which corresponds to equation (1).

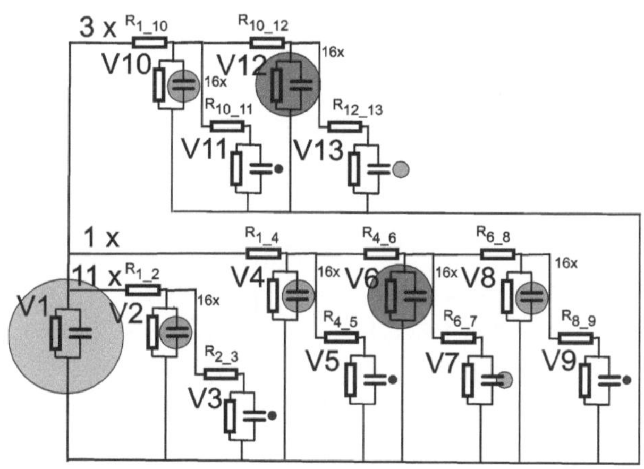

Figure 2
Electrical circuit corresponding to the pore cluster of Figure 1. Each pore class with volume V_i is represented by an RC-unit for describing surface relaxation. The connections between pores of class i and class j are represented by a resistor R_{i_j} describing the diffusion coupling.

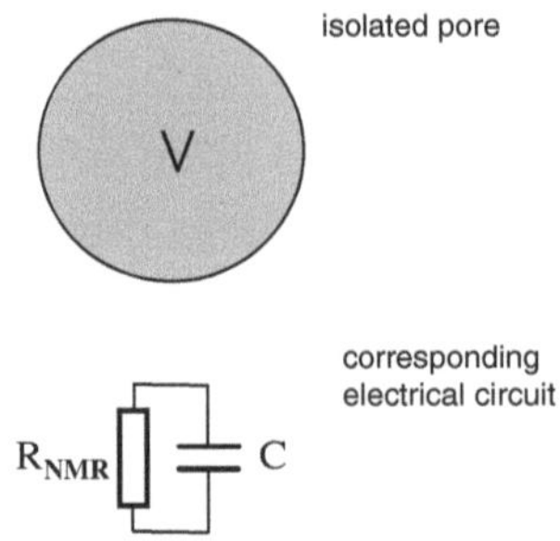

Figure 3
RC-unit corresponding to an isolated pore for describing surface relaxation.

In the next step we connect this pore to a second one with a different size (Fig. 4). The corresponding electrical circuit consists now of two RC-units connected by an additional resistor characterizing the diffusion process:

$$R_{\mathrm{diff},1\_2} = D_{\mathrm{diff}}^{-1} l_{\mathrm{diff}} (A_{\mathrm{cross}})^{-1}, \tag{14}$$

where D_{diff} is the diffusion coefficient, l_{diff} is the distance between the centers of the two pores, and A_{cross} is the cross-sectional area of the throat connecting the pores (Fig. 4).

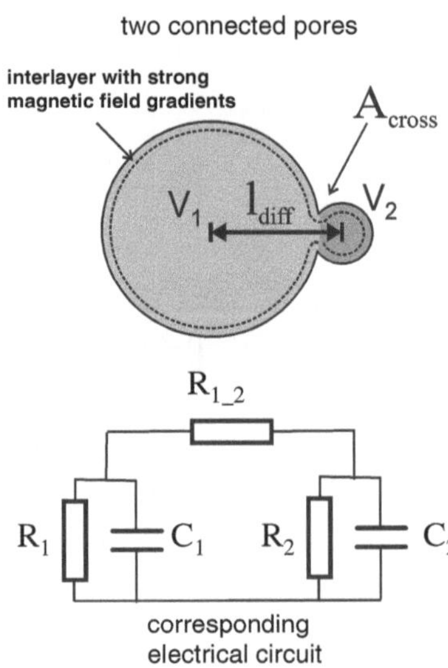

Figure 4
Electrical circuit for describing surface relaxation coupled with diffusion across the pore throat between two pores of different size. The upper cartoon shows diffusion length l_{diff} and the pore neck cross-section area A_{cross} used for numerical calculations. An interlayer of strong magnetic field gradients causes irreversible dephasing in the case of measured transverse magnetization.

If transversal magnetization is measured, the radius of the cross section is reduced by the thickness of the interlayer of strong magnetic field gradients.

On the basis of the analogous electrical circuit it is easy to establish a system of differential equations which yields n decay functions P_i related to n pore classes. The P_i's are summed up to yield the decay function of the cluster. Each function P_i itself is the sum of n exponential functions $P_{i,j}$.

$$P = \sum_{i=1}^{n} P_i = \sum_{i=1}^{n}\sum_{j=1}^{n} P_{i,j} = \sum_{i=1}^{n}\sum_{j=1}^{n} b_{i,j}\exp\left(\frac{t}{T_j}\right) = \sum_{j=1}^{n}\left(\sum_{i=1}^{n} b_{i,j}\right)\exp\left(\frac{t}{T_j}\right)$$

$$P = \sum_{j=1}^{n} b_j \exp\left(\frac{t}{T_j}\right) \quad \text{with} \quad b_j = \sum_{i=1}^{n} b_{i,j}, \tag{15}$$

where the decay functions of the pores and of the cluster share the same relaxation times T_{j}.

The relative volume V_i of a pore class of index i is given by

$$V_i = \sum_{j=1}^{n} b_{i,j}, \tag{16}$$

which corresponds to the initial condition $P_{\mathrm{i},t=0} = V_{\mathrm{i}}$.

As an example, we take a small cluster of three pore classes with the corresponding electrical circuit (Fig. 5). At each node of the circuit, the currents of the conductor branches add to zero. For instance, in the case of node 1 in Figure 5, this yields:

$$\frac{C_1 dU_1}{dt} = -\frac{U_1}{R_{\mathrm{NMR},1}} - \frac{11(U_1 - U_2)}{R_{\mathrm{diff},1\_2}}. \tag{17}$$

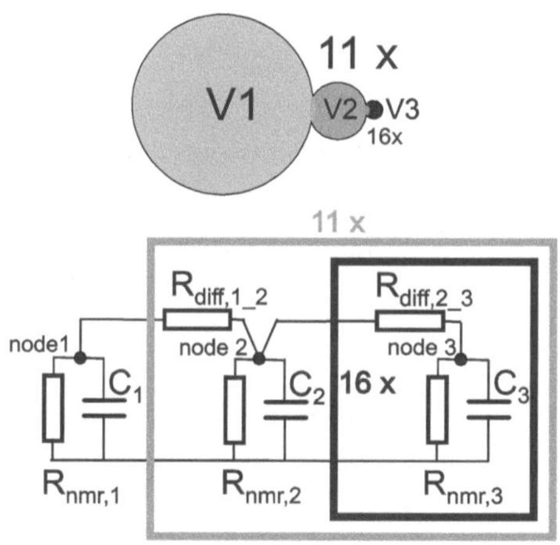

Figure 5
Cartoons of a cluster with three pore classes and the corresponding electrical circuit.

The three nodes yield a system of differential equations for voltages U_i of the capacitors C_i:

$$Y' = AY$$

$$\begin{bmatrix} U_1' \\ U_2' \\ U_3' \end{bmatrix} = A[\,U_1 \quad U_2 \quad U_3\,]$$

$$A = \begin{bmatrix} \dfrac{R^{-1}_{\mathrm{NMR},1} + 11R^{-1}_{\mathrm{diff},1\_2}}{C_1} & \dfrac{-11R^{-1}_{\mathrm{diff},1\_2}}{C_1} & 0 \\ \dfrac{-R^{-1}_{\mathrm{diff},1\_2}}{C_2} & \dfrac{R^{-1}_{\mathrm{NMR},2} + R^{-1}_{\mathrm{diff},1\_2} + 16R^{-1}_{\mathrm{diff},2\_3}}{C_2} & \dfrac{-16R^{-1}_{\mathrm{diff},2\_3}}{C_2} \\ 0 & \dfrac{-R^{-1}_{\mathrm{diff},2\_3}}{C_3} & \dfrac{R^{-1}_{\mathrm{NMR},3} + R^{-1}_{\mathrm{diff},2\_3}}{C_3} \end{bmatrix}, \tag{18}$$

where Y is the row vector of the U_i, Y' the column vector of the time derivatives of U_i and C_i corresponds to the volume (4/3) $\pi\ r_\mathrm{i}^3$.

This system is solved for the initial condition $U_{\mathrm{i},t=0} = 1$. The relaxation times T_j for the system of U_i are the same as for P_i . The coefficients $b_{\mathrm{i,j,U}}$ have to be calculated for the initial condition $U_{\mathrm{i},t=0} = 1$ which yields the decay curves

$$U_i = \sum_j b_{i,j,U} \exp\left(\frac{t}{T_j}\right). \tag{19}$$

The calculation of the decay functions P_i of equation (16) with $b_{\mathrm{i,j}} = V_\mathrm{i}\, b_{\mathrm{i,j,U}}$ follows from UC = Q and the correspondences $Q_i \hat{=} P_i$ and $C_i \hat{=} V_i$.

2. Results

2.1. Simulation of NMR Data for Longitudinal Magnetization

The simulation of NMR data was performed on the basis of the model of Figure 1 for longitudinal magnetization with geometrical parameters of Table 1 and physical parameters of Table 2. For comparison, the T_1 distribution was calculated for the case without diffusion coupling.

Each relaxation time distribution for 13 pore classes is expressed by 13 relaxation times and their coefficients. As was expected, the relaxation times T_j for diffusion coupling differs from the set for isolated pores. The coefficients $b_j = \sum_{i=1}^{n} b_{i,j}$ for "diffusion on" are different from the set $b_i = V_i$ for "diffusion off". We are accustomed that all coefficients of $T_{1,2}$ distributions derived from decay curves are positive. However, our analytical simulation yielded some of the coefficients $b_{\mathrm{i,j}}$ and b_j with a negative sign.

Table 2

Physical parameters for calculating NMR data

Diffusion coefficient D [$\mu m^2/s$]	Longitudinal magnetization	Transverse magnetization		
	Surface relaxivity ρ_1 [$\mu m/s$]	Surface relaxivity $\rho_{2,eff}$ [$\mu m/s$]	Pore radius [μm]	Surface relaxivity ρ_2 [$\mu m/s$]
2299	4	23.40	31.62	23.40
			17.78	23.40
			15.00	23.40
			11.00	23.40
			10.00	23.40
			9.00	23.98
			8.00	25.22
			5.00	30.85
			4.00	33.95
			3.50	35.95
			3.13	37.71
			2.50	41.52
			2.00	45.69

This is attributed to relaxation caused by diffusion coupling. Considering a single pore, surface relaxation alone yields a decrease of activated protons, described by an exponential function with a positive coefficient. A relaxation process based mainly on the diffusion of protons from one pore class into the other yields a decrease of activated protons for the one class characterized by a positive coefficient and an increase for the other class characterized by a negative coefficient. Thus, the receiving class i gets a negative coefficient $b_{i,j}$ of the relaxation time T_j. In Figure 5 we illustrate these relationships for the model pore cluster which we split into sub-clusters as figured in Figure 1. To each sub-cluster the decay curves $P_{i,j}$ of equation (11) are summed up for the indices of the pore classes which belong to this sub-cluster. Without diffusion coupling, each sub-cluster has as a distribution with as many relaxation times as pore classes, and each pore class has one relaxation time. With diffusion coupling, on the other hand, each pore class and each sub-cluster has the same number of relaxation times as the total cluster. However, note that lines with coefficients lower than 0.001 are not visible in the plots.

The plots of T_1 distributions of the sub-cluster are arranged such that the positive and negative coefficients of diffusion coupling between pairs of sub-clusters are highlighted for the corresponding relaxation times.

First, we discuss the T_1 distribution of the cluster with pore classes 1, 2, and 3. There are two lines for relaxation times of 121 ms and 287 ms with large coefficients of opposite sign. For the other sub-clusters in Figure 5, the coefficients of these relaxation times are very small. This implies that the diffusion process acts inside the cluster with pore classes 1, 2, and 3. The pair of corresponding exponential functions describes the diffusion coupling for short decay time t between the central pore with the pore classes 2 and 3, which are the sub-structures of the first- and second-order relative to the central pore. For $t = 0$, concentration of activated protons is equal in each pore class. According to equation (1) the relaxation of magnetization, i.e., concentration of activated protons, is faster for larger specific surface S_i/V_i, i.e., smaller pore radius r_i. Due to surface relaxation, concentration deceases from the central pore to class 2 and on to class 3. Therefore, diffusion proceeds in the same direction. In other words, a portion of the initial protons of the central pore loose their energy at the pore walls of the fractal substructures. For longer times t, the two lines with very short relaxation times can be neglected, as the corresponding exponential functions decay very fast. Subsequently, the concentrations of the three pore classes approximate equilibrium values and the exponential functions with the relaxation times 499 ms and 2089 ms predominate.

Next, we compare the relationships between the distributions of the sub-clusters (Fig. 6). For some relaxation times we find a negative coefficient in sub-clusters where the coefficient in the neighboring sub-clusters is positive. This is an indication of diffusive coupling. For instance, the sub-cluster with classes 1, 2, and 3 is coupled at relaxation time 611 ms with the sub-clusters in the right panel of Figure 6 and at the relaxation times 322 ms, 410 ms, and 1028 ms with the sub-clusters in the left panel of Figure 6.

For longer times t, the concentrations in the pore classes approximate equilibrium values and the decay functions of all sub-clusters are dominated by the four shared relaxation times at 499 ms, 884 ms, 1028 ms and 2089 ms (indicated by dotted lines in Fig. 5).

After this detailed study of the contributions of different sub-clusters we discuss the T_1 distribution of the entire cluster (Fig. 7). We compare the distribution for isolated pores with the distribution accounting for diffusion coupling. Generally, the “diffusion on” distribution (Fig. 7, center panel) is shifted towards smaller relaxation times compared to the “diffusion off” distribution (Fig. 7, top panel). Three relaxation times have negative coefficients. However, these lines are counterbalanced by lines with positive coefficients such that all lines below 700 ms disappear, apart from the line at 499 ms with a reduced coefficient (Fig. 7, bottom panel and Table 3). The result is the distribution on the bottom of Figure 7. The decay curve is composed of four exponential curves with the relaxation times which we identified in Figure 6 as the dominating shared ones. The resulting distribution is very compressed, which is referred to as “sharpening” by ARNS *et al.* (2005).

2.2. *Simulation of NMR Data for Transverse Magnetization*

Figure 8 shows the T_2 distributions for sub-clusters of the model cluster for diffusion coupling and for isolated pores. The NMR-relaxation process of transverse magnetization comprises irreversible dephasing of magnetic spins in addition to the surface relaxation at the pore walls. The irreversible dephasing is caused by large magnetic field gradients within an interlayer. Both processes act together for the decay of transverse magnetization, and therefore ρ_2 is larger than ρ_1. As a consequence, the corresponding T_2

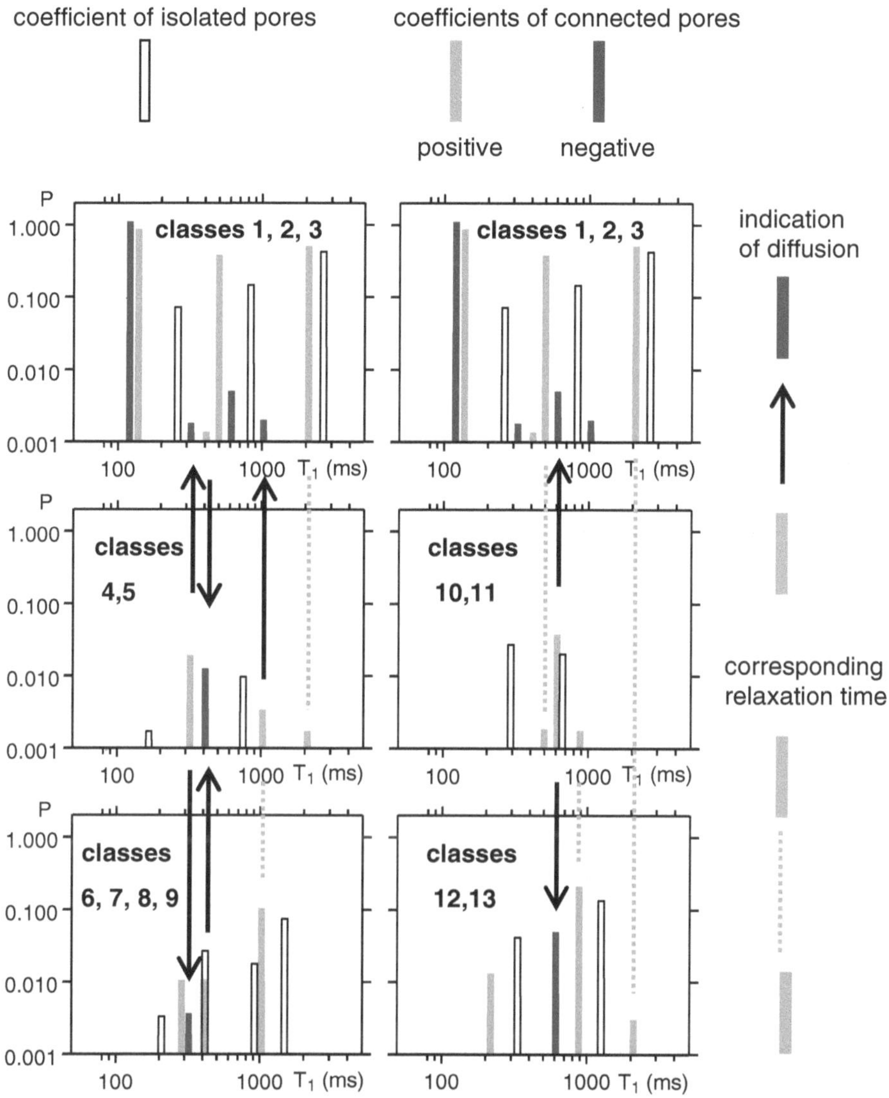

Figure 6

T_1 distributions of sub-clusters of the pore model according to Figure 1. Each sub-cluster has a relaxation time distribution comprising 13 lines with common relaxation times for all sub-clusters, but with different positive or negative coefficients. Spectral lines with coefficients smaller than 0.001 are not shown. The relationships between the distributions of sub-clusters indicated by arrows are due to diffusion coupling and are explained in the text.

relaxation times of isolated pores are smaller than the T_1 relaxation times. This shift is more pronounced for smaller pores than for larger pores. In the previous chapter, we found that diffusion coupling generally causes a shift towards shorter relaxation times.

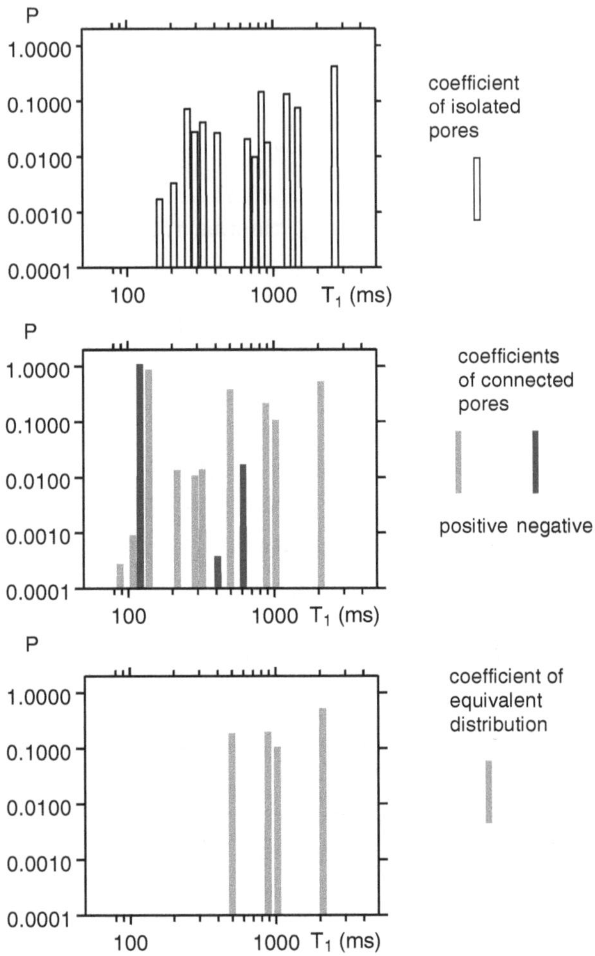

Figure 7
T_1 distributions of the entire pore cluster with connected pores according to Figure 1: Raw results from the differential equations (center) and the "sharp" distribution obtained by balancing positive and negative of neighboring spectral lines (bottom). For comparison, T_1 distribution calculated for isolated pores only (top).

For transverse relaxation, the shift due to diffusion coupling is smaller because the pores are constricted by a strong gradient interlayer. As a consequence, the cross section of the pore throats is reduced by the effective thickness of the interlayer. This reduces the effect of diffusion coupling.

We begin with the interpretation of the T_2 distribution of the sub-cluster of classes 1, 2, 3 (Fig. 7, top panel). The lines of the relaxation times 27 ms, 137 ms, and 442 ms with large coefficients correspond to the lines with 28 ms, 142, and 450 ms of isolated pores of the three pore classes of the sub-cluster. The shift between the corresponding lines is very

Table 3

Calculated T_1 and T_2 distributions

Longitudinal magnetization			Transverse magnetization		
Relaxation time $T_{1,j}$ [ms]	Coefficients b_j	Coefficients of equivalent distribution	Relaxation time $T_{2,j}$ [ms]	Coefficients b_j	Coefficients of equivalent distribution
88	0.0003		15	0.0017	0.0017
108	0.0009		20	0.0033	0.0033
121	−1.0944		27	0.0702	0.0702
139	0.8655		32	0.0262	0.0262
217	0.0134		39	0.0372	0.0372
287	0.0107		53	0.0255	0.0255
322	0.0137		96	0.0293	0.0293
410	−0.0004		121	0.0048	0.0048
499	0.3751	0.1847	137	0.1432	0.1432
611	−0.0169		153	5.5896	0.1529
884	0.2116	0.1948	208	−5.4367	0.0000
1028	0.1046	0.1046	244	0.0774	0.0774
2089	0.5159	0.5159	442	0.4283	0.4283

small. This indicates that diffusion coupling plays a minor role with respect to the fractal structure of this sub-cluster. Before protons from pore class 1 or 2 reach the pore walls of pore class 3, they become irreversibly dephased in the interlayer of pore class 2.

The additional lines in the distribution of the sub-cluster of classes 1, 2, and 3 with positive or negative coefficients are due to diffusion coupling with the sub-clusters of classes 4 to 13.

2.3. Permeability Inferred from Relaxation Time Distributions

Figure 9 shows the T_1 and T_2 distributions of the model cluster calculated for "diffusion off" and for "diffusion on". The original T_2 distribution with diffusion coupling contains a line with a negative coefficient for $T_2 = 207$ ms which was cancelled by the coefficient of the line at $T_2 = 153$ (Table 3 and Fig. 8, bottom panel).

It is common practice to use the logarithmic mean $T_{2,\mathrm{LM}}$ for inferring permeability from relaxation time distributions (KENYON, 1992, 1997):

$$k = a\phi^b T_{2,\mathrm{LM}}^2, \tag{20}$$

with the standard values a = 1, b = 4.

According to another study (PAPE *et al.*, this issue), permeability values calculated from equation (20) and b = 4 are too low compared to permeability measured in low porosity sandstone. Therefore, we prefer the Kozeny-Carman equation (3) (KOZENY, 1927; CARMAN, 1937, 1948, 1956) with the effective hydraulic pore radius r_{eff} substituted for pore radius. The effective pore radius can be calculated from $T_{2,\mathrm{LM}}$ (see equation (21) below).

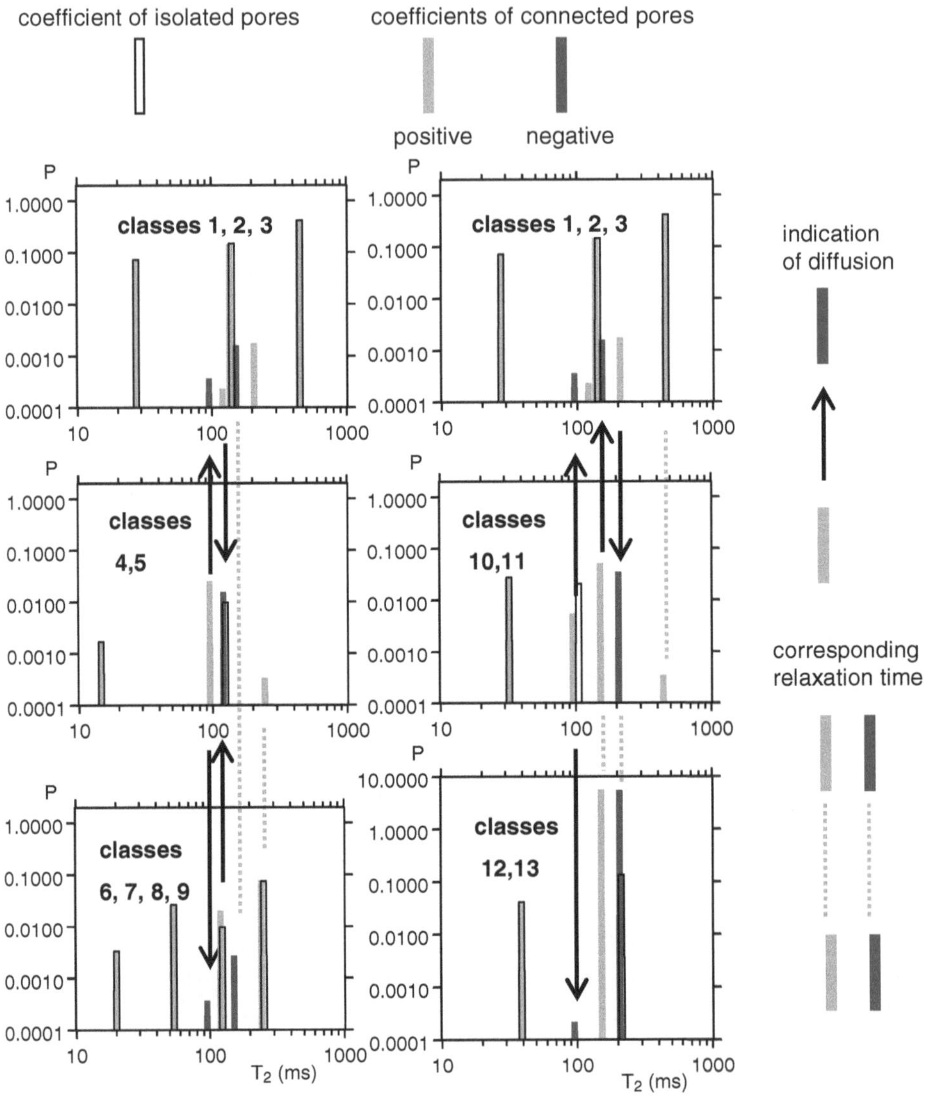

Figure 8

T_2 distributions of sub-clusters of the pore model according to Figure 1. This figure corresponds to Figure 6 for the T_1 distributions. Spectral lines with coefficients smaller than 0.0001 are not shown. The relationships between the distributions of sub-clusters compared to Figure 6 are discussed in the text.

As a result of our numerical simulation we obtained four relaxation time distributions, T_1 and T_2 distributions for isolated pores and for diffusion coupling (Fig. 9) and discussed above the differences between these distributions. It remains to assess the impact of the above considerations on predicting permeability.

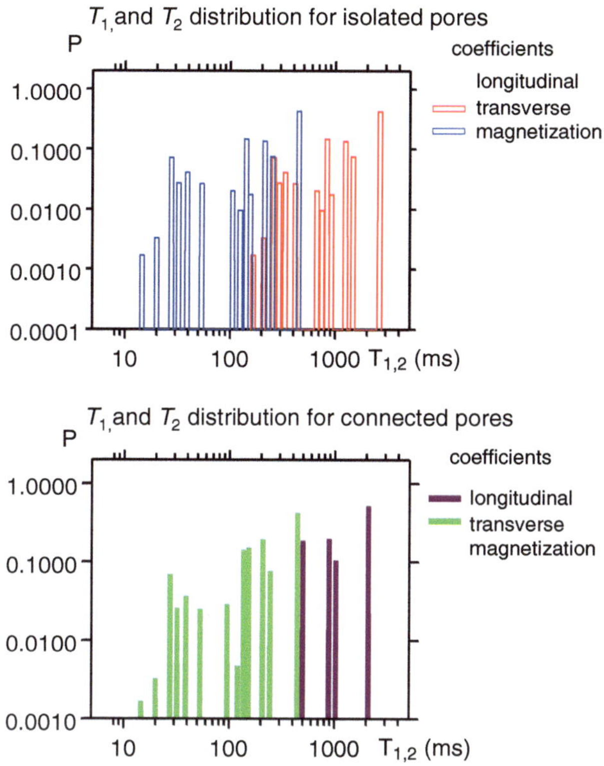

Figure 9

T_1 and T_2 distributions for the pore model with isolated pores (top) and with connected pores and diffusion coupling (bottom). The negative coefficients resulting from solving the systems of differential equations were cancelled by positive coefficients of adjacent spectral lines.

First, we calculated r_{eff} from equations (1) and (2):

$$r_{\text{eff}} = 3\rho_{1,2}T_{1,2,\text{LM}}. \tag{21}$$

For r_{eff} values larger than 9.5 we identify ρ_2 with $\rho_{2,\text{eff}}$ according to equation (4). Then, permeability k is calculated from equation (3) for a sandstone with a porosity of $\phi = 0.1$. The results are presented in Table 4. The T_1 distributions for isolated pores and for diffusion coupling are quite different due to the sharp compression in the latter case. Nevertheless, the corresponding r_{eff} values and also permeabilities are nearly identical; the values for diffusion coupling being a little larger. For transverse magnetization, the effective pore radius and the permeability values with and without diffusion coupling are also similar; the corresponding values for diffusion coupling being a little smaller. However, the r_{eff} values calculated from $T_{2,\text{LM}}$ are distinctly smaller than the r_{eff} values from $T_{1,\text{LM}}$, and therefore permeabilities derived from $T_{2,\text{LM}}$ are smaller than the

Table 4

Permeability from T_1 and T_2 distributions of the model pore cluster, calculated with equation (3) for a sandstone with porosity $\phi = 0.1$

	Longitudinal magnetization		Transverse magnetization	
	Isolated pores	Diffusion coupling	Isolated pores	Diffusion coupling
Logarithmic mean relaxation time $T_{1,2,\mathrm{LM}}$ [ms]	1249	1259	198	189
Pore radius r_{eff}, calculated from $T_{1,2,\mathrm{LM}}$ [μm]	14.99	15.11	13.92	13.25
Permeability k, calculated from r_{eff} [10^{-15} m^2]	**135**	**137**	116	105
Logarithmic mean pore radius r_{LM}, calculated from T_2 [μm]			14.99	14.27
Permeability k, calculated from r_{LM} [10^{-15} m^2]			**135**	**122**

corresponding values derived from $T_{1,\mathrm{LM}}$. The explanation comes from the decrease of ρ_2 with pore radius. In order to bring the effective pore radius derived from the T_2 distribution into agreement with the pore radius derived from the T_1 distribution, we first calculate the pore radius distribution from the T_2 distribution and equations (1), (2), and (4) and then the logarithmic mean of pore radii $r_{\mathrm{eff,LM}}$:

$$\log\left(r_{\mathrm{eff,LM}}\right) = \frac{\sum_i b_i \log(r_i)}{\sum_i b_i}, \qquad r_i = 3\rho_2 T_i \tag{22}$$

where ρ_2 is calculated from equation (4).

The r_{eff} and k values calculated in this way from the T_2 distribution for isolated pores are identical with the values of longitudinal magnetization. The values for diffusion coupling are also shifted into the proper range (Table 4).

3. *Summary and Conclusions*

We calculated the NMR relaxation of transverse and longitudinal magnetization for a fractal pore model by solving a system of differential equations. This represents the physical processes of surface relaxation, irreversible dephasing and diffusive proton exchange between pores. The analytical calculation yields the relaxation times of the system and the coefficients of the exponential functions which contribute to the decay curves of individual pore radius classes, of sub-clusters and of the entire studied pore cluster. Negative coefficients indicate diffusion coupling. A comparison of relaxation time distributions of pores and sub-clusters indicates coupling between the pore classes.

The T_1 and T_2 distributions with positive and negative coefficients were transformed into distributions of exclusively positive coefficients by balancing the coefficients of the opposite sign of neighboring relaxation times. These new distributions compare well with distributions calculated from measured decay curves. These distributions for diffusion coupling are compared to distributions for isolated pores of the same cluster. Diffusion coupling causes a "sharpening" of the distribution curve (ARNS *et al.*, 2005). In our simulations, the T_1 distribution was sharper than the T_2 distribution. Thus, for longitudinal magnetization the resolved pore radii of coupled pores are smoothed to some degree. Consequently, the fractal sub-structures of pores are not resolved by the T_1 distribution. In contrast, for transverse magnetization the resolution power is less reduced by diffusion coupling.

The "sharpening" effect can be neglected when calculating permeability from the logarithmic mean relaxation time $T_{1,2,LM}$ compared to the corresponding distribution for isolated pores. This corresponds to the results obtained by ARNS *et al.* (2005) for transverse relaxation. However, we found a discrepancy between permeability values calculated from longitudinal and transverse magnetization. Permeability calculated from the T_2 distribution is smaller, because the surface relaxivity ρ_2 depends on the pore radius. Therefore, we propose a modified way for calculating permeability which brings the results in agreement with those obtained for longitudinal magnetization. To this end, the T_2 distribution is converted into a pore radius distribution using a surface relaxivity ρ_2 which depends on the radius. Then, the logarithmic mean radius $r_{eff,LM}$ is calculated and can be inserted, e.g., into the Kozeny-Carman equation (KOZENY, 1927; CARMAN, 1937, 1948, 1956).

In conclusion, radius distributions calculated from a T_1 and from a T_2 distribution differ from one another and both deviate from the true pore radius distribution. If we compare them with the radius distribution from mercury injection, we must keep in mind that differences should arise because mercury injection attributes the entire pore volume accessible through a given pore neck to the neck radius and not to the pore size itself. Nevertheless, the methods for estimating permeability currently in use yield good results for average sandstones. Usually these methods extract one statistical parameter, for instance the logarithmic mean radius, from the distribution and use equations with parameters which are calibrated to core measurements. These methods may fail if the pore structure differs from average types of grain packing. Of special interest are rocks with low permeability compared to porosity. It was expected from NMR relaxometry that $T_{1,2}$ distributions would quantitatively yield all interesting details of fine structure such as micro-porosity. However, the volume of small pores and cavities at the pore walls may be obscured due to diffusion coupling. Another reason for underestimating the volume portion of small pore radii may lay in the way experimental NMR decay curves are inverted into $T_{1,2}$ distribution curves. Before applying an inversion program or inverse Laplace transformation, one must convert the time series of magnetization values, which is usually measured with equal linear time intervals, into a new time series with equal logarithmic time intervals. Otherwise the calculated distribution curve will be correctly adjusted for long relaxation times but not for short ones.

Carbonate rocks, in particular, reveal a large variety of pore structures, which makes it difficult to estimate permeability from NMR logs. In practice, this means that additional logs, such as gamma-ray and density logs, are required for the identification of rock type with respect to porosity-permeability relationship.

The methodology of calculating the NMR response of pore space models presented here may be applied to various pore models as a way to test their ability to explain the porosity-permeability relationship of special rock types.

References

Arns, C. H., Sheppard, A. P., Sok, R. M., and Knackstedt, M.A. (2005), *NMR petrophysical predictions on digitized core images*, Trans. SPWLA 46th Ann. Logg. Symp. 2005, paper MMM, 16 p.

Carman, P. C. (1937), *Fluid flow through granular beds*, Trans. Inst. Chem. Eng. London *15*, 150.

Carman, P. C. (1948), *Some physical aspects of water flow in porous media*, Discuss. Faraday Soc. *3*, 78.

Carman, P. C., *Flow of Gases through Porous Media* (Butterworth Scientific Publications, London 1956).

Kenyon, W. E. (1992), *Nuclear magnetic resonance as a petrophysical measurement*, Nuclear Geophysics *6* (2), 153–171.

Kenyon, W. E. (1997), *Petrophysical principles of applications of NMR logging*, The Log Analyst *38* (2), 21–43.

Kozeny, J. (1927), *Über die kapillare Leitung des Wassers im Boden (Aufstieg, Versickerung und Anwendung auf die Bewässerung)*, Sitz. Ber. Akad. Wiss. Wien, Math. Nat. (Abt. IIa) *136a*, 271–306.

Mandelbrot, B., *Fractals—Form, Chance and Dimensions* (Freeman, San Francisco 1977).

Pape, H., Arnold, J., Pechnig, R., Clauser, C., Talnishnikh, E., Anferova, S., and Blümich, B., *Permeability prediction for low porosity rocks by mobile NMR*. Pure Appl. Geophys., this issue.

Pape, H., Clauser, C., and Iffland, J. (1999), *Permeability prediction based on fractal pore space geometry*, Geophysics *64* (5), 1447–1460.

Pape, H., Clauser, C., and Iffland, J. (2000), *Permeability-porosity relationship in sandstone based on fractal pore space geometry*, Pure Appl. Geophys. *157*, 603–619.

Pape, H., Clauser, C., Iffland, J., Krug, R., and Wagner, R. (2005), *Anhydrite cementation and compaction in geothermal reservoirs: Interaction of pore-space structure with flow, transport, P–T conditions, and chemical reactions*, Int. J. Rock Mech. Min. Sci. *42*, 1056–1069.

Pape, H., Riepe, L., and Schopper, J. R. (1982), *A pigeon-hole model for relating permeability to specific surface*, The Log Analyst *23*, 1, 5-13; Errata: The Log Analyst *23* (2), 50.

Pape, H., Riepe, L., and Schopper, J. R. (1984), *The role of fractal quantities, as specific surface and tortuosities, for physical properties of porous media*, Particle Characterization *1*, 66–73.

Pape, H., Riepe, L., and Schopper, J. R. (1987a), *Theory of self-similar network structures in sedimentary and igneous rocks and their investigation with microscopical methods*, J. Microscopy *148*, 121–147.

Pape, H., Riepe, L., and Schopper, J. R. (1987b), *Interlayer conductivity of rocks — A fractal model of interface irregularities for calculating interlayer conductivity of natural porous mineral systems*, Colloids and Surfaces *27*, 97–122.

Pape, H., Tillich, J. E., and Holz, M., (2006), *Time-dependent diffusion coefficients in sedimentary rocks predicted by a fractal capillary model*, J. Appl. Geophys. *58* (3), 232–252.

(Received April 15, 2008, revised June 13, 2008, accepted June 20, 2008)
Published Online First: May 12, 2009

Pure appl. geophys. 166 (2009) 969–994
0033–4553/09/050969–26
DOI 10.1007/s00024-009-0484-3

Pure and Applied Geophysics

Oscillating Flow of a Compressible Fluid through Deformable Pipes and Pipe Networks: Wave Propagation Phenomena

Y. Bernabé[1,2]

Abstract—Similarly to blood pulse propagation in the artery system, oscillating flow can propagate as a wave in fluid-saturated pipes, networks of pipes or, by extension, in porous media, if the fluid is compressible and/or the pipes are elastically deformable. First, propagation of flow waves generated in a semi-infinite pipe by harmonic pressure oscillations at the pipe entrance is analyzed. The dispersion equation is derived, allowing determination of the phase velocity and quality factor as functions of frequency. Wave reflections at the end of a finite-length pipe and ensuing interferences between forward and backward traveling waves are then examined. Because of fluid storage in the pipe, the amplitude of the AC volumetric fluxes entering and exiting the pipe at its upstream and downstream ends are not equal. Thus, two different, upstream and downstream, frequency-dependent, AC hydraulic conductivities are introduced. Superposed on the classic viscous-inertial flow transition (controlled by the value of the pipe radius), these complex-valued parameters show another transition between an interference-free regime at low frequencies and a strong interference regime above a critical frequency that roughly scales as the pipe length. Because of attenuation, the flow wave interferences tend to gradually weaken with increasing frequencies. Finally, the single pipe model is used to investigate fluid flow waves through pipe networks with results very similar to those described above. The flow waves analyzed here are akin to the Biot's slow P waves and their propagation properties could affect seismic soundings in some geological settings.

Key words: Porous media, acoustic properties, fluid flow.

1. Introduction

Time-dependent flow of interstitial fluid can potentially provide more information about the structure of porous media than steady-state flow. For example, transient and oscillating flow methods have been used in the laboratory to measure permeability and storativity of rock samples simultaneously (e.g., Fischer, 1992; Fischer and Paterson, 1992; Hsieh *et al.*, 1981; Neuzil *et al.*, 1981; see also Song *et al.*, 2004). However, these methods rely on the assumption that the medium investigated is homogeneous. As a consequence, they fail to properly explain certain features observed in heterogeneous media. For instance, time-dependent flow through a stack of layers (i.e., *in series* arrangement of heterogeneities) is affected by the order in

[1] Earth, Atmospheric and Planetary Sciences Department, Massachusetts Institute of Technology, Cambridge, MA 02139, U.S.A.

[2] Institut de Physique du Globe de Strasbourg, Université Louis Pasteur – CNRS, France.

which the layers are placed (e.g., KAMATH *et al.*, 1992; SONG and RENNER, 2006). More generally, AC flow through a strongly heterogeneous medium depends on frequency (e.g., BERNABÉ *et al.*, 2004; SONG and RENNER, 2006). Even in homogeneous media, frequency-dependence arises from a transition from viscosity-controlled flow at low frequencies to inertia-controlled flow at high frequencies (e.g., CHARLAIX *et al.*, 1988; JOHNSON *et al.*, 1987). In most work on frequency-dependent, AC permeability (e.g., CHARLAIX *et al.*, 1988; JOHNSON *et al.*, 1987; ZHOU and SHENG, 1989), the interstitial fluid is supposed to be incompressible and the pores rigid. One important consequence is that fluid flow waves cannot propagate in the porous medium under these conditions. To clarify this statement, let us consider a semi-infinite, rigid pipe saturated with an incompressible fluid. A fluid pressure oscillation at the pipe entrance produces an oscillating flow along the pipe, which, because of mass conservation and complete absence of storage, is independent of distance from the pipe entrance. In other words, a pressure disturbance in the incompressible fluid and the associated flow are instantaneously transmitted from the pipe entrance to infinity. In contrast, studies of blood circulation through elastic arteries (e.g., ZAMIR, 2000) show that the entire body of fluid does not move in unison as in a rigid pipe. Fluid flow waves travel along the elastic arteries with a finite velocity. Note that flow wave propagation is similarly produced, even in rigid pipes, by fluid compressibility. These flow waves are dispersive and attenuated. Their existence has far-reaching consequences. Indeed, flow wave reflections and interferences greatly affect blood flow in the artery tree (e.g., DUAN and ZAMIR, 1995; WANG and PARKER, 2004; ZAMIR, 1998, 2000).

Logically, fluid flow waves must also occur in saturated pipe networks and, by extension, in porous media, provided the interstitial fluid is compressible and/or the porous medium is deformable, which is always true in natural settings. It is indeed well-known since Biot's classic work (BIOT, 1956a, b) that propagation of seismic waves is qualitatively as well as quantitatively affected in fluid-saturated porous rocks. Fluid flow waves may also occur in fractures, possibly causing resonance effects such as might be involved in volcanic tremors (CHOUET, 1996; FERRAZZINI and AKI, 1987; KORNEEV, 2008). The purpose of the present work is to derive a flow wave propagation model for a single pipe and use it to investigate flow waves through pipe networks as a first step towards application to porous rocks (note that it would be premature at this juncture to try to make realistic simulations of rocks, so unrealistically simple two-dimensional networks will be used here).

2. *AC Flow of Compressible Fluid through a Single Pipe*

2.1. *Rigid Pipe*

We consider an infinitely long, rigid pipe with a circular cross section of radius R. The radial and axial coordinates are denoted r and z, respectively, and θ is the polar angle. By reason of symmetry, the fluid velocity has no angular component and no θ dependence. The axial and radial velocity components are denoted $u(r, z, t)$ and $v(r, z, t)$, respectively, where t is time. Lastly, $p(r, z, t)$ represents the deviation with respect to an unspecified, static, fluid back-pressure (i.e., p oscillates about zero). Like $u(r, z, t)$ and $v(r, z, t)$, $p(r, z, t)$ has no θ dependence.

Since pores in rocks have very small radii, it is highly beneficial here to invoke the long-wave approximation (i.e., wavelength $\lambda >>$ pipe radius R, and/or, wave speed $c >>$ average flow velocity). As an *a posteriori* justification, it was verified that the long-wave model described in the next sections did not yield any numerical inconsistency (such as wavelengths on the order of the pipe radius) for frequencies up to 100 kHz in the case of water flowing through a rigid pipe with a radius of 100 μm or smaller.

It is best to start with the continuity equation:

$$\frac{\partial \rho}{\partial t} + \rho \frac{\partial u}{\partial z} + \rho \frac{\partial v}{\partial r} + \rho \frac{v}{r} + u \frac{\partial \rho}{\partial z} + v \frac{\partial \rho}{\partial r} = 0. \tag{1}$$

Since water is only moderately compressible and relatively low fluid velocities are considered here, the space derivatives of the density ρ can be neglected with respect to the time derivative, allowing removal of the nonlinear terms in equation (1). Note that, although not dependent on the long-wave approximation, this assumption is consistent with it. Finally, using the fluid compressibility definition $(\beta = (1/\rho)(\partial \rho/\partial p))$, we obtain:

$$\beta \frac{\partial p}{\partial t} + \frac{\partial u}{\partial z} + \frac{\partial v}{\partial r} + \frac{v}{r} = 0. \tag{2}$$

Next, the Navier-Stokes equations for u and v are written:

$$\rho \frac{\partial u}{\partial t} = -\frac{\partial p}{\partial z} + \eta \left(\frac{\partial^2 u}{\partial z^2} + \frac{\partial^2 u}{\partial r^2} + \frac{1}{r}\frac{\partial u}{\partial r} \right), \tag{3}$$

$$\rho \frac{\partial v}{\partial t} = -\frac{\partial p}{\partial r} + \eta \left(\frac{\partial^2 v}{\partial z^2} + \frac{\partial^2 v}{\partial r^2} + \frac{1}{r}\frac{\partial v}{\partial r} - \frac{v}{r^2} \right), \tag{4}$$

where η is the fluid viscosity. The long-wave approximation yields:

$$\rho \frac{\partial u}{\partial t} = -\frac{\partial p}{\partial z} + \eta \left(\frac{\partial^2 u}{\partial r^2} + \frac{1}{r}\frac{\partial u}{\partial r} \right), \tag{5}$$

$$\rho \frac{\partial v}{\partial t} = -\frac{\partial p}{\partial r} + \eta \left(\frac{\partial^2 v}{\partial r^2} + \frac{1}{r}\frac{\partial v}{\partial r} - \frac{v}{r^2} \right). \tag{6}$$

Equations (2), (5) and (6) admit axially traveling fluid flow waves as solutions (e.g., ZAMIR, 2000). Let us consider a wave traveling in the positive z direction:

$$
\begin{aligned}
u(r,z,t) &= U(r)\mathrm{e}^{-\mathrm{i}\omega(t-z/c)},\\
v(r,z,t) &= V(r)\mathrm{e}^{-\mathrm{i}\omega(t-z/c)},\\
p(r,z,t) &= P(r)\mathrm{e}^{-\mathrm{i}\omega(t-z/c)},
\end{aligned}
\tag{7}
$$

where c is the complex-valued wave speed. Plugging these expressions into equations (2), (5) and (6) leads to:

$$
-\mathrm{i}\omega\beta P + \mathrm{i}\omega\frac{U}{c} + \frac{\mathrm{d}V}{\mathrm{d}r} + \frac{V}{r} = 0 \tag{8}
$$

$$
-\mathrm{i}\omega\rho U = -\mathrm{i}\omega\frac{P}{c} + \eta\left(\frac{\mathrm{d}^2U}{\mathrm{d}r^2} + \frac{1}{r}\frac{\mathrm{d}U}{\mathrm{d}r}\right), \tag{9}
$$

$$
-\mathrm{i}\omega\rho V = -\frac{\mathrm{d}P}{\mathrm{d}r} + \eta\left(\frac{\mathrm{d}^2V}{\mathrm{d}r^2} + \frac{1}{r}\frac{\mathrm{d}V}{\mathrm{d}r} - \frac{V}{r^2}\right). \tag{10}
$$

Changing the space variable r to $\zeta = \kappa r = r\sqrt{\mathrm{i}\omega\rho/\eta}$ yields:

$$
\frac{\mathrm{d}V}{\mathrm{d}\zeta} + \frac{V}{\zeta} + \frac{\kappa\eta}{\rho c}U = \frac{\kappa\eta\beta}{\rho}P, \tag{11}
$$

$$
\frac{\mathrm{d}^2U}{\mathrm{d}\zeta^2} + \frac{1}{\zeta}\frac{\mathrm{d}U}{\mathrm{d}\zeta} + U = +\frac{P}{\rho c}, \tag{12}
$$

$$
\frac{\mathrm{d}^2V}{\mathrm{d}\zeta^2} + \frac{1}{\zeta}\frac{\mathrm{d}V}{\mathrm{d}\zeta} + \left(1 - \frac{1}{\zeta^2}\right)V = \frac{1}{\kappa\eta}\frac{\mathrm{d}P}{\mathrm{d}\zeta}. \tag{13}
$$

The radial boundary conditions are $U(\kappa R) = V(\kappa R) = 0$, and, $U(0)$ and $V(0)$ must be finite quantities. Equations (12) and (13) are Bessel equations with a right-hand term so the solutions should have the following form:

$$
U(\zeta) = A\mathrm{J}_0(\zeta) + XP(\zeta), \tag{14}
$$

$$
V(\zeta) = A'\mathrm{J}_1(\zeta) + X'\frac{\mathrm{d}P}{\mathrm{d}\zeta}, \tag{15}
$$

where A and A' are integration constants (note that they are not independent but are related in such a way as to eliminate $\mathrm{J}_0(\zeta)$ and $\mathrm{J}_1(\zeta)$ from equations 11 to 13). It must be pointed out that X and X' are not integration constants, but are (temporarily introduced) parameters that must be adjusted to reduce equations (11), (12) and (13) to a single Bessel equation for P. Indeed, if $A' = -A\,\kappa\eta/\rho c$, equations (11) to (13) give:

$$
\frac{\mathrm{d}^2P}{\mathrm{d}\zeta^2} + \frac{1}{\zeta}\frac{\mathrm{d}P}{\mathrm{d}\zeta} + \frac{\kappa\eta}{\rho X'}\left(\frac{X}{c} - \beta\right)P = 0, \tag{16}
$$

$$\frac{d^2P}{d\zeta^2} + \frac{1}{\zeta}\frac{dP}{d\zeta} + \left(1 - \frac{1}{\rho c X}\right)P = 0, \tag{17}$$

$$\frac{d^3P}{d\zeta^3} + \frac{1}{\zeta}\frac{d^2P}{d\zeta^2} + \left(1 - \frac{1}{\zeta^2} - \frac{1}{\kappa\eta X'}\right)\frac{dP}{d\zeta} = 0. \tag{18}$$

These three equations are identical Bessel equations if $X = (\kappa^2\eta^2\beta c - \rho c)/(\kappa^2\eta^2 - \rho^2 c^2)$ and $X' = X\rho c/\kappa\eta$. Thus the final solutions are:

$$P(r) = BJ_0(\gamma\kappa r), \tag{19}$$

$$U(r) = AJ_0(\kappa r) + \frac{c(\kappa^2\eta^2\beta - \rho)}{\kappa^2\eta^2 - \rho^2c^2}BJ_0(\gamma\kappa r), \tag{20}$$

$$V(r) = -A\frac{\kappa\eta}{\rho c}J_1(\kappa r) - \frac{\rho c^2}{\kappa\eta}\gamma\frac{\kappa^2\eta^2\beta - \rho}{\kappa^2\eta^2 - \rho^2c^2}BJ_1(\gamma\kappa r), \tag{21}$$

where B is an integration constant and

$$\gamma^2 = \frac{\kappa^2\eta^2}{\rho^2c^2}\left(\frac{1 - \beta\rho c^2}{1 - \frac{\kappa^2\eta^2\beta}{\rho}}\right).$$

Following ZAMIR (2000), we note that the absolute value of $\gamma\kappa r$ is very small owing to the long-wave approximation ($\gamma\kappa r$ is proportional to $\omega r/c = 2\pi r/\lambda << 1$, with a prefactor of order 1 at most). The solutions are therefore approximately equal to:

$$P(r) \approx B, \tag{22}$$

$$U(r) \approx AJ_0(\kappa r) + \frac{B}{\rho c}, \tag{23}$$

$$V(r) \approx -A\frac{\kappa\eta}{\rho c}J_1(\kappa r) + \frac{B}{2}\left(\frac{\frac{c^2}{c_0^2} - 1}{\rho^2c^2}\right)\kappa^2\eta r, \tag{24}$$

where $c_0 = 1/\beta\rho$ is the velocity of sound in the fluid. Note that, as a consequence of the long-wave approximation, P is independent on r. Then, to determine the complex-valued wave-speed c, the no-slip boundary conditions on the pipe wall, $U(R) = V(R) = 0$, are written, resulting in a homogeneous system of two linear equations in A and B, the determinant of which must be zero. This leads to the dispersion equation for c:

$$c^2 = c_0^2\left(1 - \frac{2J_1(\kappa R)}{\kappa RJ_0(\kappa R)}\right), \tag{25}$$

and to the final expression for $U(r)$:

$$U(r) = \frac{B}{\rho c}\left(1 - \frac{J_0(\kappa r)}{J_0(\kappa R)}\right). \tag{26}$$

The axial volumetric flux $q(z, t) = Q\, e^{-i\omega(t - z/c)}$ can now be calculated by integration of U over a cross section:

$$Q = \int_0^R 2\pi r U(r) \mathrm{d}r = \frac{-\pi R^2 B}{\rho c}\left(\frac{2J_1(\kappa R)}{\kappa R J_0(\kappa R)} - 1\right). \tag{27}$$

Note that, since, at any point along the pipe, the amplitude of the axial pressure gradient is ($\nabla P = i\omega B/c$, equations (26) and (27) for an infinitesimally short pipe are identical to the following expressions for harmonic flow of an incompressible fluid through a rigid pipe (e.g., BERNABÉ, 1997; CHARLAIX *et al.*, 1988; JOHNSON *et al.*, 1987):

$$U(r) = \frac{\nabla P}{\eta \kappa^2}\left(1 - \frac{J_0(\kappa r)}{J_0(\kappa R)}\right), \tag{28}$$

and

$$Q = \frac{-\pi R^2 \nabla P}{\eta \kappa^2}\left(\frac{2J_1(\kappa R)}{\kappa R J_0(\kappa R)} - 1\right). \tag{29}$$

Note also that, although not exactly zero as in the incompressible flow case,

$$V(r) = \frac{r \nabla P}{\rho c J_0(\kappa R)}\left(\frac{J_1(\kappa r)}{\kappa r} - \frac{J_1(\kappa R)}{\kappa R}\right)$$

is very small compared to $U(r)$ (by a factor of 10^{-7} at $\omega = 10^{-5}$ Hz to 10^{-4} at $\omega = 10^5$ Hz). All this seems to justify the common practice of neglecting fluid compressibility in AC flow analyses (e.g., ZHOU and SHENG, 1989). However, as pointed out by ZAMIR (2000), an entirely new class of phenomena (i.e., wave propagation, reflection and interference) arises from fluid compressibility and/or pipe deformability.

It is instructive also to visualize the frequency dependence of the complex-valued flow wave velocity c for a pipe with a radius of 100 μm (obtained as the root of equation (25) with a positive real part). As shown in Figure 1 (thin black line), $\mathrm{Re}(c/c_0)$ increases from 0 to 1 with increasing frequency, following a monotonic S curve with an inflexion point at the same frequency $\omega_{V,I}$ as the transition between the viscous and inertial flow regimes (e.g., CHARLAIX *et al.*, 1988; JOHNSON *et al.*, 1987). $\mathrm{Im}(c/c_0)$ is negative and plots on an inverted bell-shaped curve with a minimum at $\omega_{V,I}$. Another interesting representation is to plot the normalized phase velocity, $c_{Ph}/c_0 = \mathrm{Abs}(c)^2/c_0\mathrm{Re}(c)$, and the inverse quality factor, $Q_F^{-1} = -2\,\mathrm{Im}(c)/\mathrm{Re}(c)$, versus frequency (Fig. 2). We find that both parameters follow monotonic S curves with nearly identical inflexion points at $\omega_{V,I}$, c_{Ph}/c_0 increasing from 0 to 1 and Q_F^{-1} decreasing from 2 to 0. Thus, strictly speaking, flow waves are underdamped (i.e., propagating) at all frequencies except $\omega = 0$, for

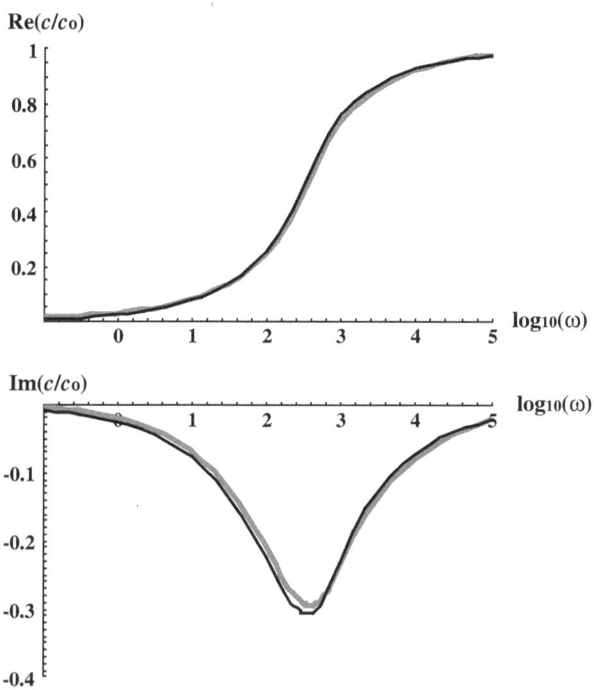

Figure 1
Real and imaginary parts of the flow wave velocity normalized to the sound velocity in the fluid as a function of frequency from equation (25) (thin black line) and from equation (44) (thicker grey line).

which they approach critical damping and zero phase velocity (i.e., non-propagating). Of course, flow wave propagation may be difficult to observe in practice (especially at frequencies below $\omega_{\mathrm{V,I}}$) owing to very high attenuation. It is also instructive to examine how the wavelength $\lambda = 2\pi c_{\mathrm{Ph}}/\omega$ varies with frequency. Above the transition frequency λ decreases as ω^{-1} whereas below $\omega_{\mathrm{V,I}}$ it varies only as $\omega^{-1/2}$. Thus, in the viscous regime, the wavelength of the flow waves is much smaller than that of sound waves in the free fluid.

2.2. *Elastic Pipe*

We now consider an elastically deformable pipe with infinitely thick walls (a finite wall thickness could, in principle, be treated in the same way but the algebra manipulations are much more complicated). By reason of symmetry, the displacement in the solid walls has no angular component and no angular dependence. The axial and radial components of the displacement vector in the solid wall are denoted x and y, respectively. A classic approach is to express x and y in terms of a scalar potential $\Phi(r, z, t)$ and an azimuthal vector potential $[0, \Psi(r, z, t), 0]$ as:

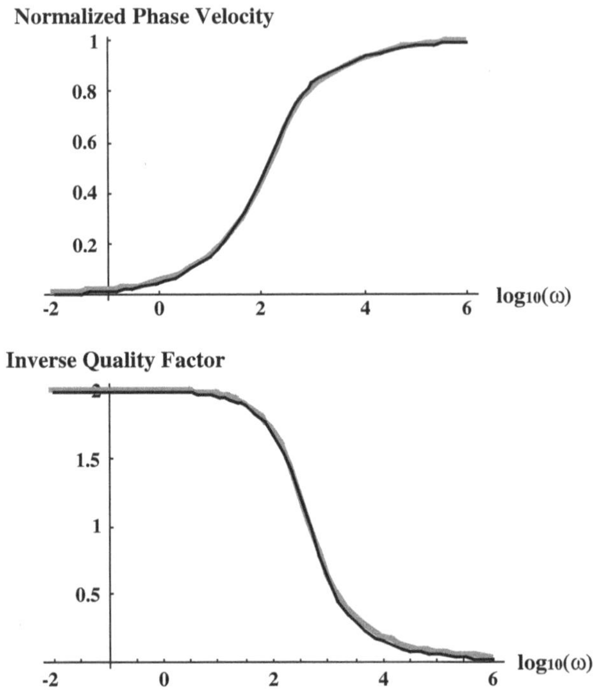

Figure 2
Flow wave phase velocity normalized to the sound velocity in the fluid and inverse quality factor as functions of frequency from equation (25) (thin black line) and from equation (44) (thicker grey line).

$$x(r,z,t) = \frac{\partial\Phi}{\partial z} + \frac{\partial\Psi}{\partial r} + \frac{\Psi}{r}, \quad \text{and} \quad y(r,z,t) = \frac{\partial\Phi}{\partial r} - \frac{\partial\Psi}{\partial z}. \tag{30}$$

The force balance equations can be rewritten in the form of two-wave equations:

$$\frac{\partial^2\Phi}{\partial r^2} + \frac{1}{r}\frac{\partial\Phi}{\partial r} + \frac{\partial^2\Phi}{\partial z^2} = \frac{1}{V_P^2}\frac{\partial^2\Phi}{\partial t^2}, \tag{31}$$

$$\frac{\partial^2\Psi}{\partial r^2} + \frac{1}{r}\frac{\partial\Psi}{\partial r} - \frac{\Psi}{r^2} + \frac{\partial^2\Psi}{\partial z^2} = \frac{1}{V_S^2}\frac{\partial^2\Psi}{\partial t^2}, \tag{32}$$

with

$$V_P = \sqrt{\frac{E(1-\nu)}{\rho_S(1+\nu)(1-2\nu)}} \quad \text{and} \quad V_S = \sqrt{\frac{E}{2\rho_S(1+\nu)}}.$$

Thus, the solutions for Φ and Ψ are also axially traveling waves:

$$\Phi(r,z,t) = F(r)\mathrm{e}^{-\mathrm{i}\omega(t-z/c)} = F(r)\mathrm{e}^{-\mathrm{i}k(ct-z)}, \tag{33}$$

$$\Psi(r,z,t) = G(r)\mathrm{e}^{-\mathrm{i}\omega(t-z/c)} = G(r)\mathrm{e}^{-\mathrm{i}k(ct-z)}, \tag{34}$$

where $k = \omega/c$ is the complex-valued axial wave-number. The functions F and G verify the following Bessel equations:

$$r^2 \frac{\mathrm{d}^2 F}{\mathrm{d}r^2} + r\frac{\mathrm{d}F}{\mathrm{d}r} - r^2 l^2 F = 0, \tag{35}$$

$$r^2 \frac{\mathrm{d}^2 G}{\mathrm{d}r^2} + r\frac{\mathrm{d}G}{\mathrm{d}r} - \left(1 + r^2 m^2\right)G = 0, \tag{36}$$

where $l^2 = k^2\ (1-c^2/V_{\mathrm{P}}^2)$ and $m^2 = k^2\ (1-c^2/V_{\mathrm{S}}^2)$ are the squares of the radial wave-numbers. These waves must have zero amplitude at infinity in the r direction. Therefore, the solutions are modified Bessel functions of the second kind:

$$F(r) = C\,\mathrm{K}_0(lr), \quad \text{and,} \quad G(r) = D\,\mathrm{K}_1(mr), \tag{37}$$

where C and D are integration constants. Note that, if the thickness of the pipe wall was finite, the modified Bessel functions of the first kind, $\mathrm{I}_0(lr)$ and $\mathrm{I}_1(mr)$, and their associated integration constants would have to be included in the solutions $F(r)$ and $G(r)$. Then, the continuity conditions on the outer free surface of the pipe would provide the two additional independent equations needed to solve the problem.

Assuming that the radial displacement of the pipe wall is negligible compared to R (Zamir, 2000), the inner wall continuity conditions can be written at the fixed position $r = R$. Consequently, the fluid velocity and pressure fields should be well approximated by those of equations (22) to (24). Contrary to the classic model of Stoneley waves in boreholes (e.g., Biot, 1952; Cheng and Toksoz, 1981; Paillet and White, 1982), the shear stress is not zero in the fluid and the tangential displacement is continuous across the solid-fluid interface. Therefore, the complete set of continuity conditions is:

$$\frac{\partial x}{\partial t}(R,z,t) = u(R,z,t), \tag{38}$$

$$\frac{\partial y}{\partial t}(R,z,t) = v(R,z,t), \tag{39}$$

$$\sigma_r(R,z,t) = \rho_{\mathrm{S}}\left(1 - \frac{2V_{\mathrm{S}}^2}{V_{\mathrm{P}}^2}\right)\frac{\partial^2\Phi}{\partial t^2} + \rho_{\mathrm{S}}2V_{\mathrm{S}}^2\left(\frac{\partial^2\Phi}{\partial r^2} - \frac{\partial^2\Psi}{\partial r\partial z}\right) = p(R,z,t), \tag{40}$$

$$\tau_{rz}(R,z,t) = \rho_{\mathrm{S}}\frac{\partial^2\Psi}{\partial t^2} + \rho_{\mathrm{S}}2V_{\mathrm{S}}^2\left(\frac{\partial^2\Phi}{\partial r\partial z} - \frac{\partial^2\Psi}{\partial z^2}\right) = \eta\frac{\partial u}{\partial r}(R,z,t). \tag{41}$$

Note that, according to the long-wave approximation, the spatial derivative of the radial velocity component v is neglected with respect to that of the axial component u in equation (41). Thus, using equations (37) and (22) to (24), we obtain the following homogeneous system of linear equation in the integration constants A, B, C and D:

$$-\mathrm{J}_0(\kappa R)A - \frac{1}{\rho c}B + k^2 c\mathrm{K}_0(lR)C + \mathrm{i}kmc\mathrm{K}_0(mR)D = 0, \tag{42}$$

$$\frac{1+\mathrm{i}}{\mathrm{c}}\sqrt{\frac{\omega\eta}{2\rho}}\mathrm{J}_1(\kappa R)A + \frac{\mathrm{i}\omega\mathrm{R}}{2\rho c^2}\left(1-\frac{c^2}{c_0^2}\right)B + \mathrm{i}klc\mathrm{K}_1(lR)C - k^2 c\mathrm{K}_1(mR)D = 0, \tag{43}$$

$$-B+\rho_S\left\{(2V_S^2-c^2)k^2\mathrm{K}_0(lR)+\frac{2V_S^2}{R}l\mathrm{K}_1(lR)\right\}C+\mathrm{i}\rho_S 2V_S^2 k\left\{m\mathrm{K}_0(mR)+\frac{\mathrm{K}_1(mR)}{R}\right\}D=0, \tag{44}$$

$$\rho(1+\mathrm{i})\sqrt{\frac{\omega\eta}{2\rho}}\mathrm{J}_1(\kappa R)A - \mathrm{i}\rho_S 2V_S^2 kl\mathrm{K}_1(lR)C + \rho_S k^2(2V_S^2-c^2)K_1(mR)D = 0 \tag{45}$$

Writing that the determinant of this system is zero leads to the dispersion equation. According to the long-wave approximation, the products kR, lR and mR all have negligibly small absolute values since they are on the order of the ratio $R/\lambda \ll 1$. Using the physical properties of water and quartz (Table 1) and taking $R = 100$ µm, one can indeed verify that kR, lR and mR are small (absolute values lower than $7\ 10^{-3}$ for frequencies up to 10^5 Hz). They also turn out to be approximately real-valued quantities (imaginary part at least one order of magnitude smaller than the real part), although this statement would have to be reversed for lR and/or mR in materials with V_P and/or $V_S < c_0$. After tedious algebraic manipulations and using the following approximate relations, $\mathrm{K}_0(\varepsilon) \approx -\mathrm{Log}(\varepsilon)$ and $\mathrm{K}_1(\varepsilon) \approx 1/\varepsilon$ for $|\varepsilon| \ll 1$, we arrive at:

$$c^4 - c^2\left(\frac{2\mathrm{J}_1(\kappa R)}{\kappa R\mathrm{J}_0(\kappa R)}\frac{\rho}{\rho_S}c_0^2 + c_0^2 + 2V_S^2\right) + 2V_S^2 c_0^2\left(1-\frac{2\mathrm{J}_1(\kappa R)}{\kappa R\mathrm{J}_0(\kappa R)}\right) = 0, \tag{46}$$

where terms of order $\varepsilon\mathrm{Log}(\varepsilon)$ were neglected. Note that this equation reduces to equation (25) in the limit of a perfectly rigid solid (i.e., infinite V_S). Equation (46) is a second degree equation on the form, $c^4 - \xi_1 c^2 + \xi_2 = 0$. Of the two pairs of roots of Equation (46), only the one given by $c^2 = \tfrac{1}{2}\left(\xi_1 - \sqrt{\xi_1^2 - 4\xi_2}\right)$, where the square-root symbol refers to the root with a positive real part, leads to physically acceptable values of c. As seen in Figures 1 and 2, equations (25) and (46) give nearly identical

Table 1

Properties of the fluid (water) and solid (quartz)

Density of fluid	$\rho = 1000$ kg m^{-3}
Viscosity	$\eta = 0.001$ kg m^{-1} s^{-1}
Compressibility	$\beta = 4.9\ 10^{-4}$ MPa^{-1}
Density of solid	$\rho_S = 2650$ kg m^{-3}
Young's modulus	$E = 95600$ MPa
Poisson's ratio	$\nu = 0.08$

results when using the values of Table 1. Since it is easier to use, only equation (25) will be used hereafter.

Finally, A is calculated as a function of B and plugged into $U(r)$:

$$U(r) = \frac{B}{\rho c}\left(1 - \frac{\kappa R J_0(\kappa r)}{2J_1(\kappa R)}\left[1 - \frac{c^2}{c_0^2} + \frac{c^2}{V_S^2}\frac{\rho}{\rho_S}\right]\right), \tag{47}$$

and the volumetric flux Q can then be inferred:

$$Q = -\frac{\pi R^2 B}{\rho c}\left(\frac{\rho c_0^2}{\rho_S V_S^2} - 1\right)\frac{c^2}{c_0^2}. \tag{48}$$

Again, one finds that, in the limit of infinite V_S, equations (47) and (48) reduce to the rigid pipe equations (26) and (27), respectively (note that deriving this statement requires using equation (25)). Consistent with the choice made earlier, equations (26) and (27) (rather than (47) and (48)) will be used in the numerical applications hereafter. This choice amounts to considering the infinitely thick quartz pipe as fully rigid. This is reasonable for a pipe with a circular cross section but would probably not be true for a pipe with a low aspect-ratio ellipsoidal cross section. However, flow wave propagation in ellipsoidal pipes is out of the scope of this paper.

2.3. *Flow Wave Interferences in a Single Pipe*

We now consider a pipe of length L. The complex-valued amplitudes of the pressure oscillations at the "upstream" and "downstream" pipe ends (i.e., $z = 0$ and $z = L$) are denoted P_U and P_D, respectively. The complex-valued amplitude of the global pressure gradient is thus $(P_D - P_U)/L$. The pressure inside the pipe is the sum of two waves traveling in opposite directions:

$$p(z,t) = B^+ e^{-i\omega(t-z/c)} + B^- e^{-i\omega(t+z/c)}. \tag{49}$$

We must have:

$$B^+ = \frac{P_D - P_U e^{-i\omega L/c}}{e^{i\omega L/c} - e^{-i\omega L/c}}, \quad \text{and,} \quad B^- = \frac{P_U e^{i\omega L/c} - P_D}{e^{i\omega L/c} - e^{-i\omega L/c}}. \tag{50}$$

Similarly, the volumetric flux $q(z,t)$ is the sum of two traveling waves:

$$q(z,t) = Q^+ e^{-i\omega(t-z/c)} + Q^- e^{-i\omega(t+z/c)}. \tag{51}$$

The complex-valued amplitudes Q_U and Q_D of the volumetric flux at $z = 0$ and L, respectively, are unequal and given by:

$$Q_U = Q^+ + Q^-, \quad \text{and,} \quad Q_D = Q^+ e^{i\omega L/c} + Q^- e^{-i\omega L/c}. \tag{52}$$

By applying equation (27) (or equation (48) in the limit of infinite V_S), one obtains:

$$Q_U = \frac{-i\pi R^2}{\rho c}\left(\frac{P_U\cos(\omega L/c) - P_D}{\sin(\omega L/c)}\right)\left(\frac{2J_1(\kappa R)}{\kappa R J_0(\kappa R)} - 1\right), \tag{53}$$

$$Q_D = \frac{-i\pi R^2}{\rho c}\left(\frac{P_U - P_D\cos(\omega L/c)}{\sin(\omega L/c)}\right)\left(\frac{2J_1(\kappa R)}{\kappa R J_0(\kappa R)} - 1\right). \tag{54}$$

These two quantities are generally different, expressing the effect of storage inside the pipe. They become identical only in the limit of an infinitesimally small pipe length.

2.4. *Apparent Hydraulic Upstream and Downstream Conductivities*

Because Q_U and Q_D are unequal, it is awkward to introduce the (usually very useful) concept of hydraulic conductivity here. In steady-state flow conditions, the hydraulic conductivity, $K_{SS} = -\eta Q/(P = \pi R^4/8$, can be considered an intrinsic property of the pipe. In the case of the oscillating flow of an incompressible fluid through a rigid pipe, a natural generalization consists in incorporating the frequency dependence into the so-called AC hydraulic conductivity,

$$K_{AC} = \frac{\pi R^2}{\kappa^2}\left(\frac{2J_1(\kappa R)}{\kappa R J_0(\kappa R)} - 1\right)$$

(e.g., BERNABÉ, 1997; CHARLAIX *et al.*, 1988; JOHNSON *et al.*, 1987). It is helpful for interpreting the results in the following sections to visualize the real and imaginary parts of K_{AC} for water AC flow through a pipe with $R = 100$ μm. The real and imaginary parts of the normalized AC hydraulic conductivity, K_{AC}/K_{SS}, are drawn on semi-logarithmic plots in Figure 3. These two curves clearly reveal the existence of two distinct, limiting flow regimes (i.e., viscosity- and inertia-controlled regimes at low and high frequencies, respectively). The sharp transition between these two regimes precisely defines a transition frequency $\omega_{V,I}$, which scales as $1/R^2$ (e.g., BERNABÉ, 1997; CHARLAIX *et al.*, 1988; JOHNSON *et al.*, 1987).

When flow waves occur owing to fluid compressibility and/or pipe deformability, the approach mentioned above is no longer possible. The easiest way is to proceed formally and define two complex-valued, (upstream and downstream) hydraulic conductivities, K_U and K_D, as:

$$K_U = \frac{-\eta L Q_U}{P_D - P_U} \quad \text{and} \quad K_D = \frac{-\eta L Q_D}{P_D - P_U}. \tag{55}$$

Using equations (53) and (54), it is found that $K_U = K_{AC}\ \Xi_U$ and $K_D = K_{AC}\ \Xi_D$, where the functions Ξ_U and Ξ_D are given by:

$$\Xi_U = \frac{\omega L}{c}\frac{P_U\cos\left(\frac{\omega L}{c}\right) - P_D}{\sin\left(\frac{\omega L}{c}\right)(P_U - P_D)}, \quad \text{and,} \quad \Xi_D = \frac{\omega L}{c}\frac{P_U - P_D\cos\left(\frac{\omega L}{c}\right)}{\sin\left(\frac{\omega L}{c}\right)(P_U - P_D)}. \tag{56}$$

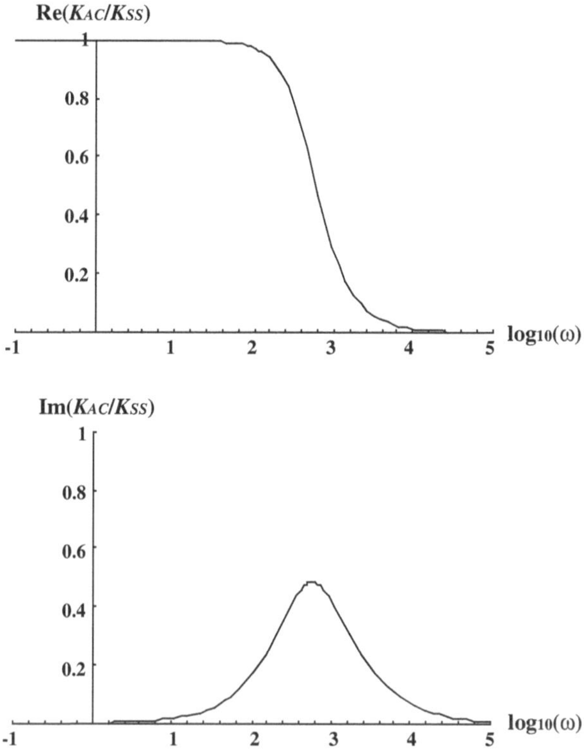

Figure 3
Real and imaginary parts of the normalized AC hydraulic conductivity for oscillating flow of an incompressible flow through a rigid pipe.

One interesting feature of these functions is that, in the limit of an infinitesimally short pipe, $\Xi_U \approx \Xi_D \approx 1$ (therefore, $K_U \approx K_D \approx K_{AC}$) for all frequencies. For finite values of L, inspection of Ξ_U and Ξ_D reveals two regions: 1) at low-frequencies, $\Xi_U \approx \Xi_D \approx 1$; 2) above a critical frequency $\omega_{1,2}$, the functions Ξ_U and Ξ_D show wide irregular oscillations, the magnitude of which increases to a maximum, then gradually dies out (note that $\mathrm{Im}(\Xi_U)$ remains negative at all frequencies). In the limit of high frequencies, $\mathrm{Re}(\Xi_U) \sim \omega^{1/2}$ and $\mathrm{Im}(\Xi_U) \sim -\omega$ while Ξ_D vanishes. In region (1), the flow waves effectively see the pipe as infinitesimally short and wave interferences do not occur. In region (2), the wavelength is sufficiently short for constructive and destructive interferences to occur. The gradual weakening of the interferences at very high frequencies is caused by the strong attenuation of the flow waves. The transition frequency $\omega_{1,2}$ separating (1) and (2) appears to correspond to a wavelength about one order of magnitude larger than L. Thus, when $\omega_{1,2}$ falls in the inertial regime (i.e., $\omega_{1,2} > \omega_{V,I}$), it should scale with $1/L$, but this scaling should breakdown in the viscous regime, inside which $\omega_{1,2}$ should vary as $1/L^2$. Thus, the products of Ξ_U and Ξ_D with K_{AC}

must yield very different results depending on the relative values of the transition frequencies $\omega_{1,2}$ and $\omega_{V,I}$.

To illustrate the last statement, Figures 4 and 5 show the real and imaginary parts of the normalized conductivities, K_U/K_{SS} and K_D/K_{SS}, calculated for $P_D = 0$ (i.e., open-ended pipe) and different values of L (in all cases, $R = 100$ μm). One important feature of K_U/K_{SS} is that its real part does not vanish in the high-frequency limit, but approaches a finite positive value K_∞/K_{SS}. This observation is easily explained by remarking that the pipe appears effectively infinite to the flow waves at very high frequencies. Thus, only the forward wave propagates and Q_U approaches a finite value Q_∞ independent of fluid viscosity and pipe length, as should be expected. Using equation (25), it can be demonstrated that $Q_\infty = \pi R^2 P_U/\rho c_0$ for $P_D = 0$ as in Figures 4 and 5, and, based on the definition of K_U, that $\mathrm{Re}(K_U/K_{SS}) = 8\eta L/R^2\rho c_0$. Thus, K_∞ increases linearly with L and eventually becomes larger (possibly orders of magnitude larger) than the steady-state value K_{SS}. To the contrary K_D/K_{SS} vanishes in the high-frequency limit. This very strong difference in high-frequency behavior between K_U and K_D is a consequence of fluid storage inside the pipe.

It is also instructive to explore a more general type of downstream boundary condition, i.e., that used in the oscillating flow method for measuring permeability (e.g., FISCHER, 1992; FISCHER and PATERSON, 1992). However, the analysis of this situation is secondary to the main purpose of this paper. Accordingly, it will be reported in Appendix A.

3. Harmonic Flow through a Two-dimensional Network of Pipes

As shown in BERNABÉ (1997), the standard method for solving a stead-state flow problem on a pipe network can be also used for an oscillating flow with a given frequency, except that a complex-valued matrix has to be inverted. The Kirchoff's laws are written for every node i in the network, $\sum_j Q_U(P_i, P_j) = 0$, where P_i and P_j denote the complex-valued amplitude of the pressure oscillation at the (connected) nodes i and j, respectively, and the function Q_U is given by equation (53) (by convention, the central node i is considered upstream and the connected nodes j downstream). This leads to a system of linear equations, which depends on the network topology (i.e., two- or three-dimensional, triangular, square, hexagonal, and so forth), the node-indexing scheme and the boundary conditions. I emphasize here that my goal was not to make realistic simulations of rocks but merely to illustrate the effect of wave reflections and interferences in relatively complex pipe systems. Hence, I decided to limit the study to two-dimensional square networks.

Semi-periodic boundary conditions were used here. In other words, unitary and zero pressure amplitudes were respectively specified at nodes on opposite sides of the network in the flow direction (i.e., the upstream and downstream sides) while periodic boundary conditions were set along the lateral sides (i.e., the network is wrapped around, forming a cylinder with its axis in the nominal flow direction).

Once the flow problem is solved, the effective upstream and downstream conductivities for the network can be calculated using definitions similar to equation (55), in which Q_U and Q_D now denote the amplitudes of the total volumetric flux into/out of the upstream and downstream network sides, P_D–$P_U = -1$ is the global pressure amplitude difference, and, L refers to the total length of the network in the flow direction.

Establishing and solving the appropriate matrix as described above must be carried out numerically for each frequency considered. Here, 71 frequencies ranging between 10^{-2} and 10^5 Hz, with a constant logarithmic increment, were used, resulting in a resolution in $\log_{10}\omega$ considerably lower than that of Figures 4 and 5. For the sake of limiting the computational cost to a minimum, I considered rather small networks (i.e., 10×10, 20×20 and 40×40). Also, it seemed preferable to limit the level of complexity arising from heterogeneity. Accordingly, the values of the pipe radius were assigned according to a rather narrow uniform distribution (equal probability between a

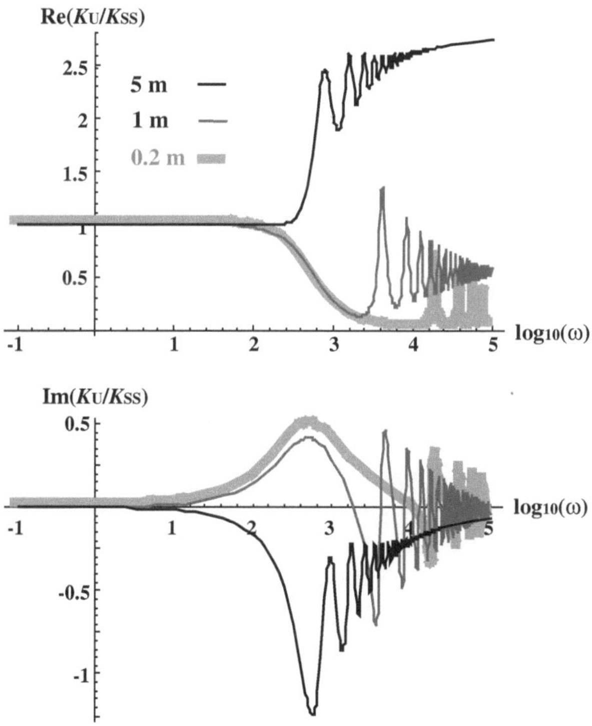

Figure 4
Real and imaginary parts of the normalized apparent conductivity K_U/K_{SS} for pipes with different lengths L (thick grey line, $L = 0.2$ m; thin grey line, $L = 1$ m; thin black line, $L = 5$ m; the pipe radius is equal to 100 μm). The frequencies are given in Hertz.

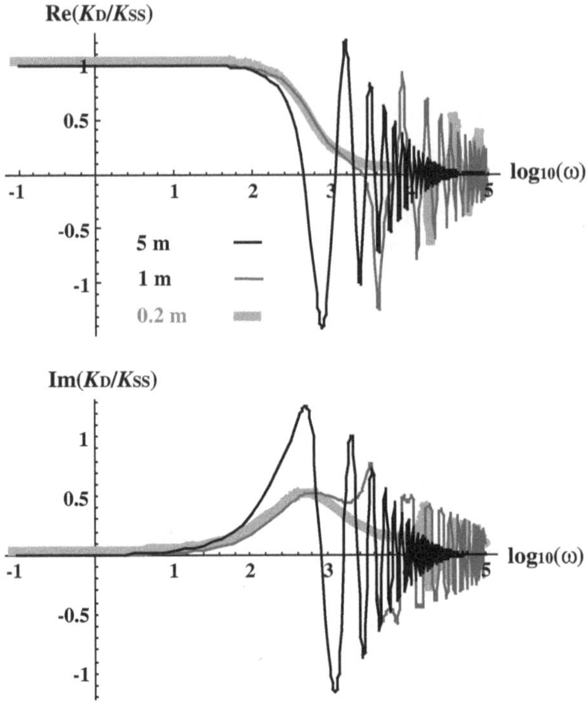

Figure 5
Real and imaginary parts of the normalized apparent conductivity K_D/K_{SS} for pipes with different lengths L (thin black line, $L = 5$ m; thin grey line, $L = 1$ m; thick grey line, $L = 0.2$ m; the pipe radius is equal to 100 μm).

minimum radius of 90 μm and a maximum of 110 μm). With such a low level of heterogeneity, 10 realizations were sufficient to estimate the ensemble average accurately even at frequencies as high as 10^5 Hz.

Figures 6 and 7 show the real and imaginary parts of K_U/K_{SS} as a function of frequency for 40 × 40 and 10 × 10 networks with different total length L. These curves look very similar to those of Figure 4. In particular, a transition frequency similar to $\omega_{1,2}$ in section 2.4 can be identified. This transition frequency appears to be a monotonically increasing function of the network length L and, therefore, is likely related to wave reflection from the upstream and downstream ends of the network. The internal reflections (i.e., from the ends of individual bonds) are visible at high frequencies although note that, at high frequencies, the resolution used here is insufficient to accurately delineate the fluctuations of K_U/K_{SS}.

In order to confirm the role of wave reflections from the upstream and downstream ends, the real and imaginary parts of K_U/K_{SS} were calculated for 40 × 40 and 10 × 10 networks with an identical total length. These networks are expected to have identical

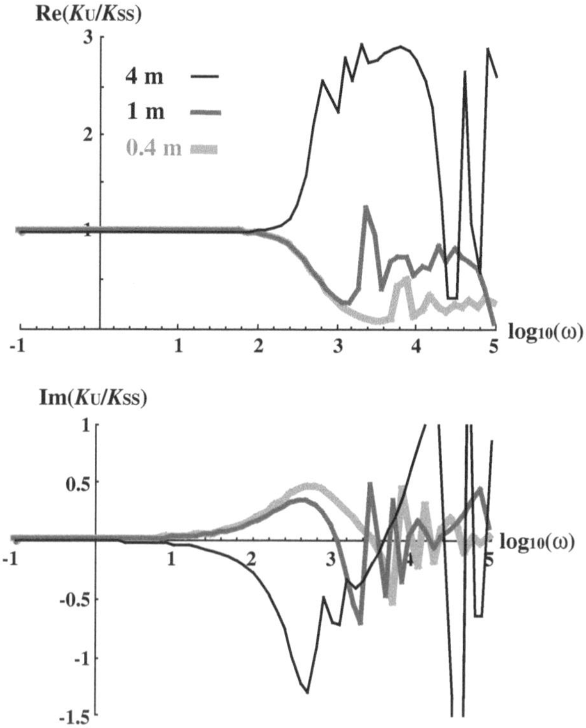

Figure 6
Real and imaginary parts of the normalized apparent permeability K_U/K_{SS} for 40 × 40 networks with different total lengths (thin, black line, $L = 4$ m; thicker, dark grey line, $L = 1$ m; thick, light grey line, $L = 0.4$ m; the mean radius of the pipes is 100 μm). Note that the diagrams show results of discrete calculations with an increment of 0.1 in log 10(ω). The discrete data points are joined by straight lines for visibility.

reflections from the upstream and downstream ends but different internal ones. Indeed, the 40 × 40 and 10 × 10 curves (thin black and grey lines, respectively) are nearly coincident in the low-frequency range up to point A in Figure 8 (i.e., somewhat above the transition frequency $\omega_{1,2}$) but large differences appear at higher frequencies. I also considered 20 × 20 networks containing a sharp discontinuity in the pipe radius distribution at mid-length in the flow direction. The mean radius was increased by a factor of 3 in the downstream half of the network, thus forming a reservoir with a large storage and high permeability. It can therefore be expected that the pipes ending at the mid-length discontinuity should act almost like open pipes and generate the same wave reflections as before. The corresponding Re(K_U/K_{SS}) and Im(K_U/K_{SS}) curves are plotted as thick light grey lines in Figure 8. The 20 × 20 curves are sufficiently similar to the other two in the frequency range near A to conclude that they reveal the existence of a strong reflection at the mid-length discontinuity. At higher frequencies, the 10 × 10 and

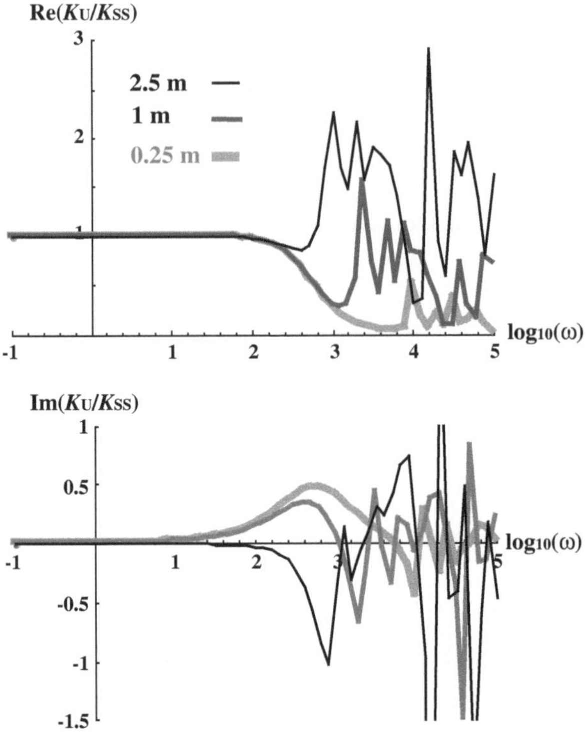

Figure 7
Real and imaginary parts of the normalized apparent permeability K_U/K_{SS} for 10×10 networks with different total lengths (thin, black line, $L = 2.5$ m; thicker, dark grey line, $L = 1$ m; thicker, light grey line, $L = 0.25$ m; the mean radius of the pipes is 100 μm).

20×20 curves are nearly coincident, indicating that the internal reflections were identical in both types of networks, as expected since the bond length was the same.

The interference of forward and backward traveling flow waves, owing to wave reflection at the ends of a network, can be intuitively predicted based on Figures 4 and 5. The effect of lateral reflections from transverse bonds (i.e., perpendicular to the nominal flow direction) is considerably more difficult to foresee. However, these reflections do exist. To demonstrate it, let us consider networks in which the radii of the transverse pipes were reduced by a factor of 100 and transverse reflections are expected to be negligible. Indeed, such modified networks are nearly disconnected in the transverse direction and, therefore, approximately equivalent to an *in parallel* bundle of almost identical pipes. A direct comparison with Figure 4 is then possible, as illustrated in Figure 9. Within the frequency resolution used here, the normalized response of a single open-ended pipe with $L = 1$ m and $R = 100$ μm (thin black lines) is identical to that of modified 10×10 networks with the same total length and mean radius in the longitudinal direction (thick light grey lines). On the other hand, the $\mathrm{Re}(K_U/K_{SS})$ and

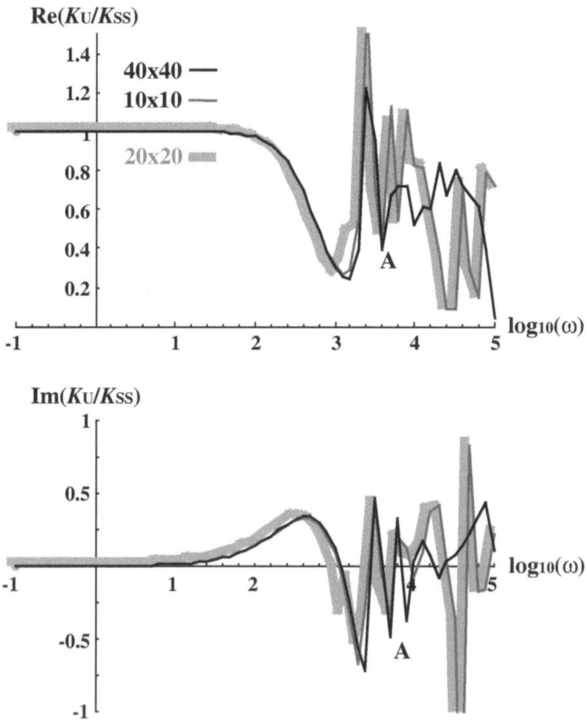

Figure 8

Real and imaginary parts of the normalized apparent permeability K_U/K_{SS} for networks with the same total length ($L = 1$ m) and the same mean radius ($R = 100$ μm) but different number of nodes (thin, black line, 40×40; thin, dark grey line, 10×10). The thick, light grey line represents the response of 20×20 networks, in which the mean radius was suddenly increased by a factor of 3 at mid-length in the flow direction. The black and dark grey curves are nearly coincident up to the point labeled A and the light grey curve is fairly similar. Above point A the grey curves become nearly coincident while the black one is quite different (see text for more details).

$\mathrm{Im}(K_U/K_{SS})$ curves for the isotropic 10×10 networks considered before (thin grey lines) display large differences with the previous ones, in particular, a slightly lower transition frequency and some large amplitude fluctuations at high frequencies. These differences must be associated with transverse internal reflections.

4. Discussion

The main point of this paper is that harmonic flow of a compressible fluid through an elastic pipe (or pipe network) propagates as a wave. Here, this type of wave was called "fluid flow wave" (or, more concisely, "flow wave") to emphasize that significant fluid displacement occurs in response to fluid pressure changes. The characteristics of these

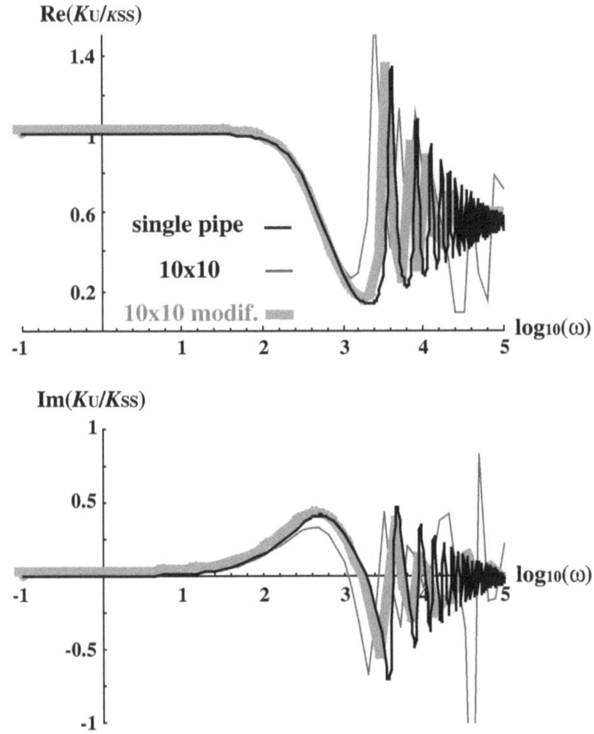

Figure 9
Real and imaginary parts of the normalized apparent permeability K_U/K_{SS} for a single pipe (thin black line) and networks with the same total length ($L = 1$ m) and the same mean radius ($R = 100$ μm) in the flow direction. The thin, dark grey line corresponds to networks with an identical mean radius in longitudinal and transversal directions. The thick, light grey line represents the response of networks, in which the mean radius in the transversal direction was reduced by a factor of 100, thus approximately removing all transversal bonds and resulting in a parallel bundle of pipes with nearly constant cross section (see text for more details).

waves were described in section 2. Among other things, it was shown that the displacement amplitude in the solid pipe wall decays exponentially with distance from the pipe axis. The flow waves are therefore interface waves similar to Stoneley waves in boreholes (e.g., BIOT, 1952; CHENG and TOKSOZ, 1981; PAILLET and WHITE, 1982). However, it must be emphasized that, despite this similarity, large differences in behavior exist. For example, the fluid pressure varies radially in borehole Stoneley waves while it does not in the flow waves presented here. Also, the dispersion equations are different. It is interesting to identify the origin of these discrepancies. In the classic model of Stoneley waves in boreholes, fluid viscosity is neglected, allowing use of a scalar elastic potential to describe the fluid displacements. This approach is equivalent to equations 2, 5 and 6 with $\eta = 0$, and, it can indeed be verified (using properties such as $\gamma\kappa = ik\sqrt{1 - c^2/c_0^2}$) that the classic solutions of BIOT (1952) for Stoneley waves in an impermeable borehole are identical to those of equations (19) to (21) in the limit of a vanishing viscosity.

Nevertheless, neglecting fluid viscosity leads to important qualitative differences because it implies a qualitatively different set of continuity conditions at $r = R$ (i.e., zero shear stress at the solid-fluid interface and discontinuous tangential displacement). Another crucial difference is that the Stoneley waves model does not assume the long-wave approximation. If the long-wave approximation is applied to the dispersion equation for Stoneley waves in an impermeable borehole (BIOT, 1952), we obtain

$$c^2 = c_0^2\left(1 - \frac{\rho c_0^2}{\rho_S V_S^2}\right)^{-1},$$

whereas equation (46) reduces to $c^2 = c_0^2$ in the limit of zero viscosity. These two equations do not coincide except in the case of a fully rigid pipe. This discrepancy is caused by the use of different sets of continuity conditions at $r = R$.

The flow waves described in this paper can, in fact, be better identified with Biot's slow P wave. BIOT (1956a,b) established the equations of propagation of mechanical waves in elastic, fluid-saturated, porous solids, from heuristically constructed potential energy and dissipation functions. He directly worked at the macroscopic scale and only considered macroscopically defined parameters and constants (Biot's equations were later rederived from microscale equations using various upscaling schemes, e.g., BURRIDGE and KELLER, 1981, or, PRIDE *et al.*, 1992). Most importantly, BIOT (1956a,b) demonstrated the existence of two compressional waves, often called the fast and slow P waves. The slow P wave is characterized by solid-fluid relative displacements much greater than the displacements of the solid grains (the opposite is true for the fast P wave) and is therefore akin to the fluid flow waves studied here. Biot's slow P waves were actually observed in the laboratory in various porous materials including natural rocks (KELDER and SMEULDERS, 1997; NAGY *et al.*, 1990; PLONA, 1980), although the presence of clay particles tended to destroy the signal (KLIMENTOS and MCCANN, 1988). Unusually slow compressional waves, possibly interpreted as slow P waves, were also observed *in situ* in natural sediments (BOYLE and CHOTIROS, 1992).

One important similarity is that both slow P wave and flow wave models use the long-wave approximation. One important difference is that Biot formulated the dissipation function assuming incompressible fluid flow through rigid pores (i.e., equations (28) and (29) in the case of pipe-like pores), conditions incompatible with flow wave propagation. However, Biot considered that poro-elastic waves were generated by elastic waves impinging on the saturated porous solid (here, they are produced by fluid pressure oscillations at the boundaries) and simply used equations (28) and (29) (or their equivalents for slit-like pores) to approximate the local fluid flow field in an infinitesimally thin layer of porous material. As we have seen in sections 2.1 and 2.2, the flow waves equations (26), (27), (47) and (48) also reduce to (28) and (29) for infinitesimally short pipes. Nevertheless, a problem remains because Biot's theory is macroscopic and an infinitesimal volume element of porous solid is presumed to contain many pores. Therefore, flow waves may be reflected at pore intersections and such internal wave reflections are not accounted for in Biot's theory. Internal reflections are not important in practice, however, because their effect is negligible except at very high frequencies. Thus, it is reasonable to identify the flow waves investigated here and Biot's slow P waves, although a rigorous identification should entail thorough upscaling of the single pipe results. For example, at the present stage it is not possible to compare the dispersion equations for Biot's slow P waves and for the single pipe flow waves. All that can be said is that they do not appear obviously inconsistent (e.g., the slow P-wave phase velocity increases with increasing frequency from 0 to near c_0 when the material solid skeleton is fairly rigid and the pores tortuosity approaches 1, conditions appropriate for comparison with the single pipe results).

Since flow waves were here envisioned as generated by fluid pressure oscillations at the boundaries of the solid, the present work is more akin to AC permeability studies (e.g., CHARLAIX *et al.*, 1988; JOHNSON *et al.*, 1987; ZHOU and SHENG, 1989), than to those based on Biot's theory. One important finding of the above-referred studies was that the AC permeability should be controlled by only one length scale. For a pipe, this length scale is equal to the radius. For a porous medium, the relevant length scale is well approximated by a dynamic volume-to-surface ratio proposed by JOHNSON *et al.* (1987) and, in most cases of interest, is comparable in order of magnitude to the mean pore radius. Here, another geometric length scale is introduced, the pipe length. In the case of a macroscopic poro-elastic solid, this length scale should correspond to the distance between major flow wave reflectors. As already discussed, reflections at pore junctions are not important in practice since their effects can only be seen at very high frequencies. Nonetheless, reflections at interfaces separating rock formations with very different physical properties should be observable. Indeed, simulations of poro-elastic waves propagation through layered media do show large oscillations of the wave attenuation (or, equivalently, wave amplitude) above a certain transition frequency within or lower than the seismic range (e.g., GELINSKY *et al.*, 1998; GUREVICH *et al.*, 1997; PRIDE *et al.*, 2002; SHAPIRO and MULLER, 1999; but note that this phenomena was not the main focus of these studies). One interesting consequence is that discontinuities in hydraulic permeability alone (all other poro-elastic parameters remaining constant) could have a seismic signature (MULLER *et al.*, 2007; SHAPIRO and MULLER, 1999). The effect of such a pure poro-elastic reflector can indeed be verified in network simulations by constructing a 40×40 network and dividing in two *in series* 40×20 sub-networks, one with pipe length and radius, L and R, and the other with $2L$ (this can be done by removing some bonds and nodes) and $R/2^{1/2}$ so the porosity remains unchanged across the interface while permeability decreases by a factor of 8. Plots of K_U/K_{SS} show the same kind of effects as in Figure 8, indicating that flow wave reflections do occur at the interface.

Flow wave interferences (or, if my interpretation is correct, Biot's slow wave interferences) also can be important in fracture networks. The relevant length should then be the distance between intersections of major fractures. Of course, in the case of fractures, the pipe geometry considered here (cylindrical pipes with a circular cross section) is quite unrealistic. Unfortunately, solving the same problem analytically in pipes with an elliptic cross section of arbitrary aspect ratio might not be possible and one might have to construct

K_U and K_D numerically. Also, for very low aspect ratios, the displacement in the solid in the direction normal to the pipe wall may no longer be negligible and writing the continuity conditions on a deforming surface will yield non-linearity.

5. Conclusions

1) Harmonic flow of a compressible fluid through an elastic pipe propagates as a wave.
2) These flow waves are dispersive and attenuated. The dispersion equation was derived, showing that the phase velocity increases from zero at low frequencies to the fluid sound velocity at high frequencies while the inverse quality factor decreases from 2 (i.e., critical damping) to zero.
3) Thus, in principle, flow waves should always be propagative and only become diffusive asymptotically at zero frequency. However, they are probably very difficult to observe in practice owing to a very high attenuation, which could be further enhanced in natural rocks by roughness of the pore walls, coating by clay particles, and so forth.
4) Reflections and interferences of flow waves become important if the distance between reflectors is greater than about a tenth of the wavelength.
5) Although more work is required to confirm this conclusion, it seems that all these properties are preserved at a greater scale, in relatively complex arrangements of pipes (i.e., here, two-dimensional square networks).

Acknowledgements

I am grateful to Dave Johnson and Larry Schwartz for insightful discussions. Christian David and Ian Main's reviews were very useful and are gratefully appreciated. I also want to express my deep appreciation to Nafi Toksoz, who always provided me with encouragement and inspiration. This work was partially funded by the U.S. Department of Energy under grant DE-FG01-97ER14760.

Appendix A

As mentioned at the end of section 2.4, we will now analyze the effect of the downstream boundary conditions used in the oscillating flow method for measuring permeability (e.g., FISCHER, 1992; FISCHER and PATERSON, 1992). In such measurements, the downstream end of the pipe is connected to a reservoir with a known storage β_D (we retrieve the open pipe case if β_D is infinite while $\beta_D = 0$ corresponds to a closed pipe). P_U is prescribed at the upstream end of the pipe and P_D measured. In addition to the flow equations written for points inside the pipe, we need to express conservation of mass at the downstream end of the pipe, i.e., at any time t the downstream volumetric flux is equal to β_D times the time-rate of change of the downstream pressure. For a given frequency ω, this can be written as $Q_D = -i\omega\beta_D P_D$. Combining this expression with equation (54) yields the value of the ratio P_D/P_U:

$$\frac{P_D}{P_U} = \frac{\pi R^2\left(\frac{2J_1(\kappa R)}{\kappa R J_0(\kappa R)} - 1\right)}{\omega c \beta_D \rho \sin(\omega L/c) + \pi R^2\left(\frac{2J_1(\kappa R)}{\kappa R J_0(\kappa R)} - 1\right)\cos(\omega L/c)}. \tag{A1}$$

We see that, for an open pipe (i.e., infinite β_D), $P_D = 0$, as it should be. For a closed pipe, we obtain $P_D = P_U/\cos(\omega L/c)$, which, as expected, reduces to $P_D \approx P_U$ in the low-frequency limit and/or for an infinitesimally short pipe.

How can we interpret this result in terms of an apparent hydraulic conductivity? We could use the formal approach of section 2.4, but K_U and K_D are not easily measurable in actual oscillating flow tests. Thus, it is more practical to follow an operational approach. The first step is to derive P_D/P_U for an incompressible fluid and a rigid pipe:

$$\frac{P_D}{P_U} = \frac{\pi R^2\left(\frac{2J_1(\kappa R)}{\kappa R J_0(\kappa R)} - 1\right)}{\omega^2 L \beta_D \rho + \pi R^2\left(\frac{2J_1(\kappa R)}{\kappa R J_0(\kappa R)} - 1\right)}. \tag{A2}$$

Note that equation (A2) is equivalent (to first order) to (A1) for an infinitesimally short pipe. In the next step, a relation between P_D/P_U and K_{AC} can be written:

$$K_{AC} = \frac{-i\omega L \beta_D \eta P_D/P_U}{1 - P_D/P_U}. \tag{A3}$$

This relation remains valid even for a closed pipe since $1-P_D/P_U \sim \beta_D$ for vanishing β_D. Note that an identical relation can be derived for a porous rock sample with negligible intrinsic storage, provided K_{AC} in equation (A3) is replaced by the product, $A_s k_{AC}$, of the sample cross-section area by the AC permeability (see BERNABÉ *et al.*, 2006). It should be emphasized here that K_{AC} (or, equivalently, $A_s k_{AC}$) can be explicitly expressed in function of P_D/P_U only because the pipe storage (or, equivalently, the rock sample storage) is negligible. Otherwise, hydraulic conductivity and storage (or, equivalently, permeability and storage) are inextricably entangled (e.g., BERNABÉ *et al.*, 2006).

Finally, the right-hand side of equation (A3) can be used as an operational definition of the apparent AC hydraulic conductivity of the pipe K^*_{AC}. Plugging (A1) into (A3), we obtain:

$$K^*_{AC} = \frac{K_{AC}}{\frac{c\sin\left(\frac{\omega L}{c}\right)}{\omega L} - \frac{iK_{AC}}{\omega L \beta_D \eta}\left(1 - \cos\left(\frac{\omega L}{c}\right)\right)}. \tag{A4}$$

We verify easily that, for any non-zero β_D, $K^*_{AC} \approx K_{AC}$ in the limit of an infinitesimally short pipe. In the open pipe case (i.e., infinite β_D), we find that K^*_{AC} is identical to K_D for all frequencies, while it is equal to zero for a closed pipe (i.e., zero β_D).

For intermediate values of β_D, K^*_{AC} is quite similar to K_D, except that, for vanishing frequencies, the ratio K^*_{AC}/K_{SS} approaches a real-valued limit between 0 and 1. However, for a given β_D, this value remains very close to unity unless the pipe is very long. Thus, we conclude that the operational definition (A4) should be largely satisfactory in laboratory experiments since the ratio L/R cannot reach sufficiently high values, given the usually rather low frequency range accessible. However, flow wave interferences in the tubing of the pore pressure system could become a problem at frequencies on the order of 500 Hz and higher.

References

Bernabé, Y. (1997), *The frequency dependence of harmonic fluid flow through networks of cracks and pores,* Pure Appl. Geophys. *149,* 489–506.

Bernabé, Y., Mok, U., and Evans, B. (2006), *A note on the oscillating flow method for measuring rock permeability,* Int. J. Rock Mech. Min. Sci. *43,* 311–316.

Bernabé, Y., Mok, U., Evans, B., and Herrmann, F. J. (2004), *Permeability and storativity of binary mixtures of high- and low-permeability materials,* J. Geophys. Res. - Sol. Earth *109,* doi:10.1029/2004JB003111.

Biot, M. A. (1952), *Propagation of elastic waves in a cylindrical bore containing a fluid,* J. Appl. Physics *23,* 997–1005.

Biot, M. A. (1956a), *Theory of propagation of elastic waves in a fluid-saturated porous solid. I. Low-frequency range,* J. Acoust. Soc. Am. *28,* 168–178.

Biot, M. A. (1956b), *Theory of propagation of elastic waves in a fluid-saturated porous solid. II. High-frequency range.,* J. Acoust. Soc. Am. *28,* 179–191.

Boyle, F. A. and Chotiros, N. P. (1992), *Experimental detection of a slow acoustic wave in sediment at shallow grazing angles,* J. Acoust. Soc. Am. *91,* 2615–2619.

Burridge, R. and Keller, J. B. (1981), *Poroelasticity equations derived from microstructure,* J. Acoust. Soc. Am. *70,* 1140–1155.

Charlaix, E., Kushnick, A. P., and Stokes, J. P. (1988), *Experimental study of dynamic permeability in porous media,* Phys. Rev. Lett. *61,* 1595–1598.

Cheng, C. H. and Toksöz, N. (1981), *Elastic wave propagation in a fluid-filled borehole and synthetic acoustic logs,* Geophys. J. Int. *46,* 1042–1053.

Chouet, B. A. (1996), *Long-period volcanic seismicity: Its source and use in eruption forecasting,* Nature *380,* 309–316.

Duan, B. and Zamir, M. (1995), *Mechanics of wave reflections in a coronary bypass loop model: The possibility of partial flow cut-off,* J. Biomech. *28,* 567–574.

Ferrazzini, V. and Aki, K. (1987), *Slow waves trapped in a fluid-filled infinite crack: Implications for volcanic tremors,* J. Geophys. Res. *92,* 9215–9223.

Fischer, G. J., *The determination of permeability and storage capacity: Pore pressure oscillation method.* In *The Determination of Permeability and Storage Capacity: Pore Pressure Oscillation Method* (eds. Evans, B. and Wong, T.-F.) (Academic Press, New York, 1992) pp. 187–211.

Fischer, G. J. and Paterson, M. S., *Measurement of permeability and storage capacity in rocks during deformation at high temperature and pressure.* In *Measurement of Permeability and Storage Capacity in Rocks during Deformation at High Temperature and Pressure* (eds. Evans, B. and Wong, T.-F.) (Academic Press, New York, 1992) pp. 213–252.

Gelinsky, S., Shapiro, S. A., Muller, T. M., and Gurevich, B. (1998), *Dynamic poroelasticity of thinly layered structures,* Int. J. Solids Struct. *35,* 4739–4752.

Gurevich, B., Zyrianov, V. B., and Lopatnikov, S. L. (1997), *Seismic attenuation in finely layered porous rocks: Effect of fluid flow and scattering,* Geophys. *62,* 319–324.

Hsieh, P. A., Tracy, J. V., Neuzil, C. E., Bredehoeft, J. D., and Silliman, S. E. (1981), *A transient laboratory method for determining the hydraulic properties of 'tight' rocks - I. Theory,* Int. J. Rock Mech. Min. Sci. and Geomech. Abstr. *18,* 245-252.

Johnson, D. L., Koplik, J., and Dashen, R. (1987), *Theory of dynamic permeability and tortuosity in fluid-saturated porous media,* J. Fluid Mech. *176,* 379–402.

Kamath, J., Boyer, R. E., and Nakagawa, F. M. (1992), *Characterization of core-scale heterogeneities using laboratory pressure transients,* SPE Formation Evaluation *7,* 219–227.

Kelder, O. and Smeulders, D. M. J. (1997), *Observation of the Biot slow wave in water-saturated Nivelsteiner sandstone,* Geophys. *62,* 1794–1796.

Klimentos, T. and McCann, C. (1988), *Why is Biot slow compressional wave not observed in real rocks?* Geophys. *53,* 1605–1609.

Korneev, V. (2008), *Slow waves in fractures filled with viscous fluid,* Geophys. *73,* N1–N7.

Muller, T. M., Lambert, G., and Gurevich, B. (2007), *Dynamic permeability of porous rocks and its seismic signatures,* Geophys. *72,* E149–E158.

Nagy, P. B., Adler, L., and Bonner, B. P. (1990), *Slow wave propagation in air-filled porous materials and natural rocks,* Appl. Phys. Lett. *56,* 2504–2506.

Neuzil, C. E., Cooley, C., Silliman, S. E., Bredehoeft, J. D., and Hsieh, P. A. (1981), *A transient laboratory method for determining the hydraulic properties of 'tight' rocks, 2. Application,* Int. J. Rock Mech. Min. Sci. and Geomech. Abstr. *18,* 253–258.

Paillet, F. L. and White, J. E. (1982), *Acoustic modes of propagation in the borehole and their relationship to rock properties,* Geophys. *47,* 1215–1228.

Plona, T. J. (1980), *Observation of a second bulk compressional wave in a porous medium at ultrasonic frequencies,* Appl. Phys. Lett. *36,* 259–261.

Pride, S. R., Gangi, A. F., and Morgan, F. D. (1992), *Deriving the equations of motion for porous isotropic media,* J. Acoust. Soc. Am. *92,* 3278–3290.

Pride, S. R., Tomeur, E., and Berryman, J. G. (2002), *Biot slow-wave effects in stratified rock,* Geophys. *67,* 271–281.

Shapiro, S. A. and Muller, T. M. (1999), *Seismic signatures of permeability in heterogeneous porous media,* Geophys. *64,* 99–103.

Song, I., Elphick, S. C., Main, I. G., Ngwenya, B. T., Odling, N. W., and Smyth, N. F. (2004), *One-dimensional fluid diffusion induced by constant-rate flow injection: Theoretical analysis and application to the determination of fluid permeability and specific storage of a cored rock sample,* J. Geophys. Res. *109,* B05207, doi:05210.01029/02003JB002395.

Song, I. and Renner, J. (2006), *Experimental investigation into the scale dependence of fluid transport in heterogeneous rocks,* Pure Appl. Geophys. *163,* 2103–2123.

Wang, J. J. and Parker, K. H. (2004), *Wave propagation in a model of the arterial circulation,* J. Biomech. *37,* 457–470.

Zamir, M. (1998), *Mechanics of blood supply to the heart: Wave reflection effects in a right coronary artery,* Proc. R. Soc. Lond. B *265,* 439–444.

Zamir, M., *The Physics of Pulsatile Flow* (Springer Verlag, New York, 2000).

Zhou, M.-Y. and Sheng, P. (1989), *First-principles calculations of dynamic permeability in porous media,* Phys. Rev. B *39,* 12,027–012,039.

(Received January 29, 2008, revised July 10, 2008, accepted July 11, 2008)
Published Online First: May 7, 2009

Pure appl. geophys. 166 (2009) 995–1010
0033–4553/09/050995–16
DOI 10.1007/s00024-009-0483-4

Pure and Applied Geophysics

Microstructural Investigation of Naturally Deformed Leucogneiss from an Alpine Shear Zone (Southern Calabria – Italy)

ROSOLINO CIRRINCIONE, EUGENIO FAZIO, PATRIZIA FIANNACCA, GAETANO ORTOLANO, and ROSALDA PUNTURO

Abstract—In order to reveal any correlation between the preferred orientation of quartz and progressive intensity of strain, a suite of deformed leucogneiss collected within a ductile shear zone outcropping in Calabria (southern Italy) was investigated. Based on the microstructural approach and matrix-clast relationships, the method applied here may be useful for appraising the connection between deformation mechanisms of constituent minerals and bulk textural properties of naturally sheared rocks. Accordingly, quartz c-axis orientation patterns were determined by image-assisted analysis. Results revealed a correlation between finite strain and textural features: As strain increases, the matrix develops at the expense of the clast counterpart, which is instead progressively reduced in both size and amount.

Key words: Textural anisotropy, quartz c-axis pattern, bulk rock rheology, shear-zone, mylonite.

1. Introduction

Shear zones are particular tectonic settings in which strain is located within the Earth's crust. They act as preferential pathways accommodating the deformation which accumulates on a geological time scale in response to plate tectonics and, at the same time, operate as efficient systems in both exhumation and burial processes during orogenic phases. Due to their geological and structural significance, shear zones play an important role in geodynamic reconstructions, since they are important environments for understanding crustal rheology. As progressive deformation continues, rocks show rheological behaviour varying in time and space, and their textures may thus furnish important insights on the dependence of rock properties with varying intensive parameters such as temperature, confining pressure, strain gradient, etc.

Recent enhanced knowledge (CHRISTIANSEN and POLLARD, 1997; STIPP *et al.*, 2002; HEILBRONNER and TULLIS, 2006; KURZ and NORTHRUP, 2008), not only of microstructural deformation mechanisms but also better definition of the flow laws that control the rheological behaviour of mineral phases in varying conditions (e.g., strain rate,

Department of Geological Sciences, University of Catania, Corso Italia 57, I-95129 Catania, Italy.
E-mail: punturo@unict.it

homologous temperature, fluid availability), have given rise to new perspectives in the characterisation of fault-related rocks.

The purpose of this work is to provide better understanding of the relationship between deformation mechanisms and the textural properties of naturally deformed rocks, by monitoring rock fabrics which are related to different finite strain conditions and therefore proposing a helpful method for study of a mylonitic shear zone.

With this aim in mind, an outstanding example of clear strain gradient within a crustal-scale shear zone is provided by the Montalto Shear Zone, located in the southern sector of the Calabria-Peloritani Orogen (southern Italy; Fig. 1). This is a composite orogenic segment of the internal Alpine chain outcropping in the peri-Mediterranean

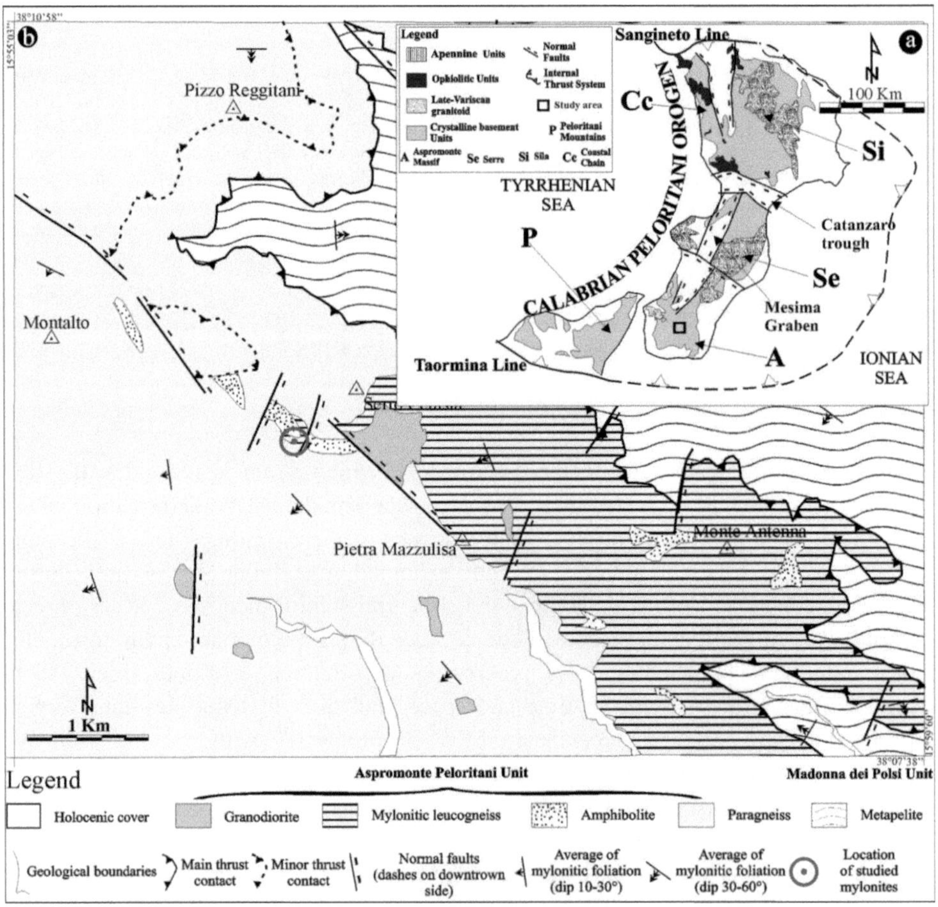

Figure 1

Geological sketch-map of study area (modified after Pezzino *et al.*, 1990). a) Simplified geological sketch of Calabrian-Peloritani Orogen; b) structural-geological map of Montalto crustal-scale shear zone, showing the location of outcrop.

region, and consists of a nappe-pile edifice enclosing Palaeozoic basement rocks and their Mesozoic sedimentary cover. In particular, the Montalto Shear Zone developed during the Alpine orogenesis (from 33 to 28 Ma; HEYMES *et al.*, 2008), at the boundary between an upper Hercynian amphibolite facies unit and an underlying early Alpine greenschist facies unit (PEZZINO *et al.*, 2008), involving several lithotypes (marble, amphibolite, micaschist, phyllite, leucogneiss, augen gneiss, paragneiss), all with differing rheological behaviour.

In order to quantify the progressive strain increment produced by this Alpine shearing event, we selected an outcrop (of approximately 6 m^3), characterised by a monoclinal attitude and located at about two kilometres ESE of Mt. Montalto (geographic coordinates: 38°9′1.13″N, 15°56′42.16″E) (Fig. 1). This site was chosen because of: **a)** the homogeneity of the constituent lithotype (leucogneiss); **b)** the relatively simple rheology of its constituents (mainly quartz and feldspars); **c)** the rare occurrence of brittle structures disturbing the continuity of the mylonitic fabric. Microstructural analysis was therefore carried out on three mylonitic rocks considered to be representative of different finite strain conditions.

Aimed at evaluating the strain gradient of this shear zone, our work included: **a)** a detailed structural and geological survey, for planning meaningful sample collection; **b)** a microstructural description of the sheared rocks, including the definition of dominant deformation mechanisms; **c)** image-assisted modal analysis of rocks, focusing on the rheological classification of minerals; **d)** determination of the preferred crystallographic orientation of mineral phases such as quartz, which control the bulk rheology of the mylonitic rock.

2. *Geological Setting*

The studied shear zone is an Alpine crustal-scale mylonitic horizon located in the southern sector of the Calabria-Peloritani Orogen (southern Italy; Fig. 1), along which two main basement tectonic slices, Madonna di Polsi Unit (MPU) and Aspromonte Peloritani Unit (APU), were stacked during late Oligocene-early Miocene syn-convergent exhumation (ORTOLANO *et al.*, 2005). The two units have different preshear tectono-metamorphic histories: High-pressure Alpine-type metamorphism exclusively affecting the MPU, and high-temperature Hercynian-type metamorphism exclusively affecting the APU. Both units show evidence of a greenschist-facies metamorphic overprint, since they were involved in the same late Alpine mylonitic shearing event that stacked them during the late Miocene (thin-skinned Alpine-Apennine thrust system).

This structural framework reveals how the southern sector of the Calabria-Peloritani Orogen is the result of a complex evolution which led to the formation of an Alpine crustal-scale mylonitic belt, the most representative outcrops of which are essentially located along the contact between the Hercynian amphibolite-facies rocks of the APU and the underlying greenschist to lower amphibolite-facies metapelites of the MPU

(Cirrincione *et al.*, 2008). This shearing event produced pervasive mylonitic foliation and stretching lineation: The former strikes averagely N45, with dip ranging between 20° and 80° SE or NW; the stretching lineation plunges 15–35°, alternatively SW or NE. The mylonitic rocks contain kinematic indicators, showing a top-to-NE sense of shear in the present-day geographic coordinates. Depending on the rheological behaviour of the various lithotypes involved, this shearing event gave rise to a series of non-coaxial structures such as σ- and δ- type porphyroclasts, S-C fabric, C'-shear bands and mica fish (Passchier and Trouw, 2005).

Evolution of the compressional thrust belt in shallower conditions produced SE-verging asymmetrical metric to hectometric folds, which involved the original mylonitic foliation together with the tectonic contact between APU and MPU, causing pronounced thickening of the original mylonitic band which currently ranges from 0.5 to 0.8 km.

In this work, attention focuses on rheological study of a single leucogneiss outcrop from the APU, considered as an ideal site for inferring the strain gradient within the shear zone.

3. Methods

The leucogneiss outcrop was selected on the basis of detailed structural field data (Pezzino *et al.*, 1990; 1992; Ortolano *et al.*, 2005; Fazio *et al.*, 2007; 2008). Sampling criteria involved rocks showing: **1)** Mesoscopically evident grain-size reduction, covering possible strain variations inside the shear zone; **2)** occurrence of abundant interconnected matrix (essentially quartz); **3)** absence of visible fractures and/or evidence of weathering.

According to the microstructural classification proposed by Schmid and Handy (1991), leucogneiss rocks were considered as composed of two rheologically different phases consisting of porphyroclasts (mostly K-feldspar) and matrix (mainly quartz). This discrimination is based both on a distinction between rheologically stronger and weaker mineral(s) and on the definition of dominating deformation mechanism(s) operating during shear strain evolution.

Following the above criteria, microstructural study focused on three oriented and progressively sheared leucogneiss specimens (samples M8, M12, M13), selected from the same outcrop and characterised by an interconnected quartz-rich matrix, more than 30% in volume, which thus governs the bulk rheology of the rock (Handy, 1990).

Clast and matrix abundances were quantified with an integrated analytical technique combining modal investigation by the point-counter method (about 2000 points for each sample) and image-assisted analysis, with the application of a density slice and phase recognition tools of SXM image analytical software (v. 1.84; Barrett, 1997) on pictures of the entire thin section acquired by a scanner (Epson Perfection 3490 Photo) at a resolution of 1200 dpi, with either single and crossed polariser sheet. Microstructure quantification was also carried out by the image analysis technique applied to thin-section pictures taken within quartz microdomains. All pictures were taken with a Nikon D70 reflex-digital camera mounted on a Zeiss Axiolab microscope with ordinary white light.

Then, within each sample, various quartz-rich domains were chosen, which were considered to be representative of the main shear plane rather than the result of the local orientation of secondary foliations (e.g., S-C fabrics and C'-shear bands). The choice of several sites (Fig. 2) was based on the statistical significance of measurements and the need to limit, or theoretically avoid, the influence of local effects on finite strain due to the potential effects of the larger porphyroclasts.

The crystallographic orientation of quartz grains was quantified by the AAVA method (Automatized Achsen Verteilungs Analyse) introduced by SANDER (1950) and automatised with AVA Generator and StereoNett software (DUYSTER, 1996), as reported in STOCKHERT and DUYSTER (1999). This software package applies the image analysis technique to calculate the azimuth and inclination of the quartz c axes at every position within the field of view by recording the changes in birefringence colours when the microscope is rotated by 90° for every pixel position. It should be noted that the azimuth determined here was bipolar (i.e., no distinction is made between a dip NE or SW).

This method yielded the crystallographic orientation of quartz, allowing us to examine the contribution of bulk rheology, with particular regard to the role of the activation of different slip systems as an effect of shearing temperature (SCHMID and CASEY, 1986; KRUHL, 1996; CAVALCANTI *et al.*, 2003).

4. *Structural and Textural Data*

The studied leucogneiss outcrop has two main mesoscopic structures: Pervasive mylonitic foliation averagely oriented 20°/230° (dip/dip direction) and marked stretching lineation at about 18°/185° (dip/dip direction). Mineral abbreviations used here are after KRETZ (1983).

The mylonitic leucocratic gneisses show a porphyroclastic texture, with 3–10 mm sized augens, mostly K-feldspar grains, in a fine-grained strongly foliated quartz-rich matrix. The common assemblage is Qtz + Wm + Pl + Kfs; accessory phases are garnet, biotite, epidote, tourmaline and chlorite. The resulting modal abundances, inferred by the point-counter method, were refined by image-assisted analysis and two main rheological categories were identified: clast and matrix (Table 1), with contrasting rheology. Because of the abundant amounts of matrix (from 64.92 to 75.90 vol%), these mylonitic rocks plot within the clast-matrix domain of the diagram of SCHMID and HANDY (1991) for genetic classification of fault-rocks, in which the interconnected matrix controls the bulk rheology of the rock.

The matrix (mostly Qtz, Wm and Pl) behaves as the weak constituent of the rock, and the porphyroclasts (mostly Kfs, Wm and Pl) as the strong ones.

The observed textural features mean that the study area exhibits the most peculiar aspects of the evolution of a complex crustal-scale shear zone. The site clearly shows that

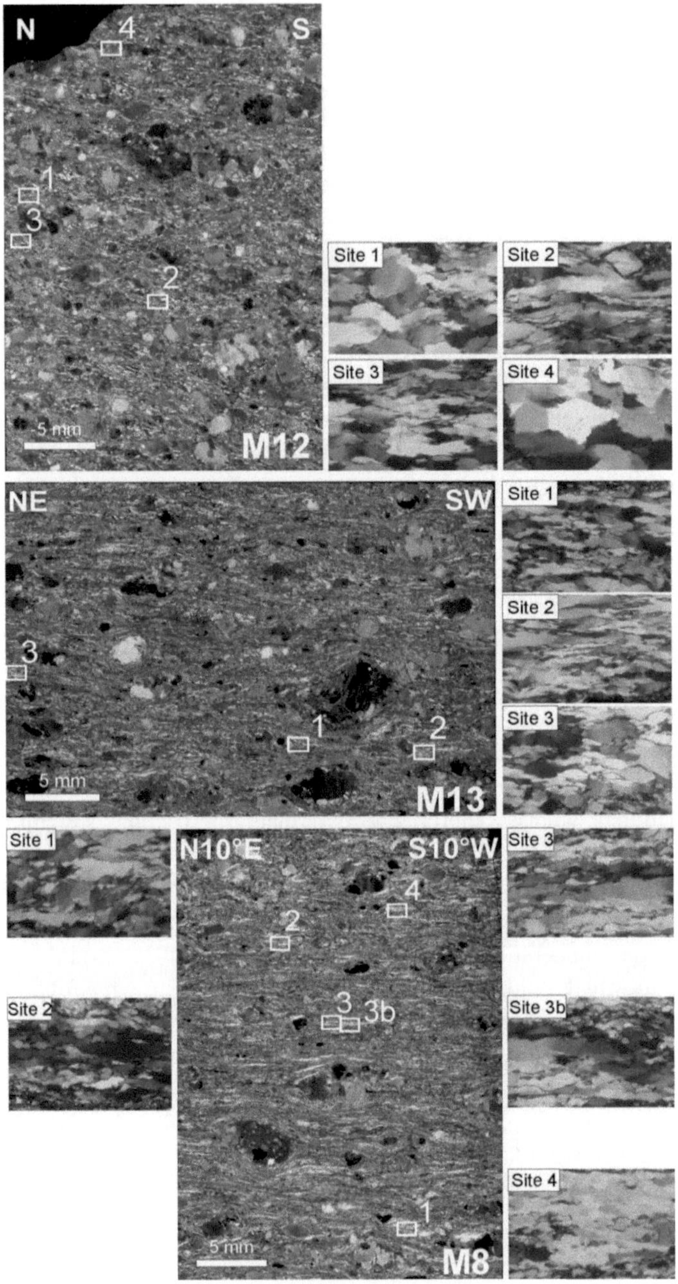

Figure 2
Location of quartz-rich domains within each thin section (samples M8, M12, M13), selected to determine quartz c-axis orientation patterns by means of image analysis techniques.

Table 1

Modal abundances of mylonites grouped into two main classes (clast and matrix) according to rheological behaviour. Brackets indicate small amounts of minerals

	Class	*Mineral Vol. %*	*Sample*		
			M12	M13	M8
polymineral aggregate	*Clast*	Kfs	21.02	9.90	10.73
		Wm + Bt	6.80	8.79	9.02
		Grt + Tur	1.31	5.40	4.03
		Pl	5.95	8.77	0.33
		Subtotal	***35.08***	***32.86***	***24.10***
	Matrix	Qtz	45.85	59.20	47.43
		Wm + Bt + (Chl)	10.98	7.41	16.13
		Pl + Ep	8.08	0.53	12.33
		Subtotal	***64.92***	***67.14***	***75.90***
	Sum	*Total*	*100.00*	*100.00*	*100.00*

various intra-crystalline deformation mechanisms operated during strain evolution, as a consequence of the rheological behaviour of each mineral phase during the evolution of the mylonitic event. In particular, the deformation behaviour of mineral phases evolves from: **a)** Solid-state flow deformation developing in a high-temperature regime, testified by syn-kinematic growth of biotite and feldspar grains showing flame perthites (VERNON, 1999), surrounded by syn-tectonically recrystallised ribbon-like quartz; to: **b)** semi-brittle behaviour, indicated by fractured sigmoidal garnet aggregates and bookshelf-sliding of feldspars, characterised by antithetic fractures with respect to the direction of shear, both surrounded by grano-xenoblastic ribbon-like quartz.

The quartz c-axis orientation patterns are shown in Figure 3. In all the pictures, the reference foliation and lineation (S_r, L_r) corresponding to mylonitic foliation and stretching lineation are horizontal and directed E-W. The quartz c-axis fabrics show specific relationships with the kinematic stretching axes during deformation, so that it is inferred that the environmental conditions of the shear zone were activated by various deformation mechanisms (LISTER and DORNSIEPEN, 1982; MAINPRICE *et al.*, 1986; HIRTH and TULLIS, 1992; STIPP *et al.*, 2002; HEILBRONNER and TULLIS, 2006). The data were then quantified by AAVA maps through the corresponding colour-coding scheme (top-right inset in all sketches of Fig. 3) and the relative quartz c-axis stereoplot. For each sample, results are then reported for the selected quartz-rich domains.

4.1. Sample M12 (Qtz + Wm + Kfs + Pl + Grt + Bt)

Three different intersecting foliations characterise this specimen; they are interpreted as S-C fabric and C'-type shear bands, indicating a top-to-NE sense of shear in the present-day geographic coordinates. Augens (major axes up to 10 mm) are essentially

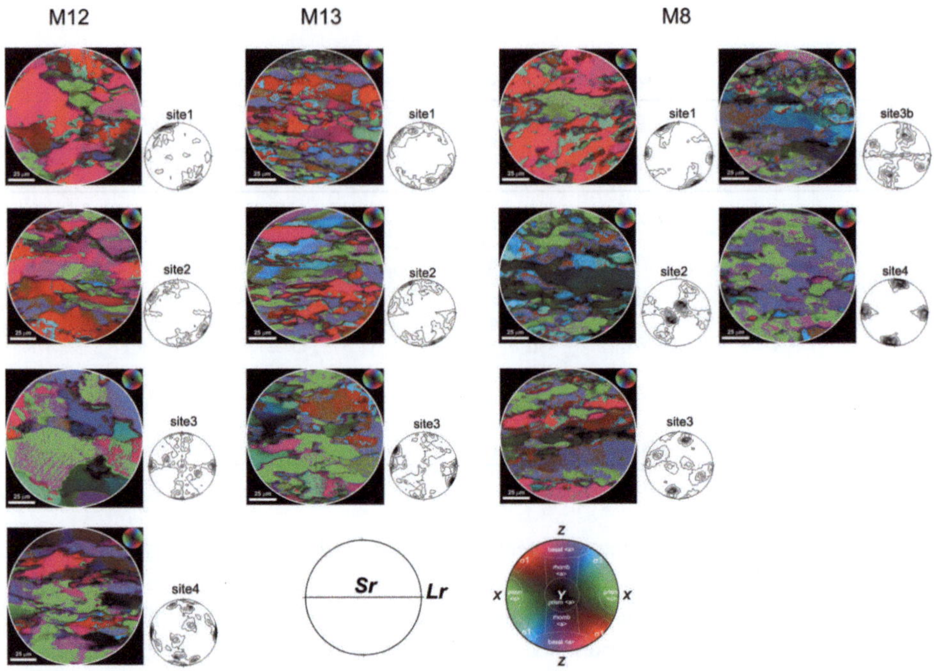

Figure 3
Various sites of quartz-rich domains selected for each sample and quartz c-axis patterns obtained by AAVA software (equal area projection, lower hemisphere; S_r is reference foliation plane; L_r is reference direction, parallel to stretching lineation). Reference stereoplot with slip systems activated at shear temperature: Modified after SCHMID and CASEY (1986) and HEILBRONNER and TULLIS (2006).

K-feldspar grains (Fig. 4a) with common plagioclase inclusions; they usually have irregular margins and sometimes form bookshelf-sliding microstructures. K-feldspar augens, frequently untwinned and showing flame perthites, reveal syn-tectonic recrystallisation, evident at the edges of porphyroclasts, where a fine mixture of white mica, quartz and epidote occurs, together with myrmekite colonies.

Garnet, which is also present as a porphyroclast, appears to be stretched along the plane of the mylonitic foliation and acts as a stronger mineral.

White mica and subordinate biotite are preferentially characterised by a sigmoidal shape, testifying to their pre-kinematic character. Rare intra-crystalline deformation produced kink banding. Quartz grains, preferentially elongated along the mylonitic foliation, form ribbon-like structures locally characterised by syn-kinematic grain boundary migration recrystallisation (Figs. 4d–f). The fine-grained quartz-rich matrix recrystallised in a dynamic regime, with major mineral axes elongated parallel to the local foliations (S and C' surfaces) and flows between porphyroclasts.

Lastly, it should be noted that another quartz population sometimes occurs at the border of large K-feldspar porphyroclasts. It was found in both pressure shadows and

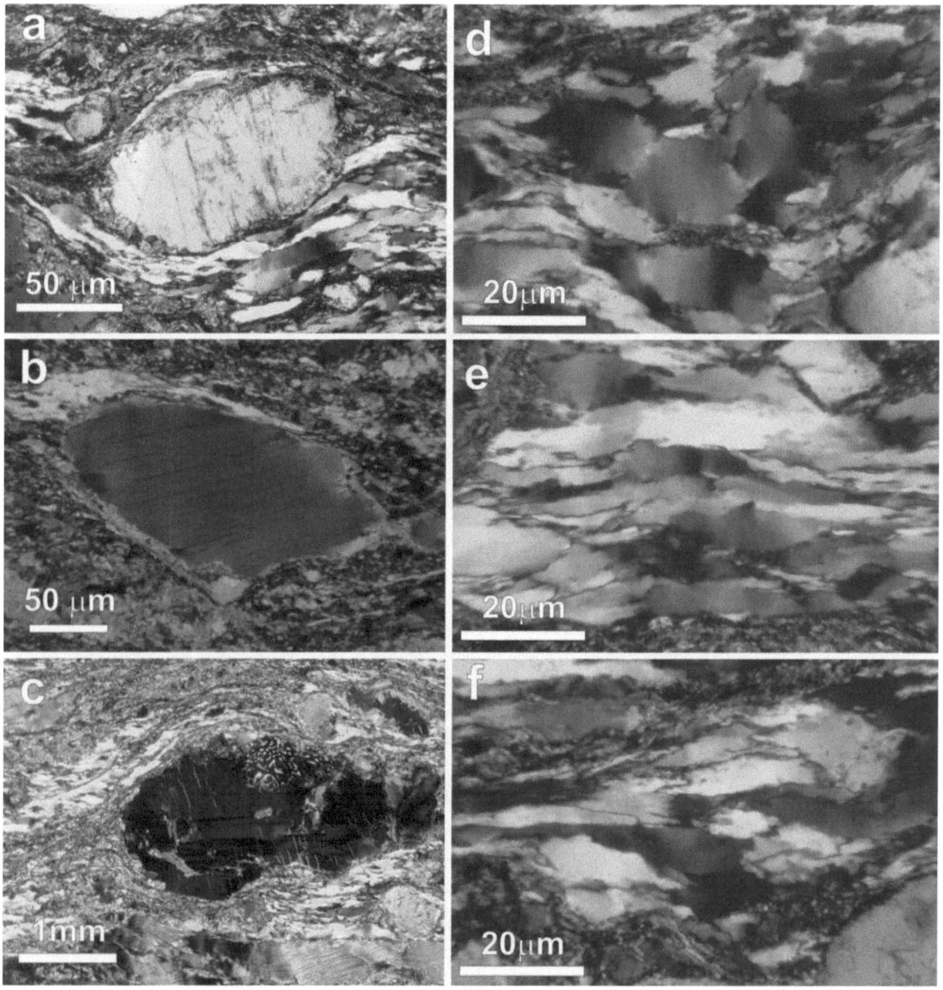

Figure 4

Photomicrographs of mylonites. a) Sigma-type feldspar porphyroclasts (sample M12) bordered by ribbon-like quartz grains, with tails mainly composed of white mica and sericite, also constituting the retrogression rim at the external edge of clast; b) pre-kinematic white mica augen (sample M8), internally deformed, as shown by curved cleavage planes and undulose extinction at clast margins; c) K-feldspar porphyroclast, with flame perthites and myrmekite at margins (sample M13) and a large mica crystal (bottom of Fig.) and other clasts defining the main near-horizontal foliation; d, e, f): Possible evolution of textures due to incremental shear strain inside quartz-rich domains. (sample M12): Note progressive elongation of grains, starting from quartz grains forming a foam texture (relic) (top), which become more elongated (center), evolving towards ribbon-like quartz in last stage of finite strain (bottom).

inside flank gulfs (Fig. 5b) and shows triple junctions, patchy extinction, and no elongation. These quartz grains were interpreted as relics of the early fabric, preserved in low strain domains near the edges of feldspar porphyroclasts. These relics form a foam

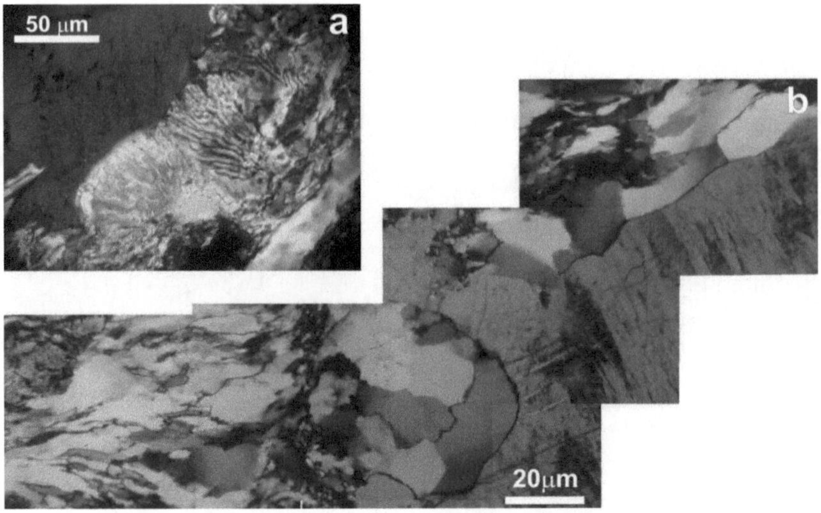

Figure 5
Sample M13: a) Myrmekites developing near a K-feldspar margin; b) polygonal quartz grains preserved inside a gulf of external border of a K-feldspar augen. These grains are very different in size with respect to matrix-forming quartz grains.

texture composed of polygonal quartz and plagioclase with linear grain boundaries and interfacial angles of about 120°.

The results of c-axis pattern analysis of this sample show how the prevalent activated slip systems are the prism ⟨c⟩ (sites 2, 3, 4) and the basal ⟨a⟩ (site 1). In addition, secondary rhomb ⟨a⟩ and prism ⟨a⟩ slip systems are visible in sites 3 and 4 (Fig. 3, Table 2).

Table 2

Quartz c-axis patterns and related active slip systems. Temperature ranges after Cavalcanti *et al. (2003)*

Metamorphic facies	Approx. T (°C)	C-axis pattern	Active slip system	M12	M13	M8
Lower greenschist + Faster strain rates	400°–450°	type 1 near Z maximum	basal ⟨a⟩	moderate	moderate	dominant
Mid-greenschist	450°–500°	type 2	rhomb ⟨a⟩	weak	weak	weak
Amphibolite	500°–650°	type 3 near Y maximum	prism ⟨a⟩	weak	nearly absent	moderate
Upper amphibolite	650°–700°	type 4 near X maximum	prism ⟨c⟩	dominant	moderate	moderate
Low shear strain	–	type 5 c-axes close to $\sigma 1$	Unfavorable slip	weak	moderate	Nearly absent

4.2. *Sample M13 (Qtz + Wm + Kfs + Pl + Grt + Bt + Ep + Tur)*

This rock is characterised by dominant S-C fabric, defined by the alignment of mica and quartz-rich layers, whereas the C'-type shear bands are poorly developed.

Porphyroclasts, up to 7 mm in diameter, are mostly composed of K-feldspar with variably developed cross-hatched twinning. White mica, plagioclase and rare garnet sometimes form porphyroclasts, all adding the effects of the stronger phases to the whole rheological behaviour.

Instead, perthites, frequent within large grains (Figs. 4c, 5a), are poorly developed within the smaller K-feldspar porphyroclasts. Large K-feldspar porphyroclasts are often poikiloblastic, due to the occurrence of idiomorphic to allotriomorphic plagioclase with common albite rims; biotite and subordinate small muscovite flakes may also be present. The enclosed plagioclase either show magmatic features such as euhedral tabular habit, simple twinning and idiomorphic zoning, or may result from the textural evolution of myrmekites (VERNON and COLLINS, 1988). The latter are mostly visible in the strain caps of the large porphyroclasts and are bulbous, with convex boundaries towards the K-feldspar (Figs. 4c, 5a). Quartz-muscovite symplectites may also be found in the same textural position. K-feldspar porphyroclasts are mantled by white mica and a fine-grained quartz-rich matrix.

Widespread ribbon-like quartz levels are well developed. They have variable grain size, indicating multistage syn-shear crystallisation, as highlighted by large relic grains with diffuse undulatory extinction, surrounded by small new ones forming an incipient oblique foliation parallel to the S surfaces. Relic polygonal quartz is also shown by the foam texture inside feldspar flank gulfs (Fig. 5b). Analysis of quartz c-axis patterns reveals activation of basal $\langle a \rangle$ slip at differing intensities. It is weak in sites 1 and 3, but becomes dominant in site 2. Prism $\langle c \rangle$ slip is nearly absent in site 1, discernible in site 2, and moderately developed in site 3. Rhomb $\langle a \rangle$ and prism $\langle a \rangle$ are minor. A maximum of c-axis density is present about 45° from X and Z positions, close to the great circle (Fig. 3, Table 2).

4.3. *Sample M8 (Qtz + Wm + Kfs + Pl + Grt + Ep + Tur)*

Like the other two mylonites, sample M8 has three intersecting foliations (S-C fabric C'-type shear bands). Augen boundaries are often more rounded and smoothed. Micas (Fig. 5b) and subordinate garnet, together with feldspars, also constitute large porphyroclasts. Mica is often fish-shaped and bookshelf-sliding structures are sometimes present.

K-feldspar augens (up to 5 mm) are mantled by a fine-grained white mica-quartz aggregate surrounding the porphyroclast edges. Relics of pre-kinematic K-feldspar are commonly involved in retrograde crystallisation of white mica, quartz, plagioclase and epidote, probably due to widespread fluid circulation during the last exhumation phase; diffuse sericitisation and saussuritisation were also detected within plagioclase.

Ribbon-like quartz layers constitute most of the rock; large relic quartz grains with undulatory extinction are almost entirely replaced by recrystallised syn-tectonic smaller grains, which sometimes give rise to an oblique foliation.

Within quartz domains, grain boundary migration recrystallisation and incipient subgrain rotation recrystallisation were also observed, probably reflecting strain accommodation by a dislocation creep mechanism. The c-axis pattern of sample M8 has various activated slip systems. Basal ⟨a⟩ slip is dominant in sites 1, 3 and 4. Prism ⟨c⟩ is weakly developed, except for sites 1 and 4, in which it otherwise prevails. Prism ⟨a⟩ slip was observed in sites 2, 3 and 3b, whereas rhomb ⟨a⟩ slip occurs exclusively in sites 2 and 3b (Fig. 3, Table 2). There is no evidence of relic polygonal texture.

5. Discussion

Microstructural study of naturally deformed mylonitic rocks (samples M12, M13 and M8; Fig. 2) allowed us to evaluate the shear-strain gradient as well as the thermal shearing evolution of a leucogneiss level within the Alpine crustal-scale shear zone of Montalto (Fig. 6). The bulk rheology of these samples was ascertained according to the fault-rock genetic classification criteria of Schmid and Handy, (1991). Determination of the LPO patterns of the most abundant matrix-forming mineral (i.e., quartz) also revealed the environmental conditions operating during mylonitic shearing evolution.

Although the leucogneiss rocks share several microstructural features (e.g., σ- and δ-type feldspar porphyroclasts; pre-kinematic mica fish; ribbon-like quartz), they were discriminated on the basis of the relative difference in the percentage amount and dimension of porphyroclasts, and by means of quartz c-axis pattern analysis, which are both meaningful controlling factors of the bulk rheology of naturally deformed rocks.

Sample M12, with the lowest shear strain, has the highest content in the stronger phases (35.08 vol %; Table 1) and less grain size reduction of porphyroclasts, especially K-feldspar (up to 10 mm sized). K-feldspar frequently shows flame perthites and

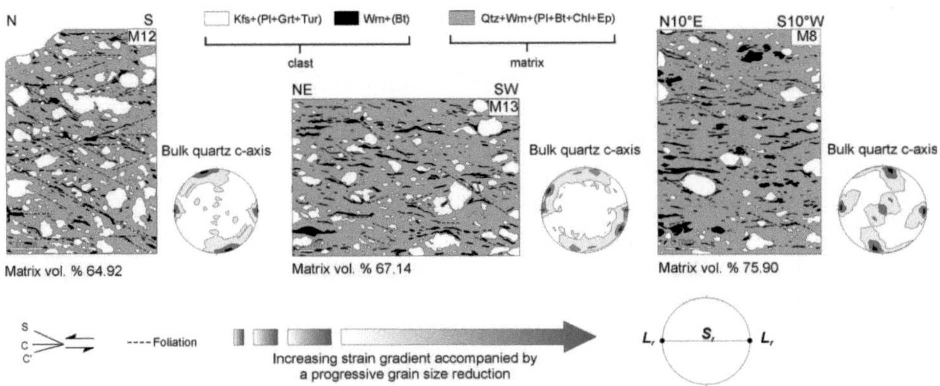

Figure 6
Classified images of thin sections (samples M8, M12 and M13), mineral modal abundances, and bulk quartz crystallographic preferred orientation. Matrix amount in different samples is positively correlated to increasing strain gradient within shear zone.

myrmekites at its borders, both indicating solid-state flow (Vernon, 2000). Syn-tectonic crystallisation of fine-grained aggregates of white mica and quartz was also observed at edges of porphyroclasts. High-temperature microstructural relics of quartz, like coarse and polygonal grains, are also preserved in low-strain domains at the edges of feldspar porphyroclasts (Fig. 5b). Quartz c-axis pattern analysis confirms the presence of several generations, with activation of moderate basal ⟨a⟩ and dominant prism ⟨c⟩ slip systems, consistent with high-temperature shearing conditions (Fig. 6, Table 2).

Sample M13 contains intermediate amounts of the strong constituent (32.86%; Table 1), the dimensions of augens being 4 mm on average. Myrmekites and flame perthites are widespread both at the edges and within K-feldspar grains, which are further mantled by a very fine-grained layer of quartz and white mica. Although ribbon-like quartz is abundant, early polygonal quartz forming a foam texture is still preserved in low strain domains (e.g., inside the gulfs of augen boundaries). All these features suggest that, compared with sample M12, this rock was affected by higher shear strain. The quartz c-axis pattern of sample M13 shows moderate activation of three distinct slip systems (Fig. 6; Table 2), corresponding to shearing evolution over a wide thermal range.

Lastly, sample M8 has the lowest percentage of strong constituent (24.10%; Table 1). Porphyroclasts, essentially K-feldspar and white mica, are more rounded than those occurring within the other two mylonites (Fig. 6). K-feldspar forms large porphyroclasts and white mica small ones, depicting both σ- and δ-type objects. Quartz grains (dominantly ribbons) have their maximum elongation parallel to the main mylonitic foliation, whereas no relics of polygonal quartz are present; here, shear strain attained its maximum value. The quartz c-axis pattern of this sample exhibits a dominant basal ⟨a⟩ slip system, with subordinate prism ⟨c⟩ and prism ⟨a⟩ slip systems (Table 2). The LPO patterns were consistent with greenschist-facies shearing conditions.

According to temperature and strain rate, the quartz c-axis patterns of all samples indicate the activation of several slip systems during shearing deformation history (Table 2). This complex shear evolution is further confirmed by the occurrence of various quartz microtextures, reflecting thermal ranges of syn-shear crystallisation (Stipp *et al.*, 2002). Quartz textures visible in the differing domains of the rock do indicate different recrystallisation conditions: low temperature is indicated by bulging microstructure (BLG), intermediate temperature by subgrain rotation recrystallisation microstructure (SGR), and high temperature by grain boundary migration recrystallisation microstructure (GBM). It is worth noting that the most frequent microstructure (SGR) is due to intermediate-temperature conditions, whereas evidence of high-temperature microtexture (GBM) is only visible within the pressure shadows of large porphyroclasts.

On the whole, quartz c-axis patterns indicate that the main active slip systems are basal ⟨a⟩ and prism ⟨c⟩, with subordinate rhomb ⟨a⟩ and prism ⟨a⟩. According to Schmid and Casey (1986), the prevailing slip systems were generated in very different thermal shearing conditions, as revealed by the dominant prism ⟨c⟩ pattern in sample M12 (upper amphibolite-facies conditions) and the dominant basal ⟨a⟩ in sample M8 (lower greenschist-facies conditions).

These apparently contradictory data may be interpreted as the result of: **a)** The influence of factors which shift the patterns usually associated with high temperatures to lower thermal shearing conditions; **b)** the "memory" of a former, distinct, high-temperature event, followed by the low-temperature Alpine event; **c)** a complex tectono-thermal shear history, evolving from high amphibolite-facies conditions at deep crustal levels towards greenschist-facies conditions at shallow levels, as a response to exhumation processes.

As regards the first hypothesis, particular fluid pressure conditions may enhance reactions and generate quartz c-axis patterns typical of higher temperatures than the true shearing climax. These occurrences are generally due to high partial fluid pressure (e.g., reactive pore fluids; SMITH and EVANS, 1984) or to a water weakening process (BLACIC, 1975; MAINPRICE *et al.*, 1986).

The second hypothesis involves the possibility that former quartz LPO relic patterns are maintained over a considerable period of time. However, it must be pointed out that intense shear events are often accompanied by a pervasive metamorphic overprint, which obliterates pristine microstructures and assemblages.

The third hypothesis is supported by examination of quartz c-axis patterns. Those related to high- temperature estimates (600-650°C) were observed in sample M12, that is a mylonite which, deformed to a lesser extent, therefore preserved the earliest deformation texture better. This "high-temperature" texture is rarely preserved in the other two samples, M8 and M13, in which the subsequent exhumation process caused re-equilibration of high-T fabrics (i.e., foam texture), as testified by the quartz slip systems ranging from ⟨c⟩-prism to ⟨a⟩-rhomb and ⟨a⟩-prism. In addition, the activation of a dominant basal ⟨a⟩ slip system in sample M8 (i.e., the most deformed; Fig. 6), due to the lowest shearing temperature (400-450°C; Table 2), is associated with the latest stage of shearing evolution which developed in greenschist-facies conditions.

On the whole, the three samples show a gradual decrease in clast amounts (from 35.08% to 24.10%) and an increase in interconnected matrix content (from 64.92% to 75.90%), which are due to increased finite strain within our rock suite (Fig. 6). This implies that the mylonitic micro-augen leucogneiss is characterised by matrix-controlled rheology, essentially composed of quartz, highlighting non-linear viscous creep deformational behaviour.

6. *Final Remarks*

This microstructural study allowed the textural properties of naturally sheared rocks to be determined, thanks to a rheological classification of rock mineral constituents based on a simplified subdivision into two categories: clast and matrix, respectively.

The lattice-preferred orientation of quartz defines two separate patterns due to the activation of slip systems representing high-temperature (⟨c⟩ prism) and low-temperature

(⟨a⟩ basal) conditions, respectively, and providing information for reconstruction of the tectono-thermal evolution of a crustal-scale shear zone.

In the future, we aim to examine the relationship between textural and seismic anisotropy, in order to reveal the influence of bulk rock features on the elastic properties of naturally deformed rocks as a consequence of increasing shear strain.

Acknowledgements

We are deeply indebted to Antonino Pezzino, whose valuable suggestions and constructive criticism helped us to improve the original manuscript greatly. We appreciated the review work of Hartmut Kern and Federico Rossetti, whose criticisms and useful suggestions significantly enhanced the clarity and organisation of the initial version. Editing carried out by Sergio Vinciguerra is appreciated. Finally, we thank G. Walton for revising the English expressions. Financial support comes from MIUR (PRIN 2007 project: "Strain rate in mylonitic rocks and induced changes in petrophysical properties across the shear zones").

References

Barrett, S. D. (1997), *Image analysis and the internet*, Scientific Data Management *1*, 18–25.

Blacic, J. D. (1975), *Plastic-deformation mechanisms in quartz: The effect of water*, Tectonophysics *27*, 271–294.

Cavalcanti de Arau'jo, M. N., Alves da Silva, F. C., Jardim de Sá, E. F., Holcombe, R. J., and De Vasconcelos, P. M. (2003), *Microstructural evolution of the Serido' Belt, NE-Brazil: the effect of two tectonic events on development of c-axis preferred orientation in quartz,* J. Struct. Geol. *25*, 2089–2107.

Christiansen, P. P. and Pollard, D. D. (1997), *Nucleation, growth and structural development of mylonitic shear zones in granitic rock,* J. Struct. Geol. *19*, 1159-1172.

Cirrincione, R., Ortolano, G., Pezzino, A., and Punturo, R. (2008), *Poly-orogenic multi-stage metamorphic evolution inferred via P–T pseudosections: An example from Aspromonte Massif basement rocks (Southern Calabria, Italy)*, Lithos *103*, 466–502.

Duyster, J. (1996), *StereoNett 2.0*, University of Bochum.

Fazio, E., Cirrincione, R., and Pezzino, A. (2008), *Estimating P–T conditions of Alpine-type metamorphism using multistage garnet in the tectonic windows of the Cardeto area (southern Aspromonte Massif, Calabria)*, Min. Petr. *93*, 111–142.

Fazio, E., Heilbronner, R., and Punturo, R. (2007), *Microstructural study and petrophysical characterization of naturally deformed leuco-gneisses from the Montalto shear zone (Aspromonte Massif, southern Italy)*, Rend. Soc. Geol. It. Nuova Serie *5*, 87–88.

Handy, M. R. (1990), *The solid-state flow of polymineralic rocks*, J. Geophys. Res., Solid Earth and Planets *95* (B6), 8647–8661.

Heilbronner, R. and Tullis, J. (2006), *Evolution of c-axis pole figures and grain size during dynamic recrystallisation: Results from experimentally sheared quartzite*, J. Geophys. Res. *111*, B10202, doi: 10.1029/2005JB004194.

Heymes, T., Monié, P., Arnaud, N., Bouillin, J. P., Pêcher, A., and Compagnoni, R. (2008), *Shortening vs. extension in the Aspromonte nappes-pile (Calabria, Italy). New insights from structural and geochronological data,* Geophys. Res. Abstr. *10*, SRef-ID: 1607–7962/gra/EGU2008-A-06173.

Hirth, G. and Tullis, J. (1992), *Dislocation creep regimes in quartz,* J. Struct. Geol. *14*, 145–159.

Kretz, R. (1983), *Symbols for the rock-forming minerals,* Am. Min. *68*, 277–279.

KRUHL, J. H. (1996), *Prism- and basal-plane parallel subgrain boundaries in quartz: A microstructural geothermobarometer*, J. Met. Geol. *14*, 581–589.

KURZ, G. A. and NORTHRUP, C. J. (2008), *Structural analysis of mylonitic rocks in the Cougar Creek Complex, Oregon-Idaho using the porphyroclast hyperbolic distribution method, and potential use of SC'-type extentional shear bands as quantitative vorticity indicators*, J. Struct. Geol. *30*, 1005–1012.

LISTER, G. S. and DORNSIEPEN, U. L. (1982), *Fabric transitions in the Saxony granulite terrane*, J. Struct. Geol. *4*, 81–92.

MAINPRICE, D., BOUCHEZ, J. L., BLUMENFELD, P., and TUBIA, J. M. (1986), *Dominant c slip in naturally deformed quartz: Implications for dramatic plastic softening at high temperature*, Geology *14*, 819–822.

ORTOLANO, G., CIRRINCIONE, R., and PEZZINO, A. (2005), *P–T evolution of Alpine metamorphism in the southern Aspromonte Massif (Calabria, Italy),* Schweiz. Min. Petr. Mitt. *85*, 31–56.

PASSCHIER, C. W. and TROUW, R. A. J., *Microtectonics* (Springer, Berlin, Heidelber 2005).

PEZZINO, A., ANGÌ, G., FAZIO, E., FIANNACCA, P., LO GIUDICE, A., ORTOLANO, G., PUNTURO, R., CIRRINCIONE, R., and DE VUONO, E. (2008), *Alpine metamorphism in the Aspromonte Massif: implications for a new framework of the southern sector of the Calabria-Peloritani Orogen (Italy)*, Int. Geol. Rev. *50*, 423–441.

PEZZINO, A., PANNUCCI, S., PUGLISI, G., ATZORI, P., IOPPOLO, S., and LO GIUDICE, A. (1990), *Geometry and metamorphic environment of the contact between the Aspromonte-Peloritani Unit (Upper Unit) and Madonna dei Polsi Unit (Lower Unit) in the central Aspromonte area (Calabria)*, Boll. Soc. Geol. It. *109*, 455–469.

PEZZINO, A., PUGLISI, G., PANNUCCI, S., and IOPPOLO, S. (1992), *Due unità cristalline a grado metamorfico diverso in Aspromonte centrale. Geometria dei loro rapporti, ambientazione metamorfica del loro contatto e caratteri petrografici delle metamorfiti,* Boll. Soc. Geol. It. *111*, 69–80.

SANDER, B., *Einführung in die Gefügekunde der geologischen Körper, Band II: Die Korngefüge* (Springer, Wien 1950).

SCHMID, S. M. and CASEY, M., *Complete fabric analysis of some commonly observed quartz c-axis patterns.* In *Mineral and Rock Deformation: Laboratory Studies* (the Paterson Volume) (eds. Heard, H.C., Hobbs, B.E.) (American Geophysical Union Monograph 36, 1986) pp. 263–286.

SCHMID, S. M. and HANDY, M. R., *Towards a genetic classification of fault rocks: geological usage and tectonophysical implications.* In *Controversies in Modern Geology* (eds. Muller, D.W., McKenzie, J.A., and Weissert, H.) (Academic Press, London 1991) pp. 339–361.

SMITH, D. L. and EVANS, B. (1984), *Diffusional crack healing in quartz*, J. Geophys. Res. *89*, 4125–4136.

STIPP, M., STÜNITZ, H., HEILBRONNER, R., and SCHMID, S. M. (2002), *The eastern Tonale fault zone: A 'natural laboratory' for crystal plastic deformation of quartz over a temperature range from 250 to 700°C*, J. Struct. Geol. *24*, 1861–1884.

STOCKHERT, B. and DUYSTER, J. (1999), *Discontinuous grain growth in recrystallised vein quartz – implications for grain boundary structure, grain boundary mobility, crystallographic preferred orientation, and stress history*, J. Struct. Geol. *21*, 1477–1490.

VERNON, R. H. and COLLINS, W. J. (1988), *Igneous microstructures in migmatites*, Geology *16*, 1126–1129.

VERNON, R. H. (1999), *Flame perthite in metapelitic gneisses at Cooma, SE Australia*, Am. Min. *84*, 1760–1765.

VERNON, R. H. (2000), *Review of microstructural evidence of magmatic and solid-state flow*, Electronic Geoscience.

Received March 12, 2008, revised September 25, 2008, accepted February 2, 2009)
Published Online First: May 12, 2009

Pure appl. geophys. 166 (2009) 1011–1036
0033–4553/09/051011–26
DOI 10.1007/s00024-009-0497-y

Pure and Applied Geophysics

Laboratory Characterization of Permeability and Its Anisotropy of Chelungpu Fault Rocks

TZU-MO N. CHEN,[1] WENLU ZHU,[2] TENG-FONG WONG,[1] and SHENG-RONG SONG[3]

Abstract—In Taiwan an international project to drill into the Chelungpu fault (TCDP) was initiated after the M_w 7.6 Chi-Chi earthquake in 1999. At Takeng, two vertical holes (A and B) to depths of about 2 km have been drilled through the northern portion of the Chelungpu fault system. In this study, we conducted systematic hydromechanical tests on TCDP drillcores collected from Hole-A at various depths above and below the major slip zone of the Chelungpu fault. We focus on the measurements of permeability as function of pressure and the brittle failure behavior. Evolution of permeability as a function of pressure and porosity was measured using either steady-state flow or a pulse transient technique. When subjected to an effective pressure reaching 100 MPa, permeability values of shaly siltstone samples range from 10^{-16} to 10^{-19} m^2. In comparison, permeability values of porous sandstones are at least an order of magnitude higher, ranging from 10^{-14} to 10^{-18} m^2. To characterize permeability anisotropy associated with the bedding structure of the rocks of the Chelungpu fault, cylindrical samples were taken from the TCDP drillcores along three orthogonal directions, denoted X, Y and Z respectively. Direction Z is parallel to the TCDP core axis, and the other two directions are perpendicular to the core axis, with X (N105°E) perpendicular and Y (N15°E) parallel to the strike of the bedding. In shaly siltstones, permeability values of samples cored along the strike of bedding (direction Y) can be up to 1 order of magnitude higher than those cored perpendicular to the strike of bedding (direction X). These observations indicate that permeability anisotropy is controlled by the spatial distribution of bedding in Chelungpu fault host rocks. Permeability evolution of fault rocks plays an important role in dynamic weakening processes, which are particularly pertinent to large earthquakes such as the Chi-Chi earthquake. Our experimental data on permeability and its anisotropy of TCDP core samples provide necessary constraints on fault models and proposed weakening mechanisms.

Key words: Permeability and porosity, Chelungpu fault, bedding anisotropy, deformation tests, dynamic weakening.

1. *Introduction*

The 1999 M_w 7.6 Chi-Chi earthquake in Taiwan was induced by the slip of Chelungpu thrust fault (Fig. 1) at a focal depth of 10 km (KAO and CHEN, 2000). The surface rupture of the Chi-Chi earthquake extended over 100 km, with scarps as high as $\sim$8 m in some locations. Synthesis of the teleseismic, strong motion and GPS data for

[1] Department of Geosciences, Stony Brook University, Stony Brook, NY 11794, U.S.A.
[2] Department of Geology, University of Maryland, College Park, MD 20742, U.S.A.
E-mail: wzhu@umd.edu
[3] Department of Geosciences, National Taiwan University, Taipei 106, Taiwan.

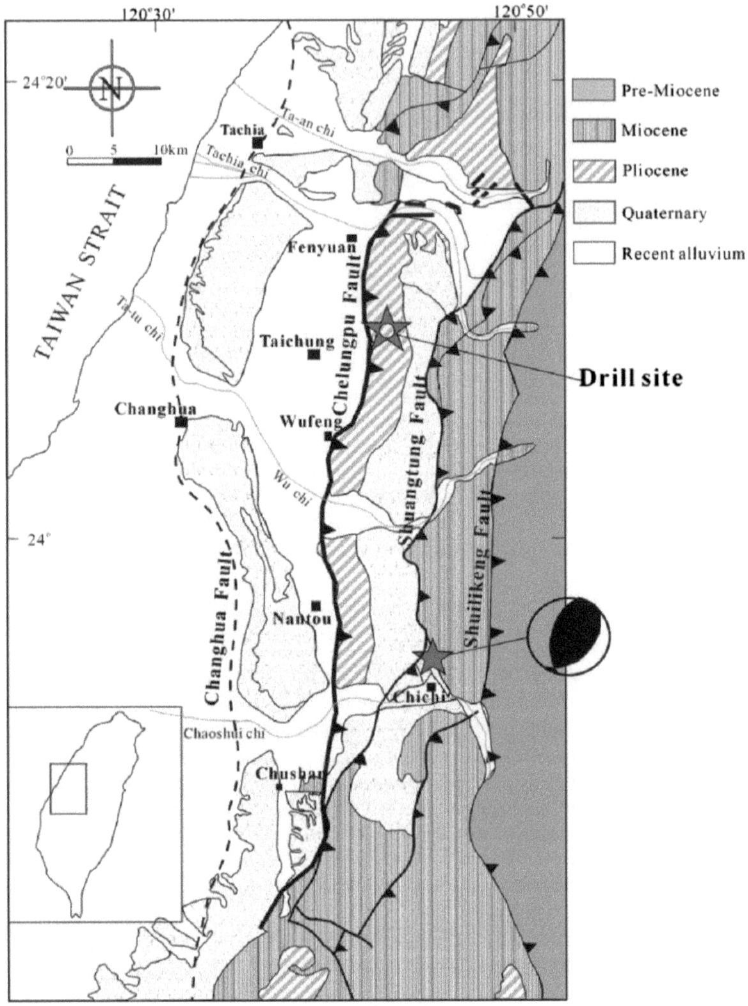

Figure 1
Geological map near the TCDP drill site, showing the surface rupture of the Chelungpu fault and the focal mechanism of the 1999 event. The star denotes the epicenter of this event which is near Chi-Chi (Song *et al.*, 2007).

the Chi-Chi earthquake indicates significant spatial heterogeneity in the slip displacement, velocity and acceleration. From Chushan (at the southern tip of the surface rupture) to Fengyuan the surface break extends over 80 km along the previously recognized Chelungpu fault, and then it deviates eastward as the rupture extended north for another 20 km (e.g., Johnson *et al.*, 2001). Most analyses show an overall trend for the slip to be appreciably higher in the northern region of the Chelungpu fault (Wang *et al.*, 2001).

While the ground velocity also follows a somewhat similar trend, surprisingly the ground acceleration tends to be higher in the southern part of the fault (SHIN and TENG, 2001; MA *et al.*, 2003).

The large slip displacement and the spatial-temporal heterogeneity of rupture of the Chelungpu fault during the Chi-Chi earthquake provide a unique opportunity to understand styles of faulting, earthquake cycles and wedge tectonics. To interpret the relatively low level of high-frequency radiation, large slip displacements and high slip velocities (up to 4.5 m/s) in the northern portion of the fault, several dynamic weakening mechanisms including frictional melting, elastohydrodynamic lubrication and thermal pressurization have been postulated (e.g., KANAMORI and HEATON, 2000; BRODSKY and KANAMORI, 2001; ANDREWS, 2002; RICE, 2006). Testing of these models requires systematic measurements on physical and chemical properties and detailed microstructural observations of Chelungpu fault rocks.

To better understand physical and chemical processes operative within the plate-bounding faults, an international project to drill into the Chelungpu fault was initiated in January 2004 (Taiwan Chelungpu-fault Drilling Project, TCDP). Two vertical boreholes (Hole-A and -B) to depths of about 2 km and one side-track (Hole-C) have been drilled through at Takeng (Fig. 2), the northern portion of Chelungpu fault system where the

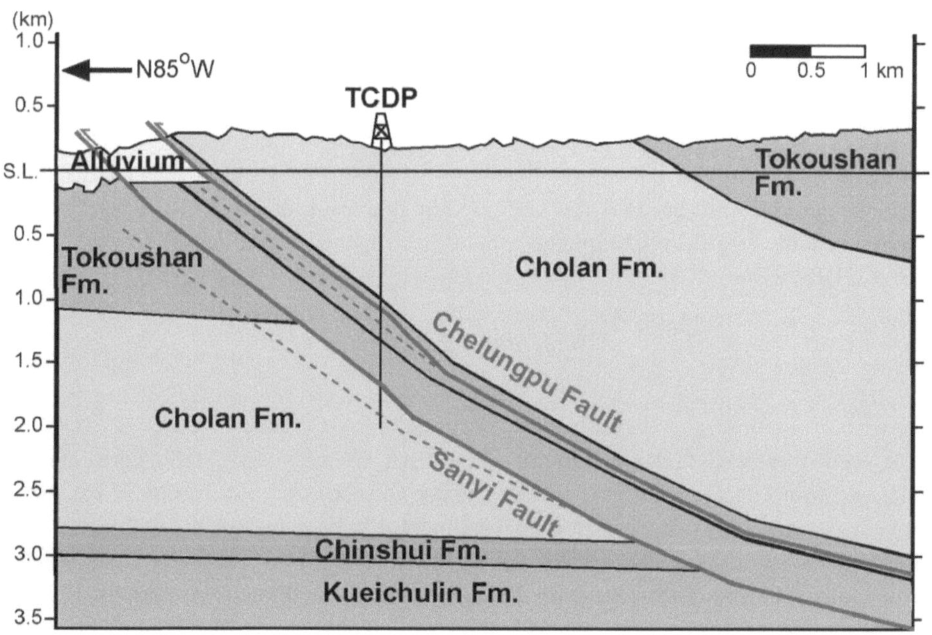

Figure 2
Schematic cross section near the TCDP drill site (modified from YEH *et al.*, 2007; HUNG *et al.*, 2007), showing the formation distribution and major faults. The borehole encountered the Chelungpu fault within the Chinshui Formation and the Sanyi fault near the contact of the Kueichulin and Cholan formations.

larger displacement occurred during Chi-Chi earthquake. The objective of TCDP is to investigate the physical and chemical properties that control the faulting mechanisms during the seismic cycle to understand the dynamic faulting process.

We obtained TCDP cores at various depths from Hole-A. Hole-A penetrated three formation units, Cholan (0 to 1029 m), Chinshui (1029 to 1303 m) and Kueichulin (1303 to 1712 m) Formation. While the layering sequence is typical of sedimentary rocks, with the youngest Cholan Formation on top and the oldest Kueichulin Formation at the bottom (Song *et al.*, 2007), the Cholan Formation (1712–2003 m) appeared again beneath the Sanyi fault (Fig. 2). The core penetrated the Chelungpu fault at a depth of ~1100 m, and encountered another major thrust fault (the Sanyi fault) at a depth of ~1712 m (Fig. 2). The late Pliocene to the early Pleistocene Cholan formation is composed of sandstone, siltstone, shale, sandstone with silty intervals, shale with sandy intervals and siltstone with sandy intervals. The lithology of the late to the early Pliocene Chinshui formation is dark gray shale with sandy or silty intervals. Greenish-gray shaly sandstone, gray sandstone and shale are dominant in the late Miocene to the early Pliocene Kueichulin formation (Ho, 1988; Song *et al.*, 2007).

In this study, we focus on the two dominant lithologic facies: shaly siltstone and porous sandstone samples (Fig. 3). Hydromechanical tests at various confining pressures were conducted to determine the rupture mechanisms during the earthquake. Microstructural analyses were performed in the fault-related rocks to clarify how the internal texture controls the anisotropy of physical properties. Specifically, how do the strength and transport properties of the Chelungpu fault rocks vary with different lithologic units, bedding orientations and depths? What is the extent of the anisotropic behavior of the mechanical and permeability properties of the Chelungpu fault rocks? What are the factors that control the anisotropic behavior within these rocks? Answers to these questions provide insights into the deformation processes of the Chi-Chi earthquake, where unusually large slip and unique rupture heterogeneity were observed.

2. *Experimental Procedure*

2.1. *Sample Description*

A well-developed fault zone architecture generally consists of fault core, damage zone, and protolith (e.g., Sibson, 1977; Chester and Logan, 1986; Caine *et al.*, 1996; Faulkner *et al.*, 2003). During earthquakes, most displacements occur within the fault core, and the associated damage zone is characterized by the extensive fractures and small faults. The protolith represents the intact, undeformed country rocks. In Hole-A, several fault zones were identified with thick fault gouge at the depth of 1111, 1153, 1222, 1580, 1712 and 1812 m. Fault zone FZ1111 at the depth ranging from 1106.29 to 1111.79 m (Fig. 3) is characterized by extensive fractures. Based on the appearance of decreased resistivity and permeability, low density and seismic velocity V_p and V_s, high

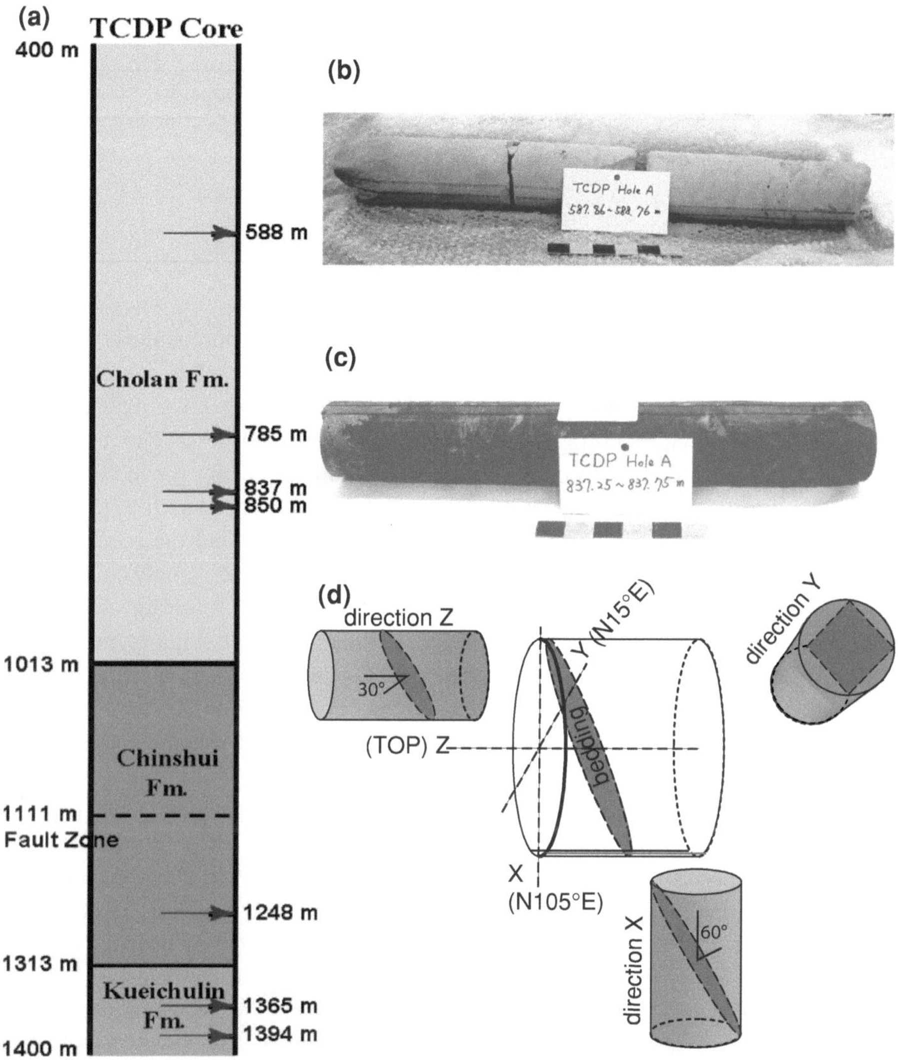

Figure 3

a) Stratigraphic column in TCDP Hole-A. The fault zone which is associated with Chi-Chi earthquake is identified by gray dashed line. Black arrows indicate the sampling locations. b) Bioturbated sandstone sample T588 which was located at the depth 587.86 to 588.76 m. c) Shaly siltstone sample T837 with a higher content of clay minerals was located at the depth 837.25 to 837.75m. The diameter of samples is ~80 mm. Gray and black lines are the reference of orientation. d) Illustration of sub-core samples (with the diameter of 1.8 mm and the length of 3.8 mm) taken from the drillcore in three orthogonal directions. Direction X samples are cored along the orientation N105°E, which results in a 60° angle between the bedding plane normal and axial direction; Direction Y samples are cored along the strike of the bedding, perpendicular to Direction X, which results in a 90° angle between the bedding normal and axial direction; Direction Z samples are cored along the axial direction of the drillcore, perpendicular to Direction X and Y, which results in a 30° angle between the bedding normal and axial direction. The bedding is indicated by dark gray color.

Table 1

Description of samples used in this study

Core Name	Depth (m)	Formation	# tested	Porosity (%)	Rock Type
T588	587.9–588.8	Cholan	3	15.9–18.7	sandstone
T785	785.4–786	Cholan	1	6.6	siltstone
T837	837.2–837.8	Cholan	14	3.5–4.3	siltstone
T850	850.3–850.8	Cholan	1	18.6	sandstone
T1248	1248.8–1249.2	Chinshui	2	4.0–5.0	siltstone

V_p/V_s, Poisson ratio and fluid contents (HUNG *et al.*, 2007), and anomalously high CO_2, CH_4 gas content (YANG *et al.*, 2005), FZ1111 is considered as the major slip zone associated with the 1999 Chi-Chi earthquake (SONG *et al.*, 2007). Clay gouges of fault core appear at the depth of 1110.37–1111.34 m, surrounded by the asymmetric damage zone. The fault zone at the depth of 1712 m represents another major thrust fault – Sanyi fault, with a drastic change of fast polarization direction (HUNG *et al.*, 2007).

Due to the limited quantity of samples from the fault zone and the requirement of larger-sized samples needed for the mechanical and permeability tests, only country rocks were used in this study. Sampling locations were decided referring to the core description, logging data, and mud gas analysis. Intact samples without macroscopic fractures were selected using the logging data and gas analysis, locations with lower V_p and V_s, anomalously high CH_4, CO_2, and radon gas concentration (HUNG *et al.*, 2007; YANG *et al.*, 2005). The samples at various depths above and below the major slip zone of Chelungpu fault were collected (Fig. 3a).

The lithological facies in Hole-A are dominantly siltstone, sandstone, and shale. The relatively tight siltstones are composed of quartz, more than 20% clay minerals, and minor feldspar. These shaly siltstones have porosities of about 4 to 6% and densities of 2590 kg/m^3. The bedding identified by the lamination of clay minerals is dipping by about 30° to the east (Fig. 3). Porous sandstones are predominated by quartz with minor feldspar. The porosities of sandstones range from 15 to 18%, and the densities are about 2120 kg/m^3. In general, bedding in porous sandstone is not visible. Both siltstone and sandstone samples at depths from 588 to 1248 m in Hole-A were investigated in this study (Table 1 and Fig. 3).

2.2. *Sample Preparation*

To understand the effect of bedding anisotropy on mechanical and hydraulic transport properties of TCDP fault rocks, cylindrical samples were taken from the drillcores along three orthogonal directions, denoted X, Y and Z, respectively (Fig. 3d). Direction Z is parallel to the TCDP core axis, and the other two directions are perpendicular to the core axis, with Direction X (N105°E) perpendicular and Direction Y (N15°E) parallel to the strike of the bedding. As such, the angle between the normal of the bedding plane and the axial direction is approximately 60°, 0°, and 30° in Direction X, Y, and Z samples respectively (Fig. 3d). The size of the samples used for our measurement is 18.4 mm in diameter and 38.1 mm in length.

Each sample was dried in the vacuum at a temperature of 80°C for at least 24 hours and then jacketed with thin copper foil. Two steel end-plugs were placed at the two ends of the sample, and then three layers of heat-shrinkable polyolefine tubing were used to keep the rocks insulated from the confining pressure medium (kerosene). Electric resistance strain gages (TML type FLA 10-11, Tokyo Sokki Kenkyujo Co.) were attached to the surface of the copper foil to measure the axial and transverse strains of the dry samples.

3. *Experimental Results*

3.1. *Mechanical Data*

Triaxial deformation tests under either dry or water-saturated conditions were conducted on TCDP siltstone and sandstone samples at room temperature (Table 1). Pressures were selected to be comparable to the *in situ* conditions. For experiments on dry samples, the confining pressure ($P_c = \sigma_2 = \sigma_3$) was fixed at 20 MPa, and pore pressure $P_p = 0$. In water-saturated experiments, pore pressure P_p was set constantly at 10 MPa and the confining pressures varied from 30, 50, to 70 MPa, therefore the corresponding effective pressures ($P_{eff} = P_c - P_p$) were 20, 40 and 60 MPa, respectively. The experiments were conducted at a nominal axial strain rate of 2×10^{-5} /s, which was sufficiently slow to ensure fully drained conditions.

The confining pressure with a resolution of 0.1 MPa was kept as a constant with fluctuation within 1% during triaxial loading. The axial stress was measured with an external loading cell with a resolution of 0.1 kN. The axial displacement was measured with the displacement transducer (DCDT) mounted outside the pressure vessel, and the displacement rate was servo-controlled at a constant strain rate. The volumetric strain for the dry sample was evaluated from the two strain gage data using the relation $\varepsilon_v = \varepsilon_{\|} + 2\varepsilon_{\perp}$, where $\varepsilon_{\|}$ and $\varepsilon_{\perp}$ are the axial and the transverse strains, respectively. The porosity change of water-saturated sample was derived from the pore volume change measured by a porosimeter with an uncertainty of 0.1%.

One of the complications in conducting water-saturated deformation experiments on TCDP shaly siltstones is that these samples would split after they had been immersed in fluids including distilled water, tap water, and also 0.01, 0.02 and 0.05 M KCl solutions. It seems that interaction of the clay minerals with the solution can significantly reduce the cohesion of the siltstone. To circumvent this problem, in a water-saturated experiment, we would first subject the oven-dried sample to a confining pressure of 15 MPa, and then evacuate the pore space by a vacuum pump before saturating the sample with distilled water at a pore pressure of 10 MPa. By subjecting samples to a confining pressure before saturation, we were able to prepare fully saturated samples that remain cohesive after deformation and permeability experiments that might last more than a week.

Both dry and water-saturated siltstone samples deformed at an effective pressure of 20 MPa show strain softening and brittle failure. The shaly siltstone sample deformed to large strain develops multiple shear faults that are aligned almost exactly with the bedding direction. Considerable water-weakening effect is observed on the brittle strength of shaly siltstone T837 samples (Fig. 4). The yield strengths of the water-saturated samples are ~2/3 of those of dry siltstone samples (Fig. 4). The Young's modulus of the water-saturated siltstone samples is only ~1/3 of those of dry siltstone samples (Fig. 4).

The strengths of the Direction X samples are slightly stronger than those of the Direction Z samples under both dry and water-saturated conditions. The anisotropy behavior in sample strength under water-saturated conditions is illustrated in Figure 5a. Porosity reduction as a function of axial strain is shown in Figure 5b. Before the onset of dilatancy *C'* prior to macroscopic fracture, porosity reduction was observed. The amount of porosity reduction decreases with increasing axial strain (Fig. 5b). Beyond *C'*, only small amount of dilatancy occurred for TCDP siltstone samples deformed at 20 MPa.

The mechanical data of saturated sandstone T588 deformed at an effective pressure of 20 MPa are shown in Figure 6. Strain softening and brittle failure in the form of a mosaic of shear bands were observed. Similar to shaly siltstone T837, the strength of sandstone T588 is anisotropic, with the sample cored in direction X being slightly stronger than that cored in direction Z. Porosity reduction as a function of axial strain is plotted in Figure 6b. Compaction was observed with increasing axial strain until *C'*, which dilatancy was observed.

Although the initial porosity of sandstone T588 (~15–18%) is considerably higher than that of siltstone T837 (3-4%), their brittle strengths are comparable under the same saturation and effective pressure condition and coring orientation (Figs. 5a and 6a). This implies that high content of clay minerals in the relatively compact siltstones significantly lowers the brittle strength of the samples.

3.2. *Permeability Data*

Permeability of both sandstone and siltstone samples were obtained by measuring the flow along the axial direction at different loading stages during hydrostatic and triaxial compression (Tables 2 and 3). Our experimental setup allows us to use two alternative techniques to measure permeability. Depending on the value of permeability of interest, the steady-state flow technique was used for rocks with relatively high permeability ($k > 10^{-15}$ m^2), and pulse transient technique was used when the samples were less permeable ($k < 10^{-15}$ m^2). The two methods could be switched if necessary during a deformation experiment (ZHU and WONG, 1997).

Shaly siltstone T837 with porosities of ~4% have permeability values ranging from 10^{-16} to 10^{-19} m^2 at effective pressures ranging from 5 to 40 MPa (Table 2).

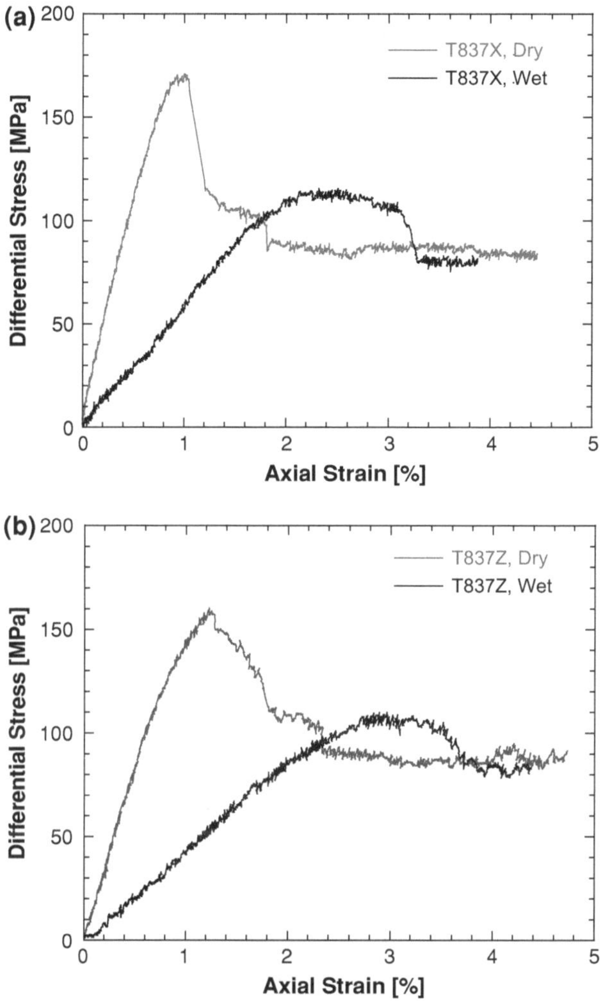

Figure 4
Mechanical data for TCDP shaly siltstone T837 samples cored in a) Direction X and b) Direction Z, deformed at an effective pressure of 20 MPa. Gray curves represent data from experiments under dry condition. Black curves represent data from experiments under water-saturated fully-drained condition. Note that both Direction X and Z samples saturated with water were significantly weaker and more compliant than the dry samples.

The permeability values of siltstone T1248 (TC#21) are comparable (Table 2). Siltstone T785 (TC#20) has the highest initial porosity (6.6%) and is most permeable among the three siltstones studied.

Permeability measurement was also conducted on a shaly siltstone T837 sample (TC#15) and a porous sandstone T588 sample (TC#12) during triaxial compression at an effective pressure of 20 MPa (Fig. 7). Both are Direction Y samples with bedding

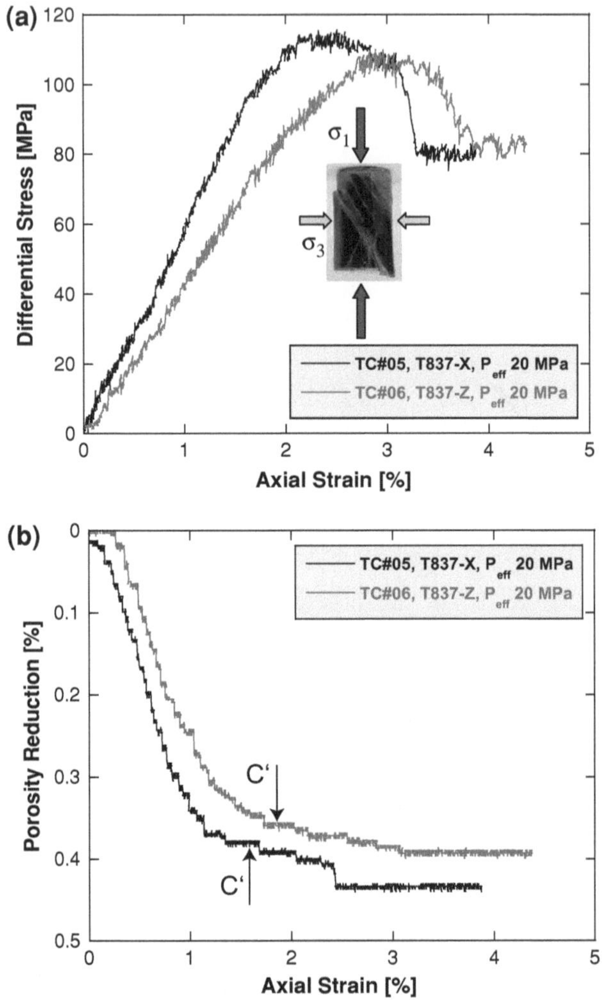

Figure 5

Mechanical data for siltstone T837 samples deformed at confining pressure of 30 MPa and pore pressure of 10 MPa. Data of Direction X samples are in black and data of Direction Z samples are in gray. a) Differential stress as a function of axial strain. Notice that T837-X is slightly stronger than T837-Z. A post-failure T837-X sample was shown as insert. Failure mode is that multiple planar shear faults align with the bedding. b) Porosity reduction as a function of axial strain. Slight dilatancy is observed.

parallel to the axial direction. Both samples failed by brittle fracture. Permeability of T837-Y decreased by two orders of magnitude as stress increased to the peak value, and then increased slightly during the stress drop accompanied by moderate dilatancy (Fig. 7a). The initial permeability of T588-Y was an order of magnitude higher than that of T837-Y. The permeability evolution of T588-Y during deformation was similar to that

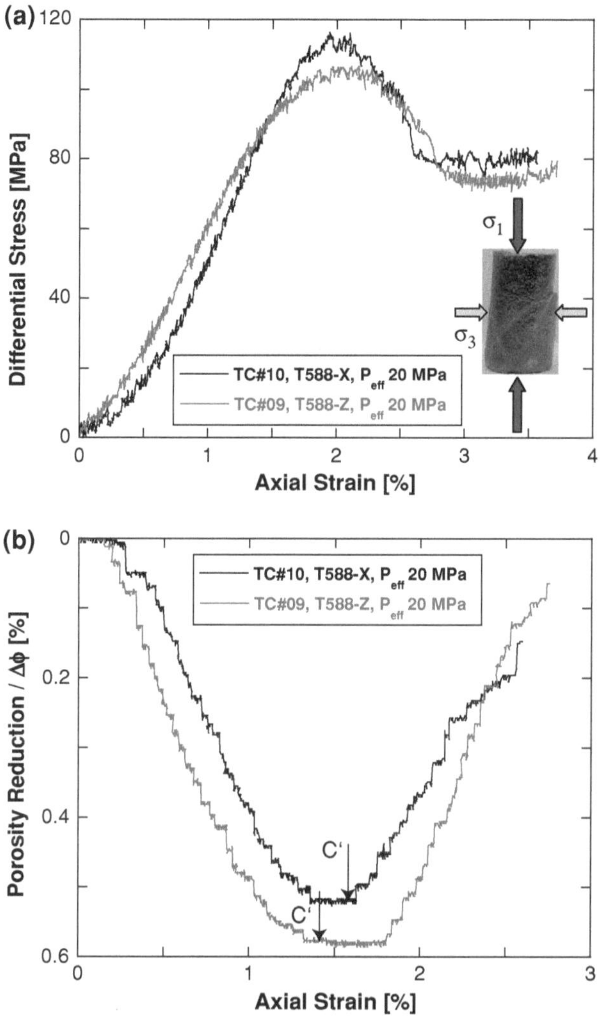

Figure 6

Mechanical data for sandstone T588 samples deformed at confining pressure of 30 MPa and pore pressure of 10 MPa. Data of Direction X samples are in black and data of Direction Z samples are in gray. a) Differential stress as a function of axial strain. Similar to siltstone samples, T588-X is slightly stronger than T588-Z. A post-failure T588-X sample was shown as insert. Failure mode consists of conjugated shear bands at different angles. b) Porosity reduction as a function of axial strain. Considerable dilatancy was observed beyond the onset of shear-induced dilatancy C' in both Direction X and Z samples.

in siltstone T837-Y. The post-failure permeability increase in T588-Y was slightly larger (Fig. 7b), consistent with the larger dilatancy observed in this sandstone sample (Fig. 6).

Permeability evolution as a function of effective pressure can be empirically described by an exponential equation (e.g., DAVID *et al*, 1994):

Table 2

Compilation of permeability data during hydrostatic compaction

Sample Number	Depth and Orientation	Effective Pressure (MPa) $P_c - P_p$	Porosity ϕ (%)	Permeability k (m^2)
Siltstone				
TC#07	T837-X	5	3.99	1.3×10^{-18}
		11	3.97	1.8×10^{-18}
		21	3.85	1.4×10^{-18}
		31	3.64	5.9×10^{-19}
		41	3.53	2.3×10^{-19}
TC#08	T837-Y	7	3.99	1.7×10^{-17}
		10	3.93	2.0×10^{-17}
		20	3.69	7.6×10^{-18}
		30	3.40	3.1×10^{-18}
		41	3.20	9.6×10^{-19}
TC#11	T837-Z	6	4.29	4.7×10^{-17}
		11	4.24	2.8×10^{-17}
		21	4.09	2.8×10^{-18}
TC#17	T837-X	6	4.34	1.3×10^{-16}
		11	4.31	6.1×10^{-17}
		21	4.20	3.5×10^{-18}
		32	3.98	3.3×10^{-19}
		41	3.88	5.4×10^{-20}
TC#18	T837-X	7	4.20	4.2×10^{-17}
		11	4.17	1.9×10^{-18}
		22	3.90	9.0×10^{-19}
		31	3.75	3.1×10^{-19}
		42	3.50	1.2×10^{-19}
Siltstone				
TC#20	T1248-X	6	4.01	6.5×10^{-17}
		10	3.98	3.0×10^{-17}
		20	3.78	4.0×10^{-18}
		31	3.59	5.3×10^{-19}
TC#21	T785-X	6	6.64	2.0×10^{-16}
		11	6.58	1.6×10^{-16}
		21	6.41	1.2×10^{-16}
		31	6.31	7.9×10^{-17}
		41	6.25	5.2×10^{-17}
		51	6.23	3.1×10^{-17}
		60	6.18	1.1×10^{-17}
		71	6.08	5.3×10^{-18}
		81	5.98	1.8×10^{-18}
		91	5.86	9.8×10^{-19}
		102	5.74	3.8×10^{-19}

$$k = k_0 \exp\left[-\gamma\left(P_{eff} - P_0\right)\right], \tag{1}$$

where k is the permeability at the effective pressure P_{eff}, k_0 is the permeability at the effective pressure P_0 which is 5 MPa in this study, and γ is a pressure sensitivity coefficient. In Figure 8, we plot permeability data of 6 siltstone T837 samples during

Table 3

Compilation of permeability data during triaixal deformation

Sample Number	Depth and Orientation	Effective Pressure (MPa) $P_c - P_p$	Differential Stress (MPa) σ_1 - σ_3	Porosity ϕ (%)	Axial Strain ε (%)	Permeability k (m^2)
Siltstone						
TC#15	T837-Y	6.4	0.0	3.99		1.2×10^{-15}
		21.0	0.0	3.74	0.00	6.4×10^{-17}
		20.9	59.9	3.55	1.13	1.8×10^{-18}
		21.0	92.6	3.55	2.88	1.3×10^{-19}
		21.0	94.9	3.57	4.27	1.4×10^{-19}
Sandstone						
TC#12	T588-Y	6.3	0.0	15.88		3.1×10^{-14}
		11.3	0.0	15.76		9.7×10^{-15}
		15.4	0.0	15.75		6.8×10^{-15}
		20.4	0.0	15.42	0.00	3.2×10^{-15}
		20.5	43.8	14.82	1.14	8.3×10^{-16}
		20.8	74.3	14.66	1.87	5.0×10^{-16}
		20.3	92.3	14.64	2.31	2.2×10^{-16}
		21.5	108.6	15.00	3.65	5.7×10^{-18}
		20.7	102.7	15.32	4.83	6.4×10^{-18}
		21.0	110.2	15.43	5.89	8.1×10^{-18}
TC#19	T850-X	5.2	0.0	18.56		5.2×10^{-14}
		10.5	0.0	18.41		5.0×10^{-14}
		15.2	0.0	18.22		3.0×10^{-14}
		20.3	0.0	17.99	0.00	2.9×10^{-14}
		20.6	60.9	17.51	0.91	9.8×10^{-15}
		20.6	109.1	17.41	1.71	3.8×10^{-15}
		20.5	126.4	17.69	2.45	2.1×10^{-15}
		20.2	84.8	18.01	3.66	1.9×10^{-15}
		20.9	92.8	18.03	5.76	1.3×10^{-15}

hydrostatic compaction. The values of γ ranging from 0.07 to 0.25 (Table 4 and Fig. 8a) are comparable to values for cores from the Kola, Cajon Pass and KTB boreholes (MORROW *et al.*, 1994; MORROW and LOCKNER, 1994), as well as the East Fort thrust fault (EVANS *et al.*, 1997) and Dixie Valley normal fault (SERONT *et al.*, 1998) zones (Table 5). Because high aspect ratio cracks are more sensitive to pressure than equal dimensional pores (WALSH, 1965), higher values of γ generally indicate higher crack density in the fault rocks.

The permeability-porosity relationship during mechanical compaction can be empirically described by a power law:

$$k = k_0(\phi/\phi_0)^{\alpha}, \tag{2}$$

where ϕ and ϕ_0 are the porosity corresponding to the permeabilities k and k_0, respectively. Higher values of porosity sensitivity exponent α indicate larger permeability loss related to porosity reduction during compaction. A compilation of our data on shaly siltstone T837 in Figure 8b indicates that there is no strong correlation between the value

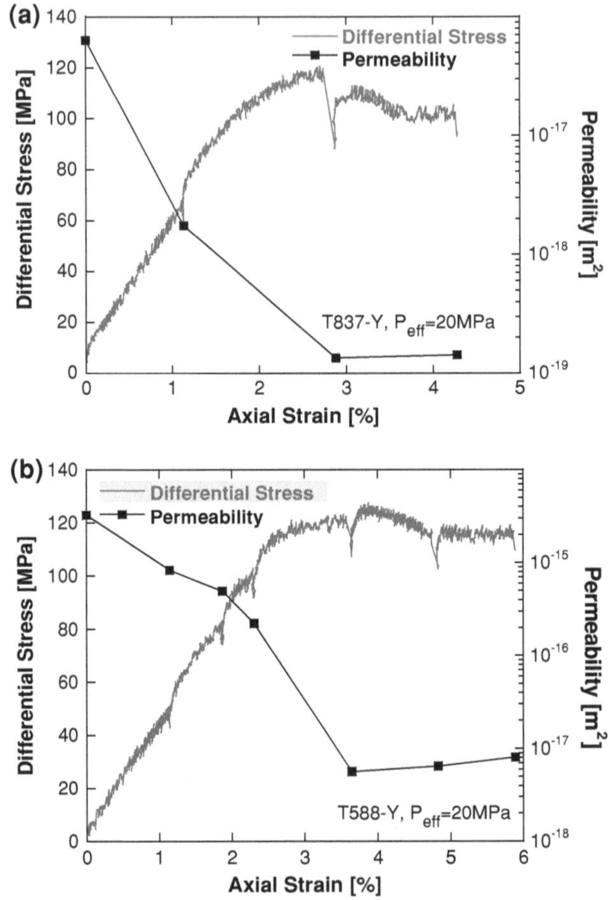

Figure 7

Stress-strain curve and permeability as a function of axial strain for a) a siltstone T837-Y sample; b) a sandstone T588-Y sample deformed at a effective pressure of 20 MPa. Before peak stress, significant permeability reduction (up to 2 orders of magnitude) was observed for both the siltstone and sandstone samples. Permeability increased slightly accompanied by strain softening and dilatancy occurred during post-failure stage.

of α and the initial porosity or bedding orientation. In general, higher α values are observed at the initial stage of compaction, which is likely caused by pore connectivity loss due to initial crack closure (e.g., WALSH, 1965).

3.3. *Permeability Anisotropy*

Permeability of a heterogeneous, anisotropic material is a second rank tensor. To understand the mechanical coupling between deformation and fluid flow, the anisotropic behavior of permeability change must be considered. In TCDP samples, the values of

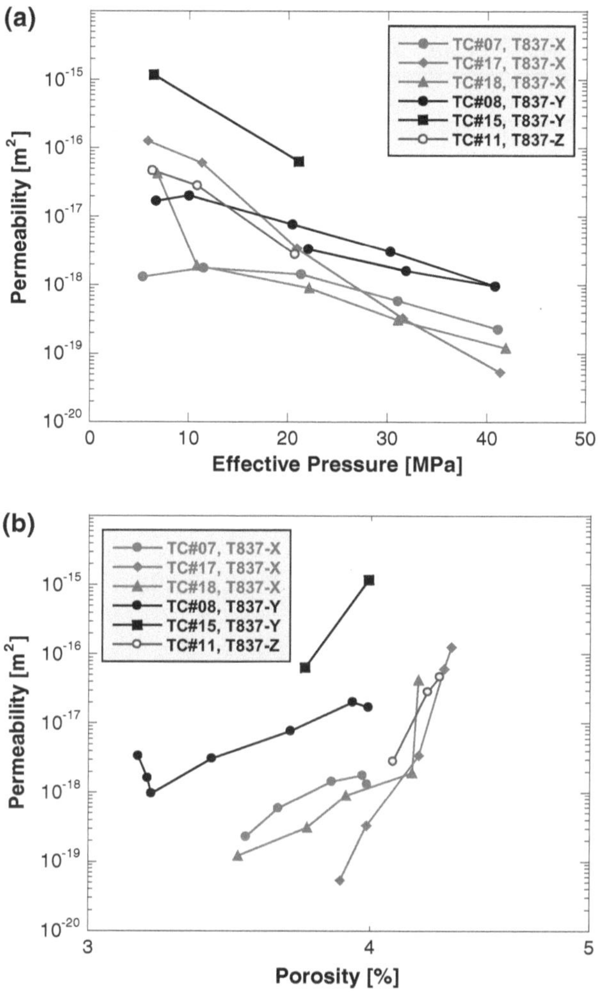

Figure 8

Permeability evolution as a function of a) effective pressure; b) porosity for 6 siltstone T837 samples during hydrostatic compaction. Data of samples cored in direction X, Y and Z are shown in gray, black and dark gray respectively. Notwithstanding the variation from sample to sample, Direction Y samples are more permeable than Direction X and Z samples.

permeability are controlled by not only the porosity change but also the spatial distribution of bedding. Figure 8b shows the permeability evolution as a function of porosity for siltstone samples T837 that were cored in three directions. Notwithstanding the variations from sample to sample, at the same porosity. Permeabilities of the Direction Y samples are significantly higher than those of the Direction X and Z samples.

Our permeability anisotropy data for the TCDP shaly siltstone samples can be interpreted using a model of a composite made up of a parallel array of bedding planes

Table 4

Effective pressure sensitivity and porosity sensitivity exponent for TCDP samples

Sample Number	$k = k_0 \exp[-\gamma\,(P_{eff} - P_0)]$		$k = k_0\,(\phi\,/\,\phi_0)^{\alpha}$		Pore Compressibility
	γ (MPa^{-1})	k_0 (m^2)	α	ϕ_o (%)	10^{-3} MPa^{-1}
Siltstone					
TC#07	0.078	1.99×10^{-18}	18.41	3.92	4.23
TC#08	0.098	2.01×10^{-17}	14.14	3.88	6.94
TC#11	0.202	4.72×10^{-17}	60.92	4.29	3.32
TC#15	0.199	1.17×10^{-15}	45.44	3.99	4.37
TC#17	0.228	1.27×10^{-16}	66.34	4.34	3.44
TC#18	0.091	1.92×10^{-18}	16.24	4.17	5.62
TC#20	0.193	6.48×10^{-17}	42.10	4.01	4.59
TC#21	0.041	1.98×10^{-16}	24.48	6.64	1.67
	0.085	3.11×10^{-17}	50.76	6.23	1.67
Sandstone					
TC#12	0.155	3.14×10^{-14}	66.14	15.88	2.34
TC#19	0.045	5.21×10^{-14}	21.70	18.56	2.06

Table 5

Compilation of pressure and porosity sensitivity of permeability for other host rocks associated with various fault zones

Material Type	γ [MPa-1]	α	Reference
gneiss (Kola)	0.032		Morrow *et al.*, 1994
basalt (Kola)	0.106		Morrow *et al.*, 1994
amphibolite (KTB)	0.058–0.116		Morrow *et al.*, 1994
granodiorite (Cajon Pass)	0.127		Morrow and Lockner, 1994
East Fort thrust fault, Wyoming protolith	0.165–0.2		Evans *et al.*, 1997
Dixie Valley normal fault protolith	0.04–0.152	53.48	Seront *et al.*, 1998

embedded in a relatively permeable matrix. With reference to Figure 9a, if δ denotes the cumulative thickness of bedding planes per unit thickness of the composite, then the effective permeability perpendicular to bedding is given by

$$k_{\perp} = \frac{1}{\delta/k_b + (1-\delta)/k_m} = \frac{k_b k_m}{\delta k_m + (1-\delta)k_b}, \tag{3}$$

where k_b and k_m denote permeabilities of the bedding and matrix materials, respectively. The effective permeability for flow parallel to bedding (corresponding to Direction Y as depicted in Figure 9b) is given by

$$k_y = k_{||} = \delta k_b + (1-\delta)k_m. \tag{4}$$

For a rock with layered heterogeneity described here, an equivalent anisotropic rock can be derived with the minimum and maximum principal directions of its permeability

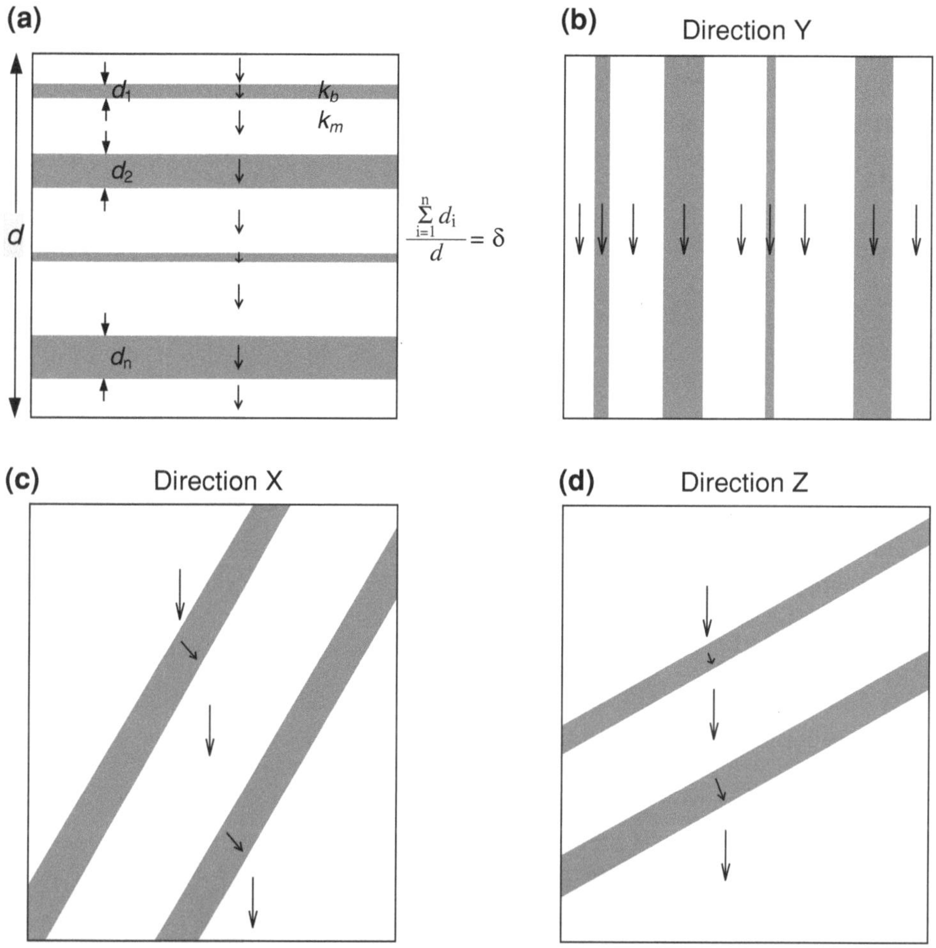

Figure 9

Schematic diagram of a composite of thickness d made up of a parallel array of bedding planes embedded in a relatively permeable matrix. The orientations of the bedding planes (dark grey stripes) are a) perpendicular; b) parallel; c) at 30°; d) at 60° with respect to the flow direction. Permeabilities of the bedding and matrix materials are k_b and k_m, respectively. The thickness of each bedding plane is d_i, and the cumulative thickness of bedding planes per unit thickness of the composite is δ. Diagrams b), c) and d) are analogs to our flow experiments in Direction Y, X and Z samples.

tensor corresponding to the effective permeabilities perpendicular and parallel to bedding planes, respectively (FREEZE and CHERRY, 1979). The symmetry intrinsic to these two end-members (with applied hydraulic gradients parallel and perpendicular to bedding) ensures that the flow is one-dimensional, if the matrix and bedding materials can be approximated as isotropic (FREEZE and CHERRY, 1979). However, if the bedding is inclined at an angle to the hydraulic gradient, it is unlikely that the flow remains one-dimensional. As illustrated in Figures 9c and 9d (for Directions X and Y), flow lines will be refracted at an interface between the matrix and bedding due to the permeability contrast. While the effective permeabilities k_x and k_z (along the direction of the applied hydraulic gradient) for Directions X and Y samples can be evaluated only if the spatial distributions of bedding thickness and spacing are prescribed, we expect them to fall between the upper and lower bounds given by (3) and (4) such that

$$k_{\perp} < k_z < k_x < k_y = k_{||}. \tag{5}$$

In the limit $k_b/k_m \to 0$, the flow lines will be axial in the matrix material and refract to become perpendicular to bedding within the bedding planes (FREEZE and CHERRY, 1979). Accordingly, the effective permeabilities of a composite with $k_b << k_m$ can be approximated by

$$k_{\perp} \approx k_z \approx k_x \ll k_y = k_{||}, \tag{6a}$$

which implies that the permeability anisotropy can be characterized by

$$\frac{k_y}{k_z} \approx \frac{k_y}{k_x} \approx \frac{k_{||}}{k_{\perp}} = 1 + \delta(1-\delta)\left(\frac{k_m}{k_b} + \frac{k_b}{k_m} - 2\right). \tag{6b}$$

We plot in Figure 10 the permeability ratio as a function of the bedding thickness parameter δ and the matrix/bedding permeability contrast k_m/k_b, as predicted by (6b). Our data for the TCDP siltstone samples (Fig. 8b) indicate a permeability anisotropy of one to

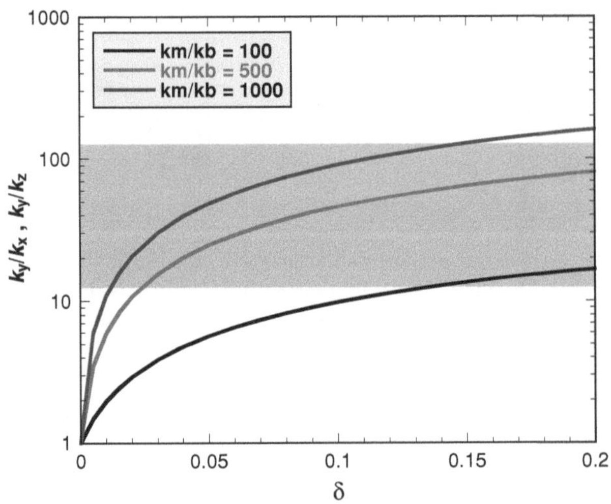

Figure 10
Bedding orientation induced permeability anisotropy as a function of the bedding thickness parameter δ and the matrix/bedding permeability contrast k_m/k_b. Shaded area highlights the permeability ratios between Direction Y samples and Direction X or Z samples obtained in this study. If the matrix/bedding permeability contrast k_m/k_b is 100, the bedding thickness parameter δ has to be at least 0.14 to generate such an anisotropy. However, the value of minimum δ is reduced to 0.01 if k_m/k_b is an order of magnitude higher.

two orders of magnitudes. With reference to this model, such an anisotropy can be explained by beddings with relative thickness δ in the range of 0.01–0.1 and permeability that is 2-3 orders of magnitude lower than the matrix, possibly due to the presence of relatively fine-grained materials. The inferred permeability contrast is sufficiently large to justify using the approximate relations (6a) and (6b).

4. Geological Implication

Our goal is to provide laboratory constraints to elucidate important questions regarding earthquake physics. Several dynamic weakening models may be pertinent to the large slip displacements and high velocities in the northern portion of the Chelungpu fault during the 1999 Chi-Chi earthquake (e.g., KANAMORI and HEATON, 2000; BRODSKY and KANAMORI, 2001; ANDREWS, 2002; RICE, 2006). Systematic laboratory measurements on the TCDP core samples provide hydromechanical constraints on the extent to which these proposed mechanisms are viable.

4.1. Pore Pressure Excess within the Chelungpu-Sanyi Thrust

One class of model hinges on the development of pore pressure excess. In an earlier study SUPPE and WITTKE (1977) recognized the potential for such pore pressure excess in the fold-and-thrust belt of northwestern Taiwan that may be controlled statigraphically by a lithological transition from relatively permeable sandstone to impermeable shale. In particular, SUPPE and WITTKE (1977) pointed out that the Chelungpu-Sanyi thrust likely rides along the transition zone at the top of the overpressured section.

RICE (1992) formulated a model for generation and maintenance of pore pressure excess in a seismogenic system. The pressure sensitivity coefficient γ plays an important role in his model. Previous studies demonstrated that significant pore pressure excess can be maintained in compact rocks and fractures that typically have relatively high values of $\gamma > 0.02$ MPa^{-1} (e.g., WONG and ZHU, 1999). In contrast, the pore pressure in porous rocks and unconsolidated materials that have low values of $\gamma < 0.02$ MPa^{-1} only slightly exceed the hydrostatic values. We obtained values of γ ranging from 0.07 to 0.25 MPa^{-1} for siltstone T837, and 0.04 to 0.15 MPa^{-1} for sandstone T588 (Table 4). These values are comparable to rocks from other boreholes and fault zones (Table 5)

It should be pointed out that the two coefficients γ and α that characterize the sensitivity of permeability to effective pressure and porosity are positively related. A linear relationship between coefficients γ and α can be derived (DAVID *et al.*, 1994):

$$\gamma = -\frac{\alpha}{\phi}\frac{d\phi}{dP} = \alpha\beta_{\phi}, \tag{7}$$

where β_{ϕ} is the compressibility of pore space. A low γ/α ratio corresponds to a relatively stiff pore space. Experimental data for porous rock and unconsolidated

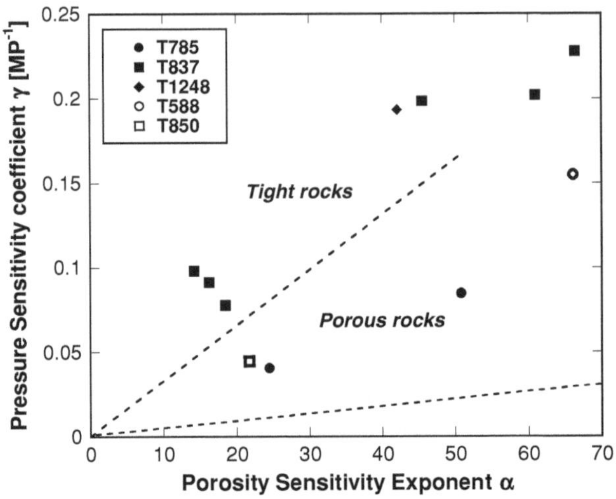

Figure 11
Correlation between the pressure sensitivity coefficient γ and the porosity sensitivity exponent α for shaly samples. Siltstones T837 and T1248 are more compressible and fall into the region of tight rocks. Siltstone T785 with a higher porosity and all sandstones are located within the range of porous rocks bracketed by these two linear boundaries. Two dashed lines represent the lowest boundaries of tight and porous rocks corresponding to the value of γ/α equal to 4.4×10^{-4} /MPa and 3.3×10^{-3} /MPa, respectively (DAVID *et al.*, 1994).

materials can be bracketed by two linear boundaries corresponding to $\beta_\phi = 4.4 \times 10^{-4}$ to 3.3×10^{-3} MPa^{-1} (e.g., DAVID *et al.*, 1994; WONG and ZHU, 1999). Compact crystalline rocks and tight sandstones have relatively high ratios of γ/α ($> 3.3 \times 10^{-3}$ MPa^{-1}), with the implications that such a compact rock is relatively compliant, possibly due to high microcrack density within the rock.

Our data show that siltstones T837 and T1248 with lower porosity 4 to 4.3% siltstones T837 and T1248 are more compressible and fall into the region of tight rocks (Fig. 11). Siltstone T785 with a higher porosity and all sandstones are located within the range of porous rocks bracketed by these two linear boundaries. These results indicate that it is possible to maintain fluid overpressure near the décollement, where an impermeable layer of shale dominates (SUPPE and WITTKE, 1977), and causes the weakening of material and localization of faulting within the wedge.

4.2. *Dynamic Weakening Mechanisms during Earthquake*

Thermal pressurization, frictional melting and elastohydrodynamic lubrication have been postulated as possible dynamic weakening mechanisms during large earthquakes (e.g., KANAMORI and HEATON, 2000; BRODSKY and KANAMORI, 2001; ANDREWS, 2002; RICE, 2006).

In elastohydrodynamic lubrication, the fault zone material can be modeled as a thin film of viscous fluid that is sheared between two rough surfaces (BRODSKY and KANAMORI, 2001). When such a viscous layer is sheared a "lubrication pressure" is induced that would lower the effective stress and hence the frictional strength. MA *et al.* (2003) adopted this model to interpret the significant difference in frequency content in displacement and acceleration between the northern and southern portions of the rupture zone for the 1999 Chi-Chi earthquake. It seems that to activate such a mechanism the effective normal stress is implicitly assumed to be so low that the gouge layer behaves like a "macroviscous" slurry, with a rheology similar to that for debris flow (WHIPPLE, 1997). However, despite the extensive physical and chemical analysis conducted on core samples including fault gouge recovered from the entire length of Hole B, no physical evidence related to the elastohydrodynamic lubrication was found (e.g., HIRONO *et al.*, 2006, 2007).

It is possible that elastohydrodynamic lubrication would not be activated unless and until another mechanism can first generate a pore pressure excess that significantly lowers the effective pressure. Frictional heating due to seismic slip can raise the pore pressure, thus lowering the effective pressure and frictional strength.

As a mechanism for generating pore pressure excess and reducing the static strength, thermal pressurization has been analyzed extensively (e.g., SIBSON, 1973; LACHENBRUCH and SASS, 1980; MASE and SMITH, 1987; LEE and DELANEY, 1987). Its potential as a dynamic weakening mechanism was elucidated by ANDREWS (2002), who explicitly analyzed the consequence of thermal pressurization on stress drop during dynamic rupture. He showed that for a planar circular crack that expands with rupture velocity v_R, stress drop induced by thermal pressurization is directly proportional to the $\sqrt{v_R r/\omega}$, where r is the propagation distance, and hydraulic diffusivity $\omega = k/(\phi\eta\beta)$, related to the permeability k, porosity ϕ, fluid viscosity η and $\beta = \beta_f - \beta_s + (\beta_b - \beta_s)/\phi$ (where β_f, β_s and β_b denote the fluid, solid and bulk compressibilities, respectively) . The stress drop is enhanced with decreasing permeability (and increasing ϕ and β). If the intrinsic failure behavior can be approximated as linear slip-weakening, then thermal pressurization can induce a significant increase of the dynamic stress drop manifested by a nonlinear slip-weakening behavior. There is a threshold distance $R \propto \omega$, beyond which thermal pressurization dominates the stress drop process. Assuming a permeability of 5×10^{-17} m^2, porosity of 2.5% and friction coefficient of 0.7 for the fault rock and plausible fluid properties, ANDREWS (2002) arrived at a value of $\omega \sim 0.02$ m^2/s which in turn implies that $R \sim 300$ m; a distance that is comparable to the fault dimension of an earthquake with $M = 3.5$. For earthquakes that involve larger slip distances, thermal pressurization is expected to dominate and results in significant weakening during dynamic rupture. The Andrews model neglects dilatancy and stress-dependent permeability changes.

A permeability of 5×10^{-17} m^2 is a reasonable value according to laboratory data for core samples from Nojima fault (LOCKNER *et al.*, 1999), Dixie Valley (SERONT *et al.*, 1998), and Median Tectonic Line (WIBBERLEY and SHIMAMOTO, 2003). Except for the Nojima samples, most existing data are for cores retrieved in proximity to the surface.

TCDP provides a unique opportunity to study cores from depths corresponding to different parts of the fault systems involved. According to our laboratory measurement of TCDP shaly siltstone samples, if we take the permeability as 10^{-18} to 10^{-17} m^2 and porosity as 0.037, then the hydraulic diffusivity would be about 10^{-4} to 10^{-3} m^2/s. This is the upper-bound of permeability in TCDP fault zone because pervasive cataclasis generally causes significant permeability reduction (e.g., ZHU and WONG, 1997). This value is lower than Andrews' (2002) critical diffusivity value by orders of magnitude. It should be pointed out that the data of LOCKNER *et al.* (2005) for TCDP fault gouge with permeability about 10^{-20} m^2 indicated even lower diffusivity value. Since the frictional strengths of the TCDP country rocks and gouge samples are comparable to Byerlee's Law (LOCKNER *et al.*, 2005; CHEN, 2008), thermal pressurization of pore fluid would be expected in the TCDP shaly rocks and may provide a mechanism for dynamic weakening of Chelungpu fault. The threshold propagation distance that is derived from our calculation of the hydraulic diffusivity for shaly samples is significantly less than 300 m which was estimated by ANDREWS (2002). This result suggests that the thermal pressurization effect could be significant in regions close to the main slip zone during the Chi-Chi earthquake.

It should also be noted that even in the absence of pore pressure excess, a mechanism such as frictional melting can also reduce the frictional strength. OTSUKI and MONZAWA (2003) recently reported field evidence of fluidization and melting in surface exposure of active faults in the Kobe-Osaka area. In core samples retrieved from boreholes penetrating the Chelungpu fault, some researches (OTSUKI *et al.*, 2005; HIRONO *et al.*, 2006; SONG *et al.*, 2007) have reported microstructural and geochemical signatures that they infer to be related to pseudotachylytes.

4.3. Bedding Related Anisotropic Behavior

Anisotropies of mechanical, permeability, magnetic and seismic properties were observed in country rocks retrieved from TCDP Hole-A. In sedimentary rocks, distributions of bedding and fabric foliation of the solid matrix and anisotropy in pore space can all contribute to the anisotropic behavior of physical properties (ZHU *et al.*, 2002).

In shaly siltstones, permeability data show that Direction Y samples cored parallel to the strike of bedding are more permeable than both Direction X and Z samples. This anisotropic behavior is primarily controlled by the distribution of low-porosity, less-permeable beds within the samples (Fig. 9).

The Chelungpu fault is located within and dips 30° to the east parallel to the strike of the bedding of the Chinshui Shale Formation (Fig. 2). Therefore, the principal stresses in triaxial compression tests on Direction X samples are analogous to the stress states in the thrust faulting regime. Our data on siltstone samples cored along the orientation N105°E (Direction X) have higher brittle strengths than those cored parallel to the TCDP core axis (Direction Z) by about 10 MPa (Fig. 5). The amount of dilatancy observed in siltstone

T837 (Fig. 5) is likely related to the strong control of bedding on its failure mode. The failure mode of siltstones is characterized by planar shear faults almost aligned with the bedding direction, which implies that bedding anisotropy may have dominant control over the style of faulting within the Chelungpu fault zone.

For TCDP sandstones, the results of mechanical strength show similar characteristics of anisotropy. Sandstone sample T588 cored along direction X is slightly stronger than that cored in direction Z (Fig. 6). However, the porous sandstone samples develop a mosaic of shear bands at different angles at brittle failure. In porous sandstones, the brittle strength difference between Direction X and Z samples may not only be dominated by the bedding orientation with respect to the maximum principal stress but the distribution of pore space as well. The microstructural observation has revealed that crack anisotropy exists in the TCDP sandstones. There are preferentially oriented microcrack planes subvertical with azimuth N105°E with poles subparallel to the orientation of minimum P-wave velocity, while the direction of the maximum P-wave velocity are along the crack planes. This anisotropy of microcrack orientation could cause the elastic and magnetic anisotropy in TCDP sandstones (Chen, 2008; Louis *et al.*, 2008).

5. *Conclusions*

Systematic measurements of the permeability and poromechanical properties were conducted on TCDP core samples recovered from Hole-A at various depths. The important conclusions are:

1) Brittle strength of clay-rich siltstones is significantly reduced at the presence of water. Comparison of data between experiments conducted under dry and water-saturated conditions show that the water-saturated samples are more compliant and their yield strength is only $\sim$2/3 of that of the dry samples.
2) The mechanical properties of both shaly siltstones and porous sandstones are anisotropic, with samples cored along the orientation N105°E (Direction X) having higher brittle strengths than those cored parallel to the TCDP core axis (Direction Z). With respect to the maximum principal stress (axial direction), the bedding orientations (normal) in Direction X, Y (cored along the strike of the bedding) and Z samples are 60, 90 and 30°, respectively.
3) Considerable permeability anisotropy related to bedding orientations is observed in siltstones samples, with Direction Y samples with bedding parallel to flow direction more permeable than both Direction X and Z samples, where beddings are at a certain angle to the flow direction.
4) For siltstones, the anisotropy of mechanical strength and permeability is primarily controlled by the distribution of bedding. The distribution of bedding as well as pore space anisotropy in forms of microcracks both likely contribute to the anisotropic behavior of porous sandstones. Our results imply that bedding anisotropy in the

Chelungpu fault rocks exerts significant control over the mechanics of faulting in the region.

5) Low permeability (10^{-16} to 10^{-19} m^2) and high pressure sensitivity values (0.07–0.25 MPa^{-1}) were obtained for shaly siltstone samples with relatively low porosities (3–4%). This indicates that pore pressure excess can be maintained and thermal pressurization of pore fluid can be a viable mechanism for dynamic weakening of the Chelungpu fault.

Acknowledgement

David Lockner kindly shared with us his friction and permeability data on TCDP samples from near the fault core. We have benefited from discussions with him, Sheryl Tembe, Daniel Davis, Laurent Louis and Christian David. Our research is partially supported by the USGS National Earthquake Hazards Reduction Program (Grants #05HQGR0065, 06HQGR0030 and 06HQGR0051) and the National Science Council of Taiwan (Grant #NSC 92-2119-M-002 –006).

References

Andrews, D.J. (2002), *A fault constitutive relation accounting for thermal pressurization of pore fluid*, J. Geophys. Res. *107*(B12), 2363, doi:10.1029/2002JB001942.

Brodsky, E.E. and Kanomori, H. (2001), *Elastohydrodynamic lubrication of faults*, J. Geophys. Res. *106*, 16357–16374.

Caine, J.S., Evans, J.P., and Forster, C.B. (1996), *Fault zone architecture and permeability structure*, Geology *24*, 1025–1028.

Chen, T.-M.N., *Experimental Constraint on Hydromechanical Properties and Microcrack Fabric of Core Samples from the Taiwan Chelungpu-Fault Drilling Project*, Master Thesis, Stony Brook Univ., New York (2008).

Chester, F.M. and Logan, J.M. (1986), *Composite planar fabric of gouge from the punchbowl fault*, California, J. Struct. Geol. *9*, 621–634.

David, C., Wong, T.-F., Zhu, W., and Zhang, J. (1994), *Laboratory measurement of compaction-induced permeability change in porous rocks: Implications for the generation and maintenance of pore pressure excess in the crust*, Pure Appl. Geophys. *143*, 425–456.

Evans, J.P., Forster, C.B., and Goddard, J.V. (1997), *Permeability of fault-related rocks, and implications for hydraulic structure of fault zones*, J. Struct. Geol. *19*, 1393–1404.

Faulkner, D.R., Lewis, A.C., and Rutter, E.H. (2003), *On the internal structure and mechanics of large strike-slip zones: Field observations of the Carboneras fault in southeast Spain*, Tectonophysics *367*, 235–251.

Freeze, R.A. and Cherry, J.A., *Groundwater* (Englewood Cliffs, Prentice-Hall, 1979).

Hirono T., Yeh, E.-C., Lin, W., Sone, H., Mishima, T., Soh, W., Hashimoto, Y., Matsubayashi, O., Aoike, K., Ito, H., Kinoshita, M., Murayama, M., Song, S.-R., Ma, K.-F., Hung, J.-H., Wang, C.-Y., Tsai, Y.-B., Kondo, T., Nishimura, M., Moriya, S., Tanaka, T., Fujiki, T., Maeda, L., Muraki, H., Kuramoto, T., Sugiyama, K., and Sugawara, T. (2007), *Nondestructive continuous physical property measurements of core samples recovered from Hole B, Taiwan Chelungpu-fault drilling project*, J. Geophys. Res. *112*, B07404, doi:10.1029/2006JB004738.

Hirono, T., Ikehara, M., Otsuki, K., Mishima, T., Sakaguchi, M, Soh, W., Omori, M., Lin, W., Yeh, E.-C., Tanikawa, W., and Wang, C.-Y. (2006), *Evidence of frictional melting within disk-shaped black materials discovered from the Taiwan Chelungpu fault system*, Geophys. Res. Lett. *33*, L19311, doi:10.1029/2006GL027329.

Ho, C.S., *An Introduction to the Geology of Taiwan: Explanatory Text of the Geologic Map of Taiwan, 2nd ed.* (Central Geological Survey, Taipei, Taiwan, 1988).

Hung, J.-H., Wu, Y.-H., Yeh, E.-C., Wu, J.-C., and TCDP Scientific Party (2007), *Subsurface structure, physical properties, and fault zone characteristics in the scientific drill holes of Taiwan Chelungpu-Fault Drilling Project*, Terr. Atmos. Ocean. Sci. *18*(2), 271–293.

Johnson, K.M., Hsu, Y.J., Segall, P., and Yu, S.B. (2001), *Fault geometry and slip distribution of the 1999 Chi-Chi, Taiwan Earthquake imaged from inversion of GPS data*, Geophys. Res. Lett. *28*, 2285–2288.

Kao, H. and Chen, W.-P. (2000), *The Chi-Chi earthquake sequence: Active, out-of-sequence thrust faulting in Taiwan*, Science *288*, 2346–2349.

Kanamori, H. and Heaton, T.H., *Microscopic and macroscopic physics of earthquakes*. In *Geocomplexity and Physics of Earthquakes* (eds. Rundle, J.B., Turcotte, D.L., and Klein, W.) (American Geophysical Union, Washington, DC, 2000) pp. 147–163.

Lachenbruch, A.H. and Sass, J.H. (1980), *Heat flow and energetics of the San Andreas Fault zone*, J. Geophys. Res. *85*, 6185–6222.

Louis, L., Chen, T.-M.N., David, C., Robion, P., Wong, T.-F., and Song, S.-R. (2008), *Anisotropy of magnetic susceptibility and P-wave velocity in core samples from the Taiwan Chelungpu-Fault Drilling Project*, J. Struct. Geol., submitted.

Lee, T.-C. and Delaney, P.T. (1987), *Frictional heating and pore pressure rise due to fault slip*, Geophys. J. R. Astron. Soc. *88*, 569–591.

Lockner, D., Naka, H., Tanaka, H., Ito, H., and Ikeda, R., *Permeability and strength of core samples from the Nojima fault of the 1995 Kobe Earthquake*. In *Proc. the International Workshop on the Nojima Fault Core and Borehole Data Analysis, Tsukuba, Japan, 22– 23 November 1999, Prelim. Rep.* (eds Ito, H., Fujimoto, K., Tanaka, H., and Lockner, D.) (U.S. Geol. Surv. Open File Rep. 00-129, 1999) pp. 147–152.

Lockner, D.A., Morrow, C., Song, S.-R., Tembe, S., and Wong, T.-F. (2005), *Permeability of whole core samples of Chelungpu fault, Taiwan TCDP scientific drillhole, EOS Trans. AGU 86*(52), Fall Meeting Suppl., Abstract T43D-04.

Ma, K.-F., Brodsky, E.E., Mori, J., Ji, C., Song, T.-R.A., and Kanamori, H. (2003), *Evidence for fault lubrication during the 1999 Chi-Chi, Taiwan, earthquake (M_w7.6)*, Geophys. Res. Lett. *30*(5), doi:10.1029/2002GL015380.

Mase, C.W. and Smith, L. (1987), *Effects of frictional heating on the thermal, hydrologic, and mechanical response of a fault*, J. Geophys. Res. *92*, 6249–6272.

Morrow, C.A. and Lockner, D. (1994), *Permeability differences between surface-derived and deep drillhole core samples*, Geophys. Res. Lett. *21*(19), 2151–2154.

Morrow, C., Lockner, D., Hickman, S., Rusanov, M., and Röckel, T. (1994), *Effects of lithology and depth on the permeability of core samples from the Kola and KTB drill holes*, J. Geophys. Res. *99*(B4), 7263–7274.

Otsuki, K. and Monzawa, N. (2003), *Fluidization and melting of fault gouge during seismic slip: Identification in the Nojima fault zone and implications for focal earthquake mechanisms*, J. Geophys. Res. *108*(B4), 2192, doi:10.1029/2001JB001711.

Otsuki, K., Uduki, T., Monzawa, N., and Tanaka, H. (2005), *Clayey injection veins and pseudotachylyte from two boreholes penetrating the Chelungpu fault, Taiwan: Their implications for the contrastive seismic slip behaviors during the 1999 Chi-Chi earthquake*, The Island Arc *14*, 22–36.

Rice, J.R. (2006), *Heating and weakening of faults during earthquake slip*, J. Geophys. Res. *111*, B05311, doi:10.1029/2005JB004006.

Seront, B., Wong, T.-F., Caine, J.S., Forster, C.B., Bruhn, R.L., and Fredrich, J. (1998), *Laboratory characterization of hydrothermal properties of a seismogenic normal fault system*, J. Struct. Geol. *20*, 865–881.

Shin, T.-C. and Teng, T.-L. (2001), *An overview of the 1999 Chi-Chi, Taiwan, earthquake*, Bull. Seismol. Soc. Am. *91*, 895–913.

Sibson, R.H. (1973), *Interactions between temperature and fluid pressure during earthquake faulting - A mechanism for partial or total stress relief*, Nature: Physical Science *243*, 66–68.

Sibson, R.H. (1977), *Fault rocks and fault mechanisms*, J. Geol. Soc. London *133*, 191–231.

SONG, S.-R., KUO, L.-W., YEH, E.-C., WANG, C.-Y., HUNG, J.-H., and MA, K.-F. (2007), *Characteristics of the lithology, fault-related rocks and fault zone structures in TCDP Hole-A*, Terr. Atmos. Ocean. Sci. *18*(2), 243–269.

SUPPE, J. and WITTKE, J.H. (1977), *Abnormal pore-fluid pressures in relation to stratigraphy and structure in the Active fold-and-thrust belt of northwestern Taiwan*, Petroleum Geol. Taiwan *14*, 11–24.

WALSH, J.B. (1965), *The effect of cracks on the compressibility of rocks*, J. Geophys. Res. *70*(2), 381–389.

WANG, W.-H., CHANG, S.-H., and CHEN, C.-H. (2001), *Fault slip inferred from surface displacements during the 1999 Chi-Chi, Taiwan, earthquake*, Bull. Seismol. Soc. Am. *91*, 1167–1181.

WHIPPLE, K.X. (1997), *Open-channel flow of Bingham fluids: Applications in debris-flow research*, J. Geol. *105*, 243–262.

WIBBERLEY, C.A.J. and SHIMAMOTO, T. (2003), *Internal structure and permeability of major strike-slip fault zones: The median tectonic line in Mie prefecture, Southwest Japan*, J. Struct. Geol. *25*, 59–78.

WONG, T.-F. and ZHU, W. (1999), *Brittle faulting and permeability evolution: Hydromechanical measurement, microstructural observation, and network modeling*. In *Faults and Subsurface Fluid Flow in the Shallow Crust* (eds. Haneberg, W.C., Mozley, P.S., Moore, J.C., and Goodwin L.B) (AGU Geophysical Monograph Series *113*) pp. 83–99.

YANG, T.F., WALIA, V., LEE, H.-F., SONG, S.-R., and WANG, C.-Y. (2005), *Compositions of on-site monitoring on dissolved gas of drilling mud flow and pore-gases of drilled cores of TCDP*, EOS Trans. AGU *86*(52), Fall Meeting Suppl., Abstract T51A–1315.

ZHU, W. and WONG, T.-F. (1997), *The transition from brittle faulting to cataclastic flow: Permeability evolution*, J. Geophys. Res. *102*, 3027–3041.

ZHU, W., MONTESI, L., and WONG, T.-F., *Effects of stress on the anisotropic development of permeability during mechanical compaction of porous sandstone*. In *Deformation Mechanisms, Rheology and Tectonics* (eds. de Meer, S., Drury, M.R., de Bresser, J.H.P., and Pennock, G.M.) (Geological Society Special Publication 200, London 2002), pp. 119–136.

(Received April 3, 2008, revised January 26, 2009, accepted March 2, 2009)
Published Online First: June 4, 2009

Pure appl. geophys. 166 (2009) 1037–1061
0033–4553/09/051037–25
DOI 10.1007/s00024-009-0495-0

Pure and Applied Geophysics

Axial and Radial Permeability Evolutions of Compressed Sandstones: End Effects and Shear-band Induced Permeability Anisotropy

JEREMIE DAUTRIAT,[1,2] NICOLAS GLAND,[1] JEAN GUELARD,[1] ALEXANDRE DIMANOV,[2] and JEAN L. RAPHANEL[2]

Abstract—The influence of hydrostatic and uniaxial stress states on the porosity and permeability of sandstones has been investigated. The experimental procedure uses a special triaxial cell which allows permeability measurements in the axial and radial directions. The core sleeve is equipped with two pressure samplers placed distant from the ends. They provide mid-length axial permeability measure as opposed to the overall permeability measure, which is based on the flow imposed through the pistons of the triaxial cell. The core sleeve is also equipped to perform flows in two directions transverse to the axis of the sample. Two independent measures of axial and complementary radial permeability are thus obtained. Both Fontainebleau sandstone specimens with a porosity of about 5.8% to 8% and low permeability ranging from 2.5 *mD* to 30 *mD* and Bentheimer sandstone with a porosity of 24% and a high permeability of 3*D* have been tested. The initial axial permeability values obtained by each method are in good agreement for the Fontainebleau sandstone. The Bentheimer sandstone samples present an axial mid-length permeability 1.6 times higher than the overall permeability. A similar discrepancy is also observed in the radial direction, also it relates essentially to the shape of flow lines induced by the radial flow. All the tested samples have shown a higher stress dependency of overall and radial permeability than mid-length permeability. The effect of compaction damage at the pistons/sample and radial ports/sample interfaces is discussed. The relevance of directional permeability measurements during continuous uniaxial compression loadings has been shown on the Bentheimer sandstone until the failure of the sample. We can efficiently measure the influence of brittle failure associated to dilatant regime on the permeability: It tends to increase in the failure propagation direction and to decrease strongly in the transverse direction.

Key words: Directional permeability, end effects, permeability anisotropy, sandstone, shear band, stress dependency.

1. Introduction

It is well known that the decrease of pore pressure due to oil production induces changes of the field stress state, and for instance it increases the effective vertical and horizontal stresses applied on the reservoir (HOLT, 1990; SCHUTJENS and DE RUIG, 1997). The compaction stemming from this stress increase modifies the structure and thus the petrophysical properties of the reservoir rocks. It is therefore of prime importance in

[1] Institut Français du Pétrole, 1&4 avenue de Bois Préau, 92852 Rueil-Malmaison Cedex, France. E-mail: nicolas.gland@ifp.fr
[2] Laboratoire de Mécanique des Solides, CNRS, Ecole Polytechnique, 91128 Palaiseau Cedex, France.

reservoir engineering to understand the interplay between stresses, porosity and permeability evolutions of the rocks. Many authors have pointed out the role of stress anisotropy for permeability and also for other structure-sensitive parameters (MORITA *et al.*, 1992; AL-HARTY *et al.*, 1999). In a recent paper, HEILAND (2003) reviews numerous experimental studies of hydromechanical coupling and stress-dependent permeability.

In order to establish the dependency of the permeability of rocks on effective pressure, the most commonly used experimental set-up is a triaxial cell, which allows investigation of the effects of different stress paths, in hydrostatic (ZOBACK and BYERLEE, 1975; WALSH and BRACE, 1984; MORROW *et al.*, 1984; and others), deviatoric (RHETT and TEUFEL, 1992a,b; WONG *et al.*, 1997; ZHU and WONG, 1997) or uniaxial conditions (TRAUTWEIN and HUENGES, 2005). Owing to the geometry of a classical triaxial cell, some authors have acknowledged the importance of frictional effects at the piston/sample interfaces. However, these effects mostly have been documented in soil mechanic (LEE, 1978; FOURIE and XIAOBI, 1991; SHENG *et al.*, 1997) rather than for consolidated geomaterials. Recently, KORSNES *et al.* (2006) have investigated the influence of these end effects on permeability of both chalks and sandstones using a modified core sleeve, which allows pore pressure sampling in a region distant from the interfaces where frictional effects take place.

The anisotropic evolutions of permeability as a result of effective stress increase are also a fundamental issue in reservoir engineering. In most cases, the oil recovery rate is closely related to horizontal permeability, which is sensitive to changes of the vertical stress. Therefore, flow experiments should be performed under deviatoric stress states. Besides, the geometry of the sample must be adapted in order to minimize end effects and to allow independent measurements of axial and radial permeabilities. Anisotropic permeabilities in porous sandstones under non-hydrostatic stress states were measured by ZOBACK and BYERLEE (1975), BRUNO (1994) and ZHU *et al.* (1997). The authors used triaxial compression and extension tests with conventional triaxial cells and measured the permeabilities along the maximum and minimum principal stresses. Conversely, SCHUTJENS and DE RUIG (1997) have utilized a true triaxial cell to perform fluid flows along the minimum principal stress.

In this paper we present new experiments performed with a special triaxial cell designed to operate under temperature and pressure conditions representative of reservoir rocks and allowing simultaneous measurements of deformation, porosity and directional permeability evolution. Two different sandstones have been studied: A tight Fontainebleau sandstone with a low permeability and a permeable Bentheimer sandstone. Axial and radial permeability measurements are presented and discussed for both materials. We focus on suspected end effects, which we investigate by FEM calculations.

2. *Experimental Set-up and Procedure*

Axial and two perpendicular radial measurements of flows and pore pressures are the most innovative features of the apparatus designed for this study (see Fig. 1). Two high

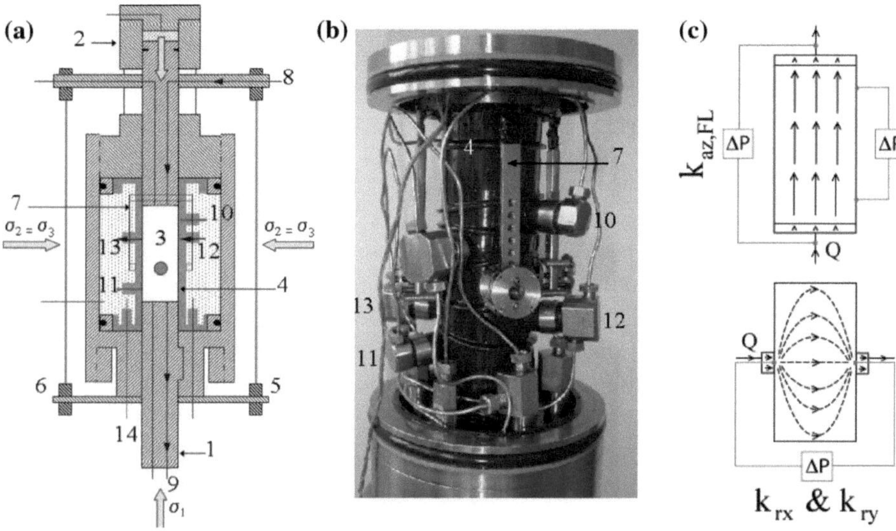

Figure 1
(a) Simplified sketch of the triaxial cell, (b) picture of the equipped core sleeve, (c) pressure differential and flow configurations for permeabilities $k_{az,FL}$, $k_{az,ML}$ and k_{rxy}; item numbers refer to: (1) fixed piston, (2) mobile piston, (3) sample, (4) sleeve, (5, 6) external LVDT, (7) internal cantilever, (8, 9) axial flow lines, (10, 11) pore fluid pressure samplers, (12, 13) radial flow ports, (14) external flow lines connectors.

pressure Isco pumps control independently axial load and confining pressure ($P_c^{\max} = 69$ MPa). An intermediate pressure Isco pump either generates a brine flow through the sample or regulates the pore pressure ($P_p^{\max} = 25$ MPa). Since temperature fluctuations affect the pore fluid viscosity, a temperature regulation is provided by placing the cell in an oven (regulated up to a maximum temperature of 200°C) and also by regulating the temperature of the pore fluid in the pump and outside lines by a thermostated bath and heating resistances, respectively. All the results presented in this paper are obtained at a temperature of 40°C, low limit for accurate temperature regulation of the oven. The triaxial cell is also equipped with deformation sensors to measure both axial and radial deformations of the sample. The axial displacement of the upper mobile piston is recorded by two external high temperature LVDT sensors from RDP. The radial deformations of the core are measured using a double cantilever sensor from ERGOTECH fastened on the core sleeve inside the cell. The dimensions of the sample are 38.1 mm in diameter and 80 mm in length.

The most important innovation of our experimental setup is the special core sleeve which allows flows generation and differential pressure measurements along the core diameter. Generally, permeability evolution of a sample subjected to deviatoric stresses is measured along the direction of the maximal stress, by sampling the pressure differential between the inlet and the outlet of the sample. Additionally, with our setup, we can sample the differential pressure between two points 60 mm apart and located at ~10 mm from each end of the sample, using two pressure samplers (items 10 and 11 on Fig. 1) placed at the contact of the lateral sample surface. This alternative way to measure axial permeability avoids end-effects linked to the contacts between the sample and pistons which will be discussed in a following section.

The radial permeability is measured by two pairs of injector-receptor ports (items 12 and 13 on Fig. 1) facing each other, positioned along two orthogonal diameters and in two distinct planes. Each injector is made of a perforated PEEK plastic head (diameter 9 mm) for the fluid injection at the surface of the sample fitted with a coaxial Hastelloy tube (diameter 1/16 in) for pressure measurement. In addition, to allow a better flow distribution at the inlet and the outlet of the sample surface, Al316 fiber discs with average mesh aperture of 25 μm are placed at the contact of the sample.

In order to measure the differential pressure induced by fluid flow through the sample, we use a set of differential pressure sensors from ABB from ± 15 mbar to ± 400 mbar adapted to measure the permeability of cores of 1 D to 1 mD. All the measurable parameters are monitored on an acquisition bench and recorded on a computer.

The cylindrical samples are cored perpendicularly to the supposed stratification according to the quarry block extraction information (no evidence of bedding planes on the block sides). They are then dried during 48 hours. Their dry weight and exact dimensions are recorded, and then they are placed under low vacuum for 5 hours. The samples are then saturated by NaCl brine of concentration 20 g . l^{-1} (commonly recommended salinity in core analysts community to prevent clays alteration), and from the saturated weight we determine the initial porosity. The saturated sample is placed into the instrumented sleeve; the cell is closed, filled with the confining oil, and the upper piston is brought in contact with the sample. At this stage, confining and axial pressures of a few MPa are applied to ensure the sealing of the sleeve. Fluid circulation is established through the system to saturate the lines and to evacuate residual gas bubbles, and the pore fluid pressure is brought at 1 MPa and regulated by a back-pressure system. This low effective pressure limits microbubbles trapping, and insures an efficient liquid permeability measurement; it also allows starting experiments at an initial low hydrostatic pressure. The oven, the pore fluid pump and the external flow lines are then regulated to 40°C. Thermal stabilization is achieved in a few hours for the cell and a few minutes for the pump. At this stage the displacement sensors are set to zero.

Two kinds of compression stress paths are performed in this study with pore pressure kept at 1 MPa. For hydrostatic loading some experiments are made stepwise, the pressure being increased by increments of 2 MPa at low pressure and 5 MPa to 10 MPa at higher pressure. Others tests consist of a continuous loading with an increase rate of confining pressure fixed between 0.05 MPa . $\min^{-1}$ and 0.1 MPa . $\min^{-1}$, low enough to be close to quasi-static conditions (HEILAND and RAAB, 2000). Deviatoric loadings are performed at a low confining pressure of 3 MPa, which insures the sealing of the sleeve, while a "uniaxial" stress state is imposed.

The porosity evolutions and the deformations of the samples are recorded during stress increase. The porous media is only connected to the thermostated Isco pump and pore pressure is regulated at 1 MPa. Hence, the variation of the pore volume of the core is directly related to the variation of the fluid volume inside the pump. The data recorded by the deformation sensors allow both the evaluation of strains and the control of the time necessary to reach compaction equilibrium. The porosity values can thus be determined precisely allowing for the total sample strains.

The axial and radial permeability measurements are based on the steady-state method. At each constant pressure step we perform 4 sets of differential pressure measurements (2 axial and 2 radial), each of them lasting about one hour. For each measurement we check for the validity of the Darcy law (laminar flow regime), by performing a sequence of 5 increasing flow rates and measuring the corresponding increasing pressure differential. Permeability is finally calculated from the slope of a linear regression as described in the following section.

All the measurements at each pressure step last about a day, consequently a complete loading sequence of one sample takes about two weeks. Unloading is implemented continuously from 65 MPa to 3 MPa at a small constant rate $\sim$0.1 MPa . min^{-1}. The permeability after unloading is measured for comparison to the initial values.

3. Permeabilities Calculation

Permeability measurements are based on the steady-state method. Darcy's law is applied for the two axial measurements:

$$\frac{Q}{A_a} = -\frac{k_{az}}{\mu}\frac{\Delta P_p}{L}, \tag{1}$$

where Q is the flow rate, A_a is the cross section of the sample, μ is the fluid viscosity, ΔP_p is the pressure differential, and L is the distance between the points of measurements. In the case of overall axial permeability measures ($k_{az,FL}$), L refers to the current length of the sample (L $\sim$ 80 mm); in the case of mid-length measures ($k_{az,ML}$), L refers to the spacing between the two pressure samplers (L = 60 mm).

For the two radial permeability measurements, a modified Darcy law is employed:

$$\frac{Q}{A_r} = -G\frac{k_r}{\mu}\frac{\Delta P_p}{D}, \tag{2}$$

where G is a geometrical factor introduced to account for the noncylindrical shape of the flow lines generated along the sample diameter D by the radial flow ports of cross-sectional area A_r. Such geometrical factor was first determined analytically by Collins (1961) for a 2D transverse flow problem on a circular core, using a conformal mapping technique. More recently, Bai *et al.* (2002) calculated the geometrical factor using a finite-element model for a three-dimensional geometry of a transverse flow in cylindrical core. This is done using the equation:

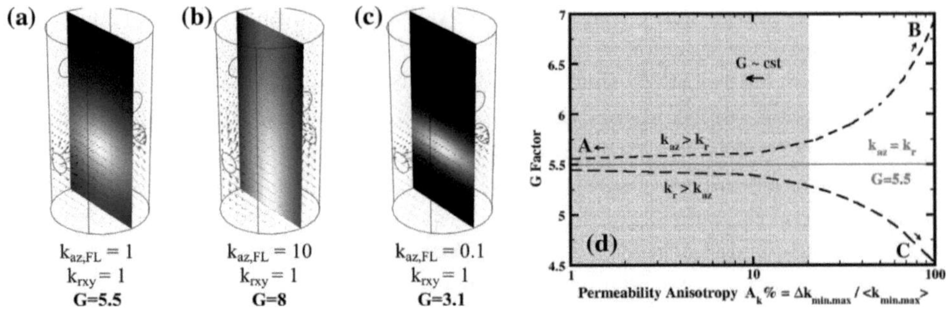

Figure 2
Geometrical model of the sleeve, radial velocity field and calculated Geometrical Factor G for different transverse permeability isotropy (a, A) $k_{az} = k_r$, (b, B) $k_{az} > k_r$, (c, C) $k_{az} < k_r$; (d) evolution of G factor with the permeability anisotropy factor.

$$G = \frac{\Delta P_{cyl}}{\Delta P_p} = \frac{A_{eq}}{A_r}, \tag{3}$$

where ΔP_{cyl} corresponds to the equivalent differential pressure generated by a flow through a cylinder with a cross-sectional area equal to A_r (conventional case for Darcy's law application) and A_{eq} is the unknown equivalent cross-sectional area of true 'ellipsoidal' flow.

For our specific experimental setup, the geometrical factor has been determined under the assumption of isotropy, using a finite-element simulation to solve the equation of flow at Darcy scale. The finite-element model is presented in Figure 2. Considering a sample of diameter $D = 38.1$ mm and length $L = 80$ mm and a radial port of diameter $d = 9$ mm, we calculated a geometrical factor $G = 5.5$ (see Fig. 2a). Nevertheless, it also remains critical to quantify the impact of permeability anisotropy on the transversal flow shape, and consequently on the geometrical factor value. Figure 2 illustrates the evolution of the geometrical factor with transversal isotropy. As clearly shown on Figures 2b and 2c, a variation of anisotropy ratio covering two orders of axial permeability magnitude induces roughly a factor 2.5 on the G value. In addition, due to the low initial anisotropy of the studied sandstones, it seems acceptable to consider a constant value of 5.5 for the geometrical factor, for materials exhibiting an anisotropy factor less than 20% (see Fig. 2d).

In order to estimate the accuracy of the permeabilities calculation, the following accuracies have been used for the different variables:

— Common variables for all measurements: Volumic flow rate $u_Q \pm 1\%$, viscosity $u_\mu \pm 2.5\%$.
— Axial permeability variables: Core length u_L (full-length case) $\pm$ 0.125% and u_L (mid-length case) $\pm$ 0.25%, injection area $u_{Aa} \pm 0.5\%$, pressure differential $u_{\Delta Pp} \pm$ 1–2% (mostly dependent of Q and the stability of flow).

— Radial permeability variables: Diameter $u_Q \pm 0.25\%$, injection area $u_{Ar} \pm 2.2\%$, differential pressure $u_{\Delta Pp} \pm$ 5–10% (mostly dependent of u_a and the stability of flow).

The accuracy calculation also integrates the error u_a related to the slope of the linear regression $Q/\Delta P_p$, which is expressed as the quadratic sum of the regression component:

$$u_a = \sqrt{a^2\left(\left(\frac{u_Q}{Q}\right)^2 + \left(\frac{u_{\Delta P_p}}{\Delta P_p}\right)^2\right) + u^2_{reg\,a} + \hat{e}_{sys}}, \tag{4}$$

with $u^2_{reg\,a}$ the uncertainty related to the linear regression and $\hat{e}_{sys}$ the systematic error related to the slope offset. Then, the estimated uncertainty of the permeability was calculated using the formula:

$$u_K^2 = \left(\frac{La}{S}\right)^2 u_\mu^2 + \left(\frac{\mu a}{S}\right)^2 u_L^2 + \left(\frac{-\mu La}{S^2}\right)^2 u_s^2 + \left(\frac{\mu L}{S}\right)^2 u_a^2. \tag{5}$$

This estimated uncertainty depends strongly on the differential pressure stability during the measurement. As discussed above, the differential pressure response of radial injection is generally noisier due to flow instability; and consequently, the radial permeability uncertainty is higher. Depending on experiments, the uncertainty on both axial permeabilities remains close to ±5%, whereas it ranges between ±10–20% for the radial direction.

4. *Characterization of Fontainebleau and Bentheimer Sandstones*

Fontainebleau and Bentheimer sandstones (FS and BS) are well documented in the literature (see for instance Klein and Reuschlé, 2003 and Sulem and Ouffroukh, 2006b). Their composition is relatively simple and they have very different permeabilities, which is a reason why they have been selected in this study. The Fontainebleau samples have been cored from a single bar obtained in the core warehouse of I.F.P. and the Bentheimer samples have been cored from a single block from the quarry of Nordhern in Germany.

The Fontainebleau sandstone is made of more than 99% quartz grains. The grain cohesion is provided by siliceous cement deposited by fluid flow during the diagenetic process (Grisoni and Thiry, 1988). Scanning Electron Microscope (SEM) observations show a good sorting of the quartz grains around a grain size value of 250 μm. Figure 3a presents a typical microstructure recorded in Electron Back-Scattering mode. This technique allows detection of secondary phases. For instance, the light gray crystal in the central part of the figure is a rutile crystal that has indented a larger quartz particle during diagenesis. Most of the macroscopic measurable porosity value is due to large pores, which are interconnected by intergranular micro-channels (or throats) at the grains boundaries. As shown on Figure 3b, the mercury injection curves reveal a narrow distribution of throat radii around a value of 7 μm, but the SEM observations and the

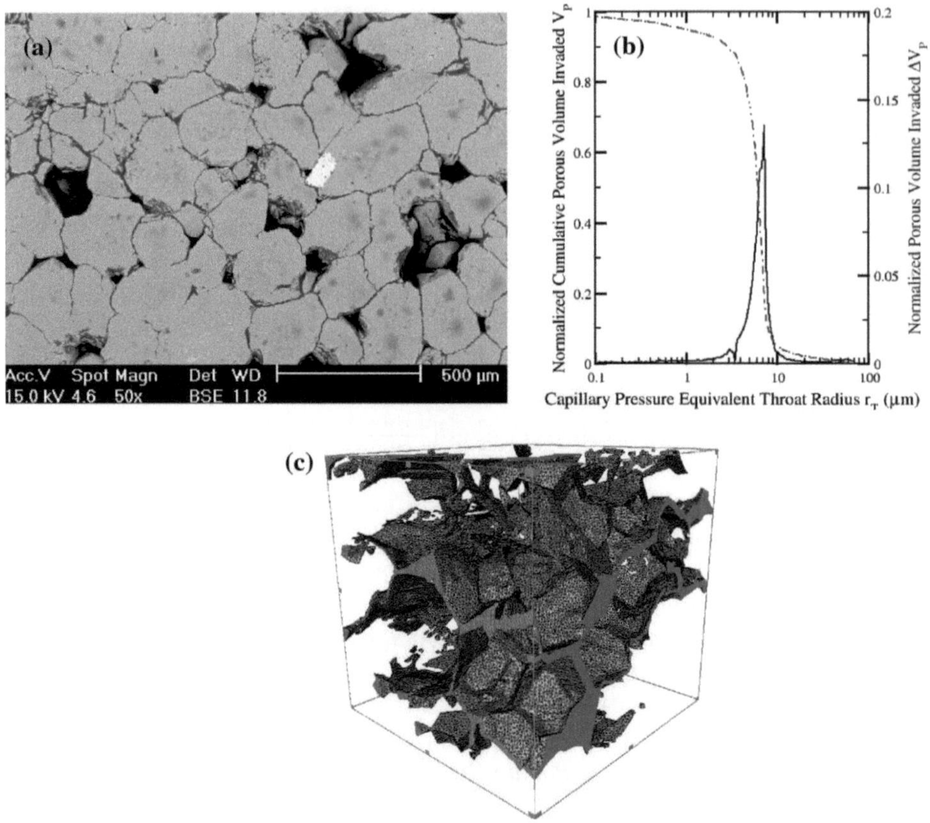

Figure 3

Microstructural characterization of Fontainebleau sandstone: (a) SEM Image in BSE mode, (b) Hg porosimetry (up to 400 MPa) and (c) pore volume extraction from a 1 mm^3 microtomography volume.

micro-tomography reconstruction (see Fig. 3c) yield a value closer to 1 μm. This discrepancy is essentially due to the shape of the throats in the Fontainebleau sandstone far from the ideal cylindrical shape model used to calculate an equivalent pore throat radius for the Mercury capillary pressure curve.

The very large range of porosity values from 2% to 30% of Fontainebleau sandstones has been previously interpreted to result from spatial heterogeneities of the diagenetic fluid flows Bourbie and Zinszner (1985). This extended range of porosity values leads to a large variation of the permeability. The two samples that we have tested present different porosities of ~5.8% and ~8%, however their permeabilities differ by one order of magnitude, since it varies between ~2.5 mD and ~30 mD (see initial porosity and axial permeabilities, values measured with our experimental setup in Table 1).

The Bentheimer sandstone is composed of 95% quartz, a small fraction of altered feldspars (less than 2%) and approximately of 3% clay homogeneously distributed in the

Table 1

Petrophysical parameters of the tested Fontainebleau samples FS1 and FS2, measured at the initial and final stages of the hydrostatic loading and after unloading. The symbol—refers to improper conditions for permeability measurement (either unstable pressure differential or plugging of the flow/pressure ports) or for porosity measurement (leak on pore fluid lines)

Sample	FS1			FS2		
Pressure Level	Initial at 5 MPa	Final at 57 MPa	Residual at 5 MPa	Initial at 3 MPa	Final at 55 MPa	Residual at 3 MPa
Porosity (%)	5.80	5.57	5.75	7.96	7.47	–
$k_{az,FL}$ (mD)	2.47	1.53	1.89	31.14	22.87	25.20
$k_{az,ML}$ (mD)	2.46	2.10	2.29	29.27	24.71	26.11
k_{rx} (mD)	–	–	–	19.60	8.11	9.64
k_{ry} (mD)	4.71	– (→3.)	3.19	–	–	–

rock matrix. The quartz grains are well sorted around a value of 200 μm. Bentheimer sandstone exhibits larger permeability values ($k_{az,ML} \sim 2.9$ D and $k_{az,FL} \sim 1.5$ D) associated with one large porosity of about 24%, and to a larger size of pore throats homogeneously distributed around 20 μm as shown by the Mercury porosimetry curve in Figure 4. The feldspar alteration also contributes to the porosity connectivity. SEM investigations show that the intergranular contacts are narrower than for the Fontainebleau sandstone, hence the radii of throats are larger and consequently the permeability is substantially higher. Because the Bentheimer sandstone is quite homogeneous, permeability measurements were also performed on additional cores on a classical permeameter giving a permeability value of $\sim$1.6 D close to the value obtained for the full-length axial permeability measured with our setup. The important discrepancy observed between the two methods of axial permeability measurements (full-length vs. mid-length) will be discussed in terms of end effects in a further section. Note that for the Fontainebleau sandstone, we did not have enough available material to plug additional samples of required dimension for measurement with the classical permeameter.

The difference of permeability between the two sandstones is in agreement with the observed shift of the pore throat radii distribution peek estimated from the Hg porosimetry. Indeed, since the local viscous dissipation varies as the power four of the characteristic dimension of the pore throats (hydrodynamic radius) and the pressure drops are located almost exclusively in the throats, the permeability is expected to follow roughly $k \propto \overline{r_t}^4$ and then $k_{BS}/k_{FS} \approx \left(\overline{r_{t,BS}}/\overline{r_{t,FS}}\right)^4$.

5. *Directional Permeability Evolutions under Hydrostatic Loading*

5.1. *Low Permeability Fontainebleau Sandstone*

The initial porosity values of the two tested samples of Fontainebleau sandstones FS1 and FS2 are respectively 5.8% and 8%. The hydrostatic compression tests have been

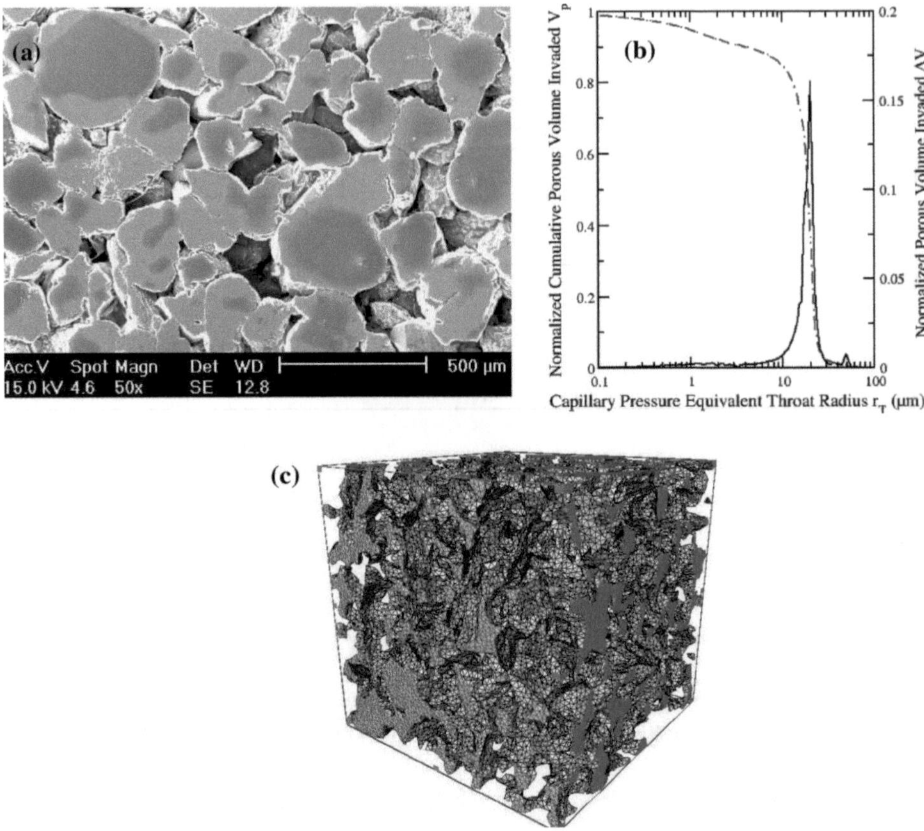

Figure 4
Microstructural characterization of Bentheimer sandstone: (a) SEM Image in SE mode, (b) Hg porosimetry (up to 400 MPa) and (c) pore volume extraction from a 6 mm^3 microtomography volume.

started at low confining pressures, $P_c = 6$ MPa for FS1 sample and $P_c = 3$ MPa for FS2 sample, in order to ensure sealing of the core sleeve. At this stage we perform a first set of permeability measurements in each direction ($k^0_{az,FL}$, $k^0_{az,ML}$, k^0_{rx} and k^0_{ry}) reported in Table 1. Both full-length and mid-length axial permeability measurements are well correlated close to 2.5 mD and 30 mD for FS1 and FS2, respectively, whereas the radial permeability exhibits a higher value close to 4.7 mD for FS1 and a lower value close to 19.6 mD for FS2. This 'apparent' permeability anisotropy is in disagreement with the known homogeneity and low structural anisotropy of the Fontainebleau sandstone (Clavaud *et al.*, 2008); this discrepancy seems mostly induced by fluctuations of radial flow conditions, as will be discussed further.

Despite the small porosity value differences between the two samples, we observe a high permeability variability in this porosity range, as previously described by Bourbie and ZINSZNER (1985). The hydrostatic compression tests have been performed on the Fontainebleau Sandstone in a stepwise manner from the initial confining pressure up to 65 MPa. For each pressure step, the variation of porous volume and the strains of the sample are monitored. After an equilibration time for deformation, we measure the permeability in each direction using the protocol previously described.

In Figure 5, we present the porosity evolution inferred from the variation of porous volume measurements and also the absolute and normalized directional permeability evolutions with the mean effective pressure ($P_{eff} = P_c - P_p$) obtained on the two Fontainebleau sandstone cores FS1 and FS2. In the following we consider the effective pressure nearly equal to the isostatic pressure, since the pore pressure remains constant at 1 MPa. By convention, the solid lines and open symbol represent the permeability in the axial direction z, measured stepwise on the full sample length $k_{az,FL}$ (circle) and on the middle length $k_{az,ML}$ (square), whereas the dashed lines and full symbols represent the results obtained for the radial permeability k_{rx} (triangle up) and k_{ry} (triangle down).

The porosity decreases slightly from 5.80% to 5.57% for FS1 and from 7.96% to 7.47% for FS2. In a first stage of confining pressure up to 20 MPa the porosity response is nonlinear, and then varies almost linearly up to 60 MPa in both experiments. After unloading back to the initial pressure, the porosity of FS1 exhibits an hysteresis with a residual value of 5.75%. The response of the Fontainebleau sandstone is therefore almost elastic with a nonlinear behavior in the early stage of the compression, characterized by a slight nonreversible deformation of the porous volume.

For the FS1 sample, the full-length axial permeability $k_{az,FL}$ evolution shows a rapid decrease of 30% of the initial value 2.47 mD until a pressure of 22 MPa. Above this pressure, which is commonly assumed to be the closure pressure of pre-existing microcracks (DAVID *et al.*, 1994), the general trend becomes close to a slow linear decrease with a slope of 2.1 E–3 MPa^{-1}. The minimum permeability reaches a value of 1.53 mD. After unloading, we measure a residual permeability of 1.89 mD. These results contrast with the evolution of the mid-length permeability $k_{az,ML}$. Indeed, the trend of permeability loss remains linear with a slope of 3.1E−3 MPa^{-1} during the complete loading. The absolute radial permeability exhibits a more pronounced linear response with a slope of 6.7 E–3 MPa^{-1} without evidence of a nonlinear response at the early stage of compression to reach a value of 3.27 mD at 50 MPa. The last measurement at 57 MPa could not be achieved.

Considering the results obtained on FS1, the hydrostatic compression test performed on FS2 is started at lower pressure $P_c^0 = 3$ MPa and we increased the number of loading steps to better identify the transition between the pseudo-elastic regime characterized by strong and nonlinear permeability drop and the elastic regime characterized by linear behavior.

As previously observed for FS1, axial full-length permeability $k_{az,FL}$ strongly decreases from 31.14 mD to17.41 mD at 30 MPa. However above this pressure the general trend is very difficult to define, because unexpectedly the permeability increases at 35 MPa to reach 23.71 mD and later follows a linear evolution with a

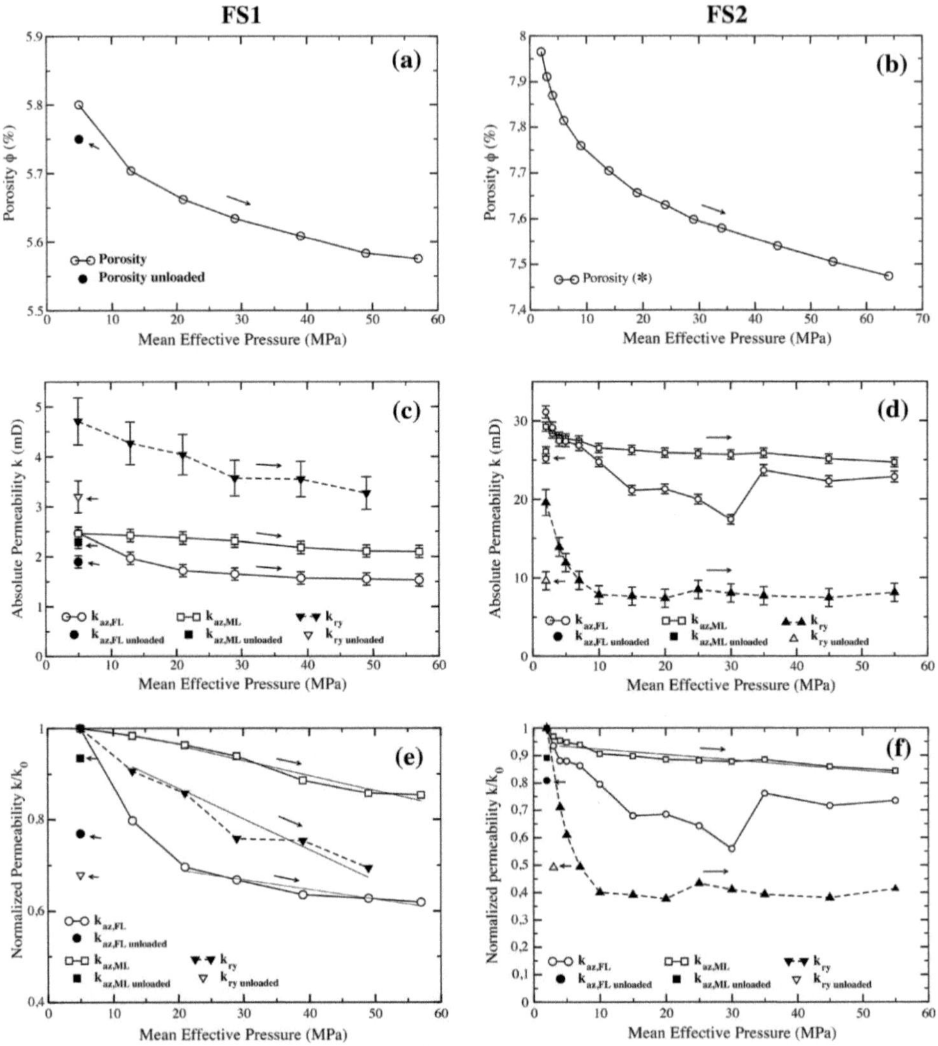

Figure 5
Porosity, absolute directional permeabilities and normalized directional permeabilities evolutions with the mean effective pressure for the FS1 (respectively (a), (c) and (e)) and FS2 (respectively (b), (d) and (f)) samples. Arrows show the loading/unloading paths; inverted filling properties of symbols refer to values at unloaded states (pressure decreased back to the initial value, $P_c = 6$ MPa for FS1 and $P_c = 3$ MPa for FS2); symbol * stands for: unloaded porosity value of FS2 sample could not be determinate due to a pump leak during the unloading.

slope of 1.95 E–3 MPa^{-1} up to 55 MPa. The mid-length axial permeability $k_{az,ML}$, of initial value 29.27 mD, exhibits a weak diminution of almost 10% until 20 MPa, and records a total decrease of less than 15% at the final effective pressure of 55 MPa, with a permeability value of 24.71 mD. After unloading, the residual full-length and mid-length axial permeabilities show close values of respectively 25.20 mD and 26.11 mD.

The radial permeability k_{rx} drops drastically from 19.60 mD to ~8 mD during the first pressure steps up to 10 MPa. Above this pressure the permeability keeps more or less a constant value, given the error estimate. After unloading back to $P_c = 3$ MPa, the residual radial permeability value is equal to 9.64 mD.

5.2. High Permeability Bentheimer Sandstone

With Bentheimer sandstone we perform both stepwise and continuous hydrostatic loading experiments. The initial axial permeabilities $k^0_{az,FL}$ and $k^0_{az,ML}$ measured at 3 MPa are respectively 1774 mD and 2891 mD. Both axial measurements are expected to be approximatively the same, nonetheless we find the mid-length value to be ~1.6 time higher than the full-length value. A similar discrepancy also has been observed by KORSNES *et al.* (2006), who used a corresponding experimental apparatus. The radial permeability k^0_{rx} calculated with a Geometric Factor $G = 5.5$ (equation (3)) yields an intermediate value of 2335 mD.

Figure 6 shows the evolution of the porosity with the increasing effective pressure. The porosity decreases from 24.4% to 23.6%. As previously observed for the Fontainebleau sandstone, the porosity response is nonlinear until 20 MPa and then evolves linearly up to 65 MPa. After unloading at 2 MPa, a pronounced hysteresis has been observed and the new residual porosity value is equal to 24.0%. The superimposed continuous curve has been obtained from continuous loading cycles performed after unloading to $P_c = 3$ MPa.

On Figure 6 we also present the evolution of directional permeabilities acquired during both stepwise and continuous loading. The results obtained from stepwise loading show a moderate response of the permeabilities with the confining pressure increase, essentially due to rather high porosity and large pore throats of the Bentheimer sandstone. Indeed, the mid-length axial permeability only shows a slight decrease, less than 10% below 10 MPa. The full-length permeability shows a more pronounced decrease of almost 50% up to 20 MPa. From 20 MPa to 65 MPa this permeability decreases linearly with a slope of 2.14E −3 MPa^{-1}. The general trend of the radial permeability is not clear, nonetheless its evolution presents an intermediate behavior with high fluctuations. After unloading, we have measured the residual permeabilities $k^r_{az,FL}$, $k^r_{az,ML}$, k^r_{ry} respectively equal to 1185 mD, 2816 mD and 2151 mD.

To limit differed local mechanisms of inelastic deformation for stress accommodation, which could influence the permeability measurements at constant confining pressure, we perform cycles of continuous loading with a rate of 0.1 $\text{MPa} \cdot \text{min}^{-1}$. For instance, a similar procedure of continuous permeability measurements during triaxial loading is discussed by HEILAND and RAAB (2000). We have measured each continuous directional permeability during two cycles of loading/unloading, taking into account the axial and radial strains of the sample. As shown on Figure 6, the results obtained by this

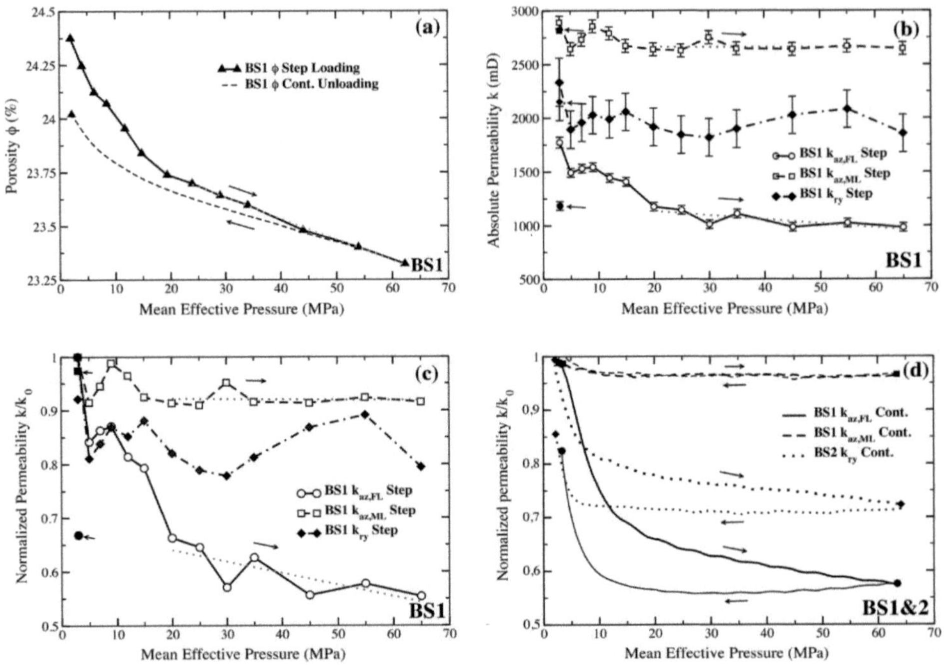

Figure 6

Evolutions of (a) porosity, (b) absolute directional permeabilities, (c) and (d) normalized directional permeabilities' evolutions during respectively ((a), (b), (c)) stepwise and (d) continuous hydrostatic compression tests. Arrows show the loading / unloading paths; inverted filling properties of symbols refer to values at the unloaded states (here, $P_c = 3$ MPa).

method are in good agreement with those obtained step by step and are less disrupted by experimental fluctuations. Those results confirm the loss of the permeability values $k_{az,FL}$, $k_{az,ML}$ and k_{ry} of almost 40%, 10% and 25% observed during the first loading cycle. As previously pointed out, the permeability drop does not occur in the same range of pressure. Indeed, the most important drop is recorded up to 25 MPa for $k_{az,FL}$, 10 MPa for k_{ry} and is negligible for $k_{az,ML}$. Independently of the measurement direction, we also observe a delay in the permeability response during the sample unloading, most probably due to frictional effects.

6. Directional Permeability Evolutions under "Uniaxial" Loading

Transport properties of rocks are strongly influenced by the type of deformation regime. In this section we aim at characterizing the influence of shear-band generation and the importance of the evolutions of grain scale microstructures on the permeability anisotropy. We also highlight the ability of our apparatus to follow the directional response with regard to strong modification of microstructures. The uniaxial compressive strength of Fontainebleau sandstone exceeds the maximum axial stress accessible with our experimental setup, therefore we did not observe brittle failure of the sample and thus any major effects on the directional permeabilities. Hence, we only present the results of three experiments obtained on the Bentheimer sandstone.

We have performed deformation and directional permeabilities' measurements under continuous triaxial loading at low confining pressure. The principal objective is to be close to the true uniaxial conditions (i.e., without confining pressure), although the application of a low confining pressure ($P_c^0 = 3$ MPa) is necessary to insure an initial sealing of the core sleeve. The axial load is then continuously increased at a rate of 0.5 MPa . min^{-1} until failure of the samples. During loading, pressure differential is monitored for each experiment in a given measurement direction and the directional permeabilities $k_{az,FL}$, $k_{az,ML}$ and k_r are calculated on three different samples.

Prior to axial loading, a first set of permeabilities measurements is taken in each direction $\left\{k^0_{az,FL}, k^0_{az,ML} \text{ and } k^0_r\right\}$ (see Table 2) at hydrostatic condition. Again we observe differences within factors from 1.9 to 2.4 for both axial permeability measurements. Regarding the axial and radial permeability measurements, the full-length axial permeability is almost 1.9 times less than the radial permeability value.

As illustrated on Figure 7a, at first, axial full-length and radial permeabilities decrease steeply. Next, a nearly linear regime is observed up to 43 MPa with slopes equal to 7.8 E–4 MPa^{-1} and 1.15 E–3 MPa^{-1} for axial full-length and radial permeability respec-

Table 2

Petrophysical parameters of the three Bentheimer samples BS1, BS2 and BS3, measured before and after the brittle failure at confining pressure of 3 MPa. For BS1 permeabilities' values, () refer to measurements before hydrostatic cycling. The symbol - refers to improper conditions for permeability measurement (either unstable pressure differential or plugging of the flow/pressure ports). The symbols (↗, ↘, →) refer respectively to the increase, the decrease or the stability of the permeability in the investigated direction

Sample	BS1 (k_{rx}^{cont}) (Dautriat *et al.*, 2009)			BS2 ($k_{az,FL}^{cont}$)			BS3 ($k_{az,ML}^{cont}$)		
Porosity (%)	24.4			24.4			23.8		
Initial State	Hydrostatic cycles Pre-strained			No preliminary hydrostatic cycling			No preliminary hydrostatic cycling		
Permeability	k_{init} at 3 MPa		k_{C*} at 3 MPa	k_{init} at 3 MPa		k_{C*} at 3 MPa	k_{init} at 3 MPa		k_{C*} at 3 MPa
$k_{az,FL}$ (mD)	(1774) 1185	↗	1560	1347	→	1315	1234	↘	1089
$k_{az,ML}$ (mD)	(2891) 2816	↗	2985	-	-	-	2826	↘	2120
k_{rx} (mD)	(2335) 2139	↘	631	2571	↘	2244	2363	↘	1667
k_{ry} (mD)	(2811) -	-	-	2620	↘	2086	2097	↘	1408

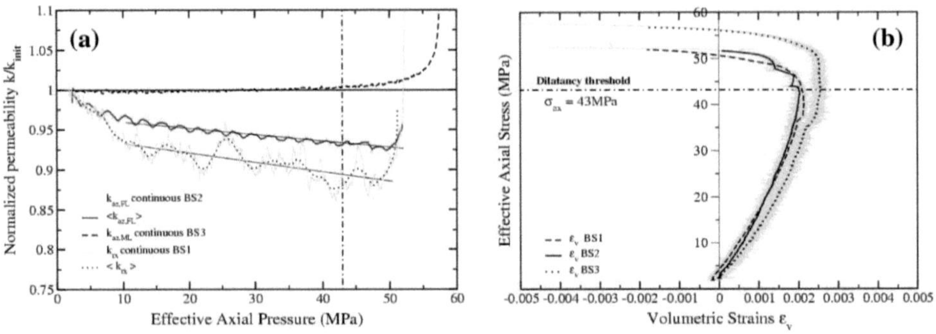

Figure 7
Evolutions of (a) directional permeability and (b) volumetric strains during continuous triaxial loading performed at low confining pressure ($P_c = 3$ MPa) on three Bentheimer sandstone samples BS1, BS2 and BS3. The average dilatancy threshold occurring at axial stress of ~43 MPa is highlighted by dashed lines on the two figures.

tively. This stress corresponds to the dilatancy threshold (see Fig. 7b), and beyond this value one records an important increase of the permeability independently of the measurement directions. Note that the radial injection conditions might change locally, leading to fluctuations of the pressure differential between the radial inlet and outlet and hence to a noisier signal than the one obtained on the evolution of axial full-length permeability. After the failure point, the pressure differential strongly diverges and can no longer be related to permeability. The mid-length axial permeability follows a different trend without significant evolution until the dilatancy threshold, and increases sharply after this point.

The critical failure pressure P_{C*} is reached at 52.5 MPa for the two first tested samples BS1 and BS2, whereas critical failure P_{C*} occurs at 57 MPa for the last sample BS3; this higher strength could be explained by the fact that the last tested sample has not been subjected to hydrostatic loading cycles.

After failure, the axial stress is decreased back to its initial value and a new set of directional permeability is performed. The permeability values measured after unloading are presented in Table 2. Note that the geometrical factor G used for the radial permeability is kept constant to 5.5, and consequently the absolute permeability value could be significantly influenced by this approximation. In fact, it is the evolution of the differential pressure which actually provides qualitative information pertaining to the flow properties along the fracture. The full-length axial permeability of BS1 increases slightly by a factor of 1.3, the mid-length is almost unaffected, and the radial permeability decreases strongly by a factor 3.4. From the measurements performed on BS2 and BS3, the failure induces a slight decrease of the axial permeability. Therefore, the radial permeability measurements present about 12% to 32% decrease from their initial values, according to the measurement direction.

In order to understand how the failure mode could affect the directional permeability values, the sample has been delicately taken out of the core sleeve and oriented with

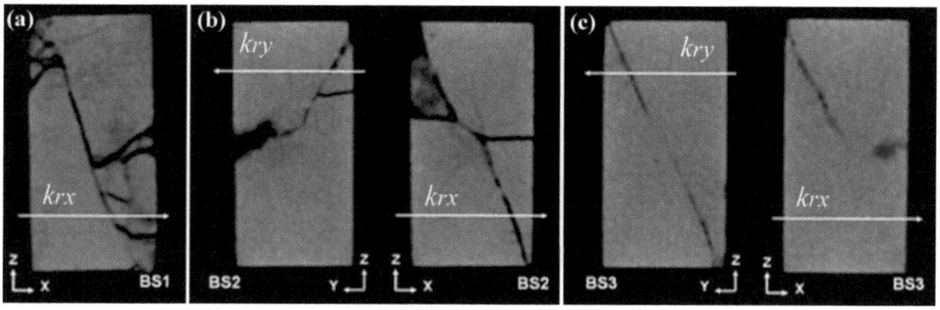

Figure 8
Tomography-RX slices of the Bentheimer sandstone samples (a) BS1, (b) BS2 and (c) BS3 subjected to 'uniaxial' compressions leading to brittle failure. The presented slices correspond to the planes in which the radial permeability measurement was performed.

respect to the injection directions and local pore pressure samplers positions. While the failure direction is clearly identified on the sample surface, we have performed CT-scan profiles of the sample along the three directions of performed flows. As shown on Figure 8, the principal shear band extends from the top to the bottom of the sample with a propagation angle close to 30°. The failure is also sub-perpendicular to the one of the radial injection direction. Several conjugated fractures are visible by tomographic analysis and their formation is probably related to the pressure control operation mode of our setup leading to the inability to control the post-peak behavior (i.e., after the dilatancy initiation). Our experimental setup does not allow the control of the strain rate applied on the sample, thus after brittle failure the radial axial stress still remains sufficient to further activate shear and to increase grain scale damage within the band.

It is known that the response in terms of porosity is highly related to the microstructure of the shear band (VAJDOVA *et al.*, 2004). In a recent study, SULEM and OUFFROUKH (2006a) have shown by image processing analysis an increase of the porosity inside the band and minor grain crushing at low confining pressure. The authors related this observation to a dilating shear banding. The increase of the porosity in the band can explain the axial permeability increase, whereas the high permeability decrease recorded in the transverse direction of the band could be related to the important sliding of sample blocks along the band, associated with intensive grain crushing after post-peak. As clearly shown on Figure 8, we observe that the radial permeability decreases mostly where the flow lines intersect an intensively damaged zone.

7. Discussion of the Experimental Results

The permeability of numerous rock types has been extensively studied under hydrostatic conditions (ZOBACK and BYERLEE, 1975a; WALSH and BRACE, 1984; DAVID *et al.*, 1994) and triaxial conditions (VAJDOVA *et al.*, 2004) in order to establish effective pressure dependency of the permeability. From those studies, the authors have shown a decrease of the permeability with the effective pressure depending on three distinct phases. During the first increments of effective pressure a large reduction of permeability is observed and described by DAVID *et al.* (1994) as a probable microcracks closure. The rate of permeability reduction decreases at high confining pressure and the authors showed an exponential evolution of the permeability with the effective isostatic pressure. When the effective pressure is further increased, the permeability exhibits a drastic drop at the critical pressure, indicating the onset of inelastic deformation characterized by pore collapse commonly associated with intensive grain crushing (ZHANG *et al.*, 1990a). The average critical pressure is about 200 MPa in sandstones, although it depends essentially on the initial porosity value of tested material. Due to the limitations of our experimental setup, we are not able to reach the pore collapse pressure of the Fontainebleau and Bentheimer sandstone. However, we mostly focus our interest on the first two regimes of permeability reduction mechanisms. In most cases, the pressure dependencies of the permeability have been determined using the overall pressure drop over the complete length of the sample. The results obtained with the additional sampling of pressure drop on the middle length of the sample provide a complementary explanation for the drastic permeability drop occurring at low effective pressure.

7.1. Evidence of End Effects on Axial Permeability

The presented results on Bentheimer sandstone show that axial permeabilities measured by the two different methods differed strongly at the beginning of loading. The comparable discrepancy factor has been observed by KORSNES *et al.* (2006) under deviatoric stresses on both chalk and sandstone samples. The authors used a modified core sleeve, allowing the measurement of mid-length permeability as in our study; they found a discrepancy factor of 1.2 to 2.4 between the two axial measurements, as we do on the Bentheimer sandstone. They concluded that end effects occurring at the sample/piston interface were responsible for these initial discrepancies.

Conversely to the Bentheimer sandstone, the initial axial permeabilities measured on the Fontainebleau sandstone samples show strictly the same value (according to the estimated error). Because of different microstructures, the two sandstones may exhibit different responses to end effects in terms of permeability damage on the first 3 MPa of loading. Mechanical preparation, such as sample cutting, may also induce some near-surface damage. Such effects may be more pronounced for the loosely packed Bentheimer sandstone.

In addition, KORSNES *et al.* (2006) did not report different evolution divergences with increasing pressure; conversely we have observed different trends depending on the measurements method. Indeed, for the three samples subjected to hydrostatic loading, the preliminary nonelastic regime response of the permeability is more pronounced for the full-length measurement (30 to 50% of the initial value), than for the mid-length measurement (less than 10% of initial value if it occurred). The porosity response also has manifested the same nonlinear evolution in the range of low effective pressure. Furthermore, the porosity of the Bentheimer sandstone showed a hysteresis and the residual value at 3 MPa is 24%. These observations suggest closure of a limited part of the pore volume, however, this mechanism alone cannot explain the discrepancies of the permeabilities evolution. We also suspect end effects to induce higher deformation of the sample at the contact of the pistons. FORTIN *et al.* (2005) have shown that during compression experiments on the Bentheimer sandstone, the acoustic emission relocalization highlighted a compressive damage area close to the interface between the pistons and the sample at an early stage of compression. This early localized compaction could explain pronounced inelastic behavior of the full-length axial permeability, while the mid-length permeability remains unaffected due to the distance of the pressure samplers from the core ends.

7.2. FEM Modelization of End Effects on Strain Localization

We have performed FEM computations in order to estimate the effects of boundary conditions and sample geometry on the stress distribution within the sample. The samples are assumed to be homogeneous and to behave as isotropic linear elastic materials.

The ideal test assumes lubrication between piston and sample; in practice the condition is closer to perfect adherence without slip, and this is even more so when there is a groove in the piston head which may indent the sample. The pistons are made of a titanium alloy with a Young's modulus at least ten times larger than the modulus of sandstone $E_{sample} = 10.3$ GPa and $E_{pistons} = 114$ GPa and also a different Poisson's ratio $\nu_{sample} = 0.2$ and $\nu_{pistons} = 0.24$. At the interfaces (top and bottom) between piston and sample, the strain incompatibilities induce local stress gradients. A FEM computation cannot exactly reproduce the experimental end conditions, but allows the check of assumptions which provide bounds to the actual conditions. A mesh of 9-noded quadrilateral elements with initial square sections of 1 mm^2 is utilized. The exact dimensions of the cylindrical samples are given and one replaces the actual pistons by two additional cylinders of 20 mm height and the same diameter. A perfect link is assumed at the two interfaces piston/sample assembly. The behavior remains elastic, and the load is applied by using pressure elements at the top and around the assembly pistons-sample. One can thus easily perform the two virtual tests of hydrostatic loading (same pressure, chosen as 10 MPa) and uniaxial loading with small confining pressure (10 MPa on top and 3 MPa around).

Figure 9 presents the results of the simulation of both configuration tests with regard to axial, radial, shear and equivalent stress inhomogeneities. The simulation shows that the ends of the sample are regions where high stress gradients appear; in particular, there is a departure from hydrostatic conditions, although the middle of the sample remains under hydrostatic stress state. The axial stress in both loading cases (Figs. 9a and 9f) is larger in magnitude near the interface and also at the outside edges of the samples. Its variation along the sample axis is not uniform but after a region where it remains

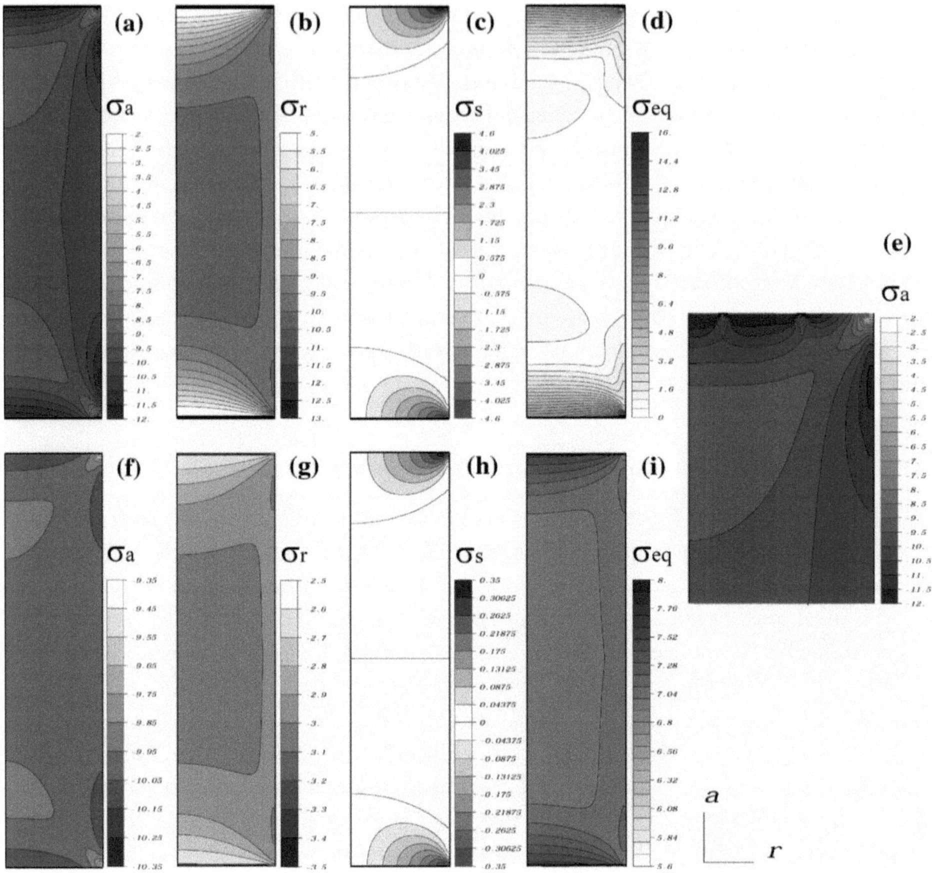

Figure 9
Results of FEM simulation for both hydrostatic at 10 MPa (top) and axial at 10 MPa with low 3 MPa confining pressure (bottom) loading, and representations of the axial (a, f), radial (b, g), shear (c, h) and equivalent (d, i) stress fields within the sample (unit in MPa). (e) Focus on the influence of the grooved piston on the axial stress at the interface between the sample and the upper piston, for the hydrostatic configuration.

constant, it decreases at first and then increases again near the pistons. There is a lens-shaped region near each end where the axial stress is slightly smaller than everywhere else. In addition, the radial stress (Figs. 9b and 9g) is fairly uniform in the middle region and compressive. It is however rapidly varying very near the interfaces and a shear stress also develops at the interfaces, more pronounced in the hydrostatic case (Fig. 9c). In the grooved pistons case (Fig. 9e) there is no evidence of a larger zone of heterogeneity than in the simple configuration. The effect remains local but induces a larger perturbation of the isocontours of stresses.

Keeping in mind the strong assumptions of homogeneity and purely elastic behavior, it remains to check whether these computed variations are responsible for early damage of the samples. Indeed, the simulations exhibit a high inhomogeneity of the stress states near the sample ends, especially during the hydrostatic loading. Due to their structural heterogeneities, the sandstones may be very sensitive to this not purely hydrostatic stress field, which could be responsible for stress localization within the grains boundaries and local grain rearrangements near the pistons' contacts. The results of the simulations support the experimental observations, and would explain the high permeability drops observed on the full-length axial permeability.

7.3. Radial Injection Conditions' Fluctuations during Compression Experiments

We have showed and simulated how the mechanical response of the samples could generate a strong loss of the axial permeability during the beginning of pressure increase. Similar experimental results indicate end effects on the radial permeability measurements, but these are related to the experimental setup and data retrieval procedure. Firstly, we find a systematic weak discrepancy factor between initial axial and radial permeability values. This factor varies from $+1.9$ and -1.4 for the FS1 and FS2 samples respectively, and of 1.4 for the Bentheimer sandstone, while LOUIS *et al.* (2003) and BENSON *et al.* (2005) have shown slight permeability anisotropy for the Bentheimer sandstone. This discrepancy is most likely related to the Geometric Factor G introduced in equation (2) to calculate the radial permeability; experimental radial fluid injection could indeed differ from the simple geometry used in our FEM model. Secondly, we have observed an important loss of the radial permeability up to a confining pressure of 10 MPa from 20% for the Bentheimer sandstone of the initial value to 60% for the FS2 sample. Those strong evolutions are not observed on the FS1 sample that is subjected to higher initial confining pressure of 6 MPa. The important loss of radial permeability could not be explained only by the response of the sandstone sample, but also by an intensive deformation of fiber within the Al316 spacers placed between injectors and sample surface. To understand how the discs are deformed during the compression tests, we have observed the microstructure of as-received and deformed discs by microtomography imaging analysis (see Fig. 10). The undeformed disc exhibits two superposed layers with averaging mesh aperture of 10 μm (transect A–B) and 25 μm (transect C–D) for an averaging thickness of 600 μm, whereas the deformed disc is inhomogeneously compacted by up to 200 μm. The most porous layer is the most intensively strained.

These high deformations of the discs might have reduced drastically their permeabilities and certainly induced important perturbation of the flow conditions at the inlet and the outlet of the injectors. The deformation of the discs could have been more pronounced during the first steps of confining pressure increase, which could explain efficiently the drastic drop of radial permeabilities. The impact of the discs deformations on their permeabilities is very difficult to quantify; we suggest that experimental measurements performed above 10 MPa might be representative of sample behavior. Nevertheless, the discs are needed to perform radial flows; indeed, compression

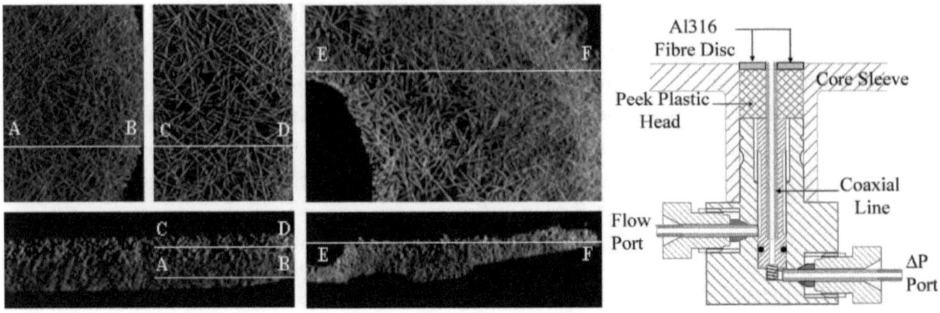

Figure 10
Microtomography slices of Al-316 fiber discs in (left) underformed (0.5 mm thickness) and (center) deformed states; (right) drawing showing the fiber disc position on a radial flow port.

tests performed without the discs lead to clogging of the injectors and the loss of the radial measurements capacity.

8. *Conclusions*

Our first experiments validate a new triaxial cell specifically designed for multi-directional permeability measurements in reservoir rocks under hydrostatic and deviatoric stresses. We have used two types of sandstones covering a wide range of porosity (from 5.8 to 24.4%) and permeabilities (from 2.5 to 3000 mD). Classical axial permeability measurements on Fontainebleau sandstone with our setup are in very good agreement with other experiments and models from the literature.

However, our classical axial (full-length) and radial permeability measurements are very sensitive at the beginning of the hydrostatic loading both for Fontainebleau and Bentheimer sandstones. The evolutions of permeabilities are characterized, in most cases, by an initial strong decrease followed by a slight linear variation. In contrast, the innovative axial mid-length measurements present a more moderate and steady evolution. Our experimental results and FEM simulations suggest stress inhomogeneities (variation in magnitude and triaxiality) in the vicinity of the piston/sample interfaces which may induce local microstuctural evolutions (damage at the grain scale or pore collapse), which in turn might strongly reduce the full-length axial permeability. Such end effects would not affect the mid-length axial permeability measurements, which are what we observe.

Our work clearly shows that hydromechanical end effects can alter significantly the standard permeability measurements using classical permeameter, even on consolidated rocks, like sandstones, thought to be less sensitive. This may lead to erroneous micromechanical interpretation of permeability reduction and also to an overestimation of the permeability decrease with overburden stress or effective stress evolution, potentially critical for field case developments. Experimental data of pressure dependent permeability, reported in the literature, should therefore be considered with care. Recent experiments performed on carbonates showed less pronounced end effects, suggesting mineralogical and/or microstructural influences.

As for the radial permeability evolutions, we have clearly observed, after loading, microstructural evolutions of the Al316 spacers inserted between the sample and the flow lines that could explain the initial loss of radial permeability.

Additional uniaxial loading experiments do confirm the influence of shear-band formation on radial permeability which is strongly reduced owing to grain crushing and densification within the shear band. These motivating results show the interest of directional measurements during a single loading experiment to investigate more efficiently the influence of high deformation on generated anisotropy. With regards to the pressure limit of our experimental setup (69 MPa) and the high strength of sandstone samples, we are only able to investigate the influence of the brittle failure regime on the permeability evolution. Nevertheless, preliminary experiments performed on soft carbonates reach higher deformation regimes, such as the shear-enhanced compaction regime, and provide a new set of data with which to understand the influence of stress paths on permeability anisotropy evolutions.

Acknowledgements

The authors thank Eric Ferne at I.F.P. and Serge Chanchole at L.M.S. for valuable technical assistance with the directional flow triaxial setup. We also are grateful to Elisabeth Bemer for helpful suggestions on the geomechanical aspects of this work. We extend thanks to Daniel Caldemaison for access to SEM at L.M.S., to Elisabeth Rosenberg and Souhail Youssef for microtomography acquisition and volume reconstruction at I.F.P. The support of Olga Vizika and Jean-Marc Lombard in this project is also acknowledged.

References

Al-Harty, S.S., Jing, X.D., Marsden, J.R., and Dennis, J.W. (1999), *Petrophysical properties of sandstones under true triaxial stresses I: Directional transport characteristics and pore volume change*, SPE *57287*.

Bai, M., Meng, F., Roegiers, J.C., and Green, S. (2002), *Improved determination of stress-dependent permeability for anisotropic formations*, SPE *78188*.

Benson, P.M., Meredith, P.G., Platzman, E.S., and White, R.E. (2005), *Pore fabric shape anisotropy in porous sandstones and its relation to elastic wave velocity and permeability anisotropy under hydrostatic pressure*, Int. J. Rock. Mech. *42*, 890–899.

Bourbie, T. and Zinszner, B. (1985), *Hydraulic and acoustic properties as a function of porosity in Fontainebleau sandstone*, J. Geophys. Res. *90*, 11524–11532.

Bruno, M.S. (1994), *Micromechanics of stress-induced permeability anisotropy and damage in sedimentary rock*, Mech. of Mat. *18*, 31–48.

Clavaud, J.-B., Maineult, A., Zamora, M., Rasolofosoan, P., and Schlitter, C. (2008), *Permeability anisotropy and its relations with porous medium structure*, J. Geophys. Res. *113*, B01202.

COLLINS, R.E., *Flow of Fluids through Porous Materials* (Reinhold Publishing Co., New York. 1961).

DAUTRIAT, J., GLAND, N., YOUSSEF, S., ROSENBERG, E., BEKRI, S., and VIZIKA, O. (2009), *Stress-dependent directional permeabilities of two analog reservoir rocks: A prospective study on contribution of μ-tomography and pore network models*, SPE Res. Eval. Eng. (in press).

DAVID, C., WONG, T.F., ZHU, W., and ZHANG, J. (1994), *Laboratory measurements of compaction-induced permeability change in porous rock: Implications for the generation and maintenance of pore pressure excess in the crust*, Pure Appl. Geophys. *143*, 425–456.

FORTIN, J., SCHUBNEL, A., and GUEGUEN, Y. (2005), *Elastic wave velocities and permeability evolution during compaction of Bleurswiller sandstone*, Int. J. Rock Mech. and Min. Sci. *42*, 873–889.

FOURIE, A.B. and XIAOBI, D. (1991), *Advantages of midheight measurements in undrained triaxial testing*, ASTM Geotech. Test. J. *14*, 138–145.

GRISONI, J.C. and THIRY, M. (1988), *Répartition des grès dans les sables de Fontainebleau: implications géotechniques des études récentes*, Bull. Liaison Labo P. et Ch. *157*, Réf. 3333.

HEILAND, J. and RAAB, S. (2000), *Experimental investigation of the influence of differential stress on permeability of a Lower Permian (Rotliegend) sandstone deformed in the brittle deformation field*, Phys. Chem. Earth (A) *26*, 33–38.

HEILAND, J. (2003), *Laboratory testing of coupled hydro-mechanical processes during rock deformation*, Hydrogeol. J. *11*, 122–141.

HOLT, R.M. (1990), *Permeability reduction induced by a non-hydrostatic stress field*, SPE Form. Eval. *5*, 444–448.

KLEIN, E. and REUTSCHLÉ, T. (2003), *A model for the mechanical behaviour of Bentheim sandstone in the brittle regime*, Pure Appl. Geophys. *160*, 833–849.

KORSNES, R.I., RISNES, R., FALDAAS, I., and NORLAND, T. (2006), *End effects on stress dependent permeability measurements*, Tectonophys. *426*, 239-251.

LEE, K.L. (1978), End *restraint effects on undrained static triaxial strength of sand*, J. Geotech. Eng. Div. *104*, 687–704.

LOUIS, L., DAVID, C., and ROBION, P. (2003), *Comparison of the anisotropic behaviour of undeformed sandstones under dry ad saturated conditions*, Tectonophys. *370*, 193–212.

MORITA, N., GRAY, K., SROUJI, F.A.A., and JOGI, P.N. (1984), *Rock property changes during reservoir compaction*, SPE *13099*.

MORROW, C.A., BO-CHONG, Z., and BYERLEE, J.D. (1984*), Effective pressure law for permeability of Westerley Granite under cyclic loading*, J. Geophys. Res. *91*, 3870–3876.

RHETT, D.W. and TEUFEL, L.W., *Stress path dependence of matrix permeability of North Sea sandstone reservoir rock*. In Rock Mechanics (ed. Tillerson and Wawersik) (Balkema, Rotterdam 1992a) pp. 345–353.

RHETT, D.W. and TEUFEL, L.W. (1992b), *Effect of reservoir stress path on compressibility and permeability of sandstones*, SPE *24756*.

SCHUTJENS, P.M.T.M. and DE RUIG, H. (1997), *The influence of stress path on compressibility and permeability of an overpressured reservoir sandstone: Some experimental data*, Phys. Chem. Earth. *22*, 97–103.

SHENG, D., WESTERBERG, B., MATTSSON, H., and AXELSSON, K. (1997), *Effects of end restraint and strain rate in triaxial tests*, Comp. and Geotech. *21*, 163–182.

SULEM, J. and OUFFROUKH, H. (2006a), Shear *banding in drained and undrained triaxial tests on a saturated sandstone: Porosity and permeability evolution*, Int. J. Rock Mech. and Min. Sci. *43*, 292–310.

SULEM, J. and OUFFROUKH, H. (2006b), *Hydromechanical behaviour of Fontainebleau sandstone*, Rock Mech. and Rock Eng. *39*, 185–213.

TRAUTWEIN, U. and HUENGES, E. (2005), *Poroelastic behaviour of physical properties in Rotliegend sandstones under uniaxial strain*, Int. J. Rock Mech. and Min. Sci. *42*, 924–932.

VAJDOVA, V., BAUD, P., and WONG, T.-F. (2004), *Permeability evolution during localized deformation in Bentheim sandstone*, J. Geophys. Res. *109*, B10406.

WALSH, J.B. and BRACE, W.F. (1984), *The effect of pressure on porosity and the transport properties of rock*, J. Geoph. Res. *89*, 9425–9431.

WONG, T.-F., DAVID, C., and ZHU, W. (1997), *The transition from brittle faulting to cataclastic flow in porous sandstones: Mechanical deformation*, J. Geophys. Res. *102*, 3009–3026.

ZHANG, J., WONG, T.-F., and DAVIS, D.M. (1990a), *Micromechanics of pressure-induced grain crushing in porous rocks*, J. Geoph. Res. *95*, 341–352.

ZHU, W. and WONG, T.-F. (1997), *The transition from brittle faulting to cataclastic flow: Permeability evolution*, J. Geophys. Res. *102*, 3027–41.

ZOBACK, M.D. and BYERLEE, J.D. (1975), *Permeability and effective stress*, Am. Assoc. Petrol. Geol. Bull. *59*, 154–158.

(Received March 31, 2008, revised July 21, 2008, accepted February 27, 2009)
Published Online First: May 26, 2009

Pure appl. geophys. 166 (2009) 1063–1087
0033–4553/09/051063–25
DOI 10.1007/s00024-009-0486-1

Pure and Applied Geophysics

Microstructural Inhomogeneity and Mechanical Anisotropy Associated with Bedding in Rothbach Sandstone

LAURENT LOUIS,[1] PATRICK BAUD,[2] and TENG-FONG WONG[3]

Abstract—This study present the result of conventional triaxial tests conducted on samples of Rothbach sandstone cored parallel, oblique (at 45 degrees) and perpendicular to the bedding at effective pressures ranging from 5 to 250 MPa. Mechanical and microstructural data were used to determine the role of the bedding on mechanical strength and failure mode. We find that samples cored at 45 degrees to the bedding yield at intermediate level of differential stress between the ones for parallel and perpendicular samples at all effective pressures. Strain localization at high confining pressure (i.e., in the compactive domain) is observed in samples perpendicular and oblique to the bedding but not in samples cored parallel to the bedding. However, porosity reduction is comparable whether compactive shear bands, compaction bands or homogeneous cataclastic flow develop. Microstructural data suggest that (1) mechanical anisotropy is controlled by a preferred intergranular contact alignment parallel to the bedding and that (2) localization of compaction is controlled by bedding laminations and grain scale heterogeneity, which both prevent the development of well localized compaction features.

Key words: Sandstone, mechanical strength, anisotropy, strain localization, heterogeneity.

1. *Introduction*

Significant anisotropy in mechanical behavior and failure strength can arise from planar rock fabrics such as bedding in sedimentary rocks, cleavage in slates, as well as preferred orientation or arrangement of minerals and cracks in igneous and metamorphic rocks. The anisotropy of tensile and compressive strengths may be associated with different failure modes and deformation mechanisms, depending on how stress is applied relative to the anisotropy planes (BAUD *et al.*, 2005).

Extensive studies have been conducted on the brittle strength of anisotropic rocks as a function of the angle θ between the maximum principal stress and the foliation or bedding. In relatively compact rocks (such as slate, phyllite, schist, gneiss and amphibolite) as well as in more porous rocks such as shale, the brittle strength

[1] Département des Sciences de la Terre et de l'Environnement, Université de Cergy-Pontoise, CNRS UMR 7072, Bâtiment Neuville 3.1, Neuville-sur-Oise, F-95031 Cergy-Pontoise, France.
E-mail: laurent.louis@u-cergy.fr
[2] Institut de Physique du Globe (CNRS/ULP), 5 rue Descartes, 67084 Strasbourg, France.
[3] Department of Geosciences, State University of New York, Stony Brook, NY 11794-2100, USA.

anisotropy is characterized by a minimum in the peak stress at $\theta = 30$–$45°$ and maxima at $\theta = 0°$ and $90°$. Such strength anisotropy can be interpreted by a model with the foliation or clay-rich bedding acting as "planes of weakness" that promotes frictional slip and wing crack growth (e.g., WALSH and BRACE,1964; RAWLING *et al.*, 2002; JAEGER and COOK, 1979). However, the anisotropic behavior is fundamentally different in a porous sandstone, the strength of which may attain maximum and minimum values at $\theta = 90°$ (perpendicular to bedding) and $\theta = 0°$ (parallel to bedding), respectively. As reviewed by BAUD *et al.* (2005), many questions related to the mechanics of anisotropic failure in porous sandstone remain unanswered.

What are the actual microstructural attributes that govern the apparent influence of bedding on failure in an anisotropic porous sandstone? In response to an applied deviatoric stress field, the pore space in the rock may dilate or compact, and the brittle-ductile transition is sensitively dependent on the interplay of dilatancy and compaction. How does bedding influence the onset and development of compactive yield, and to what extent can bedding anisotropy influence the development of strain localization and failure modes associated with the brittle-ductile transition? We chose the Rothbach sandstone, the anisotropic behavior of which had been investigated focusing on the two end-members represented by samples cored perpendicular and parallel to bedding (WONG *et al.*, 1997; BAUD *et al.*, 2005; LOUIS *et al.*, 2007a). In this study, a series of new experiments were performed on "oblique" samples cored at 45° with respect to the bedding. Systematic microstructural investigation on undeformed and deformed samples were conducted using optical and electron microscopy, as well as X-ray CT-imaging.

2. *Characterization of Microstructural Inhomogeneity Associated with Bedding*

The Rothbach sandstone belongs to the Lower Triassic "Vosgian sandstone" formation, with extensive outcrops on either side of the Rhine Graben in northeastern France and southwestern Germany. Reddish in color, the Rothbach sandstone formed from cross-bedded, fluvial sediments and has been widely used as a building stone. It has been the focus of various studies on mechanical (DAVID *et al.*, 1994; WONG *et al.*, 1997; BÉSUELLE *et al.*, 2003; BAUD *et al.*, 2004) and other physical (ZHU and WONG, 1997; LOUIS *et al.*, 2005) properties.

Our samples were cored from the block studied by BAUD *et al.* (2004), with initial porosity of 20 ± 1%. Its modal composition is 68% quartz, 16% feldspar, ~12% clays, 3% oxides and mica (DAVID *et al.*, 1994). Optical microscope observations of LOUIS *et al.* (2005) indicate a granulometric layering, that alternates between zones with significant contrasts in porosity and grain size. Whereas the more porous layers have coarser grains (with diameters in the range 200–250 μm), the more compact layers have finer grains (with diameters in the range 100–150 μm). The more compact layers (with thicknesses down to 1 mm) are significantly thinner than the more porous zones (with thicknesses up to 1 cm). LOUIS *et al.* (2005) observed that a majority of grains in Rothbach sandstone were preferentially elongated along the bedding plane, and that the intergranular contacts were as well aligned preferentially with bedding. They suggested that these microstructural attributes may collectively contribute to elastic anisotropy (on the order of 5%), as manifested by a relatively high P-wave velocity measured perpendicular to the bedding. An objective of this study is to characterize quantitatively such microstructural inhomogeneities.

2.1. X-ray-computed Tomography: Density and Porosity Contrasts

To characterize quantitatively the density and porosity contrasts associated with bedding and its spatial distribution, we used the High-Resolution CT Facility at the University of Texas at Austin to scan the Rothbach sandstone samples. The facility there employed a 200 kV microfocal X-ray source capable of a < 10 μm focal spot size, and our data were typically acquired with the source at a peak energy of 180 kV. The methodology for data acquisition and analysis was described in KETCHAM and ITURRINO (2005) and KETCHAM and CARLSON (2001).

We show in Figure 1a an image (with voxel dimension 50 μm) for an undeformed sample cored oblique to the bedding plane. The grey level of a voxel in the image corresponds to the CT-number, which is related to the attenuation of X-ray passing through the porous material. While the attenuation in a solid mineral is predominantly controlled by the atomic number of its elements, its value in a porous material is also dependent on the density, and hence sensitive to the local porosity in a voxel (KETCHAM and CARLSON, 2001). In Figure 1a, the sedimentary layering is manifested by relatively light narrow zones corresponding to dense laminations. In this sandstone sample, we visually isolated 11 zones of alternating high and low porosity.

If the spatial distribution of minerals in the solid matrix is relatively homogeneous, then the average CT-number in a given area can be approximated as linearly related to the porosity (VINEGAR *et al.*, 1991; DESRUES *et al.*, 1996). In a plot of the grey level versus density (Fig. 1b), this linear relation can be established by connecting a straight line between the two points corresponding to the grey level of the void space (with zero density) and the mean grey level of the sample image (with an average density of 2070 kg/m^3 for our sandstone). Using this linear relation, the density of a subset of the sample can then be inferred from the mean grey level of the voxels embedded in this subset. As illustrated in Figure 1c, the 11 zones identified in Figure 1a have distinct mean values of grey level, implying distinct values of average local density (Fig. 1c). The inferred density variation among the 11 zones in the sample is about ± 5% with reference to the bulk density of 2070 kg/m^3 for the entire sample. If we assume a porosity value of 20.4% (which corresponds to the arithmetic mean for > 30 samples cored in the same block), we can infer the local porosity of the 11 zones from these density estimates (Fig. 1a). We thus infer that the presence of bedding in this sample of Rothbach sandstone is associated with significant contrasts in porosity, ranging from 15.8% to 23.7%.

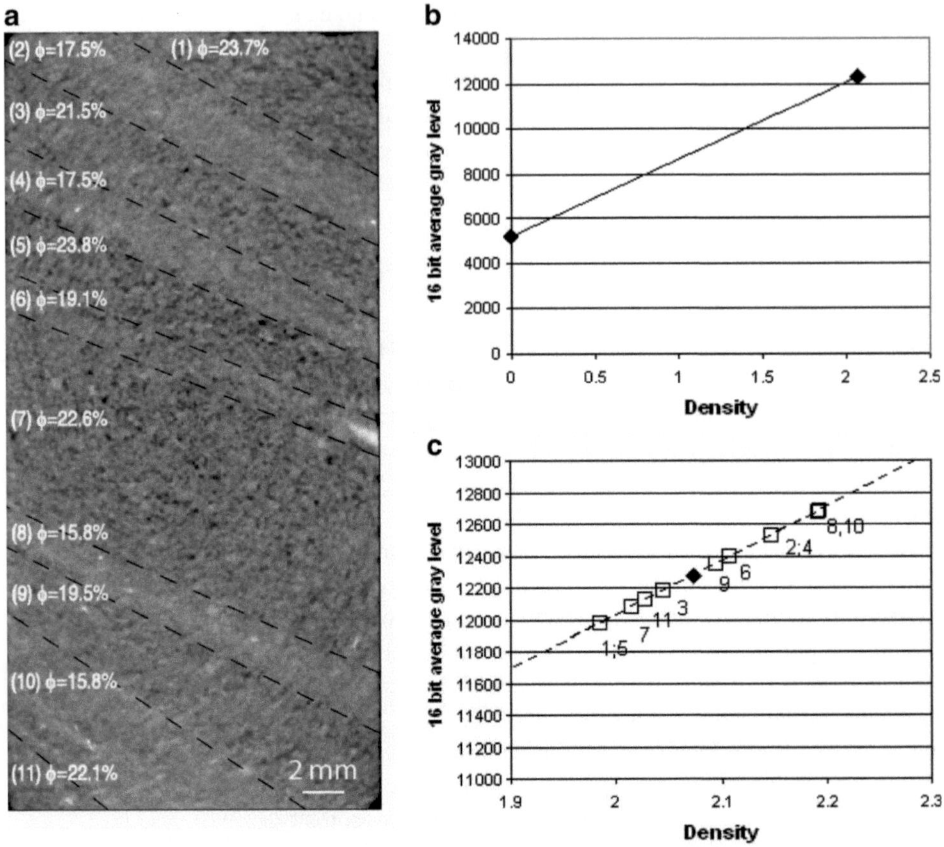

Figure 1

a. X-ray CT scan of a sample of Rothbach sandstone cored at ~45 degrees to the bedding and porosity values in visually identified regions derived directly from density in Figure 1c assuming a porosity of 20.4% for the entire sample image. b. *Ad hoc* calibration line retrieved from the CT scan image. c. Position of the zones isolated in a. on the calibration line assuming a linear relationship between average grey level and effective density.

2.2. *Scanning Electron Microscopy: Anisotropy of Intergranular Contact Normal*

A backscattered scanning electron microscope (SEM) image of the grain and pore structure is presented in Figure 2a. The bedding of this sample was sub-horizontal. It can be observed that the quartz grains and lithic fragments of quartz and K-feldspar are well cemented and coated by clay minerals (mostly illite as determined by semi-quantitative X-ray-based elemental mapping under the SEM). Similar features were observed under the laser scanning confocal microscope (Louis *et al.*, 2005). Details of an intergranular contact in one of the more porous (and relatively thick) zones as revealed by backscattered SEM (Fig. 2b) is contrasted with an image of the same area observed by cathodoluminescence (CL) SEM (Fig. 2c). Since CL-SEM can distinguish different

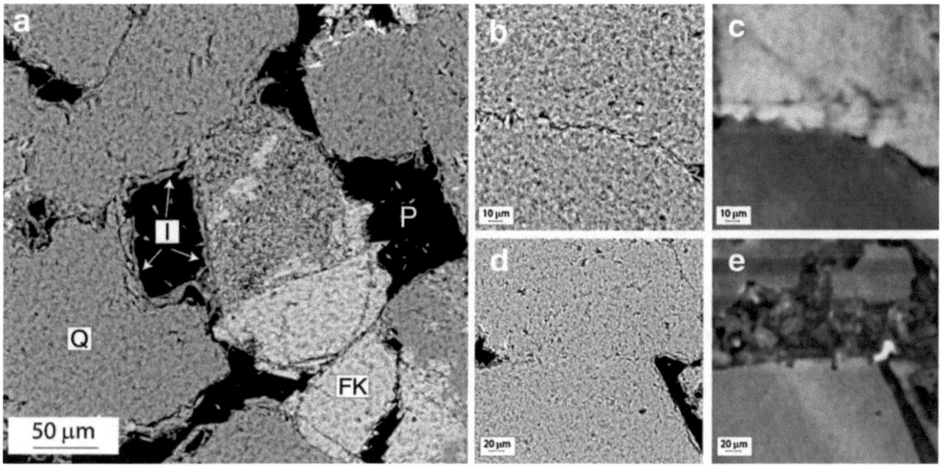

Figure 2
a. Scanning electron microscope image (BSE – backscattered electrons) of the Rothbach sandstone. Grains and lithic clasts of quartz (Q) and K-feldspar (FK) are coated with clays (mostly illite, denoted by Il). The pore space is denoted by P. b. BSE image of a grain contact in relatively high porosity zone. c. Cathodoluminescence image of the grain contact in b. d. BSE image of a grain contact in relatively low porosity zone. e. Cathodoluminescence image of the grain contact in d.

diagenetic generations of crystallization, healed cracks and crushed grains that were subsequently lithified which were not resolved by backscattered SEM are imaged in Figure 2c. Similarly, Figures 2d and 2e present backscattered and CL images, respectively, of an intergranular contact in one of the more compact (and relatively thin) zones. For either intergranular contact, the geometry of the contact and damage zone suggest that relatively intense compaction occurred after deposition and that the effective contacting surface was enhanced sub-parallel to bedding.

To characterize the contact length distribution and its anisotropy, ~1700 intergranular contacts located in both low and high porosity areas were randomly selected from a mozaic of SEM images of a sample with sub-horizontal bedding. Bedding is described here as sub-horizontal due to its crosscut nature and fuzzy boundaries between layers. The total surface investigated was about 6*20 mm^2. Each grain contact was approximated by a linear segment, the length and orientation of which were measured. Figure 3a shows the angular variation (in bins of 5°) of the contact normals so measured. The most populated angular classes of contact normal are located close to the top direction of the image that was used for the analysis. In Figure 3a we also plot the contact length as a function of its orientation (represented by an angle that is complementary to the normal direction). Compared with the contact normal distribution, even though the anisotropy of contact length is not as pronounced, it still shows an overall trend for longer contacts to be close to horizontal.

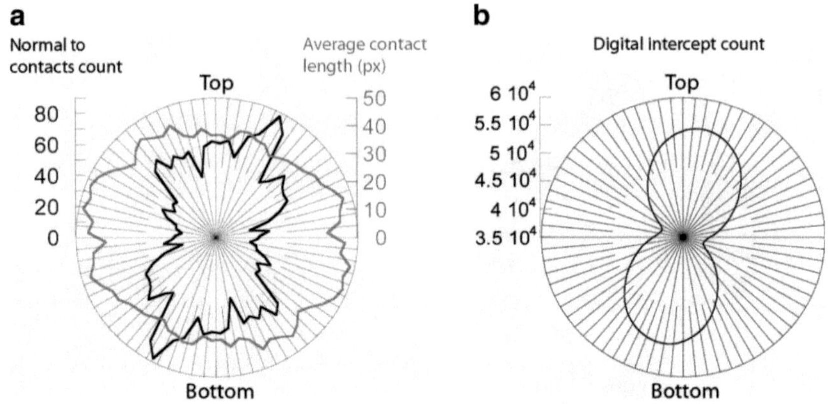

Figure 3
a. Preferred intergranular contact orientation and average length as a function of the observation angle from a SEM mosaic with nominally horizontal bedding. b. Digital intercept density taking into account both length and orientation of the intergranular contacts. Larger effective contacting surface dips 10–15 degrees to the right.

We also conducted measurement of the angular distribution of linear intercept density for this population of $\sim$1700 grain contacts. The linear intercept density is a stereological parameter defined as the number of intersections between a test array of parallel lines and a population of objects, normalized by the total length of the test array (UNDERWOOD, 1970). The anisotropy can be characterized by using arrays in different orientations. On our digital image showing all the grain contacts, an automated procedure was developed to count the number of intercepts made by the intergranular contacts with 180 test arrays at angles varying from 0° to 179°. Such stereological measurements map out the anisotropic distribution of contact length per unit area as imaged on the two-dimensional section. If the distribution in the cylindrical sample can be approximated as axisymmetric, then the stereological data can also be related to the anisotropic distribution of contact area per unit volume in the sandstone. Our data in Figure 3b indicate a maximum intercept density at 11° to the vertical, which corresponds to a maximum specific surface area for intergranular contacts oriented at 11° to horizontal. Since the bedding in this sample is sub-horizontal, this implies an overall trend for the intergranular contacts to align in orientations sub-parallel to bedding.

3. Mechanical Deformation and Failure Mode

To complement the data sets of WONG *et al.* (1997) and BAUD *et al.* (2005) for Rothbach sandstone samples cored parallel and perpendicular to bedding, in this study we focused our mechanical tests on samples cored at an oblique angle of $\sim$45° to bedding. In addition, selected tests were conducted on samples cored parallel and perpendicular to the bedding to acquire a more complete data set for all three orientations at several effective pressures (Table 1). The samples were cylinders with diameter of 18.4 mm and length of 38.1 mm. They were saturated with deionized water and deformed between lubricated end caps in conventional triaxial tests under drained conditions at a constant pore pressure of 10 MPa. The tests were performed at room temperature, axial strain rate of 1.3×10^{-5} s^{-1} and effective pressures ranging from 5 to 250 MPa. The volume change was approximated by the ratio of the pore volume change (given by the pore volumometer DCDT) to the initial bulk volume of the sample (WONG *et al.*, 1997). The acoustic emission (AE) activity was pre-amplified and discriminated from electric noise using signal amplitude and frequency characteristics (ZHANG *et al.*, 1990).

In the following, we will use the convention that compressive stresses and compactive strains are positive. Maximum and minimum compressive stresses will be denoted by σ_1 and σ_3, respectively. We will denote the pore pressure by P_p, the confining pressure by P_c ($P_c = \sigma_2 = \sigma_3$) and the difference between the confining pressure and the pore pressure ($P_c - P_p$) by P_{eff} (effective pressure). The effective mean stress $(\sigma_1 + 2\sigma_3)/3\text{-}P_p$ will be denoted by P and the differential stress $(\sigma_1 - \sigma_3)$ by Q.

We will refer to samples that were cored perpendicular, parallel and at ~45° degrees to bedding as "perpendicular", "parallel" and "oblique", respectively.

3.1. *Stress-strain Data and Acoustic Emission Activity*

Figure 4 presents the data of a hydrostatic compaction test for an oblique sample, showing the effective pressure as a function of the porosity reduction. In such a hydrostat for a porous rock, one typically observes an inflection point P^* that corresponds to the critical effective pressure for the onset of grain crushing and pore collapse. In a porous sandstone, this grain crushing pressure is primarily controlled by the porosity and grain size (ZHANG *et al.*, 1990). For this particular experiment, we obtained a value of 239 MPa for P^*. A total of eight tests performed on perpendicular, oblique and parallel samples gave an average value of 237 MPa for P^* (see also Table 1).

We also present in Figure 4 the results of two conventional triaxial tests on oblique samples deformed at 5 MPa and 150 MPa effective pressure. If the porosity change was solely controlled by the effective mean stress, then the triaxial curves should coincide with the hydrostat, and any deviation would indicate that the differential stress also contributes to porosity change. In the test at 150 MPa effective pressure, we observed an accelerated porosity reduction that initiated at the critical stress state indicated by C^*, which thus marked the onset of "shear-enhanced compaction" (WONG *et al.*, 1997). In contrast, in the test at 5 MPa, the porosity reduction was observed to decelerate with respect to the hydrostat at the critical stress state indicated by C', which thus marked the onset of dilatancy.

In Figures 5a-b and Figures 5c-d we compare the mechanical data for perpendicular, oblique and parallel samples deformed at these two effective pressures. At relatively low confinement (5 MPa effective pressure), the mechanical responses for

Table 1

Compilation of all mechanical data for the three coring directions investigated

	Effective stress σ_3–P_p (MPa)	Regime (B):brittle (C):compactive + (d):dilatancy	Mean and differential stresses at failure ( ) Onset of shear-induced dilation C′	
			$(\sigma_{1+2}\sigma_3)/3$-P_p (MPa)	$\sigma_1-\sigma_3$ (MPa)
Perpendicular	5	(B) + (d)	26 (13)	61 (24)
	10	(B) + (d)	35 (15)	74 (15)
	20	(B) + (d)	52 (38)	95 (54)
	35	(B) + (d)	77.5 (55)	127 (60)
	40	(B)	77.5	111
	55	(C)	97	124
	55	(C)	99	130
	90	(C)	131	122
	110	(C)	161	137
	130	(C)	172.5	123
	130	(C)	173	125
	150	(C)	193	126
	165	(C)	202	112
	240	(C)	240	0
	245	(C)	245	0
	245	(C)	245	0
Oblique (~45 deg)	5	(B) + (d)	24 (8)	54 (9)
	10	(B) + (d)	31 (15)	62 (15)
	20	(B)	50	86
	30	(B)	65	103
	55	(C)	95	116
	90	(C)	127	108
	130	(C)	163	98
	150	(C)	182	93
	170	(C)	199	82
	190	(C)	214	70
	238	(C)	238	0
	239	(C)	239	0
Parallel	5	(B) + (d)	22 (16)	48 (33)
	10	(B) + (d)	30 (21)	58 (33)
	20	(B) + (d)	47 (28)	80 (24)
	30	(B)	57	82
	50	(C)	80	90
	55	(C)	88	97
	90	(C)	124	99
	115	(C)	150	102
	130	(C)	161	91
	130	(C)	162	94
	140	(C)	168	84
	150	(C)	177.5	79
	190	(C)	204	38
	200	(C)	210	30
	220	(C)	220	0
	234	(C)	234	0
	238	(C)	238	0

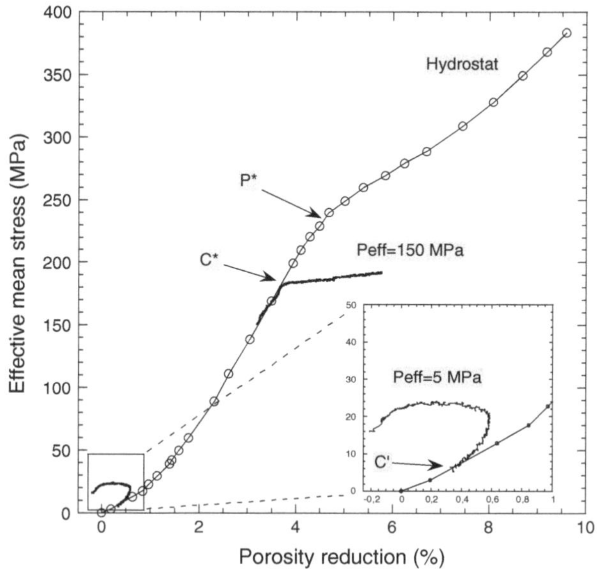

Figure 4
Hydrostatic compaction test ($\sigma_1 = \sigma_2 = \sigma_3$) for a sample of Rothbach sandstone and results of two triaxial experiments in an effective mean stress vs. porosity reduction plot. The effective stress corresponding to the onset of grain crushing and pore collapse in the hydrostatic tests ($\sim$239 MPa) is denoted by P^*. For the triaxial test at $P_{\mathrm{eff}} = 5$ MPa, porosity change decelerates with respect to the hydrostat at a critical stress state marked by C' (shear-induced dilation). For the triaxial test at $P_{\mathrm{eff}} = 150$ MPa, an accelerated porosity decrease occurs at the critical stress state C* (shear-enhanced compaction).

all three orientations are typical of the brittle faulting regime. The differential stress as a function of axial strain attains a peak, and then the sample strain softens while the differential stress drops to a residual level (Fig. 5a). With reference to the hydrostat, the data for effective mean stress as a function of porosity reduction (Fig. 5b) indicate the onset of dilatancy in the pre-failure stage before reaching the peak stress. Our data indicate that the maximum, intermediate and minimum values of peak stress were associated with the perpendicular, oblique and parallel samples, respectively.

At relatively high confinement (150 MPa effective pressure), the mechanical responses for all three orientations are typical of the cataclastic flow regime. The differential stress continues to increase as a function of axial strain, characteristic of strain hardening behavior (Fig. 5c). With reference to the hydrostat, the data of effective mean stress as a function of porosity reduction (Fig. 5d) indicate the development of shear-enhanced compaction in all three orientations. Our data indicate that the maximum, intermediate and minimum values of the critical stress for the onset of shear-enhanced compaction were associated with the perpendicular, oblique and parallel samples, respectively.

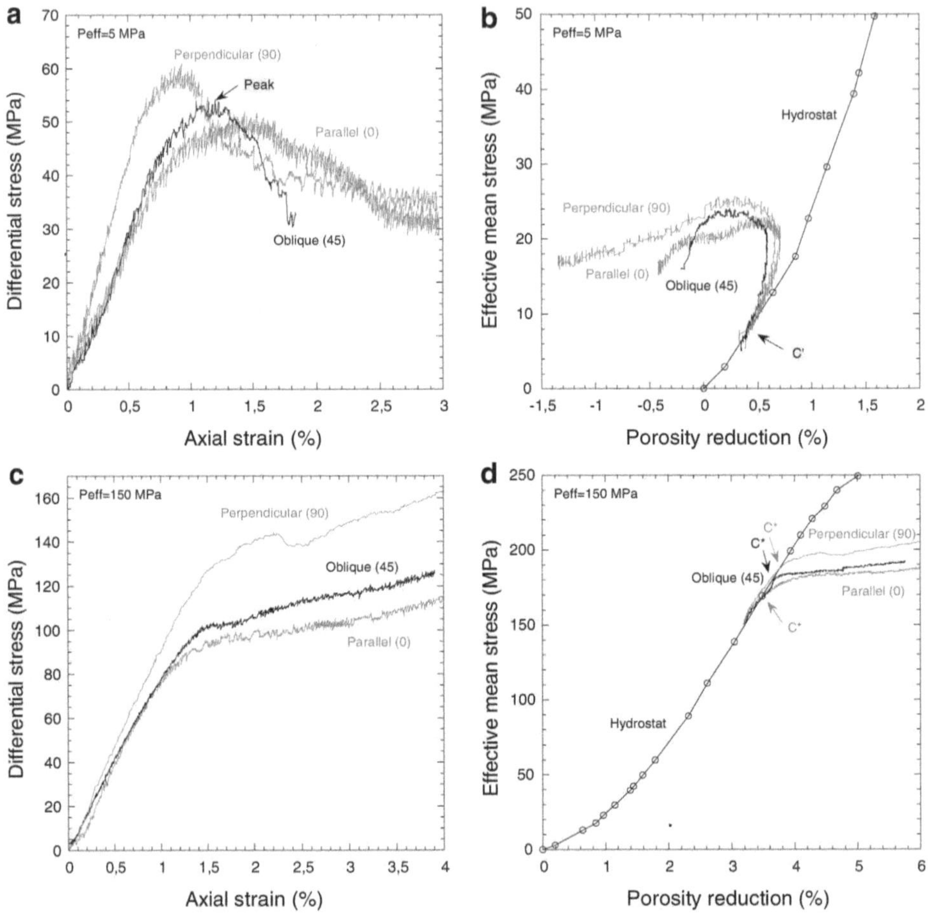

Figure 5

Example of mechanical data obtained for perpendicular, oblique and parallel samples at low (5 MPa) and high (150 MPa) effective pressure. a. Stress vs. strain curve at 5 MPa effective pressure. b. P vs. porosity reduction at $P_{\mathrm{eff}} = 5$ MPa. c. Stress vs. strain curve at $P_{\mathrm{eff}} = 150$ MPa. d. P vs. porosity reduction at $P_{\mathrm{eff}} = 5$ MPa. For both values of P_{eff}, oblique samples have intermediate mechanical strength between perpendicular and parallel samples.

We compile in Figure 6 the stress-strain curves for all 6 oblique samples deformed at effective pressures ranging from 55 MPa to 190 MPa. The AE activities recorded in these experiments (except for the one at 55 MPa) are also presented. The data for 130 MPa effective pressure were previously presented by Louis *et al.* (2007a), who compared the AE activity of this oblique sample with a perpendicular and a parallel sample deformed at the same effective pressure. They noted subtle differences in AE activity among these three samples when stressed to beyond the onset of shear-enhanced compaction. Whereas

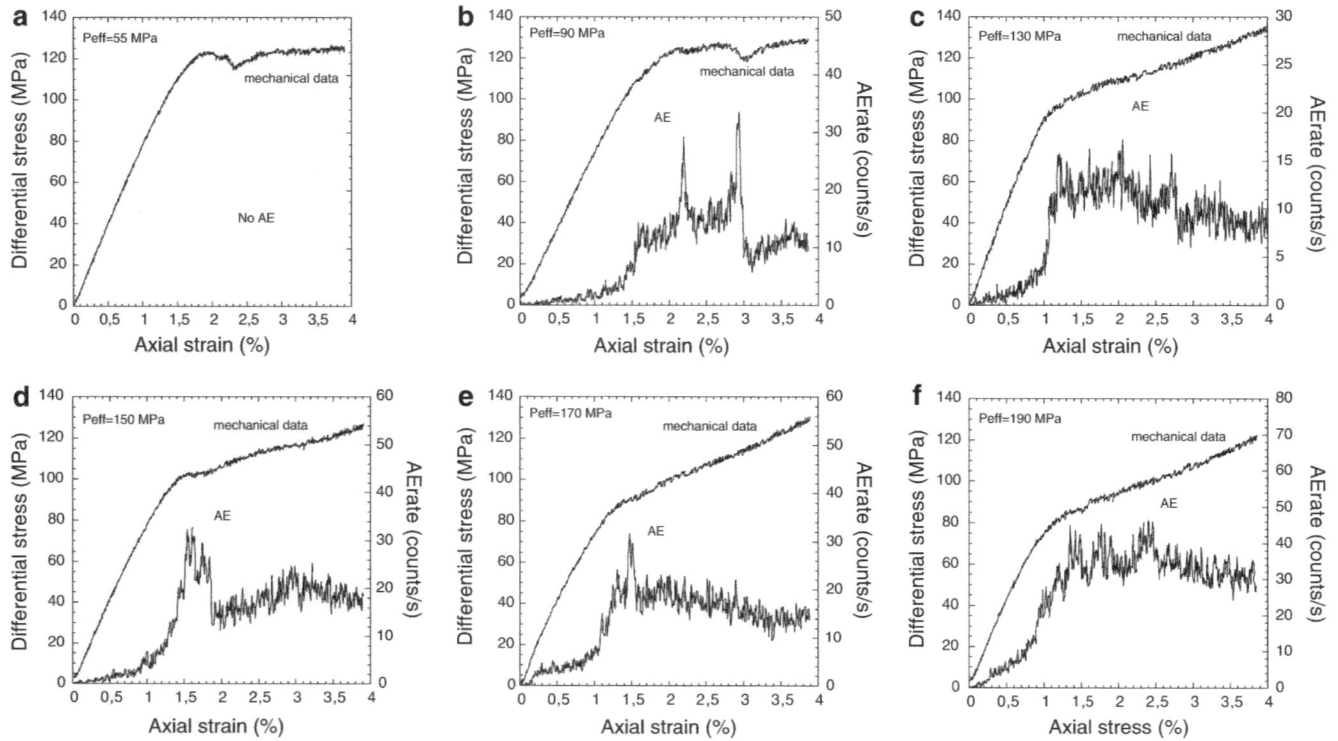

Figure 6

a.-f. Differential stress Q and acoustic emission rate vs. axial strain at $P_{\mathrm{eff}} = 55$ MPa, 90 MPa, 130 MPa, 150 MPa, 170 MPa and 190 MPa for oblique samples.

distinct upsurges in AE activity were observed in both the oblique and perpendicular samples, AE rate in the parallel sample seemed almost constant. In light of the conclusion of BAUD *et al.* (2004) that such differences in AE activity may signify differences in failure mode, the data therefore suggest the possibility that these samples cored in different orientations may have failed by different modes of strain localization. The data in Figure 6 show that the AE activities in oblique samples deformed at effective pressures of 90, 150, 170 and 190 MPa were qualitatively similar to the 130 MPa sample, in that the development of shear-enhanced compaction was manifested by at least one AE upsurge, which was sometimes accompanied by a small stress drop on the order of 10 MPa.

3.2. Brittle Strength and Compactive Yield Stress

In Figure 7a, we compiled in the *P-Q* space data on brittle strength and the stress state at the onset of shear-enhanced compaction C* of Rothbach sandstone for the three coring directions investigated (cf., Table 1). We plot the peak stress for samples that failed by brittle faulting and the differential stress at the onset of shear-enhanced compaction. We also plot in Figure 7b these critical stresses as functions of angle θ between the maximum principal stress and the bedding. There appears to be an overall trend for the maximum, intermediate and minimum values of both the brittle strength and compactive yield stress (at the same effective mean stress) to be associated with the perpendicular, oblique and parallel samples, respectively. This result contrasts chiefly with the frequent observation that strength is minimum at 30–45 degrees with respect to the bedding plane in a layered material.

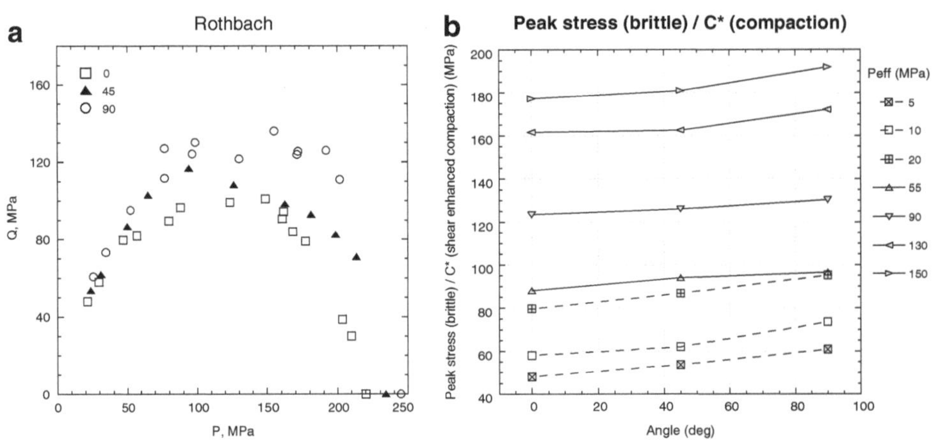

Figure 7
a. Q vs. P plot at the peak of the stress/strain curve (brittle regime) and at C^* (compactant regime) for all samples studied. Oblique samples yield at intermediate values of Q. b. Peak stress (square symbols) and Q at C* (triangles) as a function of the sample orientation for all triplets available.

Our brittle strength data for Rothbach sandstone are in basic agreement with the more systematic investigation on Adamswiller sandstone by MILLIEN (1992), who deformed nominally dry samples cored in seven bedding orientations (0°, 15°, 30°, 45°, 60°, 75° and 90°) at six fixed confining pressures ranging from room pressure to 50 MPa. They observed a general trend for the brittle strength to gradually decrease with decreasing angle, from a maximum for the perpendicular sample (90°) to a minimum for the parallel sample (0°). The brittle failure envelope for our oblique samples seems to fall very close to that of the perpendicular samples. However, it should be noted that duplicate experiments indicate significant sample-to-sample variability in the brittle strength. In contrast, the data for the compactive yield stresses were not as variable. Although previous studies (e.g., WONG *et al.*, 1997) have investigated the yield stress contrast between samples cored perpendicular and parallel to bedding, to our knowledge this is the first systematic measurement of the yield stress of porous sandstone samples cored in an oblique direction. Our data demonstrate that in general, the compactive yield stress of an oblique sample is intermediate between those of the parallel and perpendicular samples. It is likely that the compactive yield stress follows a trend similar to that for the brittle strength, showing a progressive decrease with decreasing angle, from a maximum for the perpendicular sample (90°) to a minimum for the parallel sample (0°).

3.3. Failure Mode and Spatial Distribution of Damage

Thin sections of the failed samples were prepared along a surface parallel to the sample axes. For the parallel and oblique samples, sections were prepared so that the strike of the bedding would be perpendicular to the observation plane. The oblique sample that failed at 10 MPa effective pressure was sent to the High-Resolution CT Facility at the University of Texas at Austin for a full scan at 50 μm resolution, then a thin section cut parallel to the sample axis was prepared for observation under the microscope. To characterize the spatial distribution of damage, we adopted the technique of MENÉNDEZ *et al.* (1996). The thin section of a failed sample was divided into elemental squares and each square was then assigned a damage index on the basis of optical microscope observations. If n denotes the average number of cracks per grain, then index values of I, II, III and IV correspond to $n < 2$, $2 \leq n < 5$, $5 \leq n < 10$, and $n > 10$, respectively. The damage was determined for two oblique samples compacted at effective pressures of 130 MPa and 190 MPa, using elemental squares with widths of 0.75 mm and 1.00 mm, respectively.

3.3.1 Brittle failure. At effective pressures of 5 MPa, 10 MPa and 30 MPa, shear bands were observed to develop in the failed samples for all three orientations we investigated. Our microstructural observations focused on the oblique samples.

Figure 8a shows a X-ray scan of the oblique sample that failed at 10 MPa effective pressure. The scan corresponds to an axial plane that cut through the sample. Slight bulging of the bottom part of the sample was observed, together with an open shear band

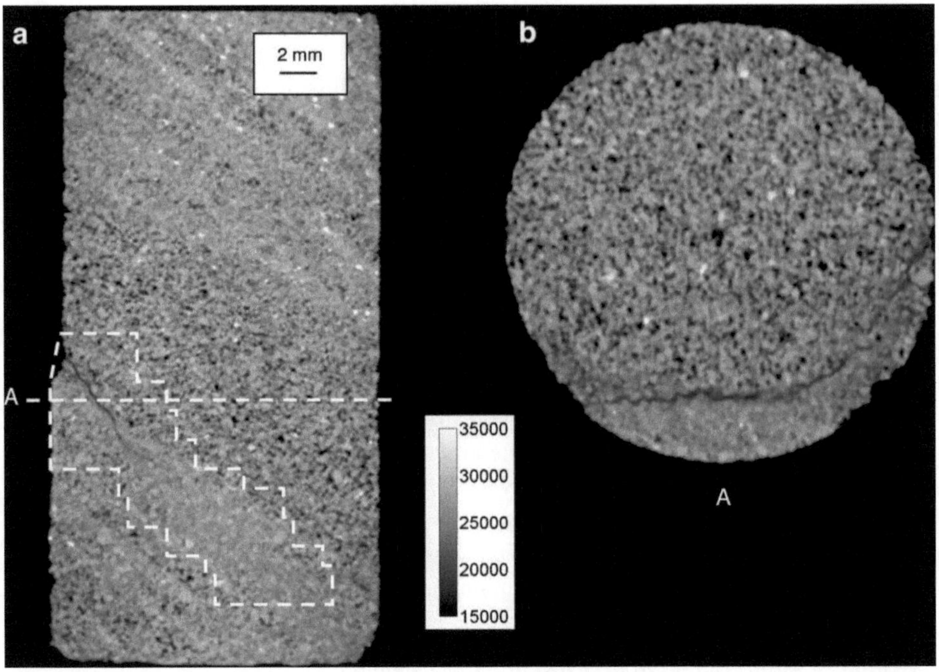

Figure 8
X-ray scan of an oblique sample deformed at $P_{\mathrm{eff}} = 10$ MPa. a. Vertical section containing the sample axis. b. Basal plane section at height ~16 mm.

emanating from the sample surface. The grey level contrast between nominally undeformed parts on either side of the open fracture suggests that the shear band propagated close to an interface between two zones of contrasting porosities and/or average grain sizes. In Figure 8b we present a transverse cross section (cut parallel to the basal plane of the sample) at the height of ~16 mm, as marked by the dotted line in Figure 8a. The shear band maps out a somewhat tortuous arc that extended across the section, and again the contrast in grey level suggests a propagation in the vicinity of a strong porosity gradient (porosity estimations using the method presented in section 2.1. provided values of respectively 19.7% and 25.0% for the lower and higher porosity zones).

A mosaic (overlapping with the region enclosed by the dotted contour in Figure 8a) was acquired under the optical microscope in reflected light (Figs. 9a-c). Damage in the form of intensely cracked grains was observed in the shear band, which cut through the section along a somewhat tortuous path with apparent thickening near the bottom of the sample. The angle of the shear band with respect to the vertical axis of the sample is ~40°, which is appreciably higher than the ~30° angle for perpendicular Rothbach sandstone samples that was reported by BÉSUELLE *et al.* (2000). It should also be noted

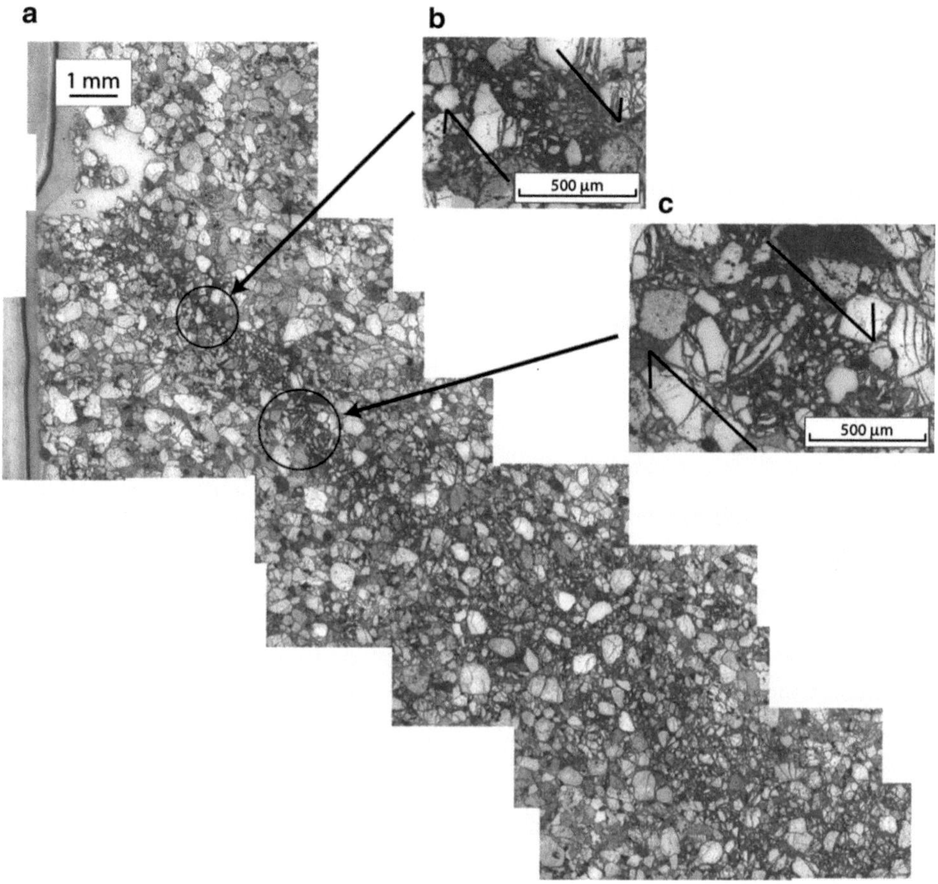

Figure 9
a. Optical microscope image mosaic corresponding to the area traced in Figure 8a. b,c. Intense damage and indications of shear are found across the localized band.

that in comparison to their perpendicular samples, the shear band analyzed here is not as geometrically complex. This difference is in basic agreement with MILLIEN's (1992) observations on Adamsviller sandstone. She concluded that when samples were cored along a direction oblique to the bedding plane, the strike of the shear band would almost systematically lie parallel to bedding.

3.3.2 Shear-enhanced compaction. Previous investigations of the damage evolution related to compactive yield are restricted to perpendicular and parallel samples. In a perpendicular sample of Rothbach sandstone failed at 130 MPa effective pressure, the microstructural observations and stereological data of BÉSUELLE *et al.* (2003) demonstrate the development of several diffuse compaction bands with intense damage that are subparallel to bedding. BAUD *et al.* (2004) suggested that the development of each of these diffuse compaction bands was probably manifested by a distinct upsurge in AE activity, and that the localized damage first initiated near an interface and then propagated laterally to form a relatively diffuse band within the more porous zone between neighboring bedding planes.

In contrast, the damage in a parallel sample was observed to be more diffuse, and as noted earlier, the AE activity did not show any discrete surge. In a parallel sample of Rothbach sandstone failed at 130 MPa effective pressure, BAUD *et al.* (2005) documented relatively short segments of elongate damage subperpendicular to σ_1. The overall damage was distributed homogeneously in that although elongate clusters had developed, they did not propagate all the way across the sample, possibly because the bedding planes inhibited their continuous growth.

Such fundamental differences in the spatial pattern of damage between a perpendicular and a parallel sample were further documented by LOUIS *et al.* (2007a). In particular, they studied an oblique sample failed at effective pressure of 130 MPa, and here we extended this previous study to also analyze an oblique sample failed at 190 MPa. In Figures 10a-c, we present the spatial distributions of damage in the perpendicular, oblique and parallel samples failed at 130 MPa. In contrast to the delocalized pattern of damage in the parallel sample, a number of diffuse, elongate zones of intense damage subparallel to bedding can be identified in both the perpendicular and oblique samples. As noted earlier, distinct upsurge in AE activity were observed in both perpendicular and oblique samples, which may be related to the propagation of these diffuse bands. Taking into account their kinematics relative to the σ_1 orientation, these diffuse bands can be categorized as compaction bands and compacting shear bands, respectively.

Our data for the oblique sample failed at 190 MPa effective pressure (Fig. 11a) show a damage pattern qualitatively similar to the compacting shear band in the 130 MPa sample. We also performed a more refined damage mapping of this sample. Under relatively high magnification, small elemental areas ($\sim$50 μm wide) were analyzed and those associated with intense grain crushing (with damage index = IV) were pinpointed on the image. In Figure 11b the dark areas correspond to these intensely damaged clusters, which are heterogeneously distributed within what appears as diffuse bands in Figure 11a under lower magnification. This observation at higher resolution reveals the presence of short subhorizontal elongated patches, what suggests possible band propagation on short distances in higher porosity zones.

4. Discussion

4.1. Microstructural Origin of the Mechanical Anisotropy

For several decades, numerous studies have been conducted on the brittle strength of anisotropic rocks as a function of the angle θ between σ_1 and the foliation or bedding.

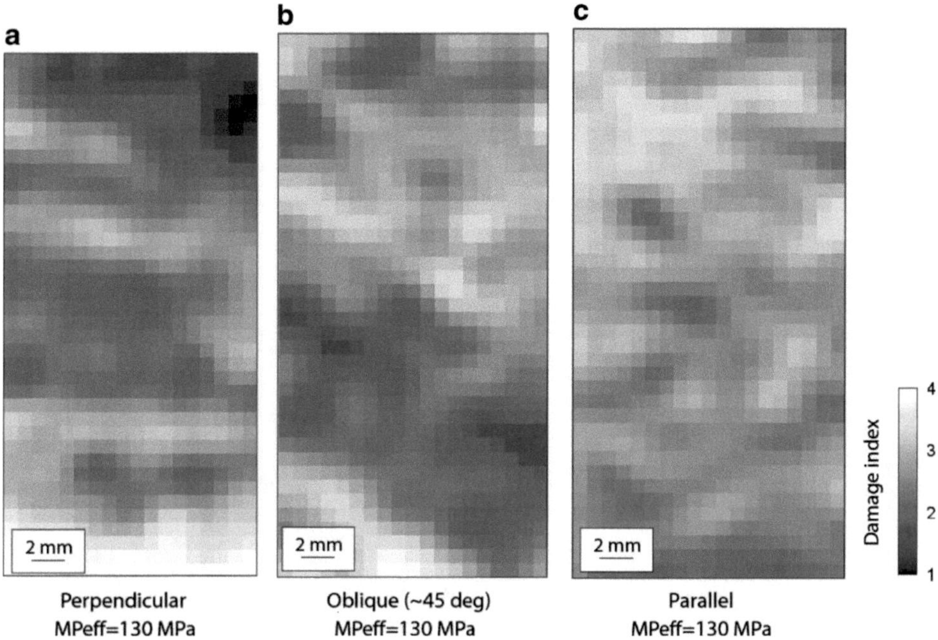

Figure 10
a.-c. Damage maps obtained by Louis *et al.* (2007a) on thin sections of samples deformed at $P_{\mathrm{eff}} = 130$ MPa. Each cell is attributed a damage index value between 1 and 4 based on the average number of cracks per grain (cf., section 3.3).

Baud *et al.* (2005) noted that two types of anisotropic behavior should be distinguished. The first type of brittle strength anisotropy manifested by a minimum in the peak stress at $\theta = 30$–$45°$ and maxima at $\theta = 0°$ and $90°$, has been widely observed in relatively compact rocks such as slate, phyllite, schist, gneiss and amphibolite, as well as in more porous rocks such as shale. Such strength anisotropy can be interpreted by a model with the foliation or clay-rich bedding acting as "planes of weakness" that promotes frictional slip and wing crack growth (e.g., Walsh and Brace, 1964, Rawling *et al.*, 2002).

The second type of behavior commonly observed in porous sandstones is manifested by a progressive decrease of brittle strength with bedding angle, with a maximum at $\theta = 90°$ (perpendicular to bedding) and minimum at $\theta = 0°$ (parallel to bedding). Rothbach sandstone has this type of strength anisotropy, and our microstructural observations have provided some insights into the micromechanics. According to our CT data (Fig. 1), the thin bedding planes are relatively compact and fine-grained. Therefore one would expect them to be stronger than the more porous and thicker zones outside. Hence the bedding in a porous sandstone may actually represent planes of relative strength (and not weakness), and accordingly the strength anisotropy in a porous sandstone is fundamentally different from that in a foliated metamorphic rock or shale.

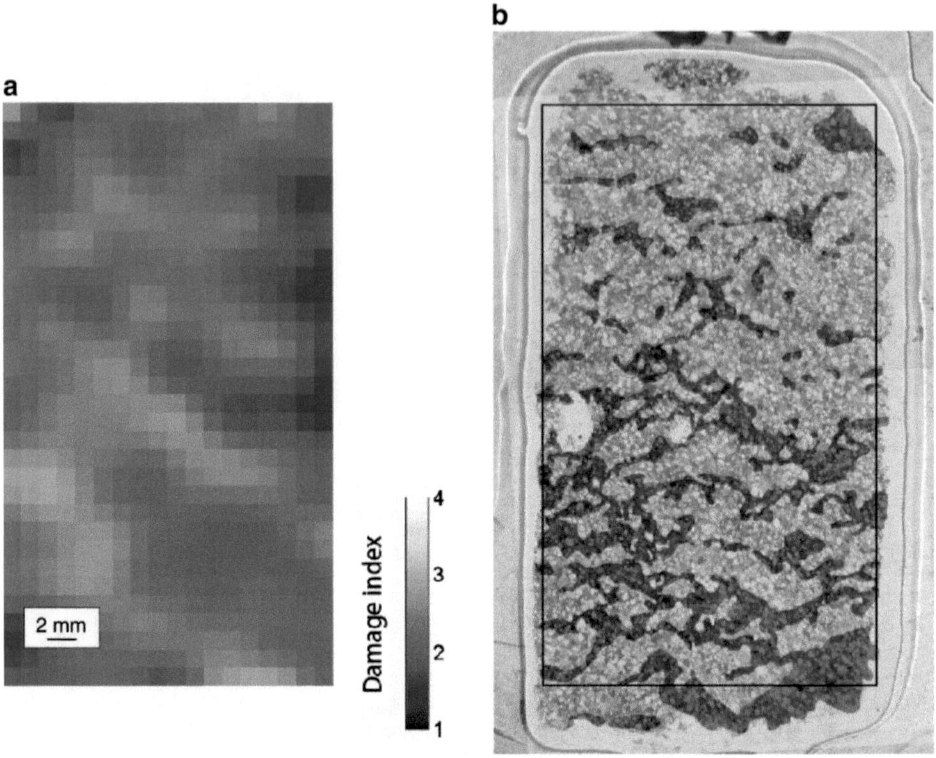

Figure 11
a. Damage map obtained from a thin section of the oblique sample deformed at $P_{eff} = 190$ MPa. b. High resolution damage mapping based on direct microscope observation for the sample deformed at $P_{eff} = 190$ MPa. Damage zones are represented with their actual extent. Deformation appears more localized than in Figure 11a with highly tortuous compaction structures.

Given the porosity (and inferred strength) contrast, a plausible model for the strength anisotropy is that of a composite with strong layers bonded to a relatively weak matrix. This class of model has been widely analyzed in the mechanics of composite materials (e.g., HULL, 1981), and it typically predicts the composite to have maximum and minimum strengths at $\theta = 0°$ and $\theta = 90°$, respectively. This model prediction is diametrically opposite to laboratory observations on porous sandstones (Fig. 7).

There are at least two microstructural attributes that are not being considered in this idealized model. First, in fluviatile sandstone such as the one studied here, the boundaries between relatively compact and more porous zones are likely to be rather diffuse and irregular. Moreover, mechanical heterogeneities inherent to granular assemblies are believed to play an important role in damage nucleation and propagation. In the case of parallel samples, which are the ones that would be expected to be mechanically stronger according to the conceptual layered model, failure may in fact occur through the solicitation of mechanically weaker zones outside the denser layers. Second, the development of grain-scale damage in a porous sandstone is strongly influenced by geometry and cementation of the intergranular contacts (e.g., MENÉNDEZ *et al.*, 1996; WONG and WU, 1995). Our new data on the anisotropic distributions of grain contact number and length (Fig. 3) suggests that these features could be for a significant part responsible for the observed mechanical anisotropy. This potential influence on brittle strength anisotropy should probably be incorporated into a more realistic micromechanics model.

In this study we have demonstrated that the anisotropies of compactive yield stress and brittle strength are qualitatively similar, and it is therefore likely that the two microstructural attributes discussed above can exert analogous control over both modes of failure. Our data also underscore the important influence of bedding on mode of strain localization. Whereas alignment of the maximum compressive stress with bedding would inhibit the development of compaction localization (and thus promote distributed cataclastic flow), diffuse shear or compaction bands readily developed in Rothbach sandstone samples with σ_1 applied oblique or perpendicular to bedding.

LOUIS *et al.* (2007b) recently suggested that a diffuse mode of compaction localization is promoted by grain-scale heterogeneity, as characterized by relatively broad distributions of CT-number and grain contact diameter. They compared the data for Berea sandstone samples (cored perpendicular to bedding) with those for Bentheim and Diemelstadt sandstones. Whereas Berea sandstone tends to develop diffuse shear or compaction bands (analogous to what we observed in Rothbach sandstone), Bentheim and Diemelstadt sandstones readily develop relatively thin and discrete compaction bands when deformed to failure under high confinement (BAUD *et al.*, 2004). To further investigate this question, we analyzed the distributions of CT-number (grey level) and grain contact length of an undeformed Rothbach sandstone sample. In Figure 12a, normalized grey level distributions of X-ray images acquired in Bentheim, Diemelstadt and Rothbach sandstones are represented. The distribution for the Rothbach sandstone was obtained by selecting in the original image higher porosity zones where compaction was observed to preferentially occur. In the same manner as for Berea sandstone (LOUIS *et al.*, 2007b), the distribution for the Rothbach sandstone is significantly broader than for Bentheim and Diemelstadt sandstones, and therefore more heterogeneous. In Figure 12b, intergranular contacts were traced from optical microscope images, then measured in length with the automated function provided in ImageJ freeware. Apart from differences in average contact length, which varies with grain size, composition and degree of cementation, one can again clearly distinguish Bentheim and Diemelstadt sandstones with nearly identical narrow contact length distribution, from the Rothbach sandstone which shows significantly broader distribution. As proxies for grain-scale heterogeneity, these statistics of CT-number and grain contact length strongly support the idea that discrete compaction bands are promoted in a relatively homogeneous granular material, in agreement with recent numerical simulations using network (KATSMAN *et al.*, 2005) and discrete element (WANG *et al.*, 2008) models.

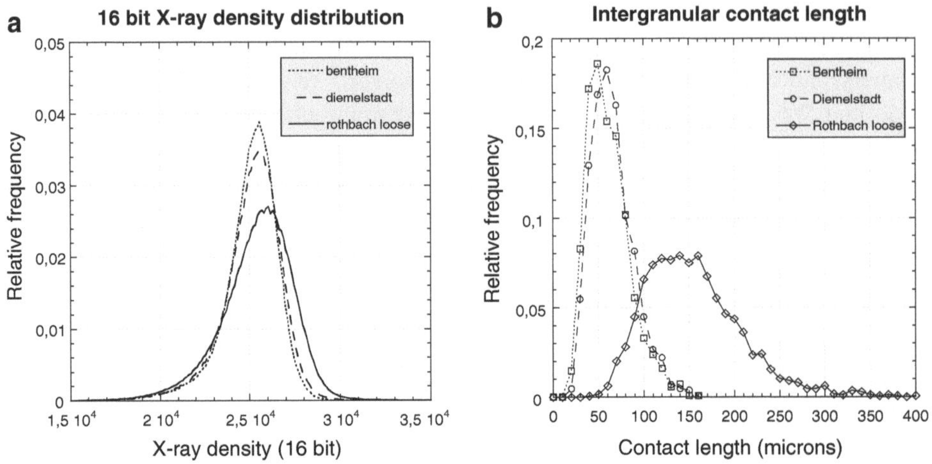

Figure 12

a. Grey level distribution of X-ray CT images from Bentheim, Diemelstadt and Rothbach sandstones. The distribution of the Rothbach is noticeably more spread than for the two other sandstones. b. Distribution of the intergranular contact length for Bentheim, Diemelstadt and Rothbach sandstones. As in a, the Rothbach sandstone is significantly different from the two others with larger intergranular contact length and broader distribution.

4.2. Compactive Yield Caps and Inelastic Deformation

BAUD *et al.* (2006) fitted the yield envelopes of four sandstones of porosity ranging from 13% to 23% with an elliptical cap of the form:

$$\frac{(P-C)^2}{A^2}+\frac{Q^2}{B^2}=1$$

in which the ellipse is anchored at P* (A + C = P*) and B/A represents the aspect ratio of the cap. For Bentheim, Berea and Darley Dale cored perpendicular to bedding as well as Adamswiller cored parallel to bedding, they consistently found B/A > 1. Figure 13 shows that such elliptical caps provide a decent fit of our yield envelopes for Rothbach sandstone cored in the three studied orientations. The caps for the parallel and oblique orientations have comparable aspect ratios slightly smaller than 1 (Figs. 13a and 13b). The cap for the samples cored perpendicular to bedding has a significantly different shape with an aspect ratio of 1.27 (Fig. 13c) in the range of the values found by BAUD *et al.* (2006) for other sandstones.

Figures 13a-c also present contours in the stress space corresponding to specific values of the volumetric strain. Since in a sandstone plastic deformation of the solid grains is negligible at room temperature under the pressure conditions, in our experiments, the plastic porosity change $\Delta\phi^p$ represent the bulk of the plastic volumetric strain and to the first approximation we can take $\varepsilon_v^p = \Delta\phi^p$. Because a lot of our experiments were stopped after a few percents of axial strain for microstructural studies, we could only draw those contours over a relatively modest range of plastic volumetric

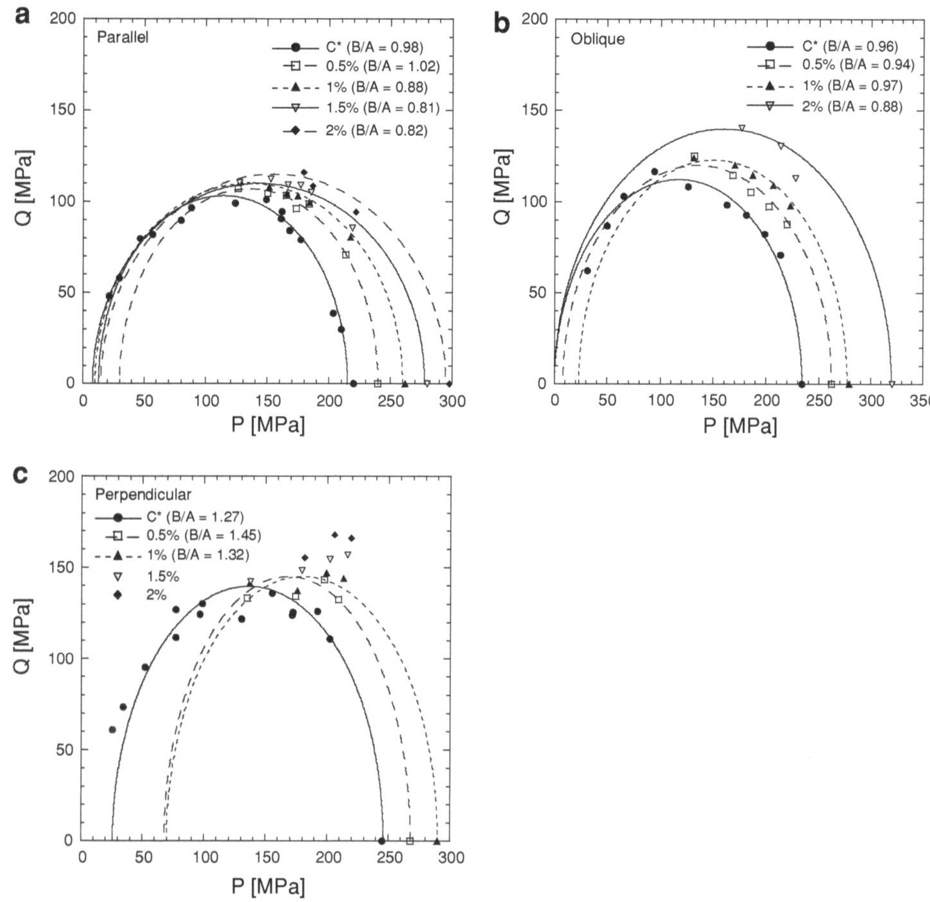

Figure 13
Yield envelopes, (Q, P) values at specific volumetric strains (0.5%, 1%, 1.5%, 2%) and corresponding elliptical fits for samples parallel (a), oblique (b) and perpendicular (c) to the bedding.

strain. Despite this limitation, Figures 13a-c show two important features: First, the shape of the contours varies significantly with increasing plastic strain for the three orientations. Our set of data on Rothbach sandstone presents therefore an overall poor agreement with both CAROLL's (1991) critical state model and the DIMAGGIO SANDLER (1971) cap model, which assumed a constant width and constant aspect ratio of the cap, respectively (BAUD *et al.*, 2006). The main observation of a cap aspect ratio significantly larger for the perpendicular samples is however still valid as plastic volumetric strain increased (Fig. 13c). Second, more strain hardening is observed for perpendicular and oblique samples than for parallel samples and therefore, since 0° (i.e., parallel) to bedding is the weakest orientation, the mechanical anisotropy quantified at the yield stress remains important as plastic volumetric strain increases.

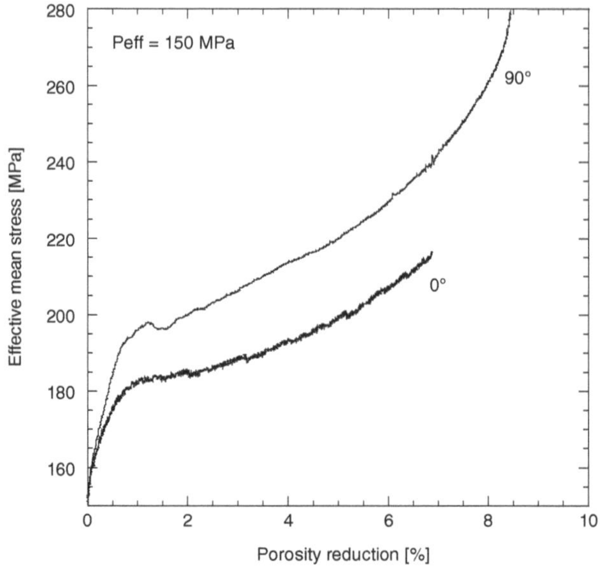

Figure 14
Effective mean stress P vs. porosity reduction for triaxial experiments at $P_{\mathrm{eff}} = 150$ MPa on samples parallel and perpendicular to the bedding.

Figure 14 presents two experiments performed at 150 MPa of effective pressure to large levels of strain for parallel and perpendicular samples. Indeed, after 5% of plastic volumetric strain ($\sim$10% axial strain), the gap between the two extreme orientations remains approximately the same at this pressure. We have seen in the previous sections that compaction localization is the dominant feature of compaction in oblique and perpendicular samples. Our data suggest that from a macroscopic point of view, the porosity reduction is comparable whether compactive shear bands, compaction bands or homogeneous cataclastic flow develop.

Because we wanted to compare the post-yield behavior against the critical state and cap models aforementioned, we assumed as a first-order approximation an elliptical yield cap for the three directions investigated, as if each of them were to reflect the behavior of a distinct isotropic material. The anisotropy of the Rothbach sandstone should be accounted for with a unique anisotropic model allowing for delineating three distinct yield envelopes for the same material. Such investigation is the objective of a future study.

5. Summary and Conclusion

The aim of this study was to investigate further the influence of sedimentary lamination in Rothbach sandstone on the anisotropy of yield stress and failure mode at various effective pressures from 5 MPa to 250 MPa. The deformed samples were cored perpendicular, oblique (45 degrees) and parallel to the bedding. For both brittle strength and compactive yield stress, we measured for each effective pressure, maximum, intermediate and minimum values for perpendicular, oblique and parallel samples, respectively. These results could not be accounted for with the conceptual model of a layered medium for that the samples cored parallel to the bedding should have been mechanically stronger than the ones cored perpendicular to the bedding. Instead, the observed mechanical anisotropy was found consistent with the preferential orientation of the intergranular contacts subparallel to the bedding, according to the principle that an increase of the total contact surface alleviates stress on each individual contact, and therefore provides more strength to the overall structure.

At low effective pressure (5 MPa to 30 MPa), shear bands were observed to develop for all sample orientations investigated. In the shear enhanced compaction domain, AE activity and distribution of damage showed that compaction localization had occurred in perpendicular and oblique samples, while parallel samples had undergone more homogeneously distributed compaction. In perpendicular and oblique samples, compaction preferentially took place in relatively more porous and presumably weaker zones. Within these zones, localization was diffuse, which was accounted for by significant grain scale heterogeneity that prevented discrete compaction bands from propagating across the sample.

Analysis of the yield loci and post-failure deformation showed that elliptical envelopes provided a reasonable fit with our experimental data, and that mechanical anisotropy persisted through compaction, with comparable amounts of porosity reduction for the extreme directions investigated. A future study will be dedicated to improving the approach on anisotropic yield by using a simple methodology such as the one described by Oda (1993).

In conclusion, the mechanical anisotropy and failure modes of the Rothbach sandstone, which was studied for its characteristic layered structure, were found to be only partly controlled by it. In experiments where samples failed by shear-enhanced compaction, the macroscopic heterogeneity materialized by the layering certainly played a role in that, when favorably oriented with respect to the maximum compressive stress, higher porosity zones were preferentially compacted (diffuse compaction bands in perpendicular samples and diffuse compactive shear bands in oblique samples). However, compaction localization as indicated by AE signatures and punctual stress drops, may rather be related to grain scale mechanical heterogeneity within these zones, as attested by X-ray CT image and thin section analyses. As far as the mechanical anisotropy is concerned, we inferred from statistics on the preferred orientation of intergranular contacts that these features were likely responsible for the maximum, intermediate and minimum mechanical strength in samples cored perpendicular, oblique and parallel to the bedding, respectively. The development of a more realistic micromechanics model relating our macroscopic and microstructural observations should aim in particular at explaining how the same microstructural features may impact in a similar way the anisotropy of brittle strength and compactive yield stress.

Acknowledgements

X-ray scanning was conducted by R. Ketcham at the High Resolution X-ray Computed Tomography Facility of the University of Texas at Austin, an NSF supported multi-user facility. The SEM images were acquired thanks to Omar Boudouma at Paris VI University. The authors would like to thank two anonymous reviewers for their thorough comments.

References

Baud, P., Klein, E., and Wong, T.-F. (2004), *Compaction localization in porous sandstones: spatial evolution of damage and acoustic emission activity*, J. Struct. Geol. *26* (4), 603–624.

Baud, P., Louis, L., David, C., Rawling, G.C., and Wong, T.-f. (2005), *Effects of bedding and foliation on mechanical anisotropy, damage evolution and failure mode*, Geol. Soc. London Special Publ. *245*, 223–249.

Baud, P., Vajdova, V., and Wong, T.-F. (2006), *Shear-enhanced compaction and strain localization: Inelastic deformation and constitutive modeling of four porous sandstones*, J. Geophys. Res. *111* (B12), doi: 10.1029/2005JB004101.

Bésuelle, P., Baud, P., and Wong, T.-F. (2003), *Failure mode and spatial distribution of damage in Rothbach sandstone in the brittle-ductile transition*, Pure Appl. Geophys. *160* (5–6), 851–868.

Bésuelle, P., Desrues, J., and Raynaud, S. (2000), *Experimental characterization of the localisation phenomenon inside a Vosges sandstone in a triaxial cell*, Int. J. Rock Mech. *37*, 1223–1237.

Carroll, M. M. (1991), *A critical state plasticity theory for porous reservoir rock*, Recent Adv. Mechan. Struct. Continua 117.

David, C., Wong, T.-F., Zhu, W., and Zhang, J. (1994), *Laboratory measurement of compaction induced permeability change in porous rocks: Implication for the generation and maintenance of pore pressure excess in the crust*, Pure Appl. Geophys. *143*, 425–456.

Desrues, J., Chambon, R., Mokni, M., and Mazerolle, F. (1996), *Void ratio evolution inside shear bands in triaxial sand specimens studied by computed tomography*, Geotechnique *46* (3), 529–546.

DiMaggio, F. L. and Sandler, I. S. (1971), *Material model for granular soils*, J. Eng. Mech. Div. Am. Soc. Civ. Eng. *97*, 935–950.

Hull, D., *An Introduction to Composite Materials* (Cambridge University Press, Cambridge 1981).

Jaeger J. C., *Fundamentals of Rock Mechanics* (Chapman and Hall, London 1979).

Katsman, R., Aharonov, E., and Scher, H. (2005), *Numerical simulation of compaction bands in high-porosity sedimentary rock*, Mechan. Mat. *37*, 143–162.

Ketcham, R. A. and Carlson, W. D. (2001), *Acquisition, optimization and interpretation of X-ray computed tomographic imagery: Applications to the geosciences*, Comp. and Geosci. *27* (4), 381–400.

Ketcham, R. A. and Iturrino, G. J. (2005), *Nondestructive high-resolution visualization and measurement of anisotropic effective porosity in complex lithologies using high-resolution X-ray computed tomography*, J. Hydrol. *302* (1-4), 92–106.

Louis, L., Wong, T.-f., and Baud, P. (2007a), *Imaging strain localization by X-ray radiography and digital image correlation: Deformation bands in Rothbach sandstone*, J. Struct. Geol. *29* (1), 129–140.

Louis, L., Baud, P., and Wong, T.-f. (2007b), *Characterization of pore-space heterogeneity in sandstone by X-ray computed tomography*, Geol. Soc. London Special Publ. *284*, 127–146.

Louis, L., David, C., Metz, V., Robion, P., Menendez, B., and Kissel, C. (2005), *Microstructural control on the anisotropy of elastic and transport properties in undeformed sandstones*, Int. J. Rock Mech. *42* (7–8), 911–923.

MENENDEZ, B., ZHU, W. L., and WONG, T. F. (1996), *Micromechanics of brittle faulting and cataclastic flow in Berea sandstone*, J. Struct. Geol. *18* (1), 1–16.

MILLIEN, A., *Comportement anisotrope du grès des Vosges: élasto-plasticité, localisation, rupture* (Ph.D. Thesis, Université Joseph Fourier - Grenoble I, Grenoble 1992).

ODA, M. (1993), *Inherent and induced anisotropy in plasticity theory of granular soils*, Mechan. Mater. *16* (1–2), 35–45.

RAWLING, G. C., BAUD, P., and WONG, T.-F. (2002), *Dilatancy, brittle strength, and anisotropy of foliated rocks: Experimental deformation and micromechanical modeling*, J. Geophys. Res.-Sol. Earth *107* (B10), ETG 8-1.

UNDERWOOD, E. E., *Quantitative Stereology* (Addison-Wesley, MA, 1970).

VINEGAR, H. J., DE WAAL, J. A., and WELLINGTOM, S. L. (1991), *CT studies of brittle failure in castlegate sandstone*, Int. J. Rock Mech. *28* (5), 441–448.

WALSH, J. B. and BRACE, W.F. (1964), *A fracture criterion for brittle anisotropic rock*, J. Geophys. Res. *69* (3), 449–456.

WANG, B., CHEN, Y., and WONG, T.-F. (2008), *A discrete element model for the development of compaction localization in granular rock*, J. Geophys. Res. *113*, doi:10.1029/2006JB004501.

WONG, T.-f., DAVID, C., and ZHU, W. (1997), *The transition from brittle faulting to cataclastic flow in porous sandstones : Mechanical deformation*, J. Geophys. Res. *102*, 3009–3025.

WONG, T.-f. and WU, L. C. (1995), *Tensile stress concentration and compressive failure in cemented granular material*, Geophys. Res. Lett. *22*, 1649–1652.

ZHANG, J., WONG, T.-F., and DAVIS, D. M. (1990), *Micromechanics of pressure-induced grain crushingin porous rocks*, J. Geophys. Res. *95*, 341–352.

ZHU, W. and WONG, T.-F. (1997), *The transition from brittle faulting to cataclastic flow in porous sandstones: Permeability evolution*, J. Geophys. Res. *102*, 3027–3041.

(Received July 24, 2008, revised December 9, 2008, accepted January 28, 2009)
Published Online First: June 2, 2009

Pure appl. geophys. 166 (2009) 1089–1106
0033–4553/09/051089–18
DOI 10.1007/s00024-009-0482-5

Pure and Applied Geophysics

Pressure-dependent Production Efficiency of an Enhanced Geothermal System (EGS): Stimulation Results and Implications for Hydraulic Fracture Treatments

GÜNTER ZIMMERMANN,[1] TORSTEN TISCHNER,[2] BJÖRN LEGARTH,[1,3] and ERNST HUENGES[1]

Summary—A series of stimulation experiments were carried out at the geothermal research well in Groß Schönebeck (EGrSk 3/90) located in the northeastern part of Germany. The intended purpose of these experiments was to develop concepts for a productivity increase of the geothermal well to create an Enhanced Geothermal System (EGS). Two different kinds of stimulation types were performed. Hydraulic gel-proppant stimulations were conducted in sandstone sections with high initial permeability. Then a different fracturing concept was applied injecting high amounts of water. This waterfrac stimulation was realized in the entire open section including sandstones and volcanic rocks. Evidence of the creation and properties of a very long vertical fracture was retrieved from pressure response analyses demonstrating a bilinear flow regime. The production efficiency of the produced artificial fractures shows a strong dependence on reservoir pressure. At increased reservoir pressure the artificial fractures of all stimulated intervals are highly conductive and subsequently become less conductive during pressure decline. Hence the range of a suitable reservoir pressure is constrained by this fracture efficiency and limits the usage of this well as an injection well for geothermal power production.

Key words: Geothermal, gel-proppant-fracs, water-fracs, reservoir stimulation.

1. Introduction

Sustainable and environmentally friendly energy can be generated from the conversion of Earth's heat (from formation fluids) into electricity. The preconditions for an economic generation of geothermal electricity are sufficiently high temperatures and flow rates of more than 50 m^3/h and 150°C (KÖHLER and SAADAT, 2003). The required temperature for this purpose can be found in the North-German Basin at 4000 m to 5000 m depth. At this depth the initial permeability of the rocks is generally insufficient for the necessary flow rates. However, stimulation operations to improve the near-well bore regions can lead to a sufficient productivity increase. These systems are generally

[1] Helmholtz Centre Potsdam, GFZ German Research Centre for Geosciences, Telegrafenberg D-14473 Potsdam, Germany. E-mail: zimm@gfz-potsdam.de
[2] Bundesanstalt für Geowissenschaften und Rohstoffe, Stilleweg 2, D-30655 Hannover, Germany.
[3] now at: Shell Exploration and Production, 28000, 9400 HH Assen, The Netherlands.

called Enhanced Geothermal Systems (EGS) and can become commercially suitable after stimulation treatments to enhance the productivity.

Geothermal power production is based on (at least) two deep boreholes (a doublet), a sustainable thermal water cycle and a power plant on the surface (e.g., Köhler, 2005). In the first borehole (production borehole), thermal water is produced from a deep reservoir and conducted to the power plant where its heat is transferred. Afterwards, the thermal water is injected in the second borehole (injection borehole) and returned to the reservoir. Concepts have to be developed to enhance the existing flow. This can be summarized by the term *hydraulic fracturing*. During stimulation experiments, fluids under high pressure penetrate into the rock and generate or extend fractures. These procedures are well known in hydrocarbon industry (e.g., Economides and Nolte, 1989) as well as in the Hot Dry Rock (HDR) technology (Hettkamp *et al.*, 2004; Baumgärtner *et al.*, 2004; Schindler *et al.*, 2008). However, the objective for using hydrothermal reservoirs requires a special stimulation technique to enable production of considerably higher amounts of fluids compared to hydrocarbon reservoirs. In contrast to the HDR technology, it was not the aim to create an underground heat exchanger but to gain access to formation fluids in the reservoir. The most important parameters in these experiments include fracture fluids volume, injection rate, viscosity (water with added polymers), the composition of chemical variants or added proppants, and the selection of the depth interval to initiate new fractures. In the following, we evaluate the stimulation experiments and their impact on the production efficiency of the reservoir, with special emphasis on productivity and injectivity of the well as a function of reservoir pressure. The outcome of this evaluation will have a direct implication for the further use of this well.

2. *Geology and Well History*

The site used for this study Groß Schönebeck is located 50 km northeast of Berlin. This former gas well EGrSk 3/90 drilled in 1990 was re-opened and deepened to 4294 m at the end of the year 2000 to gain access to the Rotliegend formation (Fig. 1). Here, hydrothermal aquifers can be accessed with formation fluids of 150°C and porosities of up to 10% (Huenges and Hurter, 2002). The well was therefore selected as an *in situ* laboratory for scientific investigations of geothermal power production. Experiments in this *in situ* geothermal laboratory should lead to a reliable technology for sufficient production of deep fluids in such reservoirs.

The well encounters the typical sequence of various geological formations known in the North-German Basin. A series of 2370 m of Quarternary to Triassic sediments is underlain by 1492 m of the Zechstein salinar (evaporatic rocks). The following section of this well, which was foreseen for testing, comprises 400 m of Rotliegend formation (siltstones, Dethlingen sandstones, conglomerates and 60 m of underlying volcanic rocks) up to the final depth of 4294 m (Huenges *et al.*, 2002; Holl *et al.*, 2004). This section below 3874 m was an open-hole section at times of the intended stimulation

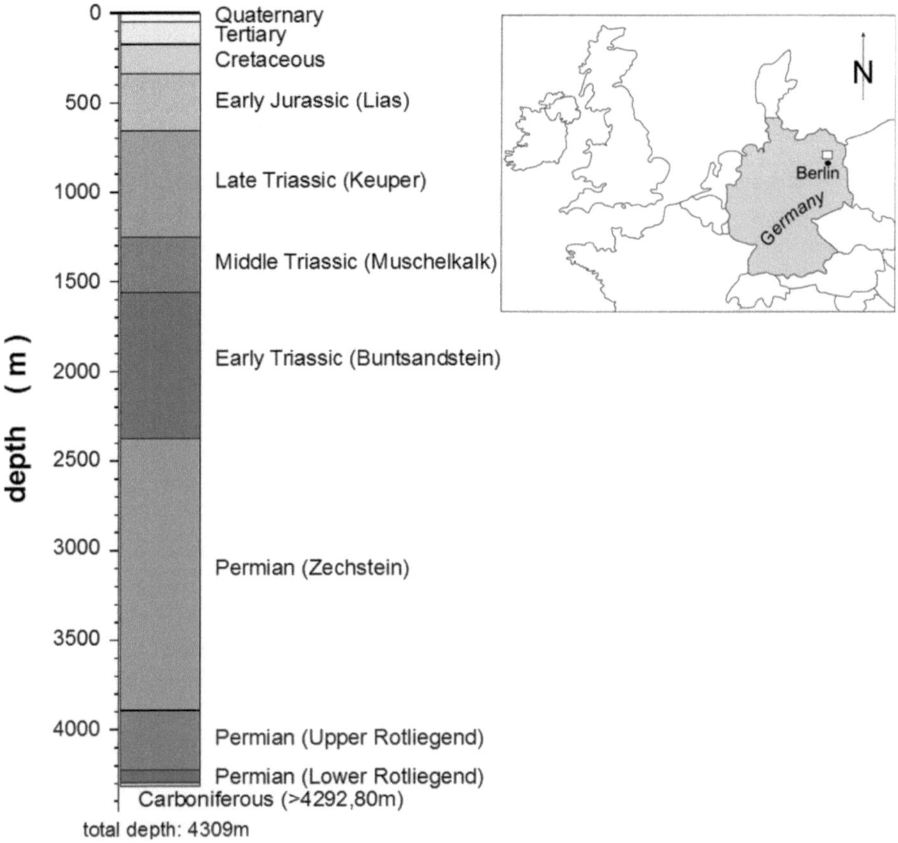

Figure 1
Location and geological profile of the well EGrSk 3/90.

treatments of the well with access to the reservoir rocks. The main targets are sandstones of the Upper Rotliegend (Dethlingen Formation/Lower Elbe subgroup) as well as the volcanic rocks (andesites) of the Lower Rotliegend, where permeability is mainly due to connected fractures. The Dethlingen sandstones represent an effective reservoir horizon with a porosity of 8–10% and a permeability of 10–100 mD (Trautwein and Huenges, 2005). The Elbe-Basis-sandstone as the lower part of the Dethlingen Formation exists in NE Brandenburg as a well-sorted middle-grained to fine-grained sandstone, which has been deposited in a fluviatile setting. The effective reservoir thickness is approximately 80 m.

In 2003 the open hole section had to be cased with a perforated liner due to instabilities of the borehole wall in the siltstone layers. During this treatment the well was deepened to 4309 m and reached the top of the carboniferous.

3. Development of Productivity and Injectivity of the Well

The productivity index and injectivity index are defined as the ratio between flow rate and pressure change in the reservoir during water production, respectively water injection. The production efficiency of a stimulation treatment can be quantified by the change of this index. However one must consider that this ratio generally depends on pressure changes in the reservoir and the total fluid volume involved. Therefore, a comparison of hydraulic test results to quantify the stimulation achievement should be drawn under similar test conditions. For all following tests these conditions are summarized in Table 1.

Two different stimulation techniques were applied, which were specially designed according to the different rock properties. In the sandstone layers with their high permeability, hydraulic fracture stimulations in isolated intervals were performed using high viscous gel and proppants. In the low permeability volcanic rocks, waterfrac stimulations with high flow rates were carried out. Due to the high flow rates necessary for the waterfrac stimulation, it had to be performed in the entire open section of the well, because frictional pressure increase due to a packer system in conjunction with a tubing of reduced diameter prevented high flow rates. Hence the waterfrac stimulation included the volcanic rocks as well as the sandstone layers, and interpretation of the success of the waterfrac stimulation can hardly distinguish between the two rock types, because fracturing took place in both rock types. Nonetheless the achieved contribution of each layer can be determined by flow and temperature logs and hence yields the individual production efficiency.

The initial hydraulic condition of the well was tested with a production test at the entire open hole section between 3874 m to 4294 m. The resulting productivity index achieved 0.97 $m^3/(h\ MPa)$ at maximum pressure drawdown. Subsequently, a flow log was run which showed outflow of the conglomerates and the volcanic sequences of the reservoir. The rocks of the Rotliegend Sandstones intended for use as the geothermal reservoir were nearly impermeable (HUENGES *et al.*, 2002; ZIMMERMANN *et al.*, 2003). The reason for this impermeable sandstone section is believed to be the mud infiltration during the long standstill period of approximately 10 years. Permeability measurements

Table 1

Data from hydraulic tests in EGrSk3/90 (PI = productivity index; II = injectivity index)

	Duration days	volume m^3	flowrate m^3/h	diff. pressure MPa	PI/II $m^3/(h\ MPa)$
casing lift test Jan 2001	0.51	167	13.5	14.0	0.97
casing lift Test Feb 2002	0.58	307	22.4	10.5	2.13
pumping test Aug 2002	37	580	1.0	1.7	0.59
Injection test Jan 2003	8.3	720	3.6	4.3	0.83
flow back test Feb 2003	0.24	338	59.0	14.7	4.03
flow back test Dec 2003	1.0	859	15.6–52.8	6.7	7.46
injection test Dec 2004	17.9	3091	7.2	3.5	2.02

on cores from this section of the well showed mean values of 10^{-14} m^2 (10 mD) and documented the general usability of these reservoir rocks (TRAUTWEIN, 2005; TRAUTWEIN and HUENGES, 2005). Transmissibility was calculated from the shut-in period of the initial test and estimated to 4–6 × 10^{-14} m^3 (0.04–0.06 Dm). This value reflects the transmissibility of the conglomerates and volcanic rocks, which are merely due to natural fractures.

The first stimulation experiments were gel-proppant stimulations in two isolated intervals of the Rotliegend sandstones (ZIMMERMANN *et al.*, 2003; LEGARTH *et al.*, 2005). Visualization by Borehole Televiewer (BHTV) and Formation Micro Imager (FMI) confirmed the creation of an open vertical fracture with a height of more than 100 m and in the direction of the maximum horizontal stress (S_H = 18.5° ± 3.7°) (HOLL *et al.*, 2003, 2004). The productivity index could be raised to 2.13 m^3/(h MPa) due to the stimulation treatments (ZIMMERMANN *et al.*, 2003). Compared to the previous test this result indicates a doubling of the productivity index of the well. LEGARTH *et al.* (2003) conclude that the limited achievement was strongly influenced by the proppant properties during the treatment and prevented a better result of the stimulation treatments.

To determine the hydraulic parameters of the stimulated reservoir in more detail and to obtain stable conditions over a longer period, a long-term production test was performed in summer 2002 (ZIMMERMANN, 2004; REINICKE *et al.*, 2005). The productivity-index was estimated at pseudo steady-state conditions to 0.59 m^3/(h MPa). Transmissibility of the productive formations was estimated from pressure build up towards the end of the shut-in to assure pseudo radial flow conditions and was calculated to 3.1 10^{-14} m^3. This low transmissibility suggests that the stimulated sections have no connection to the high permeable sandstones.

The first waterfrac treatment started in January 2003 with a moderate injection test. The observed injectivity index of 0.83 m^3/(h MPa) corresponds to the productivity derived in the aforementioned production test at similar low difference pressure. For this reason it can be assumed that the hydraulic response of the reservoir is similar for production and injection for low pressure changes (decreasing for injection as well as increasing for production).

During the first part of the waterfrac treatment a total of 4284 m^3 fluid was injected under high pressure into the reservoir. At the beginning a pressure step test with gradually increasing injection rates up to 24 l s^{-1} was performed. At the injection rate of 8 l s^{-1} pressure increase reduces due to an enhanced injectivity of the formation. This effect can be interpreted as a mechanical reaction of the rock due to an opening of the existing generated artificial fracture, as well as the extension of a pre-existing fracture in the conglomerates and volcanic rocks at the bottom of the well (HUENGES *et al.*, 2006).

In the subsequent flow-back test a significant increase of productivity could be achieved (ZIMMERMANN *et al.*, 2005), which was above 4 m^3/(h MPa) during the whole test. This is an indication that the massive water injection produced additional fractures, so that the experiment was rated successful and represented roughly another doubling of the productivity index.

The waterfrac stimulation was continued with flowrates between 108 m^3/h and 144 m^3/h and a total injection volume of 7291 m^3. The pressure step-rate test indicates multiple fracture opening events. Fracture closure pressure was determined by pressure decline analyses during shut-in at 6.4 MPa above formation pressure. In the following we discuss the result of this stimulation treatment in terms of pressure-dependent productivity efficiency.

4. Pressure Dependence of Stimulation Results

In general, the production efficiency of produced artificial fractures is dependent on reservoir pressure changes. Hence, this dependence will be discussed in this section, because it is a crucial issue for the further use of this well in a geothermal doublet. Two cases will be discussed which represent the final reservoir conditions of the well: high increased reservoir pressure during a flow-back test and low increased reservoir pressure during an injection experiment.

4.1. Production Efficiency at High Reservoir Pressure

After the last hydraulic stimulation treatment in the well a flow-back test was performed with a subsequent shut-in followed by a second flow-back to reduce the remaining well head pressure. Within the first part of the test of 24 hours a total amount of 859 m^3 of water was produced from the formation, indicating another increase of productivity in comparison with former tests (Fig. 2). The results show that the stimulation treatments yielded an increase of productivity up to 14 $m^3/(h\ MPa)$ determined at fracture equilibrium pressure of approximately 53 MPa, equivalent to a formation pressure increase of 9.8 MPa. The productivity index decreases with decreasing difference pressure giving a clear indication of a closing fracture (Fig. 2). Hence it can be concluded that a self-propping effect due to shear displacement is nonexistent for the fractures of the Rotliegend sandstones and only residual fracture conductivity remains. At the end of stable flow conditions at a flow rate of 50 m^3/h the remaining productivity index is 7.5 $m^3/(h\ MPa)$ at a formation pressure increase of 6.5 MPa, which corresponds nearly to another doubling.

The pressure response of the flow-back test, the subsequent shut-in and an additional flow-back reveal a bilinear flow signature (Figs. 3, 4) ($p \sim t^{0.25}$; according to CINCO-LEY and SAMANIEGO-V, 1981) during the first part of the test. At the end of the shut-in the derivative pressure signature achieves an almost constant level, indicating a pseudo-radial flow response ($p \sim \ln t$; e.g., HORNE, 1995). From the maximum derivative pressure the maximum transmissibility could be estimated to 6.5×10^{-14} m^3 (0.065 Dm). Fracture conductivity was calculated according to the bilinear flow analysis to 4×10^{-13} m^3 (0.4 Dm), the corresponding fracture half length is 142 m.

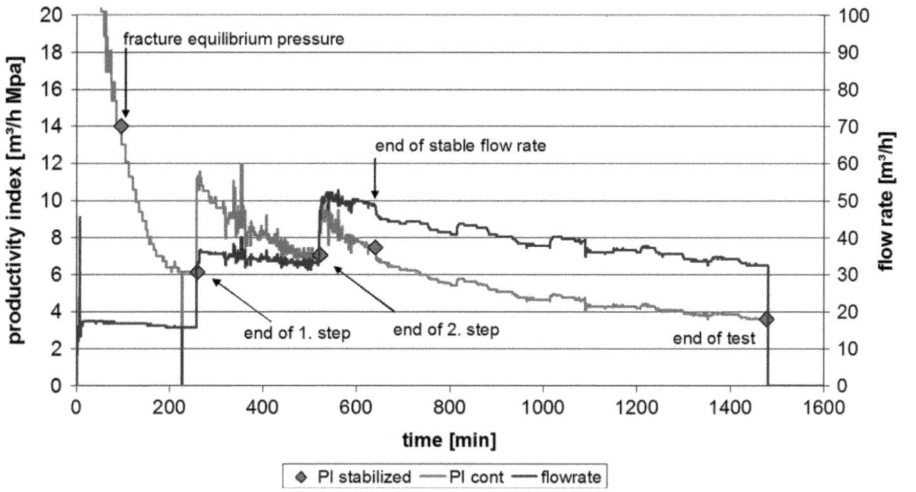

Figure 2

Trend of productivity index during first flow-back test. At the end of the stable flow rate (approx. after 11 hours of flow) the remaining productivity index is 7.5 $m^3/(h\ MPa)$ at the flow rate of 50 m^3/h.

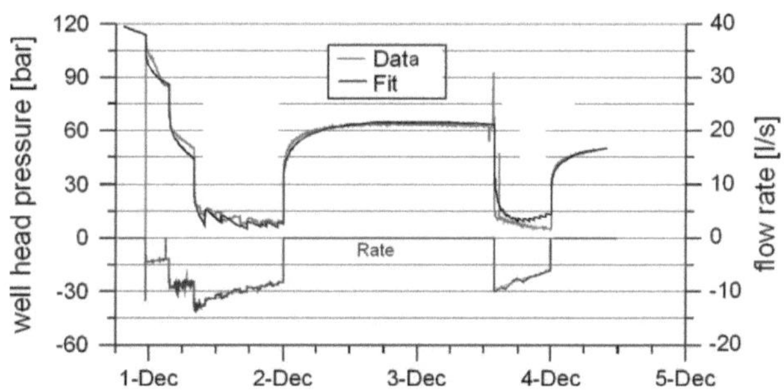

Figure 3

Pressure match of flow-back test to obtain hydraulic parameters such as fracture length and formation transmissibility.

4.2. Production Efficiency at Low Reservoir Pressure

To obtain the production efficiency of the reservoir rock at low pressure increase, a long-term injection experiment was carried out to obtain the hydraulic parameters of the reservoir and the fractures. The duration was scheduled to reach pseudo-radial flow conditions and potential boundaries of the compartment. Furthermore, the longer testing time yields results with higher accuracy and higher reliability. During 18 days 3091 m^3 of preconditioned water was injected into the reservoir with a density of 1000 kg/m^3 and a pH value lower than 5 to avoid iron scaling. The mean flow rate was set to 7.2 m^3/h over the whole time (Fig. 5). This limitation of the flow rate was due to the fact that a

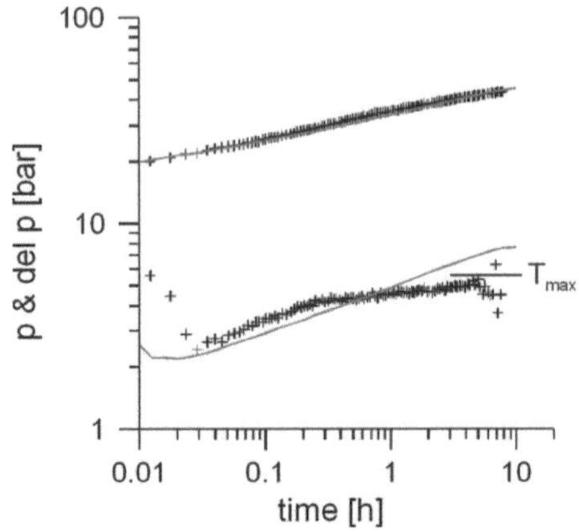

Figure 4
Log-log response of pressure and derivative as a function of superposition time of flow back test (Fig. 3). The upper curve represents pressure, the lower curve pressure derivative with maximum transmissibility derived at the end of the derivative.

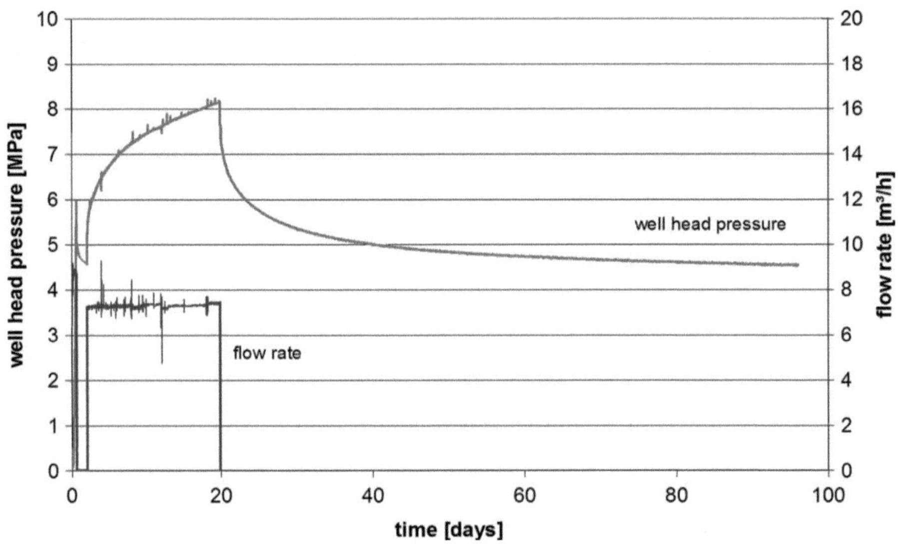

Figure 5
Flow and well head pressure during the long-term moderate injection experiment and adjacent shut-in period.

mechanical reaction of the rock, i.e., hydraulic fracturing, has to be avoided to obtain the pure hydraulic behaviour of the reservoir at low difference pressure. Furthermore, at low difference pressure hydraulic parameters of this injection test are comparable to an equivalent production test. The injectivity index was calculated to 2.02 $m^3/(h\ MPa)$ at the

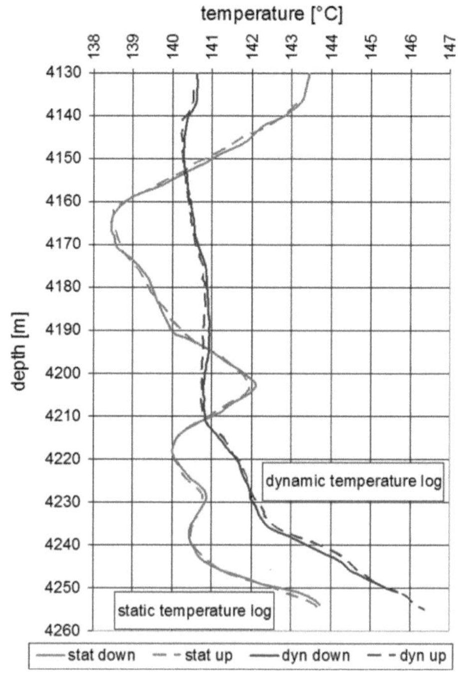

Figure 6
Temperature profiles before (static) and during production (dynamic) in the open section of the well. The temperature changes before production are due to the injection of cold water during the stimulation treatments.

end of the injection phase at an increased reservoir pressure of 3.6 MPa. After injection the well was shut in and the pressure response was monitored over additional 76 days. At the end of the shut-in a well head pressure of 4.5 MPa remained due to buoyancy of the injected water. This remaining well head pressure was used to move on with a short flow-back test in conjunction with temperature logging to obtain the inflow horizons of the reservoir. The production rate was set to 7.2 m^3/h just as before the injection rate to achieve comparable results. The temperature log, which was run during production, shows inflow from the bottom of the well appearing in the log profile as a change of slope (Fig. 6). Due to the obstruction below 4260 m the measured bottom of the interval was limited to 4255 m. The most significant change in the temperature profile and hence the most productive inflow was detected from 4212 m to the bottom at 4255 m yielding an effective inflow interval of 43 m. The limitation of this inflow to the conglomerates and volcanic rocks reveals that the contribution of the stimulated intervals of the sandstone layers due to the proppant frac treatment can almost be neglected at low difference pressures, i.e., the artificial sandstone fractures are closed at these conditions.

Well test analysis was carried out for the injection time and the subsequent shut-in time. Due to the signature of the pressure curve during injection and shut-in (Fig. 7), a

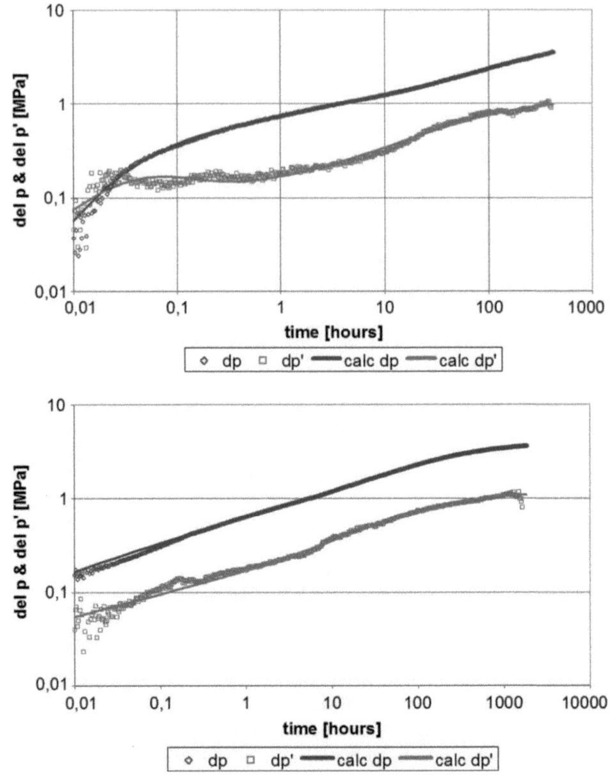

Figure 7
Log-log representation of the pressure response and its derivative during injection phase (top) and shut-in phase (bottom) of the experiment. The continuous lines represent the calculated results from well test analysis.

model with a finite conductive vertical fracture and homogeneous reservoir conditions was applied. Since no indication of any boundary effects was visible in the pressure response towards the end of the test, an infinite reservoir model was assumed. The pressure versus time curve and the derivative versus time curve showed quarter slopes in the log-log representation, indicating a bilinear flow behaviour with a finite conductive vertical fracture in the reservoir (e.g., CINCO-LEY and SAMANIEGO-V. (1981)). Toward the end the pressure derivative curve starts to converge to a horizontal line indicating the beginning of pseudo-radial flow and hence enabling the calculation of the reservoir transmissibility.

From the bilinear flow analysis the fracture conductivity was calculated to 7.8×10^{-13} m^3 (0.78 Dm) during injection and to 9.6×10^{-13} m^3 (0.98 Dm) during shut-in. These values document high fracture conductivities which can be related to the conglomerates and the volcanic rocks, and which remain stable especially during the shut-in with decreasing difference pressure. From this it follows that a self-propping effect due to shear displacement keeps the fractures open in the lower part of the well. The corresponding fracture half length of the vertical fracture was estimated to 309 m and 255 m, respectively. These half lengths are longer than expected from well test analysis after the waterfrac treatment at increased reservoir pressure with a calculated fracture half length of 142 m. Hence, it can be concluded that due to the fracture treatment a natural fracture system was connected. The transmissibility is about 4×10^{-14} m^3 (0.04 Dm) for both periods. The radius of investigation (e.g., LEE 1981) was calculated to 527 m for the injection time after 18 days of injection; for the shut-in period of 76 days the radius of investigation reached 1100 m.

5. *Results of Fracture Simulation*

The simulation of the last hydraulic stimulation in the well displays the achieved geometry of all stimulation treatments, since the last treatment was performed in the open hole which includes the previously stimulated intervals. Hence all previous stimulated sections are affected by this last treatment and the geometry reflects an equivalent model of several artificial fractures.

The fracturing process was modelled with the three-dimensional fracture simulator FRACPRO™ (CLEARY, 1994; CLEARY *et al.*, 1983). The aim was to model the fracture dimensions by matching the net treatment pressures (Fig. 8). A detailed description of this analysis can be found in LEGARTH (2003). A reasonable pressure match of the real data represents one plausible solution for the fracturing process and fracture geometry in reality. Determining fracture dimensions and geometry by modelling is important in order to set up the subsequent production schedule and as real-time modelling with the applied simulator becomes possible to optimize fracture and treatment design on site.

According to model calculation the pressure data of the stimulation treatment demonstrated the existence of an artificial fracture. Assuming one single vertical fracture, it spans vertically a height of 100 m in a north-south direction and extends horizontally at least 160 m into the formation (Figs. 9, 10). The mean fracture aperture ranges approximately 5 mm during the stimulation treatment at an injecting flow rate of 108 m^3/h.

6. *Interpretation and Discussion*

The initial production test before the stimulation treatments showed inflow only from the volcanic rock section of the reservoir. Since these rocks only have negligible matrix permeability this inflow is due to natural fractures of the conglomerates and volcanic rocks. The sandstone layers were nearly impermeable due to drilling induced near well bore damages referred to as well bore skin.

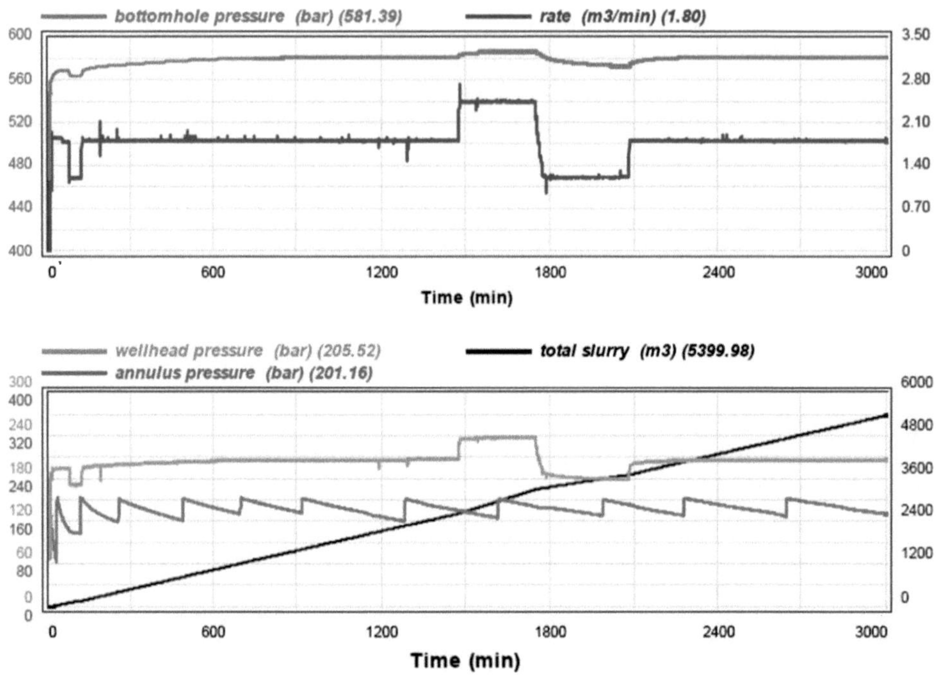

Figure 8
Pressures and flow rates during the waterfrac stimulation.

After stimulation treatments of the sandstones a flow log showed additional inflow from the sandstones. Furthermore the productivity index had increased (see Table. 1). At low difference pressure the situation is different: the inflow from the sandstones decreases and the artificial fractures close (Fig. 11; bottom). Hence the productivity index is similar to that of the initial condition. This interpretation is supported by the determination of the associated transmissibility before and after stimulation, which shows similar values. Transmissibility was calculated from shut-in after production tests and represents in both cases a value at low difference pressure.

Next two waterfrac stimulations were performed in the open section of the well which included the sandstone layers and the volcanic rocks. These stimulation treatments led to an additional access to the sandstone intervals and the volcanic rocks in the vicinity of the borehole due to the generation of additional artificial fractures.

After these waterfrac stimulations the effect in the sandstones is twofold: At high difference pressure the artificial fractures give access to the sandstone reservoir with a corresponding fracture half length of 142 m according to the hydraulic test results and 160 m according to fracture simulation. This leads to an additional pressure-dependent increase in productivity index and hence enables the access to the reservoir (Fig. 11; mid and top). During highly increased formation pressure above the equilibrium pressure all

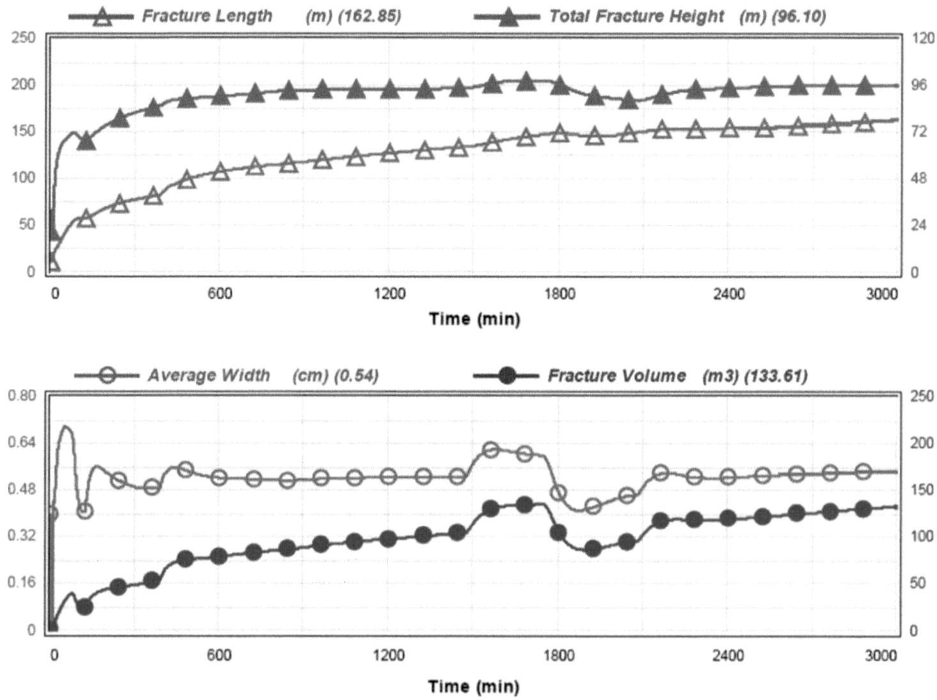

Figure 9
Calculated geometry of generated fracture during the waterfrac stimulation.

fractures including the artificial fractures in the sandstones as well as the natural and artificial fractures in the conglomerates and in the volcanics are open, yielding an injectivity index of approximately 14 m^3/(h MPa) (Fig. 11; top). At low difference pressure the fracture half lengths and the fracture apertures are reduced, consequently most parts of the fractures are not effective and the link to the reservoir rocks of the sandstone layers is disconnected again. Only in the near vicinity of the well the fractures are effective and improve the access to the borehole. However this is only a skin reduction and is limited to the near borehole environment and hence does not transcend the zone of reduced permeability to the undisturbed sandstone reservoir. This interpretation is supported by the calculated transmissibility, which has not changed substantially after the stimulation treatments and is more than ten times lower as is expected for these highly permeable sandstones.

In the conglomerates and the volcanic rocks at the bottom of the well new additional fractures were created. At low difference pressure only these sections contribute to the transmissibility. According to the low matrix permeability of these rocks this contribution is low and only the fracture system is effective. This interpretation is supported by the result of temperature logging during a production test, which showed only inflow from

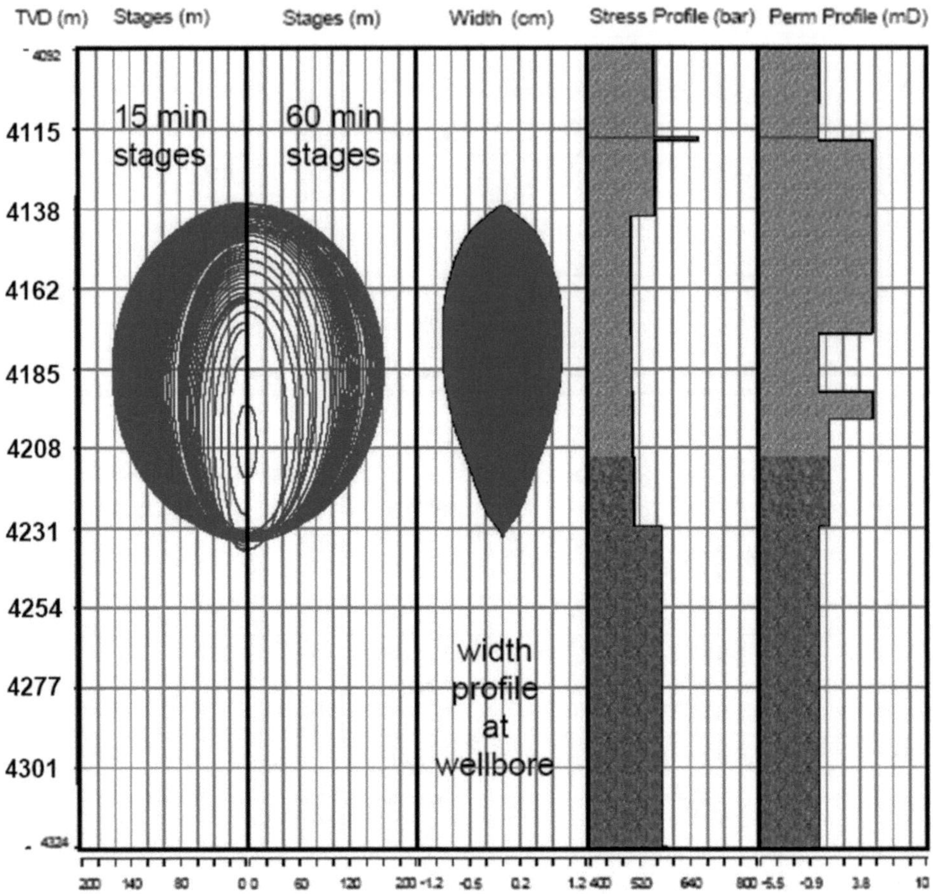

Figure 10
Final geometry of the generated fracture and stress and permeability profile of the waterfrac stimulation.

the conglomerates and the volcanic rock. The calculated extension of the fracture system is approximately 300 m and hence a nearly doubled value if compared to the previous analysis at high difference pressure. A most likely explanation of this outcome is the connection to a natural fracture system.

The results of hydraulic testing reveal that the effect of self propping of the stimulated fractures is different in the sandstone section and in the volcanic rocks. Self propping of a stimulated fracture presumes a shear displacement of the given fracture plane due to differential stress and preexisting natural fractures. The state of stress is normal to strike slip faulting with a differential stress in the order of 50 MPa in the sandstones and 30 MPa in the volcanic rocks. (S_v = 105 MPa; S_H = 90-100 MPa; S_h = 55 MPa (sandstones); S_h = 75 MPa (volcanics)). S_H was calculated for the sandstones by

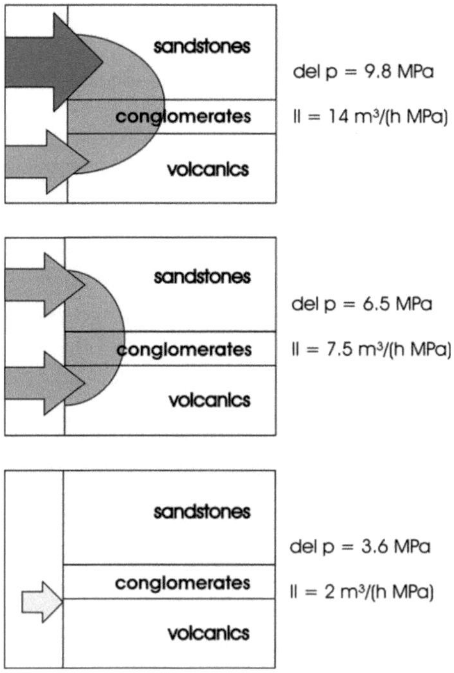

Figure 11

Pressure-dependent production efficiency of inflow into the reservoir rock (II = injectivity index). Top: All artificial fractures (from stimulation) are open at reservoir pressure equal to fracture equilibrium pressure. Mid: Residual opening of artificial fractures at fracture closure pressure. Bottom: Closure of artificial fractures at low reservoir pressure in the sandstone section; only the fractures in the conglomerates and volcanic rocks remain residually open.

fracture mechanics modelling (MOECK *et al.*, 2008) and is assumed similar in the volcanic rocks. S_h is approximately equal to fracture opening/closure pressure and was determined in the sandstone section by a leak-off test (HUENGES *et al.*, 2006). Due to operational limitations a leak-off test in the volcanics section exclusively was not possible, but it could be performed in the adjacent well located in the same reservoir compartment.

The flow characteristics indicated that the sustainability of hydraulic fracturing efficiency is different for the treated rock types. It is most likely that volcanic rocks consist of natural fractures, but we have no evidence for this, because no image logs were run in this section. Opposite to this, core samples of these sandstones do not show open natural fractures, only mineralised fissures are visible. Hence we conclude that a naturally fractured regime supports the shear displacement in the volcanic rocks during the hydraulic fracturing process. In case no natural open fractures exist like in the sandstone section, hydraulic fracturing is dominantly a tensile fracturing process and not sustainable. These considerations are in line with the aforementioned results of the hydraulic stimulation treatments, even if the differential stress in the sandstones is higher than in the volcanic rocks.

7. *Conclusions*

Enhanced geothermal systems (EGS) are engineered reservoirs that have been created to extract economical amounts of heat from low permeability and/or porosity geothermal resources. This includes all geothermal resources that are currently not in commercial production and require stimulation or enhancement. For these purposes we used the former gas exploration well in Groß Schönebeck as a down-hole geothermal laboratory.

The results reflect the learning curve from several reservoir hydraulic stimulation treatments. These experiments are major steps towards developing a procedure to increase the thermal water productivity from a prior low permeable sedimentary reservoir. The obtained values of productivity appear to indicate the feasibility of geothermal power production from a sedimentary geothermal reservoir.

The pressure dependence of productivity and injectivity of the geothermal reservoir queries the sustainability of the created artificial fractures. In volcanic rocks with natural fractures one can expect a self-propping effect due to shear displacement keeping the artificial fractures open after reservoir pressure release. In sedimentary rocks this self propping cannot be expected in general. The immediate consequence of this is that the fractures close after the injection pressure is reduced. Only a proppant package can avoid closure of the produced fractures, therefore stimulations in sedimentary rock should comprise high proppant concentration to obtain a multi layer proppant package and hence a high fracture conductivity.

The results of pressure-dependence production efficiency of the reservoir exclude the usage of this well as a production well. On the other hand this well is suitable as an injection well within a doublet system due to an increased injectivity as a result of the stimulated fractures remaining open at increased reservoir pressure.

Acknowledgement

This multidisciplinary project is a joint venture of research institutes (GFZ Potsdam, BGR Hannover, GGA Hannover), Universities (TU Berlin, RU Bochum) and industry partners (GTN Neubrandenburg, BWG Neubrandenburg, MeSy Bochum). It is funded by BMWI, BMBF, BMU, MWI Brandenburg and MWFK Brandenburg.

References

Baumgärtner, J., Jung R., Hettkamp, T., and Teza, D. (2004), *The status of the hot dry rock scientific power plant at Soultz-sous-Forêts*, Z. Angew. Geol. 2, 12-16.

Cinco-Ley, H. and Samaniego-V, F. (1981), *Transient pressure analysis for fractured wells*, J. Petro. Techn., 1749–1766.

Cleary, M.P. (1994), Critical issues in hydraulic fracturing of high-permeability reservoirs, SPE Paper 27618, Proceedings, *European Production Operations Conference and Exhibition* (Aberdeen, U.K., 15-17 March).

CLEARY, M.P., KAVVADAS, M., and LAM, K.Y. (1983), *Development of a fully three-dimensional simulator for analysis and design of hydraulic fracturing*, SPE Paper 11631.

ECONOMIDES, M.J. and NOLTE, K.G. *Reservoir Stimulation*, 2nd edn. (Schlumberger Educational Services, Houston, Texas. 1989).

HETTKAMP, T., BAUMGÄRTNER, J., BARIA, R., GERARD, A., GANDY, T., MICHELET, S., and TEZA, D. (2004), *Electricity production from hot rocks, Proc. 29th Workshop on Geothermal Reservoir Engineering* (Stanford University, Stanford, California, January 26-28, 2004), SGP-TR-175.

HOLL, H.-G., HURTER, S., SAADAT, A., KÖHLER, S., WOLFGRAMM, M., ZIMMERMANN, G., TRAUTWEIN, U., WINTER, H., LEGARTH, B. and HUENGES, E. (2003), *First-hand experience in a second hand borehole: Hydraulic experiments and scaling in the geothermal well Groß Schönebeck after reopening, Proc. Internat. Geotherm. Conf., IGC-2003 Reykjavik, Multiple integrated uses of geothermal resources*, S01, paper 060, 8–13.

HOLL, H.-G., MOECK, I., and SCHANDELMEIER, H. (2004), *Geothermal well Groß Schönebeck 3/90: A low enthalpy reservoir (Rotliegend, NE Germany), Proc. 66th EAGE Conference and Exhibition*, F032, Paris.

HORNE, R.N. *Modern Well Test Analysis*, 2nd edn. (Petroway Inc., Palo Alto, California. 1995).

HUENGES, E. and HURTER, S. (2002), *In-situ Geothermielabor Groß Schönebeck 2000/2001*, Scientific Technical Report, (GeoForschungsZentrum Potsdam) STR02/14.

HUENGES, E., HURTER, S., SAADAT, A., KÖHLER, S., and TRAUTWEIN, U. (2002), *The in-situ geothermal laboratory Groß Schönebeck: Learning to use low permeability aquifers for geothermal power, Proc. Twenty-Seventh Workshop on Geothermal Reservoir Engineering* (Stanford University, Stanford, California, January 28-30, 2002), SGP-TR-171.

HUENGES, E., TRAUTWEIN, U., LEGARTH, B., and ZIMMERMANN, G. (2006), *Fluid pressure variation in a sedimentary geothermal reservoir in the North German Basin: Case study Groß Schönebeck, Pure Appli. Geophys. 163*, 2141–2152.

KÖHLER, S. (2005), *Analysis of the combined heat and power plant Neustadt-Glewe, Proc. World Geothermal Congress 2005* (Antalya, Turkey, 24-29 April 2005), 1309.

KÖHLER, S. and SAADAT, A. (2003), *Thermodynamic modeling of binary cycles – Looking for best case scenarios, Proc. Internat. Geotherm. Conf, IGC-2003 Reykjavik, Multiple integrated uses of geothermal resources*, S01, Pape 061, 14–19.

LEE, W.J. *Well Testing*, SPE Textbook Series, Vol. 1 (Society of Petroleum Engineers, Houston, Texas 1981).

LEGARTH, B. (2003), *Erschließung sedimentärer Speichergesteine für eine geothermische Stromerzeugung*, Ph.D. Thesis, Technical University Berlin, D83, Faculty VI: Civil Engineering and Applied Geosciences, published as Scientific Technical Report STR03/09, GeoForschungsZentrum Potsdam, Germany.

LEGARTH, B., HUENGES, E., and ZIMMERMANN, G. (2005), *Hydraulic fracturing in sedimentary geothermal reservoirs*, Internat. J. Rock Mech. Mining Sci. *42*, 7–8, 1028–1041.

LEGARTH B., TISCHNER, T., and HUENGES, E. (2003), *Stimulation experiments in sedimentary, low-enthalpy reservoirs for geothermal power generation, Germany*, Geothermics *32*, 4–6, 487–495.

REINICKE, A., ZIMMERMANN, G., HUENGES, E., and BURKHARDT, H. (2005), *Estimation of hydraulic parameters after stimulation experiments in the geothermal reservoir Groß Schönebeck 3/90 (North-German Basin)*, Internat. J. Rock Mech. Mining Sci. *42*, 7–8, 1082–1087.

SCHINDLER, M., NAMI, P., SCHELLSCHMIDT, R., TEZA, D., and TISCHNER, T. (2008), *Summary of hydraulic stimulation operations in the 5-km deep crystalline HDR/EGS reservoir at Soultz-sous-Forêts, Proc. Thirty-Third Workshop on Geothermal Reservoir Engineering, Stanford University, Stanford, California, January 28-30, SGP-TR-185.*

TRAUTWEIN, U. (2005), *Poroelastische Verformung und petrophysikalische Eigenschaften von Rotliegend Sandsteinen*, Ph.D. Thesis, Technische Universität Berlin.

TRAUTWEIN, U., and HUENGES, E. (2005), *Poroelastic behaviour of physical properties in Rotliegend sandstones under uniaxial strain, Internat. J. Rock Mech. Mining Sci. 42*, 7–8, 924–932.

ZIMMERMANN, G. (2004), *Results of moderate pumping tests in the deep well Groß Schönebeck 3/90 in summer 2002*, Scientific Tech. Report, GeoForschungsZentrum Potsdam, STR04/03, 123–135.

ZIMMERMANN, G., HURTER, S., SAADAT, A., KÖHLER, S., TRAUTWEIN, U., HOLL, H.-G., WOLFGRAMM, M., WINTER, H., LEGARTH, B., and HUENGES, E. (2003), *The in situ geothermal laboratory Groß Schönebeck- stimulation experiments of sandstones in 4200 m depth, Proc. 28 Workshop on Geothermal Reservoir Engineering*, Stanford University, Stanford, California, SGP-TR-173.

ZIMMERMANN, G., REINICKE, A., HOLL, H. G., LEGARTH, B., SAADAT, A., and HUENGES, E. (2005), Well test analysis after massive waterfrac treatments in a sedimentary geothermal reservoir, *Proc. World Geothermal Congress 2005*, Antalya, Turkey, 24–29 April 2005, 1129.

(Received March 28, 2008, revised August 5, 2008, accepted September 11, 2008)
Published Online First: June 2, 2009

Pure appl. geophys. 166 (2009) 1107–1123
0033–4553/09/051107–17
DOI 10.1007/s00024-009-0475-4

Pure and Applied Geophysics

Impact of Poroelastic Response of Sandstones on Geothermal Power Production

GUIDO BLÖCHER, GÜNTER ZIMMERMANN, and HARALD MILSCH

Abstract—During geothermal power production using a borehole doublet consisting of a production and injection well, the reservoir conditions such as permeability k, porosity φ and Skempton coefficient B at the geothermal research site Gross Schoenebeck/Germany will change. Besides a temperature decrease at the injection well and a change of the chemical equilibrium, also the pore pressure p_p will vary in a range of approximately 44 MPa ± 10 MPa in our reservoir at −3850 to −4258 m depth. This leads to a poroelastic response of the reservoir rocks depending on effective pressure p_{eff} (difference between mean stress and pore pressure), resulting in a change in permeability k, porosity φ and the poroelastic parameter Skempton coefficient B. Hence, we investigated the effective pressure dependency of Flechtinger sandstone, an outcropping equivalent of the reservoir rock via laboratory experiments. The permeability decreased by 21% at an effective pressure range from 3 to 30 MPa, the porosity decreased by 11% ($p_{\mathrm{eff}} = 6$ to 65 MPa) and the Skempton coefficient decreased by 24% ($p_{\mathrm{eff}} = 4$ to 25 MPa). We will show which mechanisms lead to the change of the mentioned hydraulic and poroelastic parameters and the influence of these changes on the productivity of the reservoir. The most significant changes occur at low effective pressures until 15 to 20 MPa. For our *in situ* reservoir conditions $p_{\mathrm{eff}} = 43$ MPa a change of 10 MPa effective pressure will result in a change in matrix permeability of less than 4% and in matrix porosity of less than 2%. Besides natural fracture systems, fault zones and induced hydraulic fractures, the rock matrix its only one part of geothermal systems. All components can be influenced by pressure, temperature and chemical reactions. Therefore, the determined small poroelastic response of rock matrix does not significantly influence the sustainability of the geothermal reservoir.

Key words: Porosity, permeability, Skempton coefficient, poroelasticity, sandstone.

1. Introduction

The technical feasibility of geothermal power production will be demonstrated by means of the geothermal research wells Gross Schoenebeck (40 km north of Berlin/ Germany) using a borehole doublet. The reservoir is located in −3850 to −4258 m depth within the Lower Permian of the NE German Basin. The main targets are sandstones of the Upper Rotliegend (Dethlingen Formation/lower Elbe subgroup) as well as the volcanic rocks (andesites) of the Lower Rotliegend, where permeability is mainly due to connected fractures. The Dethlingen sandstones represent an effective reservoir horizon

Helmholtz Centre Potsdam, GFZ German Research Centre for Geosciences, Telegrafenberg, A6 104, D-14473 Potsdam, Germany. E-mail: bloech@gfz-potsdam.de; zimm@gfz-potsdam.de; milsch@gfz-potsdam.de with a porosity of 8–10% and a permeability of 1^*10^{-14} to 1^*10^{-13} m^2 (TRAUTWEIN and HUENGES, 2005). During geothermal power production using the doublet for production and injection simultaneously flow rates up to 100 m^3/h will be applied (ZIMMERMANN *et al.*, 2008). According to the productivity index $PI = 10$ $m^3/(h^*MPa)$ (ZIMMERMANN and REINICKE, 2009), which is defined as the ratio between flow rate and pore pressure change in the reservoir during water production, a pore pressure change dp_p of $\pm$ 10 MPa will be induced. The *in situ* formation pressure is $p_p = 43.8$ MPa, determined by p,T-logs at stationary conditions of the geothermal target horizon at a depth of 4100 m (LEGARTH *et al.*, 2005). In addition to a fault pattern analysis, stress ratios were calculated for the −4035 m deep Rotliegend sediments based on frictional failure theory (MOECK *et al.*, 2008). By means of the stress ratios and the vertical stress $S_V = 105$ MPa the maximum horizontal stress $S_H = 100$ MPa and the minimum horizontal stress $S_h = 55$ MPa were determined. By means of the mean stress $\bar{\sigma} = (S_V + S_H + S_h)/3 = 82.7$ MPa the effective mean $\bar{\sigma}_{eff} = \bar{\sigma} - p_p = 42.9$ MPa can be calculated (TERZAGHI, 1923). For our particular case an effective stress range of 42.9 MPa $\pm$ 10 MPa is of interest.

In order to show the influence of this effective stress change on the geothermal power production, we investigated the effective pressure dependency of the hydraulic rock properties permeability k and porosity φ as well as the hydro-mechanical property Skempton coefficient B (SKEMPTON, 1954). The permeability is a measure of the ability of the reservoir rock to conduct water and therefore influences the potential production rate. The porosity indicates the potential volume of geothermal fluid in the reservoir rock and the Skempton coefficient indicates the mechanical reservoir behavior on pore pressure change. The investigation was done by means of laboratory experiments with Flechtinger sandstone (Sventesius quarry, Germany), an outcropping equivalent of the reservoir rock.

Various previous studies still investigated the effective pressure dependency of these rock properties. CARROLL and KATSUBE (1983) developed a theory of hydrostatic poroelasticity in terms of porosity and bulk volume. By means of this theory changes in effective pressure can be related to changes in porosity (ZIMMERMANN, 1991). The effective stress dependency of the porosity also controls changes of elastic modules and the seismic velocity (SHAPIRO, 2003). The permeability can always be expressed as a function of confining p_c and pore pressure p_p (AL-WARDY and ZIMMERMAN, 2003). If the permeability follows the effective pressure law $p_{\text{eff}} = p_c - \chi p_p$, where χ denotes the effective pressure coefficient, then it can be expressed as a function of this effective pressure, $k = f\ (p_c - \chi p_p)$. Several experiments determined the effective pressure coefficient χ; MORROW *et al.* (1986) and BERNABE (1987) found that a value of 1 was the limiting value for crystalline rocks, ZOBACK and BYERLEE (1975) and AL-WARDY and ZIMMERMAN (2003) found that the effective stress coefficient depends on the clay content. Often χ is taken to be constant, yielding a linear expression for effective pressure, but KWON *et al.* (2001, and references therein) mentioned that χ may itself be a function of pressures p_c or p_p, or microstructural changes in pore structure.

The dependency of permeability on pore structure was also investigated by various studies: A general overview can be found in BEAR (1972); using hydraulic radius models

(KOZENY, 1927; CARMAN, 1937) or geometric parameters determined by a fractal approach (PAPE *et al.*, 2000) or mercury injection (KATZ and THOMPSON, 1987). It has to be taken into account, that the pore geometry parameters cannot be directly measured under pressure. Therefore, the permeability has to be correlated to other rock properties such as porosity or formation factor (BERNABE, 1988), which can be measured under pressure. Besides permeability also the Skempton coefficient was investigated under different experimental conditions. These studies showed that B is nearly independent of deviatoric stress change, but decreases as a result of the stiffening effect due to increasing confining pressure (LOCKNER and STANCHITS, 2002), under axisymmetric loading (LOCKNER and BEELER, 2003) and as a function of effective stress (ZIMMERMANN, 1991). Furthermore, an additional content of clay particles inside the pore spaces has also been shown to cause the Skempton coefficient to decrease (AL-WARDY and ZIMMERMAN, 2004). The stiffening of the framework can be explained by closure of open microcracks at high confining pressure levels (MESRI *et al.*, 1976).

The investigation of the mechanism, which leads to a change of permeability, porosity and Skempton coefficient depending on effective pressure, is beyond the focus of the present study. The aim of this paper is to show the impact of porosity, permeability and Skempton coefficient changes on the geothermal power production. For this purpose, relative changes of these parameters, depending on the expected effective pressure change in our reservoir, were measured by laboratory experiments and estimated by the mentioned relations. By means of the relative changes we derive evidence pertaining to the hydraulic behavior of the rock matrix during production and injection.

2. *Laboratory Experiments*

For our laboratory experiments three similar cylindrical-shaped samples of Flechtinger sandstone were prepared. All samples were taken from one sandstone block with a dimension of 10 cm × 10 cm × 30 cm. By means of these samples we tested permeability, porosity and Skempton coefficient under *in situ* conditions. A detailed description of these three experiments can be found in the following. In addition, mercury porosimetry was performed after the experiments on broken parts of the samples, which have a volume of approximately 1 to 2 cm^3. This independent method also yields a pore radius distribution of the samples as a function of the (effective) cumulative porosity as well as an average pore radius. In addition, the specific inner surface distribution is calculated from the mercury injection curve.

2.1. Permeability Measurement

The permeability experiment was performed in a High-Pressure-Temperature (HTP) permeameter (Fig. 1) which allows a variety of continuous petro-physical measurements at a maximum temperature, lithostatic and pore pressure of 200°C, 140 MPa and 50 MPa,

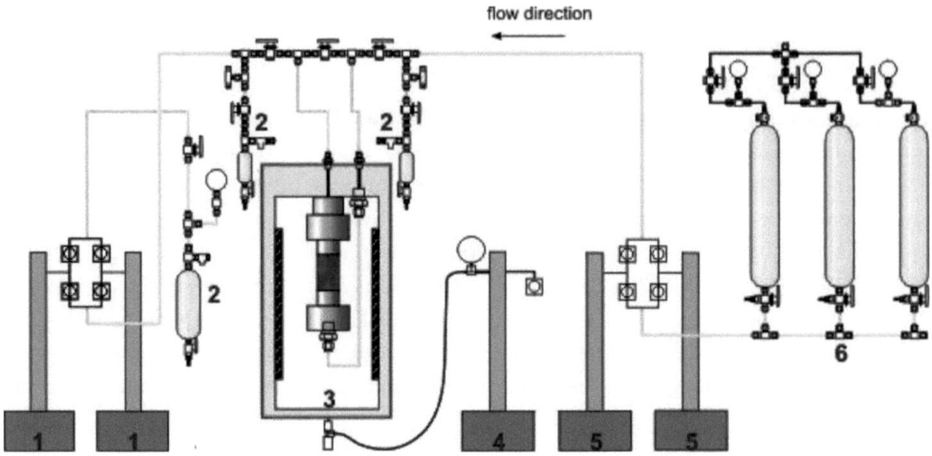

Figure 1
Schematic setup of the HTP-permeameter used for the permeability measurements: (1) downstream pump system maintains constant pore pressure (max. 50 MPa), (2) reservoirs for fluid sampling (max. 10 MPa), (3) pressure vessel with heater (max. 140 MPa and 200°C), (4) confining pressure pump (max. 140 MPa), (5) upstream pump system delivers constant fluid flow (max. 100 ml/min, max. 50 MPa) and (6) fluid reservoirs (10 l, max. 10 MPa).

respectively. A detailed description of this apparatus can be found in MILSCH *et al.* (2008a).

For investigating the permeability dependence on changes in effective pressure, a sample of Flechtinger sandstone was taken. The sample was cylindrical in shape with a diameter of 3 cm and a length of 4 cm. The permeability experiment was performed at a constant temperature of 40°C. A 0.1 molar NaCl-brine was used as the pore fluid which allowed a concurrent measurement of the electrical conductivity. The results of these measurements are published elsewhere (MILSCH *et al.*, 2008b).

The experiment with the Flechtinger sandstone was performed at a maximum confining and pore pressure of 47 MPa and 42 MPa, respectively. A constant flow rate of 0.05 ml/min was applied. We adjusted the effective pressure by increasing the confining pressure to its maximum and subsequently increased the pore pressure. Then, the pore pressure and the confining pressure were successively reduced to 2.5 MPa and 5 MPa, respectively. The increment of both pressures varied between 0.5 and 12.8 MPa. The described pressure cycling was repeated three times. During the experiment the pressure difference *dp* across the sample was measured. So as to keep the error (offset and/or drift of pressure sensors) in determining permeability as small as possible, a special measurement technique that refers to only one pressure sensor was applied. For this purpose, at each pressure step a bypass between the up- and the downstream side of the sample was opened and a reference value at the upstream side was measured. After closing the bypass, the upstream pressure increased to a steady state value. Consequently,

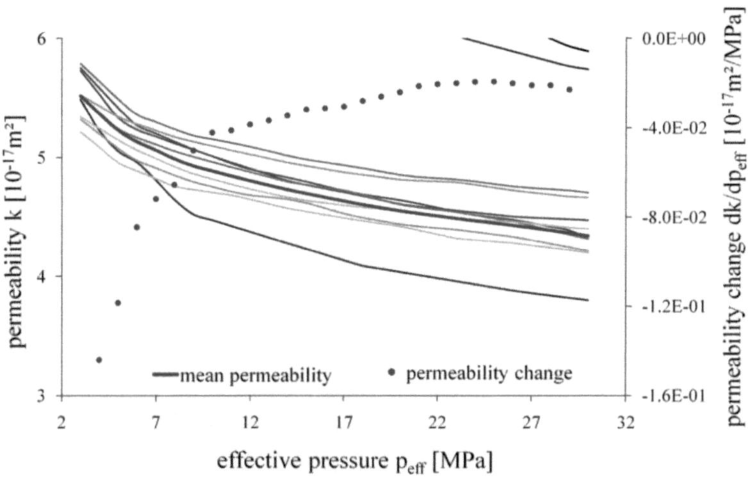

Figure 2
Permeability k measured for Flechtinger sandstone as a function of effective pressure. For each ramp we calculated the average permeability over an effective pressure interval of 1 MPa and subsequently the mean of the last nine pressure ramps. The color from black to light grey indicates the chronological sequence of the experiment. The dots indicate the permeability change over an effective pressure interval of 1 MPa.

the pressure gradient ∇p can be calculated by the measured pressure difference and the length l of the sample. By means of the determined pressure gradient, the permeability k of a rock is determined by a steady-state method using Darcy's Law, thus assuming laminar flow conditions:

$$k = -\frac{Q * \eta}{A * \nabla p}, \tag{1}$$

where Q, A and η denote flow rate, the cross section of the sample and the dynamic viscosity of the fluid.

The permeability of the Flechtinger sandstone measured with the permeameter is shown in Figure 2. There, two issues were observed: An initial decrease in permeability for the first three pressure ramps (not shown here) and a decrease in permeability due to an effective pressure increase. The former is interpreted to be due to a consolidation of the non preconditioned sample. This consolidation is irreversible and leads to a steady deformation. After the consolidation process the rock acts completely linear elastic. Therefore, the decrease in permeability due to an increase in effective pressure can be assumed as poroelastic response. The determined permeability is $5.5^{*}10^{-17}$ m^2 at 3 MPa effective pressure and decreased to $4.3^{*}10^{-17}$ m^2 at 30 MPa effective pressure.

For a better illustration of the permeability decrease due to effective pressure increase, we also calculated the effective pressure derivative of permeability (Fig. 2).

The absolute permeability change decreased strongly from $-1.4^{*}10^{-18}$ m^2/MPa to $-4.2^{*}10^{-19}$ m^2/MPa at low effective pressure ($p_{\text{eff}} = 4$ to 10 MPa). Subsequently, the permeability change decreased to a minimum value of $-2.0^{*}10^{-19}$ m^2/MPa at 21 MPa effective pressure and then it kept near constant at effective pressure above 21 MPa. The constant value of permeability change indicates a linear decrease of permeability due to effective pressure change.

2.2. *Porosity Measurement*

The porosity measurement was performed at the same High-Pressure-Temperature (HTP) permeameter as in the permeability test. Therefore, the sample was cylindrical in shape with a diameter of 3 cm and a length of 4 cm. The temperature was kept constant at 40°C and a 0.1 molar NaCl-brine was used as the pore fluid which allowed a concurrent measurement of the electrical conductivity. The experiment was carried out under hydrostatic conditions, so that the sample reacts to confining pressure only. During the first pressure ramp, the confining pressure was increased to a maximum of 50 MPa and was then reduced to the initial condition of 10 MPa. Subsequently, two further pressure cycles were performed with maximum confining pressure of 70 MPa. The confining pressure was controlled by adjusting the flow rate of the pressuring oil. For the porosity measurements we ensured an average confining pressure change of 0.13 MPa/min. During the total time of the experiment, the pore pressure p_p was kept constant at 5 MPa. Therefore, the effective pressure $p_{\text{eff}} = p_c - p_p$ ranged from 5 to 65 MPa. A direct measurement of porosity change depending on effective pressure change was impossible. Therefore, we determined the initial porosity at atmospheric pressure by means of mercury injection, 2-D image analysis and weighting method. For our particular sample the initial porosity φ_{init} of 8.54% was determined by the ratio between fluid- and bulk volume. The fluid volume $V_f = 2.415$ cm^3 can be obtained by the dry and wet sample mass, which were 67.715 g and 70.130 g. The bulk volume $V_b = 28.27$ cm^3 can be calculated by means of the sample dimensions. During the total time of the experiment, we measured the required fluid volume dV which was removed and added to the sample to ensure a constant pore pressure of 5 MPa. This fluid volume is defined negative because it indicates a loss of pore volume. By means of the initial fluid volume V_f, the sample volume V_b and the removed amount of water dV, the porosity φ can be calculated as follows:

$$\varphi(dV) = \frac{V_f + dV}{V_b}. \tag{2}$$

The calculated porosity of the three confining pressure ramps is shown in Figure 3. It is obvious, that during the first and the beginning of the second confining pressure ramp an irreversible porosity loss occurred. In contrast, the second part of the second pressure cycle as well as the complete third pressure cycle shows the linear-elastic behavior of the Flechtinger sandstone. This linear-elastic behavior leads to a decrease in porosity due to an increase of effective pressure. The determined porosity is 0.076 at 6 MPa effective pressure and decreased to 0.068 at 65 MPa effective pressure.

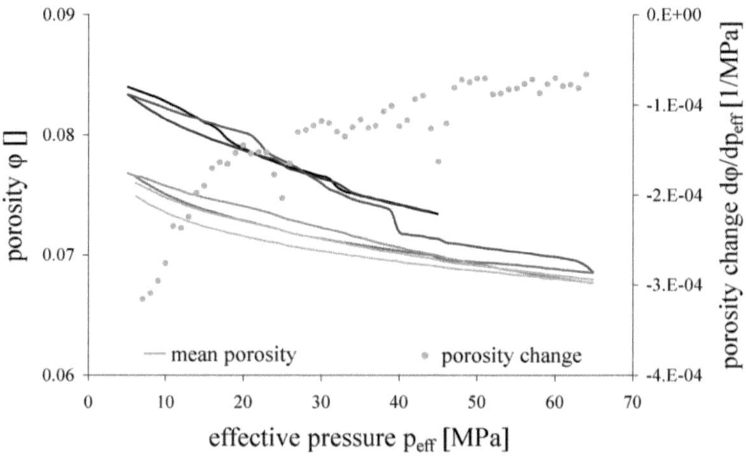

Figure 3
Porosity φ measured for Flechtinger sandstone as a function of effective pressure. For each ramp we calculated the average porosity over an effective pressure interval of 1 MPa and subsequently the mean of the last three confining pressure ramps. The color from black to light grey indicates the chronological sequence of the experiment. The dots indicate the porosity change over an effective pressure interval of 1 MPa.

As for the permeability measurement, we also calculated the effective pressure derivative of porosity (Fig. 3). There, the strongest reduction of porosity change from $3.2^{*}10^{-4}$ 1/MPa to $1.5^{*}10^{-4}$ 1/MPa was observed in a range between 7 MPa and 20 MPa effective pressure. A further reduction of porosity change to $7.0^{*}10^{-5}$ 1/MPa at 51 MPa effective pressure followed and for higher effective pressure the porosity change was constant. This indicates a linear decrease of porosity due to effective pressure.

2.3. Skempton Coefficient Measurement

For the Skempton coefficient measurement, a sample of Flechtinger sandstone identical to the permeability and porosity experiments before was examined. At the pre-experimental stage, the sandstone samples were completely saturated within the hydraulic system. The samples underwent a two-day flow-through with distilled water before the pore pressure ports of the hydraulic system were closed to ensure a maximum saturation of more than 95%. The saturation was measured following the procedure of Smith and Smith (1998).

The sample was cylinder-shaped with a diameter of 5 cm, a length of 10 cm and a porosity of ca. 8.3% was determined. The total volume of the sandstone cylinder was 196.4 cm^3 and the pore volume was ca. 16.3 cm^3. For measuring the axial and lateral strains of the sandstone samples, two axial extensometers and one radial chain extensometer were installed (MTS, 2004). The sample was mounted inside the pressure chamber (Fig. 4). This test system allows confining pressure up to 140 MPa. To apply

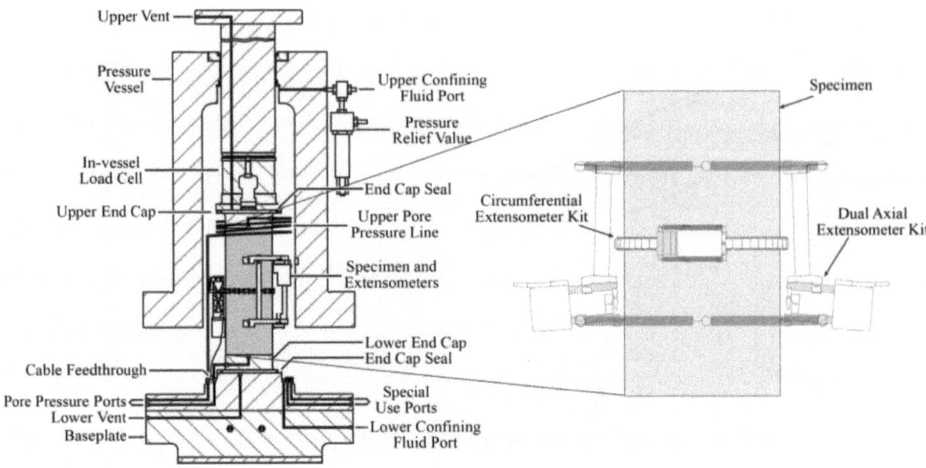

Figure 4
Laboratory test configuration showing Flechtinger sandstone sample in the pressure chamber (MTS, 2004). Pore pressure change was measured with the hydraulic shut-off valves closed to minimize external pore pressure system volume.

hydrostatic conditions, the chamber was filled with pressuring oil, and to keep the sample from being contaminated, it was surrounded by a silicon jacket. The mechanical system as well as the hydraulic system was connected to the Flechtinger sandstone. To minimize the dead volume of the hydraulic system, all capillaries were plugged by wires and the remaining dead volume was measured to 4 ml; this is equivalent to 25% of pore volume. This additional volume in the tubing exterior can significantly influence measurements of B (GREEN and WANG, 1986). WISSA (1969) suggested to minimize the ratio of exterior saturant volume to sample pore volume (0.3%), because the exterior saturant volume acts as a pore fluid reservoir and violates the undrained condition $dm_f = 0$. Therefore, it can be assumed that the measured values of B are significantly lower than for the fully undrained condition. Hence, our measurements specify a lower boundary.

To reduce the irreversible deformation and to minimize the nonlinear effects, we preconditioned the Flechtinger sandstone sample by applying a pressure cycle from 5 to 50 MPa, and repeated this procedure three times (according to HART and WANG, 1995). The raising of the confining pressure from 5 to 50 MPa and its lowering was done in 15 min. Throughout the preconditioning, the pore system was open to the atmosphere to achieve drained conditions. In the first part of the Skempton experiment we measured the undrained poroelastic properties of an intact (without visible damage) Flechtinger sandstone at confining pressures from 5 to 50 MPa. The experiment was carried out at hydrostatic conditions so that sample properties are related to changes in confining pressure p_c. To show the reversibility of the induced pore pressure change, the confining pressure was increased to a maximum of 50 MPa and was then reduced to the initial conditions five times. We measured the elastic deformation of the sandstone given by its axial and radial extension, as well as the pore pressure throughout the experiment. By means of the measured pore pressure change we determined the isotropic Skempton coefficient defined by $B = dp_p/dp_c$ (LOCKNER and BEELER, 2003). After closing the shut-off valves of the hydraulic system, the confining pressure was increased at a constant rate of 0.075 MPa/min to its highest level of 50 MPa. At this level, the confining pressure was kept constant for 4 hours and was then decreased to its initial condition, again at a constant rate of 0.075 MPa/min. At the beginning of the first pressure cycle, the isotropic Skempton coefficient increased with increasing effective pressure to 20 MPa. This could be due to a possible dissolution of gas in the fluid of an incompletely saturated sample. Therefore, an additional saturation of the sample was performed. By means of the described ramps an effective pressure range $p_{\mathrm{eff}} = p_c - p_p$ of approximately 10 MPa was applied. Therefore, two additional cycles were performed, where the confining pressure was increased to a maximum of 95 MPa and subsequently reduced to 5 MPa, resulting in an effective pressure range from 4 to 26 MPa. These two cycles started with the adjustment of the initial conditions, which were 5 MPa for confining pressure and 2.8 MPa for pore pressure. After closing the shut-off valves the confining pressure was increased at constant steps of 10 MPa to its highest level of 95 MPa. At this level the confining pressure was kept constant for 1 h and was then decreased to its initial condition, again at constant steps of 10 MPa. During the two additional pressure cycles the Skempton coefficient decreased with increasing effective pressure from 0.81 at 4 MPa to a minimum of 0.62 at 25 MPa.

There are two limiting boundaries for the isotropic Skempton coefficient: if the confining pressure affects only the grain structure it should be zero, and if the confining pressure affects only the fluid inside the pore spaces it should be one. We assume that the decrease of the Skempton coefficient results from the changing ratio of pore volume to grain volume as well as from a change of the pore-space geometry. This leads to a higher bulk modulus of the sample. On the one hand, the water is more compressible than the grains, therefore the pore volume decreases with increasing effective pressure. On the other hand, irregular pore spaces were more sensitive to effective pressure, therefore their shapes became more pressure resistant, and pore throats and microcracks were closed. This geometry change caused by closing of microcracks and elastic deformation of grains resulted in a more stable structure.

To investigate the dependence of the Skempton coefficient on effective pressure change, we differentiated the isotropic Skempton coefficient with respect to effective pressure (Fig. 5). The derivatives of these equations show that the change of the Skempton coefficient was highest at low pressure and was reduced by an increasing effective pressure. The initial value of the Skempton coefficient change was $1.3^{*}10^{-2}$ 1/MPa at $p_{\mathrm{eff}} = 5$ MPa and decreased to $7.3^{*}10^{-3}$ 1/MPa at $p_{\mathrm{eff}} = 15$ MPa. For higher effective pressure, the Skempton coefficient change was near constant at $6.6^{*}10^{-3}$ 1/MPa.

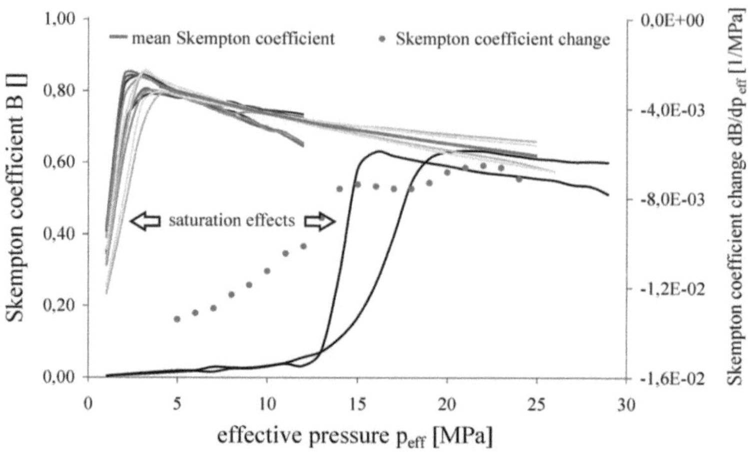

Figure 5

Skempton coefficient B measured for Flechtinger sandstone as a function of effective pressure. For each ramp we calculated the average Skempton coefficient over an effective pressure interval of 1 MPa and subsequently the mean of the last four confining pressure ramps. The color from black to light grey indicates the chronological sequence of the experiment. The vertical branches of the curves are due to a saturation deficit. The dots indicate the Skempton coefficient change over an effective pressure interval of 1 MPa.

3. *Estimation of Permeability and Skempton Coefficient at Reservoir Condition*

As mentioned in the introduction, for the geothermal power production in Gross Schoenebeck an effective pressure range of 42.9 MPa $\pm$ 10 MPa can be assumed. Only the porosity measurement included this effective pressure range and the determined values are listed in Table 1. The permeability and Skempton coefficient measurements were conducted to effective pressure values of 30 MPa and 25 MPa, respectively. Above these effective pressure levels both rock properties showed a near-linear dependence on effective pressure. Therefore a linear approximation for higher effective pressure levels can be sufficient. The permeability change was determined with $-2.0^{*}10^{-19}$ m^2/MPa and the change of Skempton coefficient was $-6.6^{*}10^{-3}$ 1/MPa. By means of the permeability

Table 1

Variation of permeability (linear extrapolated), porosity (measured) and Skempton coefficient (linear extrapolated) in the target effective pressure range between 33 MPa and 53 MPa. The relative value refers to the parameter value determined at in situ conditions

Parameter	p_{eff} = 33 MPa (rel. change)	p_{eff} = 43 MPa (in situ)	p_{eff} = 53 MPa (rel. change)
Permeability k [10^{-17} m^2]	4.28 (**+4.9%**)	**4.08**	3.88 (**−4.9%**)
Porosity φ	0.0710 (**+1.7%**)	**0.0698**	0.0688 (**−1.4%**)
Skempton coefficient B	0.566 (**+13.2%**)	**0.500**	0.434 (**−13.2%**)

value at 30 MPa effective pressure ($4.3^{*}10^{-17}$ m^2) and the Skempton coefficient value at 25 MPa effective pressure (0.619) and its pressure dependence, values within the target effective pressure range can be linear extrapolated (Table 1). These derived values are only rough estimations based on measurements below the target pressure range. Therefore, a better approximation of permeability and Skempton coefficient based on the measured porosity change should be derived in the following.

3.1. Permeability Calculation

The aim of this calculation is an extrapolation of permeability at high effective pressure by means of empirical model concepts. Various models exist, which correlate permeability and porosity based on pore space geometry attributes, such as hydraulic radius, inner surface, tortuosity, etc. In order to derive permeability values at effective pressure between 33 MPa and 53 MPa, we tested three of these models: 1) hydraulic radius model based on Kozeny-Carman-equation (KOZENY, 1927; CARMAN, 1937), 2) permeability-porosity relation of Rotliegend sandstones based on a fractal approach by PAPE *et al.* (2000) and 3) permeability calculation by mercury-injection test based on the percolation theory mentioned by KATZ and THOMPSON (1987).

For the first permeability calculation the equation $k = \varphi^{*}R^2/6$ was used, there R denotes the hydraulic radius. The hydraulic radius $R = \varphi/M$, was determined by use of mercury injection method (there M is the specific inner surface). This model led to a permeability $k = 1.37^{*}10^{-19}$ m^2 at atmospheric pressure. For the second permeability calculation the equation $k = 155\varphi + 37315\varphi^2 + 630\,(10\varphi)^{10}$ was used. This equation is based on a fractal model for Rotliegend sandstones of northeast Germany and was calibrated at several core sample measurements. It led to a permeability of $k = 2.07^{*}10^{-16}$ m^2 at atmospheric pressure. The third calculation based on the mercury injection led to a permeability of $k = 5.92^{*}10^{-16}$ m^2 at atmospheric pressure. For our specific rock sample a permeability of $k = 5.5^{*}10^{-17}$ m^2 was measured. In consideration of approximately 8% porosity, the permeability of our sample was lower than assumed. The model calculations by Pape and Katz and Thompson confirm this result. Therefore, we used the more general definition of permeability (BEAR, 1972):

$$k = f(s) * f(\varphi) * L^2. \tag{3}$$

There $f(s)$, $f(\varphi)$ and L^2 denote a function of shape, a function of porosity and a reference length square, respectively. The reference length can be the hydraulic radius or particle size as well.

In order to find a linkage between permeability and porosity, the often-used function of porosity $f(\varphi) = \varphi^3/(1-\varphi)^2$ (BEAR, 1972) was chosen. There the porosity function can be directly calculated by means of the laboratory results. The cross-plot of measured permeability and porosity function (Fig. 6) shows a near-linear dependency of the permeability k on the porosity function $f(\varphi)$. Therefore, a constant geometry term $f(s)*L^2$ over the total range of effective pressure can be assumed. To verify this assumption, the

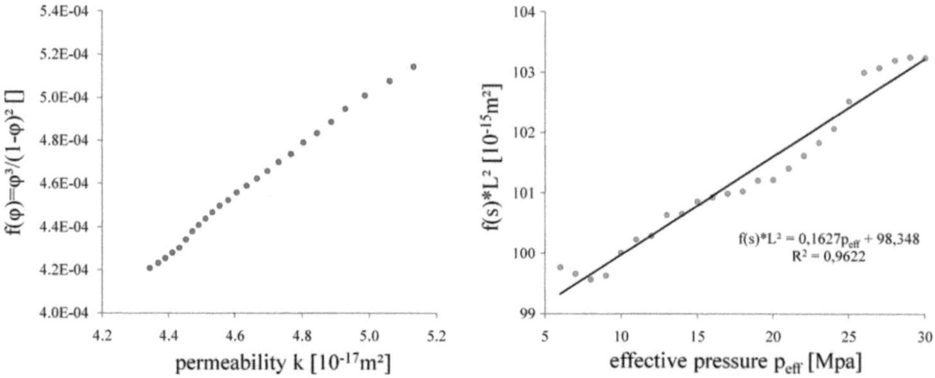

Figure 6
Cross-plot of measured permeability k and porosity function $\varphi^3/(1-\varphi)^2$ (calculated by porosity measurements) (left) and geometry function $f(s)*L^2$ depending on effective pressure (right).

geometry term was calculated over the total range of effective pressure by means of $f(s)^*L^2 = k/f(\varphi)$, as shown in Figure 6. It becomes obvious that the geometry term depends on effective pressure itself. Although the influence of the geometry term on permeability was small in comparison to the porosity function, we derived a linear dependency of the geometry term on effective pressure $f(s)^*L^2 = 0.1627p_{eff} + 98.348$. There the unit of $f(s)^*L^2$ is $[10^{-15}\ m^2]$ and thus of p_{eff} is [MPa].

By means of Equation (3) and the definitions of geometry term and porosity function, we were able to calculate the permeability at each point of effective pressure, where the porosity was known. The results of this calculation are shown in Figure 7 and Table 2. With these calculations the measured permeability values can be well reproduced and lead to a better prediction of permeability in the effective pressure range between 33 MPa and 53 MPa. Although the newly calculated values differ marginally in comparison to the linear extrapolation, the permeability values based on porosity and geometry seem more accurate. For an effective pressure range of $p_{eff} = 43$ MPa $\pm$ 10 MPa, we calculated a

Table 2

Variation of permeability (porosity-dependent), porosity (measured) and Skempton coefficient (bulk module-dependent) in the target effective pressure range between 33 MPa and 53 MPa. The relative value refers to the parameter value determined at in situ conditions

Parameter	$p_{eff} = 33$ MPa (rel. change)	$p_{eff} = 43$ MPa (*in situ*)	$p_{eff} = 53$ MPa (rel. change)
Permeability k $[10^{-17}\ m^2]$	4.29 (**+3.6%**)	**4.14**	4.03 (**−2.7%**)
Porosity φ	0.0710 (**+1.7%**)	**0.0698**	0.0688 (**−1.4%**)
Skempton coefficient B	0.653 (**+1.1%**)	**0.646**	0.630 (**−2.5%**)

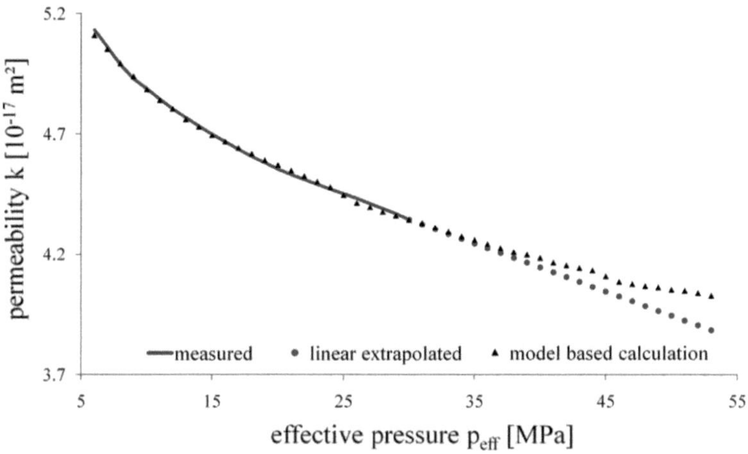

Figure 7
Permeability depending on effective pressure; measured by laboratory experiment (blue curve), extrapolated (blue dots) and calculated by means of porosity measurement (black triangles).

permeability range of $k = 4.08^{*}10^{-17}$ m^2 $\pm$ 4.9% with linear extrapolation and a maximum change of $k = 4.14^{*}10^{-17}$ m^2 $\pm$ 3.6% based on porosity measurement.

3.2. Skempton Coefficient Calculation

In contrast to permeability, the isotropic Skempton coefficient can be expressed in terms of the bulk module (K) and the porosity φ (Detournay and Cheng, 1993):

$$B = \frac{\frac{K_f}{\varphi}\left(1 - \frac{K_{fr}}{K_s}\right)}{\frac{K_f}{\varphi}\left(1 - \frac{K_{fr}}{K_s}\right) + K_{fr}\left(1 - \frac{K_f}{K_s}\right)}, \tag{4}$$

where the subscripts f, fr and s stand for fluid, framework and solid, respectively. By means of this equation we derived Skempton coefficient values in the target effective pressure range of $p_{eff} = 43$ MPa $\pm$ 10 MPa. The first step was the calculation of Skempton coefficient based on the porosity measurements. Since porosity was measured at effective pressure change, Skempton coefficient depends on effective pressure as well (Fig. 8). The results of this calculation were unsatisfying, because the porosity change showed no significant effect on Skempton coefficient. It is a well known fact that the Skempton coefficient strongly depends on the bulk modulus. Therefore, the bulk modulus of the fluid, the solid and the framework over the observed effective pressure range had to be determined. Values for the bulk modulus of the fluid considered as pure water could be taken from the program REFPROP (Lemmon *et al.*, 2007). The main fraction of the mineral content of the Flechtinger sandstones is quartz. Therefore, the bulk modulus of quartz (Levien *et al.*, 1980) was taken for the solid. Due to the fact that the bulk modulus

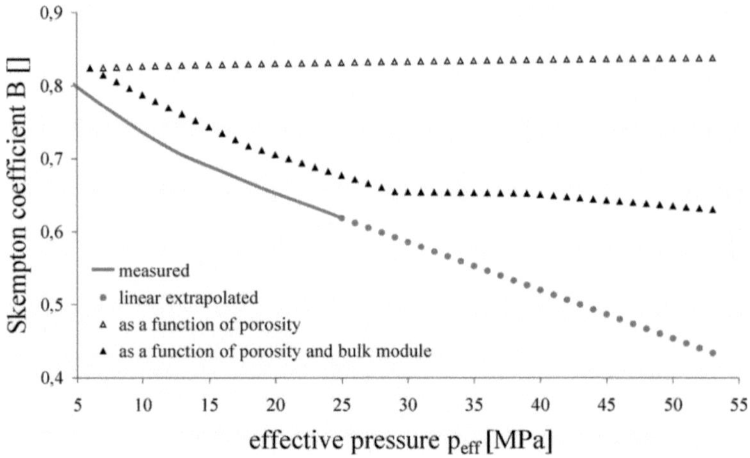

Figure 8
Skempton coefficient depending on effective pressure; measured by laboratory experiments (red curve), extrapolated (red dots), calculated by means of porosity (white triangles) and calculated by means of porosity and bulk module (black triangles).

of the framework was not determined by our laboratory experiments, we used published data as well (TIGREK *et al.*, 2005).

As porosity, calculations have shown that also the bulk modulus of the fluid and of the solid has no significant effect on the Skempton coefficient. In contrast, the effective pressure-dependent bulk modulus of the framework characterized the change in Skempton coefficient (Fig. 8). Although the absolute value of calculated and measured Skempton coefficient differs, the change of the Skempton coefficient with effective pressure can be reproduced. It has to be taken into account that the curve of the calculated Skempton coefficient shows a break at $p_{eff} = 29$ MPa. This could be a special effect of the specific sample used by TIGREK *et al.* (2005) and cannot be clarified here, because it was not discussed in the paper. For effective pressure values above 29 MPa the decrease of the Skempton coefficient is lower than expected by linear extrapolation. For an effective pressure range of $p_{eff} = 43$ MPa $\pm$ 10 MPa, we calculated a Skempton coefficient range of $B = 0.500 \pm 13.2\%$ with linear extrapolation and a maximum change of $B = 0.646 \pm 2.5\%$ based on porosity and bulk modulus of the framework (Table 2). The linear extrapolation values indicate the maximum change of Skempton coefficient over the observed effective pressure range. Due to the used bulk modulus of the framework, we assume that Skempton coefficient change calculated by porosity and bulk modulus is underestimated. The true Skempton coefficient change should be between these two calculated ranges. We expect a relative change of Skempton coefficient similar to those of porosity and permeability.

4. Conclusions

To show the impact of poroelastic response of sandstones on geothermal power production at the research site Gross Schoenebeck, we measured the effective pressure dependency of permeability, porosity and Skempton coefficient of Flechtinger sandstone, an outcropping equivalent of our reservoir rock. For the future geothermal power production flow rates will be about 100 m^3/h, resulting in an effective pressure change of $p_{eff} = 43$ MPa $\pm$ 10 MPa in the reservoir.

Only the porosity measurement included in this effective pressure interval and maximum changes of $\varphi = 0.0698 \pm 1.7\%$ were determined. By means of the porosity measurement and the permeability measurement a maximum permeability change of $k = 4.14^{*}10^{-17}$ $m^2 \pm 3.6\%$ at $p_{eff} = 43$ MPa $\pm$ 10 MPa can be calculated. With additional values of bulk module of the Flechtinger sandstone, similar relative changes in Skempton coefficient can be derived.

During the laboratory experiments, the permeability decreased by 21%, the porosity decreased by 11% and the Skempton coefficient decreased by 24%. These significant changes occurred at low effective pressures until 15 to 20 MPa. For our *in situ* reservoir conditions a change of 10 MPa effective pressure will result in a change in permeability of less than 4% and in porosity of less than 2%. Besides the rock matrix, a geothermal system consists of natural fractures and fault zones as well as induced hydraulic fractures. The hydraulic properties of all these components can be assumed as effective pressure-dependent. In addition, chemical reactions such as precipitation and dissolution can significantly influence the productivity of the reservoir. In contrast to the above-mentioned facts, we can conclude that the determined poroelastic responses did not affect the sustainability of the geothermal reservoir.

References

Al-Wardy, W. and Zimmerman, R. W., *Effective stress law for the permeability of clay-rich sandstones, 16th ASCE Engineering Mechanics Conference*, July 16-18 (University of Washington, Seattle 2003).

Al-Wardy, W. and Zimmerman, R. W., *Skempton coefficient of clean and clay-rich sandstones.* In *Proc. 17th ASCE Engineering Mechanics Conference* (University of Delaware, Newark 2004) DE, 6.

Bear, J., *Dynamics of Fluids in Porous Media*, (*Elsevier* 1972) 764 pp.

Bernabe, Y. (1987), *The effective pressure law for permeability during pore pressure and confining pressure cycling of several crystalline rocks, J. Geophys. Res.*, *92*, 649–657.

Bernabe, Y. (1988), *Comparison of the effective pressure law for permeability and resistivity formation factor in Chelmsford granite*, Pure Appl. Geophys., 127, 4.

Carman, P. (1937), *Fluid flow through granular beds*, Trans. Inst. Chem. Eng. London *15*, 150.

Carroll, M. and Katsube, N. (1983), *The role of Terzaghi effective stress in linear elastic deformation, J. Energy Resour. Tech. 105*, 509–511.

Detournay, E. and Cheng, A. H. D. (1993), *Fundamentals of poroelasticity, analysis and design methods* (*Pergamon Press*, Chapter 5 in *Comprehensive Rock Engineering: Principles, Practice and Projects*, Vol. II, *Analysis and Design Method*), pp. 113–171.

Green, H. G. and Wang, H. F. (1986), *Fluid pressure response to undrained compression in saturated sedimentary rock*, Geophys. *51*, 4, 948–956.

HART, D. J. and WANG, H. F. (1995), *Laboratory measurements of a complete set of poroelastic moduli for Berea sandstone and Indiana limestone*, J. Geophys. Res., B, Solid Earth Planets *100*, 17741–17751.

KATZ, A. and THOMPSON, A. (1987), *Prediction of rock electrical conductivity from mercury injection measurements*, J. Geophys. Res., *92*, 599–607.

KOZENY, J. (1927), *Über die kapillare Leitung des Wassers im Boden (Aufstieg, Versickerung und Anwendung auf die Bewässerung*, Sitz. Ber. Akad. Wiss. Wien. Math. Nat. (Abt. IIa), *136a*, 271–306.

KWON, O., KRONENBERG, A., GANGI, A., and JOHNSON, B. (2001), *Permeability of Wilcox shale and its effective pressure law*, J. Geophys. Res. *106*, 19.339–19.353.

LEGARTH, B., HUENGES, E., and ZIMMERMANN, G. (2005), *Hydraulic fracturing in a sedimentary geothermal reservoir: Results and implications*, *Internat. J. Rock Mech. Min.* Sciences *42*, 1028–1041.

LEMMON, E., HUBER, M., and MCLINDEN, M. (2007), *Reference fluid thermodynamic and transport properties-refprop, version 8.0*, National Institute of Standards and Technology, Gaithersburg, NIST Standard Reference Database 23.

LEVIEN, L., PREWITT, C. and WEIDNER, D. (1980), *Structure and elastic properties of quartz at pressure*, Am. Mineralogist *65*, 920–930.

LOCKNER, D. A. and BEELER, N. M., *Stress-induced anisotropic poroelasticity response in sandstone. In Proc. 16th ASCE Engineering Mechanics Conference* (University of Washington, Seattle, WA 2003), July 16–18.

LOCKNER, D. A. and STANCHITS, S. A. (2002), *Undrained poroelastic response of sandstones to deviatoric stress change*, *J. Geophys. Res.*, B, Solid Earth Planets *107*, 30.

MESRI, G., ADACHI, K., and ULLRICH, C. R. (1976), Pore pressure response in rock to undrained change in all-round stress, Geotechnique, *25*, 317–330.

MILSCH, H., SPANGENBERG, E., KULENKAMPFF, J., and MEYHÖFER, S. (2008a), *A new apparatus for long-term petrophysical investigations on geothermal reservoir rocks at simulated in situ conditions*, Transp. Porous Med. *74*, 73–85, doi 10.1007/s11242-007-9186-4.

MILSCH, H., BLÖCHER, M., and ENGELMANN, S. (2008b), *The relationship between hydraulic and electrical transport properties in sandstones: An experimental evaluation of several scaling models*, Earth Planet Sci. Lett. *275*, 355–363.

MOECK, I., SCHANDELMEIER, H., and HOLL, H. G. (2008), *The stress regime in a rotliegend reservoir of the Northeast German basin*, Internat. J. Earth Sci., in press.

MORROW, C., BO-CHONG, Z., and BYERLEE, J. (1986), *Effective pressure law for permeability of Westerly granite under cyclic loading*, J. Geophys. Res. *91*, 3870–3876.

MTS (2004), *Mts rock and concrete mechanics testing systems*, MTS Systems Corporation, 14000 Technology Drive Eden Prairie, MN USA 55344.

PAPE, H., CLAUSER, C., and IFFLAND, J. (2000), *Variation of permeability with porosity in sandstone diagenesis interpreted with a fractal pore space model*, Pure Appl. Geophys *157*, 603–619.

SHAPIRO, S. A. (2003), *Elastic piezosensitivity of porous and fractured rocks*, Geophys, *68*, 2, 482–486.

SKEMPTON, A. W. (1954), *The pore pressure coefficient a and b*, Geotechnique *4*, 143–147.

SMITH, I. and SMITH, G. N., *Elements of soil mechanics*, (*Blackwell*, Oxford 1998).

TERZAGHI, K. (1923), *Die Berechnung der Durchlässigkeitsziffer des Tones aus dem Verlauf der hydrodynamischen Spannungserscheinungen*, Sitzungsberichte der Akadamie der Wissenchaften in Wien, Mathematisch-Naturwissenschaftliche Klasse, Abteilung, IIa, *132*, 105–124.

TIGREK, S., SLOB, E., DILLEN, M., CLOETINGH, S., and FOKKEMA, J. (2005), *Linking dynamic elastic parameters to static state of stress: toward an integrated approach to subsurface stress analysis*, Tectonophysics, *397*, 167–179.

TRAUTWEIN, U. and HUENGES, E. (2005), *Poroelastic behaviour of physical properties in rotliegend sandstones under uniaxial strain*, Internat J. Rock Mechan. Min. Sci. *42*, 7-8, 924–932.

WISSA, A. E. Z. (1969), *Pore pressure measurement in satureted stiff soils*, J. Soil. Mech. *95*, 1063–1073.

ZIMMERMAN, R. W., *Compressibility of Sandstones (Developments in Petroleum Science),* volume 29, 173 pages, (Elsevier Publi. Co. 1991).

ZIMMERMANN, G. and REINICKE, A. (2009), *Hydraulic fracture stimulation of a deep sandstone reservoir to create an enhanced geothermal system* (EGS) - Laboratory testing and field experiments, Geothermics, submitted.

ZIMMERMANN, G., REINICKE, A., BRANDT, W., BLÖCHER, G., MILSCH, H., HOLL, H.-G., MOECK, I., SCHULTE, T., SAADAT, A., and HUENGES, E. (2008), *Results of stimulation treatments at the geothermal research wells in gross Schoenebeck/Germany, PROC, Thirty-Third Workshop on Geotherm Reservoir Engin.* (Stanford University, Stanford, California), January 28–30, SGP-TR-185.

ZOBACK, M. and BYERLEE, J. (1975), *Permeability and effective stress*, Am. Assoc. Petro. Geologists, Bulletin *59*, 154–158.

(Received April 1, 2008, revised September 23, 2008, accepted October 2, 2008)
Published Online First: May 7, 2009

Pure appl. geophys. 166 (2009) 1125–1163
0033–4553/09/051125–39
DOI 10.1007/s00024-009-0481-6

Pure and Applied Geophysics

Permeability Prediction for Low Porosity Rocks by Mobile NMR

H. Pape,[1,4] J. Arnold,[1,3] R. Pechnig,[1,3] C. Clauser,[1] E. Talnishnikh,[2] S. Anferova,[2] and B. Blümich[2]

Abstract—Estimating permeability from NMR well logs or mobile NMR core scanner data is an attractive method as the measurements can be performed directly in the formation or on fresh cores right after drilling. Furthermore, the method is fast and non-destructive. Compared to T_1 relaxation times, commonly measured T_2 distributions are influenced by external and internal magnetic field gradients. We performed two-dimensional T_1 and T_2 relaxation experiments on samples of Rhaetian sandstone, a rock with low porosity and small pore radii, using a mobile NMR core scanner which operates within a nearly homogeneous static magnetic field. Because small pore sizes are associated with high internal magnetic field gradients, standard methods from NMR logging in the oil industry cannot be applied for accurate permeability prediction. Therefore, a new model theory was developed, which describes the pore radius dependence of the surface relaxivity ρ_2 by both an analytical and a more practical empirical equation. Using corrected ρ_2 values, permeability can be predicted accurately from the logarithmic mean of the T_2 distribution and the physically based Kozeny-Carman equation. Additional core plug measurements of structural parameters such as porosity, permeability, specific inner surface area and pore radius distributions supported the NMR results.

Key words: Permeability, NMR, surface relaxivity, rock physics.

1. Introduction

Storage and transport properties such as porosity and permeability are essential for characterizing reservoir rocks, and different petrophysical measurements such as resistivity and NMR are used to estimate them. The common approach to porosity in NMR logging is based on polarizing the magnetic spin moments of hydrogen nuclei of fluids in porous rocks. NMR logs are considered suitable for estimating the pore size distribution and to infer the formation permeability (Kenyon, 1992). Commonly the buildup of nuclear magnetization and the irreversible signal decay are measured in terms of relaxation curves following an excitation with suitable radio-frequency (rf) pulses.

[1] Applied Geophysics and Geothermal Energy E.ON Research Center of Energy, RWTH Aachen University, Germany.
[2] Institute of Technical and Macromolecular Chemistry, RWTH Aachen University, D-52056 Aachen, Germany.
[3] Geophysica Beratungsgesellschaft mbH, D-52064 Aachen, Germany.
[4] Eichendorffstr. 10, D-30952 Ronnenberg, Germany. E-mail: h.pape@geophysik.rwth-aachen.de

The characteristic times for buildup of longitudinal magnetization and the loss of transverse magnetization are denoted as T_1 (longitudinal relaxation time) and T_2 (transverse relaxation time), respectively. Distributions of T_1 and T_2 times are obtained from experimental relaxation curves by inverse Laplace transformation or by inversion with suitable fitting functions. Assuming a fast diffusion regime, which applies to the majority of rocks, the rate-limiting step is relaxation at the surface (KLEINBERG *et al.*, 1994). As a consequence, the decay rate of magnetization decay depends mainly on the surface-to-volume ratio for measurements in a homogeneous magnetic field. Thus the relaxation time distribution reflects the pore radius distribution of the rock. Compared to T_1, T_2 measurements are preferred in NMR logging due to shorter measurement times and correspondingly a higher possible logging speed. However, the measured T_2 values may be affected not only by the properties of interest, such as the pore size, but also by the inhomogeneity of the static magnetic field and the internal magnetic field gradients (KENYON, 1992). This causes a further reduction in relaxation times. Internal magnetic field gradients arise from magnetic susceptibility contrasts between rock mineral surfaces and the pore fluid (GODEFROY and CALLAGHAN, 2003; LONNES *et al.*, 2003). The magnetic susceptibility value of water is negative (-9.26×10^{-6} SI at 20 °C), as for diamagnetic minerals such as rock salt, calcite, gypsum, quartz, and graphite. By contrast, most minerals and rocks have positive values (KLEINBERG and HORSFIELD, 1990; GEOTEK 2000, Multi-Sensor Core Logger Manual).

Compared to high-field NMR, measurements in low magnetic fields are preferred for rock studies, because the internal field gradient is proportional to the applied magnetic field strength (KLEINBERG *et al.*, 1994). However, estimating permeability only from NMR T_2 relaxation time measured in a low and homogeneous static magnetic field is difficult. This is due to the unknown nature of the surface relaxivity and the internal magnetic field gradients of each sample. Therefore, interpretation of NMR relaxation data is not as simple as it looks at first sight and should be calibrated using additional information from core analysis.

The permeability can not be measured directly by NMR relaxation technique in bore holes. Yet NMR logs yield some important parameters, which allow estimating permeability. First, NMR measurements yield a porosity log which can be used to infer permeability using poroperm crossplots (TIAB and DONALDSON, 1996). Second, NMR relaxation curves are related to distributions of pore sizes. This represents the real advantage because this information gives access to the effective hydraulic pore radius r_{eff}, which is another important parameter for calculating permeability. In practice, the conversion from T_2 distributions to pore sizes and the permeability prediction are based on empirical relationships which were initially calibrated to a data set of a different formation. However, a logging technique is less attractive if one needs core data of the same bore hole for calibration. In the present case of low Allermoehe sandstone, permeability prediction with an empirical equation of the type of the so-called Timur-Coates equation (COATES *et al.*, 1991) was not satisfactory, as it yielded much lower values than directly measured gas permeabilities. Therefore, we used a permeability equation of the Kozeny–Carman type (KOZENY, 1927, CARMAN, 1957), which involves porosity, tortuosity and effective hydraulic pore radius r_{eff}. Glover and Walker (submitted 2008) show that the applied equation is purely based on physical principles and can be derived from electro-kinetic theory. The same is true for the RGPZ model, developed by Revil, Glover, Pezard, and Zamora (see GLOVER *et al.*, 2006, and references therein), which takes the effective grain radius as the characteristic length parameter of the porous medium.

Our NMR laboratory device allowed measurements with an external homogeneous field. This made it possible to study the surface relaxivity ρ_2, which relates T_2 to pore radius. In practice, ρ_2 is taken as a constant conversion factor. We found that ρ_2 depends on pore radius in a characteristic way. In this paper, we developed a model assuming the existence of an interlayer where the internal magnetic field gradients are localized. We calibrated the model with our samples and supplemented the current method of deriving r_{eff} from T_2 distributions by adding a correction term.

In this paper we will show, how permeability prediction from NMR data can be improved: (1) by using the function of surface relaxivity ρ depending on pore radius, and (2) by using a permeability model which is fully based on physics. These principles are not restricted to mobile NMR measurement, but can be transferred to the interpretation of NMR borehole measurements.

A central question is the compatibility of downhole and portable measurements. For the comparison of both methods, a NMR log of the Allermoehe bore hole was available.

2. *Samples and Measurements*

We studied the petrophysical properties of a low-porosity sandstone formation in a hot water aquifer drilled and logged in the Allermoehe borehole, which is located in the northern German sedimentary basin near Hamburg. It was drilled into the Rhaetian hot water aquifer that is considered a geothermal resource (BAERMANN *et al.*, 2000a, b; PAPE *et al.*, 1999b, 2005). Rhaetian sandstones were cored in the depth interval between 3220 m and 3250 m. From the same depth interval, additional NMR logging data are available, which was recorded by the CMR (Combinable Magnetic Resonance) tool of Schlumberger. For laboratory NMR measurements and additional petrophysical investigations we drilled core plugs with diameters of 20 mm at 20 positions from the Allermoehe cores. The selected cores are fine-grained sandstones with high quartz content.

One-dimensional (1D) T_2 measurements were performed in the laboratory on 25 core plugs with a mobile, low-field NMR core scanner. Additionally, two-dimensional (2D) T_1-T_2 experiments were carried out on 16 of these samples. The 2D NMR measurements allowed us to compare the T_2 to the T_1 distributions, the latter not being influenced by diffusion within external and internal magnetic field gradients.

In order to characterize the samples, a set of petrophysical properties was measured including Klinkenberg corrected gas permeability, porosity, formation factor, and specific inner surface area. Moreover, we obtained the pore size distribution on identical core plugs by mercury injection. Grain sizes of the sandstones were measured on thin-sections by image analysis.

In summary, the samples have porosities ranging between 2% and 12% and low permeabilities up to 20 mD[1]. Mercury porosimetry yields pore-sizes between 0.01 μm and 100 μm. Magnetic susceptibility values were recorded by a Geotek® multi-sensor Core Logger (GEOTEK 2000, Multi-Sensor Core Logger Manual) with a range from 0×10^{-6} SI to 25×10^{-6} SI.

2.1. *Mobile NMR*

Prior to NMR laboratory testing, the Allermoehe core plugs were dried to a constant weight in an oven at a temperature of 60° C to remove all water from interconnected pores. Then they were saturated with distilled water in a vacuum exsiccator to constant weight. Additionally, porosity was measured with the Archimedes method, which was performed during the saturation procedure for the NMR measurements.

The NMR measurements were performed by a mobile Halbach sensor (Fig. 1) as described by ANFEROVA *et al.* (2007). The magnet system weighs less than 30 kg and features a sufficiently homogeneous magnetic field within the cylindrical volume of the sample. The static magnetic field B_0 is equal to 0.22 T, corresponding to a proton resonance frequency of 9.6 MHz. The device is constructed from two identical arrays, which are stacked behind each other and separated by an axial gap of 10 mm for improved homogeneity of the magnetic field. Each array consists of three magnet rings containing 16 magnet blocks of a volume of 30 mm^3. The inner diameter of the magnet is 140 mm. The homogeneity of the magnetic field in the central part of the magnet with a diameter of up to 80 mm is sufficient for non-destructive transverse relaxation measurements. Relaxation time T_2 distributions can be measured in water-saturated drill cores and plugs of diameters between 20 mm and 80 mm by means of exchangeable cylindrical rf-coils. Additionally, 2D T_1-T_2 correlation experiments can be performed in the central part of the magnetic field with the highest homogeneity on plugs with diameters of 20 mm. Whereas 1D T_2 distributions are measured by applying a CPMG pulse sequence (DUNN *et al.*, 2002), T_1-T_2 correlation experiments require a relaxation-relaxation editing pulse sequence (HÜRLIMANN and VENKATARAMANAN, 2002). A multi-modal log-normal distribution (Matlab code courtesy of Andreas Hartmann; RWTH Aachen University 2006; now at Baker Hughes INTEQ GmbH, Celle) was used to obtain 1D T_2 distribution curves. 2D T_1-T_2 distribution curves are obtained from inverse Laplace transformations. Measurement times for T_2 distributions and T_1-T_2 correlation experiments on the Allermoehe core plugs depend on the sample porosity and on the number of

[1] 1 mD = Millidarcy; darcy = 1 D = 9.87×10^{-13} m^2.

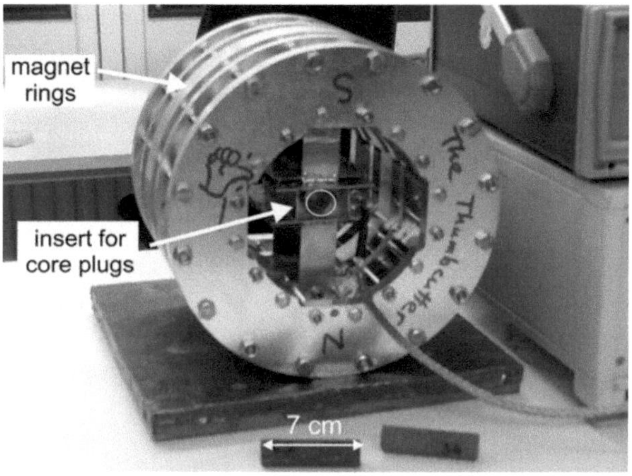

Figure 1
Mobile Halbach NMR sensor with an inner diameter of 140 mm and combined with a special insert for core plugs with a diameter of 20 mm.

Table 1

Experimental parameters and measurement times of the mobile NMR sensor

Parameter	Modified Halbach core-scanner
Frequency	9.6 MHz
Number of scans	32–80
Pulse length	9.7 μs
Inter-echo time	0.06 ms–0.15 ms
Number of echoes	1500–6400
Saturation delay time	1 s
Saturation recovery time	1 ms–3.5 s
Measurement time	1D measurements: 30 s–5 min
	2D measurements: 30 min

scans. They vary from a few minutes for CPMG experiments to 30 minutes for 2D experiments. Experimental parameters of the Halbach sensor used for 1D T_2 measurements and for 2D T_1-T_2 correlation experiments are shown in Table 1.

2.2. *Independent Methods*

The following supporting measurements were performed on corresponding core plugs: Porosity ϕ, permeability k, mercury porosimetry, and specific surface area. These measurements were necessary for accurate interpretation of NMR core data in terms of permeability. Additionally, electrical resistivity was measured to obtain the formation factor F and thus tortuosity T.

All NMR porosities were correlated to porosity values measured on corresponding core plugs with a helium gas pycnometer (Geopyc 1360 from Micromeritics). The total volume of the plugs was measured with the raw density pycnometer Accupyc 1330 (Micromeritics). For comparison, porosity was measured using the Archimedes method. The mean value of the difference $\phi_{\text{pyc}}-\phi_{\text{Archimedes}}$ was 0.006.

Permeability was determined from gas-flow measurements. For low permeability rocks, the effective gas permeability depends on pressure and may therefore deviate from water permeability. Hence, the Klinkenberg correction (RIECKMANN, 1970) was applied to account for this effect. Each sample was measured for one pair of inlet-pressure and outlet-pressure with the mean pressure between 0.119 and 1.74 MPa. The Klinkenberg constant b was derived from the empirical regression with permeability k, recommended by the API: $b = 0.777k^{-0.39}$.

Correcting for the Forchheimer effect was not necessary, because the Reynold numbers Re indicated laminar flow. We calculated $Re = (\rho/\eta)(v/\phi)d$ with density ρ, viscosity η, Klinkenberg corrected Darcy velocity v, porosity ϕ, and effective grain diameter d. For the Allermoehe samples we obtained $Re < 0.1$.

In order to estimate pore size distributions from NMR relaxation measurements, additional information was obtained by mercury porosimetry on selected core plugs. Mercury injection is a well established and widely used method for obtaining pore size information (DULLIEN, 1979). It yields the pore throat sizes weighted by the relative pore volumes to which the pore throats provide access (COATES *et al.*, 1999). Mercury is injected into a core plug with incrementally increasing pressure up to a maximum of 60,000 psi[2]. During each step, both the pressure and volume of mercury are measured after the pressure reaches equilibrium. Applying the Washburn equation (WEBB, 2001), each pressure step can be associated with a particular pore throat size which decreases with increasing pressure. The device Autopor 9520 of Micromeritics was used for the measurements.

Precise measurements of the specific inner surface were performed based on nitrogen-adsorption according to the Brunauer-Emmet-Teller (BET) method (BRUNAUER *et al.*, 1938). The specific surface $S_{\text{por,BET}}$ is defined as the ratio of the absolute surface area of a solid and its pore volume. The surface area includes all parts of accessible inner surfaces, i.e., mainly pore wall surfaces. The BET measurement determines the amount of adsorbed gas required to cover the accessible internal pore surfaces of a solid with a complete monolayer. The volume of the adsorbate at full monolayer can be calculated from the adsorption isotherm by means of the BET equation (Catalogue of Reference Procedures, Testing and Chemical Analysis, BUNDESANSTALT FÜR MATERIALFORSCHUNG UND PRÜFUNG, Germany 2005).

The formation factor F was obtained from electrical resistivity measurements. The inverse of the formation factor describes the effective porosity with respect to transport processes such as fluid flow, electrical conductance, and diffusion (PAPE *et al.*, 2000,

[2] 1 psi (pound per square inch) = 6.98×10^3 pascal.

2006). For the determination of F we used the equation of electrical conductivity σ_0 (the inverse of resistivity):

$$\sigma_0 = \frac{1}{F}\sigma_{\mathrm{w}} + \sigma_q = \frac{\phi}{T_{el}}\sigma_{\mathrm{w}} + \sigma_q, \tag{1}$$

with electrical conductivity of the electrolyte σ_{w}, interlayer conductivity $\sigma_{\mathrm{q},}$ and electrical tortuosity T_{el}.

The parameters F and σ_{q} are calculated from a series of resistivity measurements on a sample stepwise saturated with saltwater with increasing salinity. According to Archie's first law (ARCHIE, 1942), F and T_{el} can be substituted by porosity. Otherwise, the tortuosity can be expressed by the ratio of grain radius r_{grain} and the effective hydraulic pore radius r_{grain} acccording to the fractal model of pore structure, calibrated to a set of reservoir sandstones (PAPE *et al.*, 1976a, 2000):

$$F = \phi^{-m}; \quad T = \phi^{-m+1} = \left(\frac{r_{grain}}{r_{eff}}\right)^{0.8(D-2)}, \tag{2}$$

where D is the fractal dimension.

The so-called cementation or tortuosity exponent m varies between 1 and 3 depending on rock texture (SERRA and SERRA, 2004).

Table 2 presents NMR data and petrophysical parameters from independent methods, measured on the Allermoehe core plugs. Table 3 contains the calculated tortuosities T_{el} and cementation exponents m. The values of the samples AC1, AC3 and AC 24 with porosities $\phi \leq 0.03$ seem to be too low compared to a bead pack with $m \sim 1.5$ (GLOVER *et al.*, 2006). The three Allermoehe samples with very low porosity are anhydrite cemented. The exchange of fluid for several times during the measuring procedure of F causes dissolution of anhydrite. The relative increase of porosity is especially high in the case of low initial porosity. Therefore, the measured tortuosity values do not correspond to the original state of the three samples and therefore the values can not be used.

2.3. Logging Data

When we started, a commercial CMR Log (Combinable Magnetic Resonance tool recorded by Schlumberger) was available which yielded logarithmic mean values of surface relaxivity, $T_{2,\mathrm{LM}}$ and additionally three fractions of porosity: clay bound porosity with $T_2 < 3$ ms, capillary bound water with 3 ms $< T_2 <$ 33 ms and free water with $T_2 >$ 33 ms.

3. Permeability Models and Empirical Relationships

This chapter presents relationships between the physical processes such as hydraulic flow, electrical conduction, mercury injection, and NMR relaxation, which can be used to

Table 2

Petrophysical parameters measured on the Allermoehe core plugs from various depths: Logarithmic mean grain radius $r_{grain,LM}$, pycnometer porosity ϕ_{pyc}, formation factor F, specific inner surface area $S_{por,BET}$ measured with the BET method, gas permeability k_{gas}, logarithmic mean values from 1D T_2 distributions $T_{2,LM}$ and from 2D T_1 and T_2 distributions $T_{1,2LM}$.

Sample	Depth [m]	$r_{grain,LM}$ [μm]	ϕ_{pyc}	F	$S_{por,BET}$ [μm^{-1}]	k_{gas} [mD]	$T_{2,LM}$_1D [ms]	$T_{1,LM}$_2D [ms]	$T_{2,LM}$_2D [ms]
AC1	3224.45	106	0.02	53	40.42	0.16	2.95	–	–
AC3	3225.64	97	0.02	54	39.79	0.42	7.95	–	–
AC4	3235.34	102	0.09	56	2.38	11.6	82.17	510	81.84
AC6	3236.67	121	0.06	72	10.59	1.85	14.23	250	30.11
AC7	3236.79	121	0.08	67	9.40	3.59	32.93	220	29.07
AC9	3239.88	110	0.08	70	8.45	5.26	24.37	300	37.04
AC10	3240.69	102	0.11	38	5.24	20.7	30.37	330	41.34
AC12	3241.44	91	0.09	68	10.89	3.13	32.78	330	28.17
AC13	3241.75	89	0.09	58	9.21	4.42	31.79	190	24.34
AC14	3242.56	100	0.09	44	5.63	10.1	30.60	320	35.76
AC15	3243.16	110	0.09	50	4.34	19.6	34.02	440	35.03
AC16	3243.42	102	0.08	55	6.60	8.99	26.01	330	33.29
AC18	3243.78	110	0.08	56	10.89	6.16	21.95	320	33.73
AC19	3243.92	106	0.10	47	6.50	13.9	32.01	370	39.90
AC20	3244.79	116	0.08	60	5.17	7.2	29.77	320	35.35
AC21	3245.10	120	0.07	79	6.71	3.1	19.49	–	–
AC22	3245.54	116	0.07	89	9.34	0.39	17.47	150	7.68
AC23	3246.15	113	0.08	92	8.37	0.4	18.88	130	7.47
AC24	3246.65	117	0.03	65	38.67	0.91	25.65	–	–
AC25	3247.31	113	0.06	66	12.52	1.71	21.57	240	23.89

estimate permeability from NMR measurements. These relationships can be divided into two types: (1) empirical ones, which vary in their validity from formation to formation and (2) analytical ones, which are based on geometrical and physical principles (GLOVER *et al.*, 2006). The latter have the advantage that they are of general validity.

The simplest empirical method for predicting permeability is from a permeability-porosity crossplot. A power-law relationship can be established between permeability and porosity based on fractal theory (PAPE *et al.*, 1967a). A fit to a large data set of Rotliegend sandstone from northeastern Germany (Pape *et al.*, 1999a, 2000) yields:

$$k_\phi = \left(31\phi + 7463\phi^2 + 191(10\phi)^{10}\right) \times 10^{-12} \quad \left(k \,\mathrm{in}\, \mathrm{m}^2\right). \tag{3}$$

From a data set of Allermoehe sandstone, PAPE *et al.* (1999b, 2005) derived the following relationship:

$$k_\phi = 0.309(100\phi)^{4.85} \times 10^{-12} \quad \left(k \,\mathrm{in}\, \mathrm{m}^2\right). \tag{4}$$

The empirical Kenyon equation (KENYON, 1997) is commonly used for estimating permeability from porosity and the logarithmic mean $T_{2,LM}$ of the T_2 distribution from logging data:

$$k = a\phi^4 T^2_{2,\mathrm{LM}}. \tag{5}$$

Usually, a value of $a = 4$ mD/(ms)^2 is used for reservoir sandstones (MORRIS *et al.*, 1996).

The derivation of a physically justified relationship includes several steps. In the first step hydraulic flow and electrical conduction are connected via geometrical and semi-geometrical parameters: porosity ϕ, tortuosity T and a length parameter, for instance the effective hydraulic pore radius r_{eff}. In the last step, NMR relaxation time has to be related to specific surface area per pore volume S_{por}, which corresponds to an inverse length parameter. In practice, surface relaxivity ρ is used as constant conversion factor, which has to be calibrated for each formation. We will show later, that the surface relaxivity of transverse magnetization, ρ_2, increases with decreasing pore radius. It additionally depends on effective pore radius r_{eff}.

The various physical processes discussed in this paper depend on length parameters such as pore body radius r_{pore}, pore radius $r_{\mathrm{pore,mercury}}$ from mercury injection, effective hydraulic pore radius r_{eff}, and grain radius r_{grain}. Therefore, the conversion of different length parameters plays a fundamental role when we relate different processes. The geometrical relationships between length parameters become more complicated from the fact that porous rocks reveal distributions of pore and grain sizes. We can obtain a distribution of $r_{\mathrm{pore,mercury}}$ from mercury porosimetry, which measures the radius of pore entrance. From these distributions we can extract a characteristic value, for instance the logarithmic mean, in order to relate it with the effective hydraulic pore radius r_{eff}. A general complication is the dependence of the length parameters on the resolution power of the measurement method. This statement is obvious when considering surface area measurements on rough surfaces.

From these remarks it gets clear that it is difficult to derive a purely physically based model for the prediction of permeability from NMR relaxometry. For some steps in the model, empirical relationships cannot be avoided. Some of the relationships between geometrical and semi-geometrical parameters consist of power laws, which are originally calibrated to data sets of specific rock formations. However, they can also be theoretically justified by assuming a fractal pore space structure. The equations become physically based by expressing the exponents of the power laws, which are usually not integer, as a function of the unique parameter fractal dimension D. Another group of relationships result from the analytical RGPZ model (GLOVER *et al.*, 2006, GLOVER and WALKER, submitted 2008). The characteristic parameter of this theory is the cementation exponent m. The parameter D can be related to m by equation (2).

3.1. Permeability Relationships

We used a special form of the Kozeny-Carman equation (KOZENY, 1927; CARMAN 1956), where permeability is related to porosity ϕ, tortuosity T and the effective hydraulic pore radius r_{eff}:

$$k = \frac{\phi}{8T} r_{eff}^2. \tag{6}$$

The term (ϕ/T) can be replaced by the inverse formation factor F^{-1}, and by ϕ^m, according to the first Archie equation (2) (ARCHIE, 1942).

Equation (6) was analytically derived from electro-kinetic theory by Glover and Walker (submitted 2008) as a result of the RGPZ model. The original length parameter in this model is the effective grain diameter d_{eff} which is related to effective pore radius r_{eff} by the so-called Θtransformation:

$$d_{\text{eff}} = 2\Theta r_{\text{eff}} \quad \text{with } \Theta = \sqrt{\frac{a_{\text{grain pack}} m^2 F^2}{8}}, \tag{7}$$

where the logarithmic mean value from grain radius distributions can be taken for (1/2) d_{eff}. The parameter $a_{\text{grain pack}}$ is a constant in the range 2–12 and characterizes the topology of the grain pack with $a_{\text{grain pack}} = 8/3$ attributed to a pack of quasi-spherical grains.

An equivalent equation for $r_{\text{grain}} / r_{\text{eff}}$ can be derived from the fractal pore model (PAPE *et al.*, 1967a, 2000) which is based on the fractal expression for tortuosity T (see eq. 2):

$$\frac{r_{grain}}{r_{eff}} \approx \frac{a_{shape\_spere}(1-\phi)}{a_{shape\_cylinder}\left(\frac{\phi}{T}\right)} = \frac{3}{2} F(1-\phi) \approx \left(\frac{3(1-\phi)}{2\phi}\right)^{1/(1-0.8(D-2))}. \tag{8}$$

Permeability can be expressed by F, $a_{\text{grain pack}}$, and a length parameter Λ which is related to the pore throats (SCHWARTZ *et al.*, 1992; KOSTEK *et al.*, 1992; BERNABÉ and REVIL, 1995):

$$k \approx \frac{\Lambda^2}{a_{\text{grain pack}} F} \quad \text{with} \quad \Lambda \approx \frac{d_{eff}}{2mF}. \tag{9}$$

Equations (7) and (9) combine to equation (6) with elimination of m and $a_{\text{grain pack}}$.

The effective pore radius can be substituted by $S_{\text{por,hydr}}$, the specific surface normalized by pore volume. The value of $S_{\text{por,hydr}}$ is smaller than the respective $S_{\text{por,BET}}$ because the surface is smoothed by the physical process of fluid flow. For a hydraulic model with smooth pores, r_{eff} is related to $S_{\text{por,hydr}}$ by

$$r_{\text{eff}} = \frac{a_{\text{shape}}}{S_{\text{por,hydr}}}, \tag{10}$$

with $a_{\text{shape}} = 2$ for cylindrical pores and $a_{\text{shape}} = 3$ for spherical pores.

This equation is primarily based on a pore model which consists of a bundle of smooth, cylindrical capillaries of radius r_{eff} (PAPE, 2000). By contrast, the simplest model for NMR relaxation, which relates relaxation times T_1 and T_2 to S_{por} and r_{eff}, was developed for isolated smooth spherical pores (KLEINBERG *et al.*, 1994) with:

$$\frac{1}{T_{1,2}} = \rho_{1,2} S_{por,NMR}, \tag{11}$$

where $\rho_{1,2}$ is the surface relaxivity.

and

$$r_{eff} = \frac{3}{S_{por,NMR}}. \tag{12}$$

Based on a fractal model for porous rocks, several geometrical relations were established by PAPE *et al.* (1967a, 1999a, 2000) in which specific surface, pore radius, tortuosity, and porosity are connected via the fractal dimension *D*. This is the fundamental geometric parameter for the description of the structure of pore-space. We regard $D = 2.36$ as a standard value for northern German sandstones (PAPE *et al.*, 1967a). This value was first calculated for a set of 41 well investigated reservoir sandstones with the petrophysical data reported by PAPE *et al.* (1987b). The principal idea of the fractal concept is the dependence of the measure of geometrical parameters such as the area of a rough surface on the power of resolution of the measuring method. The largest value is attributed to $S_{por,BET}$ measured with the highest resolution by the smallest yardstick, which is given by the size of adsorbed nitrogen molecules d_{N2}. By contrast, the dynamic process of fluid flow in a capillary is determined by the smoothed surface of the hydraulic shear zone $S_{por,hydr}$. PAPE *et al.* (1967a) thought that the pore radius would be a suitable yardstick, and described the relationship between both specific surfaces by

$$\frac{S_{por,BET}}{S_{por,hydr}} = \left(\frac{r_{eff}}{d_{N2}}\right)^{D-2}. \tag{13}$$

The fractal dimension *D* was calculated from equation (13) with the laboratory data (PAPE *et al.*, 1987a, 1987b) and the assumed value of 4.3 nm for d_{N2}. The effective pore radius results from the measured formation factor and gas permeability according to equation (6) and $S_{por,hydr} = 2/r_{eff}$. The mean value of *D* was 2.36 with the variance 0.018, minimum value 2 and maximum value 2.7. Then $(d_{N2} \times 10^{-6})^{D-2} = 0.141$ with d_{N2} in m. The values 0.141 and 2.36 are empirically derived and calibrated to the data set used. Subsequent studies on more data sets showed that the fractal dimension $D \sim 2.4$ is in good agreement with the structure and the petrophysical parameters of many consolidated sandstones, which underwent mechanical and chemical compaction during burial over geologic time (PAPE *et al.* 1999a, 2000). Additionally, we studied the sandstones of the original data set with Scanning Electron Microscope (SEM). The samples with $D < 2.3$ showed smooth pore wall. Either, the grains had their original round form from sedimentation, or they had planar crystal faces from quartz cementation. Fractal dimensions $D > 2.4$, calculated from BET measurements and permeabitity, could be explainded by the content of clay minerals, which may considerably enlarge $S_{por,BET}$. However, this additional surface enlargement is restricted to the small size range of the grain size distribution of clay minerals, whereas fractal substructures are distributed over a large range of length scales. This interpretation could be confirmed by SEM. Therefore, we prefer to introduce a lamellar factor q into equation (13) rather than to expect a larger fractal dimension for sandstones with clay content:

$$\frac{S_{\mathrm{por,BET}}}{S_{\mathrm{por,hydr}}} = \mathrm{q}\left(\frac{r_{\mathrm{eff}}}{\mathrm{d}_{\mathrm{N2}}}\right)^{\mathrm{D}-2}. \tag{14}$$

The dependence of surface area on resolution power becomes visible when $S_{\mathrm{por,mercury}}$ is calculated from the pore radius distribution from mercury porosimetry and plotted versus radius in a log-log plot. Then, for a distance, the slope of this curve is 2-D for radii smaller than the radius of the main peak of the radius distribution curve. (See Fig. 5 with explanations below.)

Inserting the values of d_{N2} and D, equations (13) and (10) yield:

$$r_{\mathrm{eff}} = 10^{-6}\left(\frac{2}{0.141 \times 10^{6} S_{\mathrm{por,BET}}}\right)^{\frac{1}{0.64}}, \tag{15}$$

r_{eff} in m, $S_{\mathrm{por,BET}}$ in m^{-1}.

From equations (15) and (6) follows:

$$k_{S\mathrm{por,BET}} = 497 \times 10^{-6}\left(\frac{\phi}{T}\right)\left(10^{6} S_{\mathrm{por,BET}}\right)^{-3.125}, \quad k_{S\mathrm{por,BET}}\,\mathrm{in}\,\mathrm{m}^2, \quad S_{\mathrm{por,BET}}\,\mathrm{in}\,\mathrm{m}^{-1}. \tag{16}$$

Tables 3 and 4 present petrophysical parameters which are derived from data of Table 2. Permeabilities of the Allermoehe core plugs were calculated with equation (16) and $S_{\mathrm{por,BET}}$. As tortuosity values, we used the quotient of formation factor F and pycnometer porosity ϕ_{pyc}, which is strictly speaking an electrical tortuosity T_{el}. Except for one outlier, the calculated permeabilities agree within one order of magnitude with the measured gas permeabilities.

Effective hydraulic pore radii were derived from equation (6) and the calculated permeabilties $k_{S\mathrm{por,BET}}$ and the measured permeabilities k_{gas} (Table 4). We find that r_{eff} values calculated from F and k_{gas} are generally smaller than the logarithmic mean values $r_{\mathrm{eff,LM}}$ from pore size distribution curves, which we regard as representative for the hydraulic radius. In order to get better r_{eff} values from equations (6) and (16), the porosity and the tortuosity term have to be corrected. First, the bulk porosity ϕ has to be reduced by the clay bound porosity ϕ_{clay}. Second, the electric tortuosity T_{el} has to be replaced by the hydraulic tortuosity T_{hydr}. Numerical simulations on two-dimensional networks by David (1993) showed that the so-called 'network tortuosity' for hydraulic flow was 1.5 times larger than the 'network tortuosity' for electrical current. The tortuosity with respect to processes of transport depends mainly on path elongation and on the variations of pore radius, which can be important in sandstone. Interpretation of PFG-NMR measurements on Allermoehe sandstones by Pape *et al.* (2006) showed that the radius of pore bodies, between 10 and 12 μm, exceeds the derived pore throat radius with value between 0.8 and 2.7 μm.

Tortuosity values for self-diffusion of water were determined in the range between 5 and 20. The influence of path elongation is similar for electric tortuosity and for hydraulic tortuosity. However, the impact of constrictions on tortuosity is larger for fluid flow, as was shown by SCHOPPER (1966).

Therefore, equations (6) and (16) have to be extended to:

$$k = \left(\frac{1}{8}\right)\frac{(\phi - \phi_{\text{clay}})}{T_{\text{hydr}}} r_{\text{eff}}^2 \tag{17}$$

and

$$k_{\text{Spor,BET}} = 497 \times 10^{-6}\left(\frac{\phi - \phi_{\text{clay}}}{T_{\text{hydr}}}\right)\left(\frac{10^6 S_{\text{por,BET}}}{q}\right)^{-3.125}, \quad k_{\text{Spor,BET}} \text{ in m}^2, S_{\text{por,BET}} \text{ in m}^{-1}, \tag{18}$$

where q is a factor of surface enlargement corresponding to the clay content, as discussed in detail by PAPE *et al.* (1967a, 1967b, 1999a).

The expression $(\phi - \phi_{\text{clay}})$ corresponds to the porosity which can be calculated from the pore radius distribution from mercury intrusion as shown below. For further calculations we estimated T_{hydr} by ϕ^{-1}, which corresponds to Archie's first equation, but with a higher exponent than one gets from the measured electrical formation factor and porosity (see Table 3). For low porosities, T_{hydr} was restricted to a maximum value of 20 according to the results of PFG-measurements (PAPE *et al.*, 2006).

The corrections in equation (18) balance in some way because $(\phi - \phi_{\text{clay}})/T_{\text{hydr}} < \phi/T_{el}$ and $(S_{\text{por,BET}}/q)^{-3.125} > S_{\text{por,BET}}^{-3.125}$, which explains the good fit to the permeability values previously obtained from equation (16).

For comparison, effective pore radii $r_{\text{eff,RGPZ}}$ were calculated with equation (7) from the logarithmic mean grain radius $r_{\text{grain\_LM}} = d_{\text{eff}}/2$ and from the cementation exponent m. Additionally, effective pore radii $r_{\text{eff,fractal}}$ were obtained from $r_{\text{grain\_LM}}$ and fractal dimension $D = 2.36$ with equation (8). Then, permeabilities k_{RGPZ} and k_{fractal} were calculated with equation (6) (see Table 4). The permeability values estimated from the effective grain radius on the basis of the RGPZ model and of the fractal model are very close to the directly measured gas permeabilities. This results shows that the two models, which are based on very different premises, do not contradict each other. Additionally, we got a justification for assuming a fractal dimension of $D = 2.36$ for the samples of our study. We found that permeability prediction from grain sizes gave good results. However, it is more difficult to compare $T_{1,2}$ distributions with grain radius distributions than to compare them with pore radius distributions. Therefore, we studied pore throat distributions from mercury intrusions as an intermediate step.

3.2. *Interpretation of Mercury Injection Curves*

To develop a method for permeability prediction which is suitable for different formations, information on the effective pore radius derived, e.g. from mercury porosimetry should be used. The following discussion of the radius distribution obtained from mercury injection compared to relaxation time distributions is necessary to thoroughly understand the structural and physical conditions. This will help us later to interpret T_1 and T_2 distributions for estimating permeability.

The curves of mercury injection porosimetry were used firstly to obtain an estimate of the effective hydraulic pore radius and secondly to determine the factor $3\rho_{1,2}$ which relates pore radius r to $T_{1,2}$ of the NMR relaxation curves according to equations (11) and (12):

$$T_{1,2} = \frac{1}{3\rho_{1,2}}\left(\frac{3}{S_{\mathrm{por,NMR}}}\right) = \frac{1}{3\rho_{1,2}} r. \tag{19}$$

Figure 2 presents the mercury injection curve of plug AC9 as a typical example, where the relative volumes ΔV_i are plotted versus radius r. Additionally, we find four curves for different kinds of S_{por} functions, which depend on the radius. In the following, the arguments are set in square parentheses.

1) $S_{\mathrm{por,capillary}}[r] = 2/r$. This is the specific surface divided by the volume for each individual pore class.
2)

$$S_{\mathrm{por,frac}}[r] = 0.141 S_{\mathrm{por,BET}}\left(10^6 r_{\mathrm{eff}}\right)^{-0.36}, \tag{20}$$

r_{eff} in m

which represents the fractal specific surface, derived from $S_{\mathrm{por,BET}}$ and adjusted to each radius r of the distribution as resolution yardstick. Equation (20) corresponds to equation (13)

3)

$$S_{\mathrm{por,mercury}}[r_j] = \left(\frac{S_{\mathrm{por,frac}}[r_n]}{\sum_{i=1}^{n}\left(\frac{2}{r_i}\right)\Delta V_i \Big/ \sum_{i=1}^{n}\Delta V_i}\right)\left(\frac{\sum_{i=1}^{j}\left(\frac{2}{r_i}\right)\Delta V_i}{\sum_{i=1}^{j}\Delta V_i}\right), \tag{21}$$

This function describes the specific surface, calculated from the radius distribution from mercury intrusion. The resolution power increases with decreasing radius. The nominator of the second term contains the sum of the surfaces of all pore classes from the largest pore with $i = 1$ to the pore class with index $j \leq$ n. The sum in the denominator is the total volume of the same pore classes. Due to the first term, $S_{\mathrm{por,frac}}[r_n]$ is equal to $S_{\mathrm{por,mercury}}[r_n]$, where r_n is the smallest radius of the distribution.

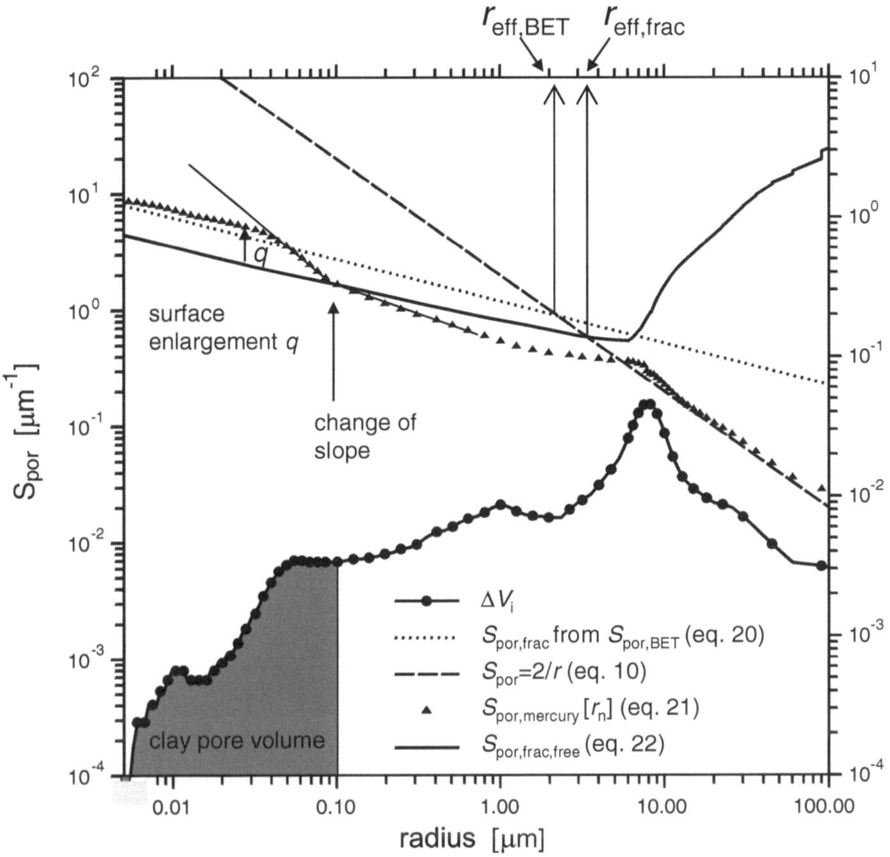

Figure 2
Radius distribution curve of sample AC9, which serves to demonstrate a method to deduce several parameters of pore structure based on the theory of fractal dimensions.

4)

$$S_{\text{por,frac,free}}\left[r_{\text{j}}\right] = S_{\text{por,mercury}}\left[r_{\text{border}}\right]\left(\frac{r_{\text{j}}}{r_{\text{border}}}\right)^{-0.36}\left(\frac{V_{\text{rel}}}{\sum_{i=1}^{j}\Delta V_{\text{i}}}\right), \tag{22}$$

$$V_{rel} = \sum_{i=1}^{n} \Delta V_i = 1.$$

The radius r_{border} marks the border between the clay volume part and the volume part of the distribution which is relevant for fluid flow. We derive r_{border} from the shape of $S_{\text{por,mercury}}$ and take the radius, where the slope changes from steep to less

Table 3

Petrophysical parameters of Allermoehe samples: Tortuosity T_{el} and cementation exponent m from electrical measurements and hydraulic tortuosity T_{hydr} estimated from porosity

Sample	ϕ_{pyc} [%]	T_{el}	m	$T_{hydr} = \phi^{-1} < 20$
AC1	2	(1.06))	(1.01))	20.0
AC3	2	(1.08)	(1.02)	20.0
AC4	9	5.04	1.67	11.2
AC6	6	4.32	1.52	16.7
AC7	8	5.36	1.66	12.5
AC9	8	5.60	1.68	12.5
AC10	11	3.80	1.65	9.1
AC12	9	6.12	1.75	11.1
AC13	9	5.22	1.69	11.1
AC14	9	3.96	1.57	11.1
AC15	9	4.50	1.62	11.1
AC16	8	4.40	1.59	12.5
AC18	8	4.48	1.59	12.5
AC19	10	4.70	1.67	10.0
AC20	8	4.80	1.62	12.5
AC21	7	5.53	1.64	14.3
AC22	7	6.23	1.69	14.3
AC23	8	7.36	1.79	12.5
AC24	3	(1.95)	(1.19)	20.0
AC25	6	3.96	1.49	16.7

steep in the direction of increasing radius (see Fig. 5). We find r_{border} equal to 1 μm. This value was later used for all samples, and is consistent with the cutoff $T_2 < 3$ ms used in the interpretation of CMR data. The parameter $S_{por,frac,free}$ is based on the value of $S_{por,mercury}$ [r_{border}]. The function $S_{por,frac,free}$ describes a fractal specific surface which has nearly the same fractal dimension as the specific surface described by $S_{por,frac}$ due to the term $(r_j/r_{border})^{-0.36}$. In the clay region of the radius distribution, we find $S_{por,frac,free} = S_{por,frac}/q$, where q is the factor of surface enlargement due to the clay content from equation (18). The last term in equation (22) causes a smaller slope of the curve in the log-log plot (Fig. 5) compared to the slope of equation (20). $S_{por,frac,free}$ is an empirical function which shall describe the free water part of the pore distribution.

The construction of four different S_{por} curves allows us to obtain three parameters which are needed in the calculation of permeability from equations (17) or (18). We just mentioned the surface enlargement factor q. Additionally, we get free porosity $(\phi - \phi_{clay})$ from the volume $V_{free} = \sum_{i=1}^{ib} \Delta V_i$ with the index ib of the class with r_{border}. For this, V_{free} is firstly transformed into the respective volume of mercury, which intruded into the sample, and secondly related to the volume of the sample. The third parameter of interest is an estimate of r_{eff}. In Figure 2 we find $r_{eff,BET}$ at the radius,

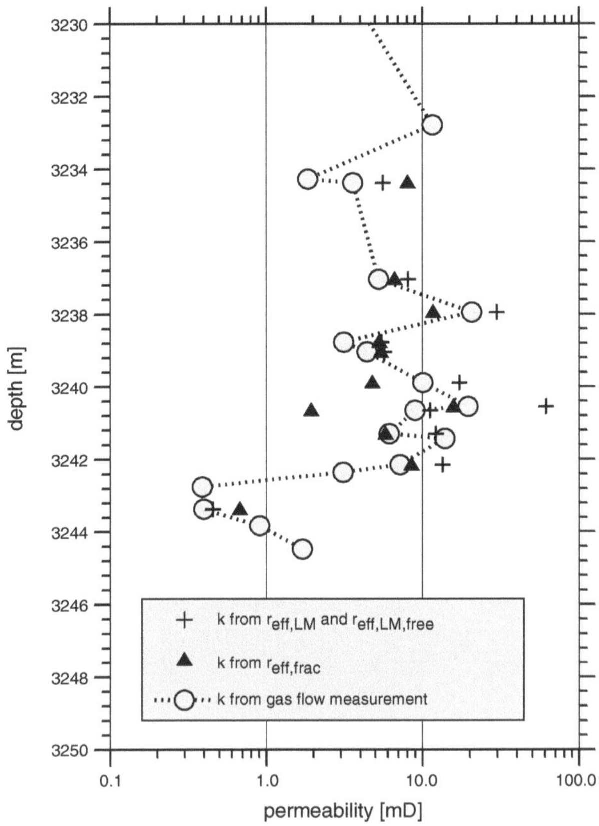

Figure 3
Variation of Permeability of Allermoehe sandstone samples with depth calculated from pore radius distributions according to equation (17) for different definitions of pore radius (crosses and triangles see legend) and compared to gas permeability (circles).

where the line of $S_{por,frac}$ crosses the line of $S_{por,capillary}$. This hydraulic radius is solely based on $S_{por,BET}$; therefore, it is exactly the same as r_{eff} from equation (15). In an analogous way we take the radius $r_{eff,frac}$ at the crossing of the curve of $S_{por,frac,free}$ and the line of $S_{por,capillary}$. This radius is based on $S_{por,BET}$ and on the radius distribution as well and gives a good estimate for r_{eff}, which was proven by comparison of measured gas permeability with calculated permeability from equation (10). The new parameters, derived from radius distributions of several Allermoehe samples, and calculated permeability values are presented in Table 5.

This method for obtaining r_{eff} values from individual pore radius distribution curves is time-consuming. However, we used it in order to compare $r_{eff,frac}$ with the logarithmic mean values $r_{eff,LM}$ (Table 5), which are defined by

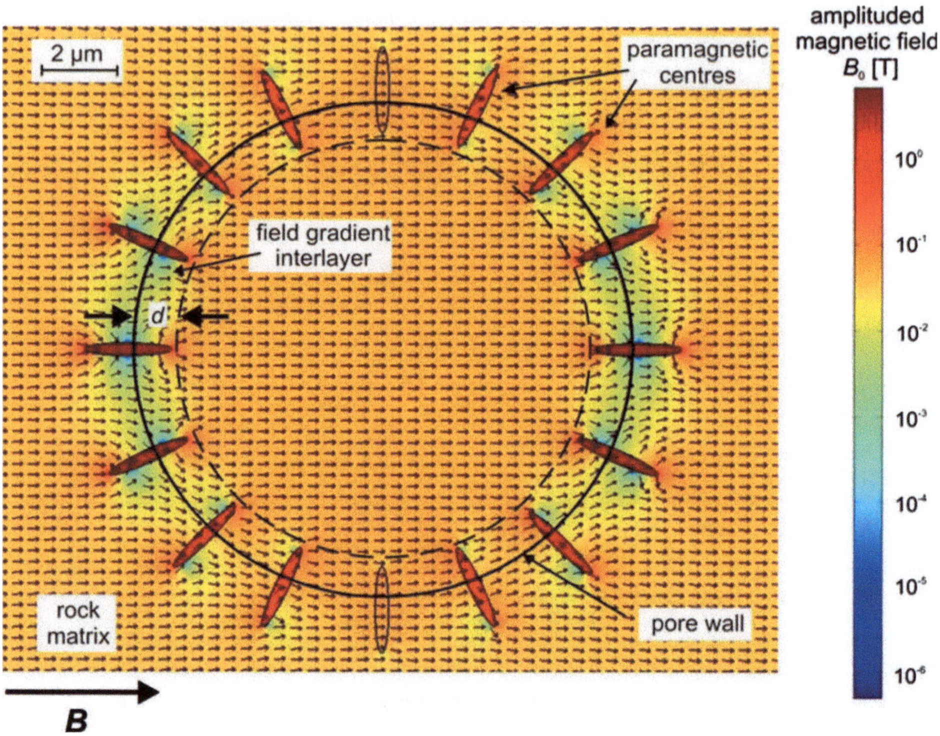

Figure 4
Schematic model of a pore in a constant, homogeneous static magnetic field of amplitude $B = 0.2$ T. Strong local magnetic field gradients caused by paramagnetic centers at the pore wall affect a layer of thickness d.

$$r_{\mathrm{eff,LM}} = \exp_{10}\left[\frac{\sum_{\mathrm{i}} \Delta V_{\mathrm{i}} \log(r_{\mathrm{i}})}{V_{\mathrm{rel}}}\right] \tag{23}$$

with the integral volume V_{rel} equal to one.

In general, the calculated $r_{\mathrm{eff,LM}}$ values were close to the $r_{\mathrm{eff,frac}}$. For sample AC23 with a high clay content, the effective pore radius $r_{\mathrm{eff,LM,free}}$ was calculated from the upper part of the pore radius distribution with radius >0.05 µm. Permeabilties were calculated from equation (17) using $r_{\mathrm{eff,frac}}$ and $r_{\mathrm{eff,LM}}$ values as effective pore radii. In the case of sample AC23 the term $r_{\mathrm{eff,LM}}$ is replaced by $r_{\mathrm{eff,LM,free}}$. Table 5 presents correction factors c_ϕ with $c_\phi \phi = (\phi - \phi_{\mathrm{clay}})$, which are used to calculate permeabilities from $r_{\mathrm{eff,frac}}$. For the permeability prediction from $r_{\mathrm{eff,LM}}$ and $r_{\mathrm{eff,free,LM}}$ we used a medium value $c_\phi = 0.7$. In all cases tortuosity was set to $T_{\mathrm{hydr}} = \phi^{-1}$. The corresponding results are presented in Table 5. Figure 3 shows the calculated permeabilities from

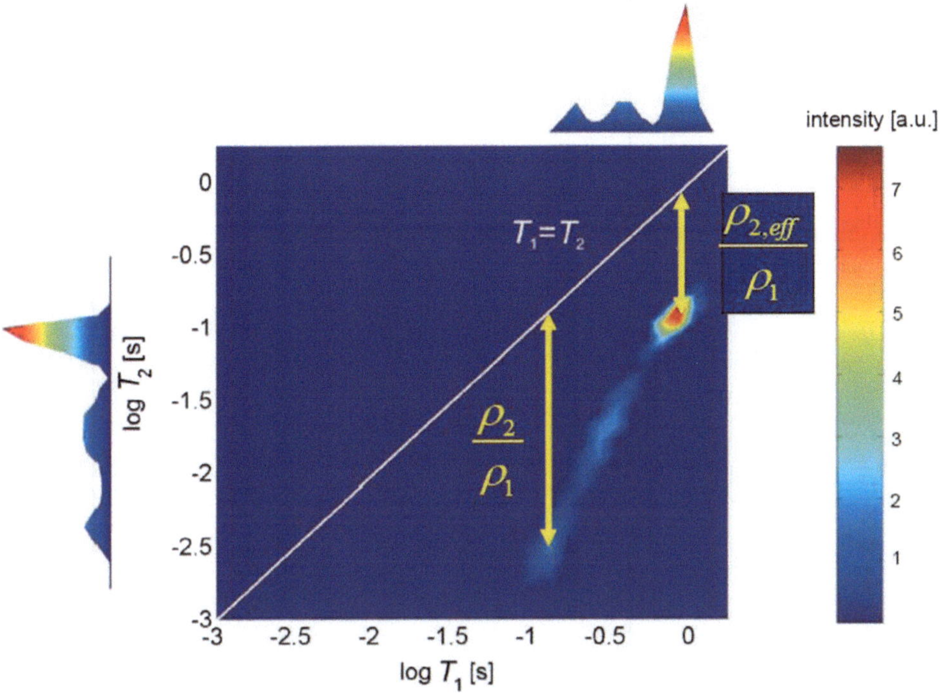

Figure 5
Two-dimensional T_1-T_2 map of measurements recorded on Allermoehe core plug AC15. Due to high internal gradients, the T_1/T_2 ratio is not constant.

$r_{eff,frac}$, $r_{eff,LM}$, and $r_{eff,LM,free}$ plotted versus depth in comparison to the measured permeabilties k_{gas}. The fit of permeabilities calculated in both ways is very good.

4. *A New Model for the Dependence of Surface Relaxation Time ρ_2 on Pore Radius*

4.1. Surface Relaxation and "Dephasing"'

As equation (19) indicates a linear relationship between NMR-relaxation times and pore radii, the $T_{1,2}$ distribution curves can be processed similarly as the mercury injection curves by calculating the logarithmic mean values $T_{1,2,LM}$. In the case of longitudinal relaxation the linearity is sufficiently fulfilled, i.e., relaxivity ρ_1 is constant with respect to pore radius although ρ_1 may vary for different samples. The physical process related to ρ_1 is called surface relaxation. It acts directly at the mineral surface of the pore walls where the magnetization of the hydrogen atoms is lost. For the same sample, the T_2 distribution curves are shifted towards smaller values. This indicates a second effective relaxation process. It is caused by dephasing of the transverse magnetization vectors due to diffusion

Table 4

Petrophysical parameters of Allermoehe samples: Permeabilities k_{gas} from gas flow measurements, $k_{Spor,BET}$ derived from T_{el} and $S_{por,BET}$, k_{RGPZ} derived from T_{el} and $r_{eff,RGPZ}$, and $k_{fractal}$ derived from T_{el} and $r_{eff,fractal}$; effective pore radii r_{eff} calculated from k_{gas}, r_{eff} from $k_{Spor,BET}$, $r_{eff,LM}$ from mercury injection curves, $r_{eff,RGPZ}$ from $r_{grain,LM}$ and m, and $r_{eff,fractal}$ from $r_{grain,LM}$ and fractal dimension $D = 2.36$

Sample	k_{gas} [mD]	$k_{Spor,BET}$ [mD] eq.(16)	k_{RGPZ} [mD] from $r_{eff,RGPZ}$ eq. (6)	$k_{fractal}$ [mD] from $r_{eff,fractal}$ eq. (6)	r_{eff} [µm] from k_{gas} eq. (6)	r_{eff} [µm] from $k_{Spor,BET}$ eq. (6)	$r_{eff,LM}$ [µm] from mercury intrusion	$r_{eff,RGPZ}$ [µm] from $r_{grain,LM}$ eq. (7)	$r_{eff,fractal}$ [µm] from $r_{grain,LM}$ eq. (8)
AC1	0.16	0.09	–	–	0.26	0.19	–	–	0.30
AC3	0.42	0.09	–	–	0.42	0.20	–	–	0.29
AC4	11.6	598.38	7.93	8.98	2.26	16.27	–	1.88	1.99
AC6	1.85	4.38	6.35	4.85	1.02	1.58	–	1.91	1.57
AC7	3.59	6.84	6.60	6.62	1.37	1.90	3.14	1.88	1.91
AC9	5.26	9.13	4.71	6.99	1.70	2.25	3.78	1.62	1.93
AC10	20.7	82.35	26.17	23.82	2.37	4.74	5.27	2.82	2.58
AC12	3.13	4.25	3.23	6.44	1.29	1.51	2.76	1.33	1.84
AC13	4.42	8.42	5.40	8.87	1.42	1.96	2.82	1.58	1.94
AC14	10.1	51.66	17.93	12.04	1.87	4.24	4.90	2.51	2.03
AC15	19.6	102.53	13.64	14.28	2.77	6.36	9.23	2.34	2.31
AC16	8.99	25.15	9.35	5.75	1.97	3.31	4.45	2.03	1.61
AC18	6.16	5.16	10.11	6.50	1.64	1.51	4.63	2.13	1.73
AC19	13.9	30.87	14.53	13.86	2.26	3.38	–	2.34	2.31
AC20	7.2	49.45	8.84	7.08	1.84	4.84	4.87	2.06	1.86
AC21	3.1	16.63	4.03	4.86	1.39	3.22	–	1.60	1.72
AC22	0.39	5.25	2.51	3.66	0.52	1.92	–	1.34	1.60
AC23	0.4	7.16	1.92	4.01	0.54	2.28	0.29	1.19	1.75
AC24	0.91	0.08	–	–	0.68	0.21	–	–	0.62
AC25	1.71	2.83	7.53	3.06	0.94	1.22	–	1.99	1.26

of water molecules within an internal, inhomogeneous magnetic field. Large field gradients are caused by paramagnetic minerals at the pore walls. They are effective within a certain distance from the minerals. For simplification, this will be represented by a thickness d of this field gradient interlayer (Fig. 4).

In this situation, transverse relaxation can be described by:

$$\frac{1}{T_2} = \frac{1}{T_1} + \frac{1}{T_{\text{dephase}}}, \tag{24}$$

where T_{dephase} is the relaxation time due to dephasing with corresponding "dephasing relaxation" ρ_{dephase} as shown in Appendix A:

$$T_{\text{dephase}} = \frac{r - d}{3\rho_{\text{dephase}}}. \tag{25}$$

Combining equations (24), (25), (10) and (11) we obtain:

$$\rho_2 = \rho_1 + \frac{\rho_{\text{dephase}}}{\left(1 - \frac{d}{r}\right)}, \tag{26}$$

$$\frac{T_1}{T_2} = 1 + \frac{\rho_{\text{dephase}}}{\rho_1\left(1 - \frac{d}{r}\right)}. \tag{27}$$

Equations (26) and (27) show that ρ_2 and T_1/T_2 decrease with pore radius r. Surface relaxivity and relaxivity of the field gradient interlayer can be separated in an analysis of a two-dimensional T_1-T_2 (2D) correlation experiment. But this is possible only if the data were measured in presence of a very homogeneous external field, as was produced by the Halbach magnet.

4.2. Calibration of the Model by aid of 2D T_1-T_2 Correlation Experiments

We performed 2D T_1-T_2 relaxation correlation experiments with the NMR core scanner in order to study the influence of diffusion on the shape of the T_2 distribution function for low porosity rocks, such as those from the Rhaetian sandstone at Allermoehe. Initial results are published in ANFEROVA *et al.* (2007). The T_1/T_2 ratio, determined from 2D T_1-T_2 correlation experiments performed on the Allermoehe samples is not constant even for short echo spacing. It decreases with pore size from values of 10 to 4. Figure 5 shows the 2D T_1-T_2 map based on measurements on the Allermoehe core plug AC15.

The T_1-T_2 correlation experiments were used to determine the relationship between ρ_1 and ρ_2. Analogous to the mercury injection curves, the differential magnetizations Δm of each $T_{1,2}$ fraction of the distribution curves are plotted as relative volumes, with the integral volume m_{rel} equal to one. First, all relaxation curves were correlated with the radius distribution curves from mercury injection by shifting the relative volume distribution relative to the pore size distribution until the maxima of both curves coincide for large radii and large $T_{1,2}$ values. Then, ρ_1 and $\rho_{2,\mathrm{eff}}$ are determined from equation (19) for each sample (Table 6). Compared to surface relaxivity values from literature, our data seem to be very low. This can be explained by the dependence of surface area (in eq. 19) on the resolution power of measurement. We calibrated the relaxation times to the radius of pore throats from mercury porosimetry with the mean value $r_{\mathrm{pore\ throat}} = 4$ µm. The ρ_1 values of HOWART *et al.*, (1990), which cluster at $\rho_1 = 50$ µm/s, correspond to the radii of pores, which were optically measured on thin-sections. HÜRLIMANN *et al.*, (1994) calibrated relaxation time T_1 to the radius of pores which was measured with the PFG-NMR method and obtained a mean value $\rho_1 = 21$ µm/s for 10 sandstones. For Allermoehe sandstone, PFG-NMR measurements (PAPE *et al.*, 2006) yielded a mean radius of pore bodies $r_{\mathrm{pore}} = 11$ µm. Our surface relaxivity values have to be multiplied with the factor $r_{\mathrm{pore}}/r_{\mathrm{pore\ throat}} = 2.75$ for comparison with the values of HÜRLIMANN *et al.* Then, our mean value $\rho_1 = 4.3$ µm/s yields a comparable value of 12 µm/s.

Core plug AC9 is considered representative for all Allermoehe samples studied in this work due to its petrophysical properties (porosity, range of pore sizes, and permeability). We used it to demonstrate the pore radius dependence of ρ_2 and calculate the thickness of the interlayer d.

For large $T_{1,2}$ values, the first maxima of the relative volume distribution occur at $T_1 = 749.9$ ms and $T_2 = 133.4$ ms. The second maxima for the next largest $T_{1,2}$ values occur at $T_1 = 177.8$ ms and $T_2 = 17.78$ ms. Correlating the T_1 curve with the radius distribution curve at the first maximum yields a conversion factor of $c_{r},T_1 = 1/3\rho_1 = 78.64$ ms µm^{-1} according to equation (19). Consequently, the first maximum corresponds to $r_1 = 9.53$ µm and the second maximum corresponds to $r_2 = 2.26$ µm. Inserting the appropriate T_1 and T_2 values corresponding to both maxima in equation (27) and dividing the two equations yields:

$$\frac{\left(\frac{T_1(r_1)}{T_2(r_1)} - 1\right)}{\left(\frac{T_1(r_2)}{T_2(r_2)} - 1\right)} = \mathrm{h} = \frac{\left(1 - \frac{d}{r_2}\right)}{\left(1 - \frac{d}{r_1}\right)}. \tag{28}$$

From this one obtains the thickness of the interlayer:

$$d = r_2 \frac{(1 - \mathrm{h})}{1 - \mathrm{h}\left(\frac{r_2}{r_1}\right)}. \tag{29}$$

Inserting the appropriate values of the parameters on the right-hand side of equation (29) yields $d = 1.25$ µm. The value seems to be rather large compared with the radius of pore throats. However, the border between the interlayer and central part of the pore is not sharp but marks a continuous increase of positive and negative gradients of magnetic field towards the pore wall. A large portion of interlayer volume within the pore throats hinders the effects of diffusion coupling between neighbored pores of different size (Pape and Clauser, this issue). The surface relaxivities of AC9 obtained from the comparison of the $T_{1,2}$ distribution curves with the mercury injection curve are $\rho_1 = 4.24$ µm/s and $\rho_{2,\text{eff}}\,[r_1] = 23.82$ µm/s. Thus, equation (26) yields the "dephasing relaxation" $\rho_{\text{dephase}} = (\rho_{2,\text{eff}}\,[r_1] - \rho_1)\,(1 - d/r_1) = 17.01$ µm/s. Starting with an assumed value for ρ_1, the value of ρ_{dephase} can be directly obtained from the two dimensional $T_1 - T_2$ map (see Fig. 8) as $\rho_{\text{ dephase}} = \rho_1\,(T_1/T_2) - \rho_1$.

The relationship between pore radius r, surface relaxivity ρ_2, and transverse relaxation time T_2 is described by equations (19) and (26). We constructed a table with ρ_2 and T_2 values calculated for a column of r values. Because the function ρ_2 according to equation (26) has a pole for $r = d$, the values for radii near the pole make no physical sense. Therefore, we truncated the function at a pore radius $r = 3$ µm which is about three times larger than d and corresponds to $T_2 = 40$ ms. To estimate correct ρ_2 values for T_2 times shorter than 40 ms, the ρ_2 curve which has been calculated for large pore radii was extrapolated by a straight line in the plot of log ρ_2 versus log T_2 with the slope at $r = 3$ µm.

Data measured on Allermoehe sample AC9 yielded a slope of -0.3 which yields:

$$\rho_2[r_i] = \rho_{2,\text{eff}}[r_1]\left(\frac{T_2[r_\text{i}]}{T_2[r_1]}\right)^{-0.3} \quad \text{for} \quad r_i < r_1, \tag{30}$$

with $r_1 = 9.54$ µm.

Figure 6 presents the dependences of surface relaxivity ρ_2 on pore radius and on T_2. For sample AC9, ρ_2 values calculated from empirical equation (30) are compared to the curves calculated with equations (26) and (19). Additionally, the fit to data of the two dimensional T_1-T_2 measurement is presented.

On the basis of equation (30) it is possible to obtain corrected pore radii from radii $r_{\rho 2,\text{eff}}$ which are calculated with the constant $\rho_{2,\text{eff}}$ and equation (19). In order to calculate permeability later, we applied this correction to the logarithmic mean values $r_{\rho 2,\text{eff,LM}}$ derived from the logarithmic mean $T_{2,\text{LM}}$ of the T_2 distribution. We obtained the correction term by multiplying equation (30) with $3T_2[r_i]$ and dividing $T_2[r_i]$ and $T_2[r_1]$ in equation (30) by $3\rho_{2,\text{eff}}$:

$$r_{\text{corr}}\left[T_{2,\text{LM}}\right] = r_{\rho 2,\mathit{eff}}\left[T_{2,\text{LM}}\right]\left(\frac{r_1}{r_{\rho 2,\mathit{eff}}\left[T_{2,\text{LM}}\right]}\right)^{0.3} \quad \text{for} \quad r_{\rho 2,\mathit{eff}} < r_1, \tag{31}$$

with $r_1 = 9.54$ µm.

Using the exponent of 0.3 yields a corrected pore radius $r_{\text{corr}}\,[T_{2,\text{LM}}]$, which is similar to the pore radius $r_{\rho 1}\,[T_{1,\text{LM}}]$. The pore radius $r_{\rho 1}\,[T_{1,\text{LM}}]$ is obtained with equation (19) and ρ_1 from the T_1 distribution curve which is measured simultaneously on sample AC9.

In Figure 7a, the shape of the T_1 and T_2 distribution curves of sample AC9 are compared. The relaxation curves are correlated by shifting the relative volume distributions until the maxima of both curves for large $T_{1,2}$ values coincide. This is achieved by multiplying the T_2 values by the term $(\rho_{2,\text{eff}}/\rho_1)$. In the next step the T_2

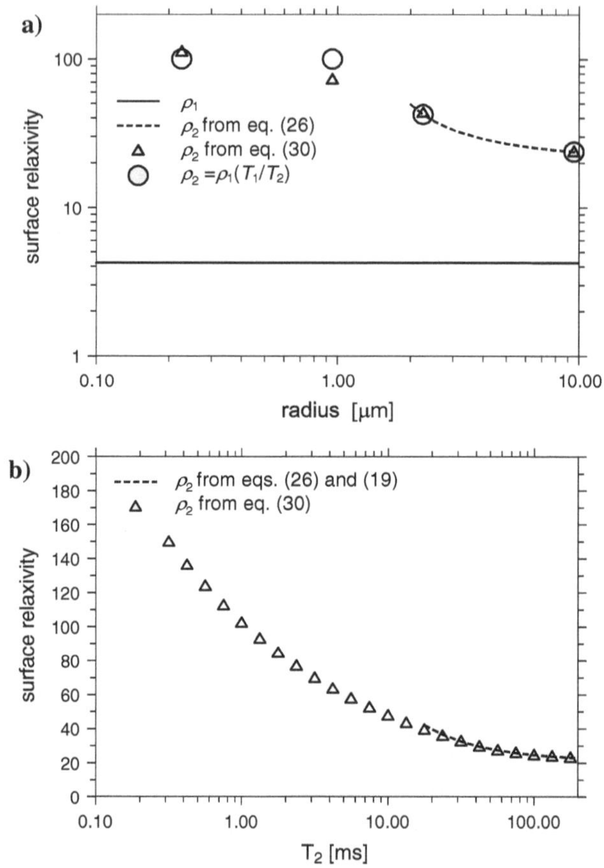

Figure 6
Dependence of surface relaxivity ρ_2 on pore radius demonstrated for sample AC9. a) The surface relaxivity ρ_2, calculated with equation (26) and the thickness d of the interlayer with large magnetic field gradients is plotted for radius >2 μm. The other surface relaxivities (spheres from the empirical equation (30) and triangles for $\rho_2 = \rho_1(T_2/T_1)$ represent the values at the maxima of the T_1 and T_2 distributions of the two-dimensional measurement (see Fig. 8). They are plotted versus the maxima of the pore radius distribution from mercury intrusion. b) The surface relaxivities ρ_2 (triangles) calculated for T_2 with the empirical equation (30) are compared with values (dashed line) calculated with the aid of a table based on eqs. (26) and (19).

distribution curve is renormalized in two different ways. First, it is corrected using the thickness of the field gradient interlayer. Accordingly, each measured T_2 value of the T_2 distribution curve is multiplied by the term (ρ_2/ρ_1), where the individual ρ_2 values are determined from equations (19) and (26). Second, the T_2 distribution curve is renormalized empirically. Therefore, T_2 values are multiplied by the correction terms $(r_1/r_{\rho 2,\mathrm{eff}})^{0.3}$ and $(\rho_{\ 2,\mathrm{eff}}/\rho_1)$. Both approaches yield a nearly perfect match with the T_1 distribution (Fig. 7b).

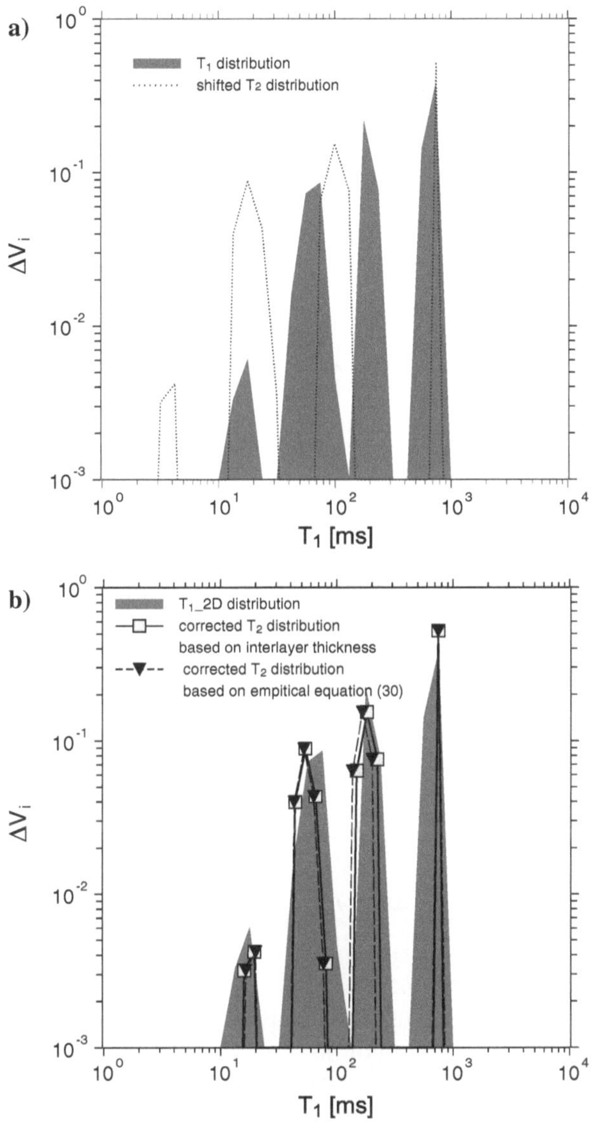

Figure 7

Relationships between T_1 and T_2 distribution curves of the T_1-T_2 correlation experiment, explained for sample AC9. (a) The T_2 distribution curve is shifted so that the first maxima for large $T_{1,2}$ values of both distribution curves coincide. (b) Two correction procedures applied for the T_2 distribution curve are demonstrated which compensate the effect of enhanced relaxation due to magnetic field gradients in the layer at the pore walls.

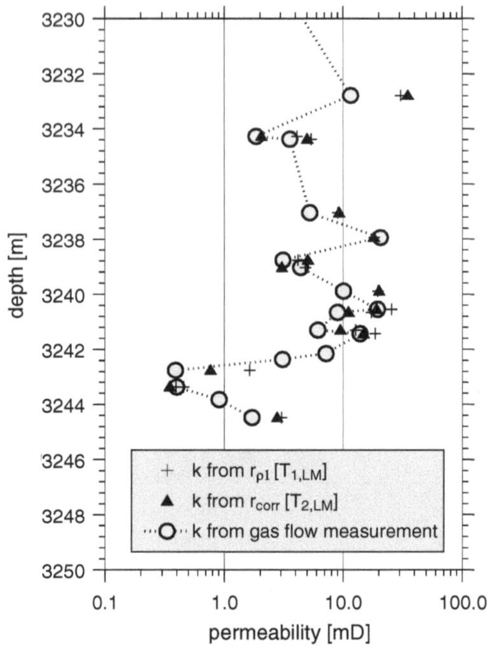

Figure 8
Permeabilities calculated from 2D T_1 and T_2 distributions according to equations (32) and (33) (Table 4) plotted versus depth and compared to gas permeabilities for several Allermoehe samples.

5. Permeability Prediction by use of the New Model

5.1. Permeability from 2D T_2 Relaxation

For all samples for which a pore radius distribution curve was available, values of $r_{\rho 1}$ $[T_{1,\mathrm{LM}}]$ and $r_{\rho 2,\mathrm{eff}}$ $[T_{2,\mathrm{LM}}]$ were calculated from the T_1 and T_2 distribution curves. Mean values of ρ_1 and $\rho_{2,\mathrm{eff}}$ (Table 6) of these samples were used to determine values of $r_{\rho 1}$ $[T_{1,\mathrm{LM}}]$ and $r_{\rho 2,\mathrm{eff}}$ $[T_{2,\mathrm{LM}}]$ for the rest of the samples. From the $r_{\rho 2,\mathrm{eff}}$ $[T_{2,\mathrm{LM}}]$ values also r_{corr} $[T_{2,\mathrm{LM}}]$ values were calculated according to equation (31). The permeability was then predicted using the following equations:

$$k = \left(\frac{1}{8}\right)\frac{c_\phi \phi}{T_{\mathrm{hydr}}} r_{\rho 1}^2 \left[T_{1,LM}\right] \tag{32}$$

and

$$k = \left(\frac{1}{8}\right)\frac{c_\phi \phi}{T_{\mathrm{hydr}}} r_{corr}^2 \left[T_{2,LM}\right] \tag{33}$$

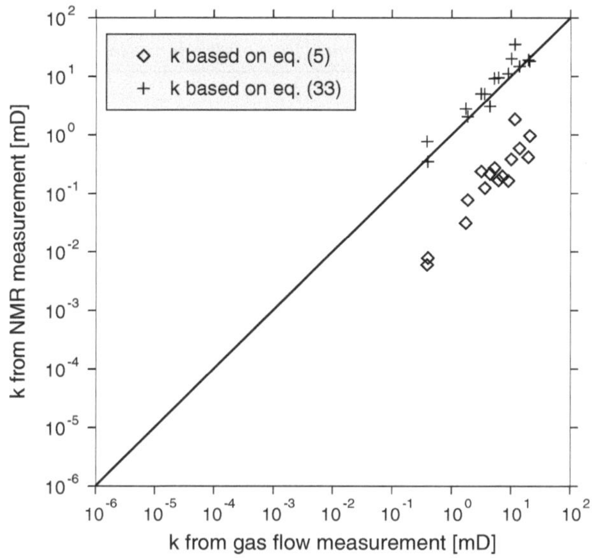

Figure 9
Correlations of Allermoehe core plug permeabilities: results from gas flow measurements are compared to calculated permeabilities based on 2D T_2 measurements: 1) Permeability calculated from $T_{2,\mathrm{LM}}$ according to the standard method in the oil industry with equation (5) using a factor $a = 4$. 2) Calculated from $r_{\mathrm{corr}\ (T_{2,\mathrm{LM}})}$ according to equation (33).

with tortuosity $T_{\mathrm{hydr}} = \phi^{-1}$, and $T_{\mathrm{hydr}} = 20$ for $\phi < 0.05$ (cf. Table 3). The factor $c_{\phi} = 0.7$ was chosen because of the high clay-content of the Allermoehe samples. Table 6 shows the resulting pore radii and permeabilities.

In Figure 8, permeabilities calculated from 2D T_1 and T_2 distributions according to equations (32) and (33), are plotted versus depth and compared to gas permeabilities. Additionally, Figure 9 presents calculated permeabilities from 2D T_2 distributions versus measured gas permeabilities: (1) permeability was calculated according to the standard method in the oil industry, based on equation (5) with a factor of $a = 4$ generally accepted for sandstones, and (2) permeability was calculated from $r_{\mathrm{corr}\ (T_{2,\mathrm{LM}})}$ according to equation (33). Whereas permeabilities derived from equations (32) and (33) agree well with gas flow measurements, results from the standard oil industry method seem to underestimate permeability systematically for the low porosity Rhaetian sandstones from Allermoehe.

5.2. *Permeability from 1D T_2 Relaxation*

Permeabilities of the Allermoehe core plugs were also derived from 1D transverse relaxation measurements performed with the NMR core scanner. For the samples with available pore size distribution, individual relaxivities $\rho_{2,\mathrm{eff}}$ were determined by equating the maxima of both frequency distributions. A mean value of $\rho_{2,\mathrm{eff}}$ was

Table 5

Results based on the mercury intrusion measurements

Sample	$r_{eff,frac}$ [μm]	q	c_ϕ	k [mD] from $r_{eff,frac}$ eq. (17)	$r_{eff,LM}$ [μm]	$r_{eff,LM,free}$ [μm]	k [mD] from $r_{eff,LM}$ eq. (17)	k [mD] from $r_{eff,LM,free}$ eq. (17)
AC7	3.62	1.15	0.77	8.05	3.14		5.59	
AC9	3.18	1.79	0.82	6.65	3.78		8.12	
AC10	3.32	1.02	0.87	11.72	5.27		29.82	
AC12	2.58	2.07	0.79	5.30	2.76		5.48	
AC13	2.60	1.72	0.80	5.43	2.82		5.72	
AC14	2.38	1.00	0.84	4.79	4.90		17.22	
AC15	4.21	1.00	0.88	15.79	9.23		61.19	
AC16	1.76	1.00	0.78	1.94	4.45		11.24	
AC18	2.98	1.00	0.82	5.82	4.63		12.15	
AC20	3.59	1.00	0.82	8.55	4.87		13.45	
AC23	1.61	3.51	0.33	0.68	0.29	0.9		0.46

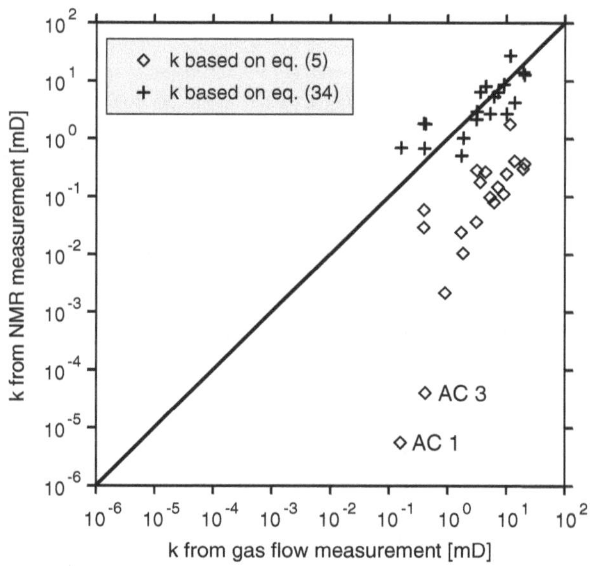

Figure 10

Correlations of Allermoehe core plug permeabilities: results from gas flow measurements are compared to calculated permeabilities based on 1D T_2 distributions measured with an NMR core scanner: 1) Permeability calculated from $T_{2,LM}$ according to the standard method in the oil industry with equation (5) using a factor $a = 4$. 2) Permeability calculated with equation (34).

calculated and used for the other samples. Pore radii $r_{\rho 2,eff}$ ($T_{2,LM}$) were calculated from logarithmic mean relaxation times $T_{2,LM}$ according to equation (19). In the case of samples AC1 and AC3 with very small pore radii, $T_{2,LM}$ values were calculated using a lower T_2 cutoff of $T_2 = 0.9$ ms. The pore radii were corrected

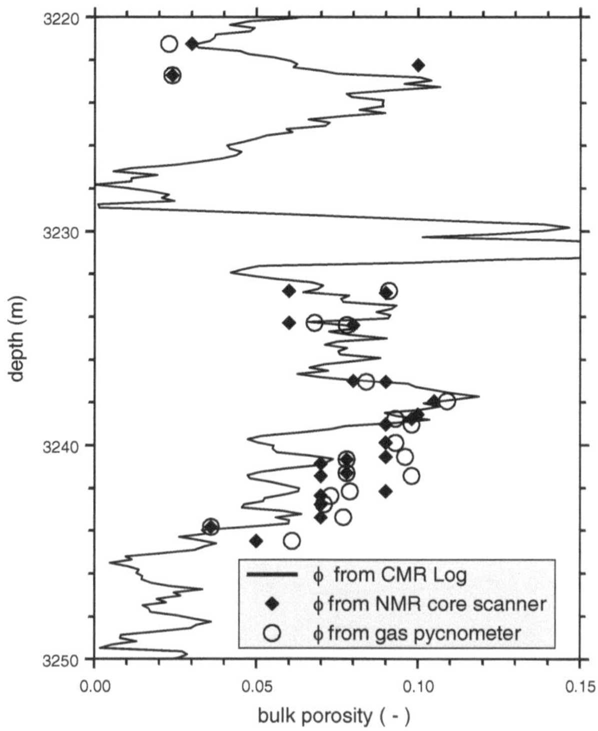

Figure 11

Comparison of NMR logging data recorded in the Allermoehe borehole and mobile NMR core scanner results: The Schlumberger CMR porosity log (full line) correlates well with core plug porosities measured with the gas pycnometer (circles) and with the NMR core scanner (diamonds).

according to equation (31) by multiplication with the term $(r_1/r_{\rho 2,\mathrm{eff}})^{0.3}$ resulting in r_{corr} $(T_{2,\mathrm{LM}})$.

Permeability was calculated from equation (33) with $c_\phi = 0.7$ and tortuosity $T_{\mathrm{hydr}} = \phi^{-1}$ for $\phi \geq 0.05$, and $T_{\mathrm{hydr}} = 20$ for $\phi < 0.05$.

Equation (33) is consistent with:

$$k = \left(\frac{1}{8}\right)\left(\frac{c_\phi \phi}{T_{\mathrm{hydr}}}\right)\left(3\rho_{2,\mathrm{eff}} T_{2,\mathrm{LM,corr}}\right)^2 \tag{34}$$

where $T_{2,\mathrm{LM,corr}}$ is the corrected logarithmic mean:

$$T_{2,\mathrm{LM,corr}} = T_{2,\mathrm{LM}}\left(\frac{r_1}{3\rho_{2,\mathrm{eff}} T_{2,\mathrm{LM}}}\right)^{0.3}, \tag{35}$$

with $r_1 = 9.54$ μm.

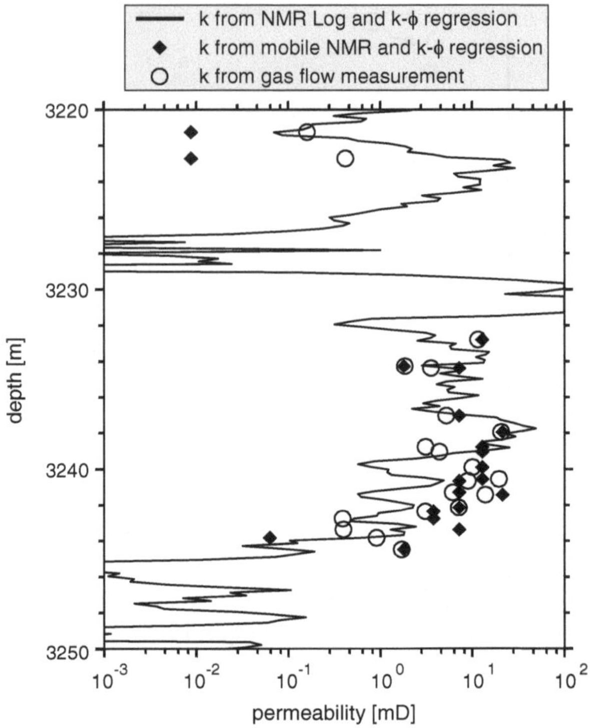

Figure 12
Correlation of core plug permeabilities: Results from gas flow measurements are compared to calculated permeabilities from a power-law relationship between permeability and porosity according to equation (4).

The corrected pore radii r_{corr} $[T_{2,LM}]$ and permeabilities are presented in Table 7. To obtain permeability values according to equation (34) the mean value of $\rho_{2,eff}$ of the set of samples was used. For comparison, permeability values were calculated by the standard oil industry method from 1D T_2 distributions (equation (5) with a factor of $a = 4$). Whereas permeabilities derived from equation (34) agree well with gas flow measurements. Results from the standard method used in the oil industry underestimate permeability systematically (Figure 10). Particularly, permeability is underestimated by four orders of magnitude for the samples AC1 and AC3 which contain a considerable amount of clay.

6. *Comparison of NMR Logging and Mobil NMR Sensor Data*

In this study, we used relatively small samples for the relaxation measurements with our laboratory NMR device. However, relaxation measurements on long cores moving stepwise through the Halbach magnet can also be performed with the mobile NMR core scanner. These measurements yield the porosity, because the initial amplitude of the

Table 6

Results based on mercury injection curves and on 2D T_1-T_2 correlation experiments performed on the Allermoehe samples: pore radii $r_{\rho 1}$ [$T_{1,LM}$] and $r_{\rho 2,eff}$ [$T_{2,LM}$] are calculated according to equation (19) using surface relaxivities ρ_1 and $\rho_{2,eff}$; the r_{corr} [$T_{2,LM}$] values are calculated according to the empirical equation (31). Permeabilities k follow from equations (32) and (33). For comparison gas permeabilities (k_{gas}) are also shown.

Sample	ρ_1 [μm/s]	$\rho_{2,eff}$ [μm/s]	$r_{\rho 1}$ [$T_{1,LM}$] [μm] eq. (19)	$r_{\rho 2,eff}$ [$T_{2,LM}$] [μm] eq. (19)	r_{corr} [$T_{2,LM}$] [μm] eq. (31)	k [mD] from $r_{\rho 1}$ ($T_{1,LM}$) eq. (32)	k [mD] from r_{corr} ($T_{2,LM}$) eq. (33)	k_{gas} [mD]
AC4			6.51	6.12	6.97	34.92	26.87	11.6
AC6			3.19	2.25	3.45	3.79	1.61	1.85
AC7	4.83	21.89	3.19	1.91	3.07	5.35	2.07	3.59
AC9	4.24	23.82	3.82	2.65	3.87	8.48	3.98	5.26
AC10	4.24	23.82	4.20	2.95	4.18	18.71	9.36	20.7
AC12	2.37	17.76	2.35	1.50	2.59	4.83	1.62	3.13
AC13	4.21	13.30	2.40	0.97	1.91	2.61	0.68	4.42
AC14	5.31	37.20	5.10	3.99	5.16	19.14	11.44	10.1
AC15	4.28	35.10	5.65	3.69	4.88	17.13	9.77	19.6
AC16	5.77	33.62	5.71	3.36	4.57	11.85	6.39	8.99
AC18	5.09	29.68	4.89	3.00	4.23	10.13	5.12	6.16
AC19			4.72	2.98	4.21	15.67	7.88	13.9
AC22			1.92	0.57	1.32	0.76	0.14	0.39
AC23	2.22	12.96	0.87	0.29	0.82	0.38	0.05	0.4
AC25			3.06	1.79	2.93	2.74	1.02	1.71
mean value	4.256	24.915						

decay curve and the area under the T_2 distribution curve are directly proportional to the number of polarized hydrogen nuclei in the pore fluid. Hence, the T_2 distribution can be directly calibrated in terms of porosity (COATES *et al.*, 1999; BLÜMICH *et. al.*, 2004; ARNOLD *et al.*, 2006). To demonstrate this on the Allermoehe samples, we calibrated the mobile NMR core scanner for porosity measurements using a glass tube of known cross section and filled with distilled water. Then, porosity was calculated from the relaxation curves of the plugs. Comparison to porosity from the gas pycnometer shows that NMR relaxation measurements yield reasonable porosity values. The correlation coefficient R^2 between NMR core scanner and pycnometer measurements is 0.8 with a standard deviation of 1.1%. The porosity values from the mobile NMR core scanner also agrees well with the porosity data of the CMR log from the Allermoehe borehole, where we used the sum of free water and capillary water for the comparison (Figure 11).

For the comparison of permeability data we used the 1D T_2 measurements with the mobile NMR core scanner and porosity measurements with the pycnometer. Permeability from the CMR log was calculated with ϕ = free water + capillary water from the CMR log itself. Figure 12 shows the comparison of permeability data measured with the mobile NMR core scanner and with the downhole CMR tool, which were calculated with the calibrated permeability-porosity relationship of equation (4). Both data sets are in good agreement and compare well with the

Table 7

Results based on 1D T_2 relaxation measurements with the improved Halbach core-scanner: $\rho_{2,eff}$ values follow from equation (19); r_{corr} [$T_{2,LM}$] values result from equation (31); and permeabilities "k from r_{corr} [$T_{2,LM}$]" are calculated with equation (29) and "k from $T_{2,LM,corr}$" result from equations (34) and (35) using the mean value of $\rho_{2,eff}$

Sample	$\rho_{2,\mathrm{eff}}$ [μm/s] eq. (19)	r_{corr} [$T_{2,\mathrm{LM}}$] [μm] eq. (31)	k [mD] from r_{corr} [$T_{2,\mathrm{LM}}$] eq. (34)	k [mD] from $T_{2,\mathrm{LM,corr}}$ and mean $\rho_{2,\mathrm{eff}}$ eqs. (34) and (35)	k_{gas} [mD]
AC1	–	1.53	0.69	0.69	0.16
AC3	–	2.49	1.80	1.80	0.42
AC4	–	6.14	27.09	27.10	11.6
AC6	–	1.79	1.02	1.02	1.85
AC7	21.60	3.32	6.26	6.42	3.59
AC9	17.37	2.31	3.01	2.71	5.26
AC10	23.46	3.33	11.87	12.78	20.7
AC12	13.10	2.33	3.89	2.97	3.13
AC13	22.05	3.29	7.76	8.06	4.42
AC14	13.10	2.22	3.53	2.70	10.1
AC15	27.85	4.07	11.87	14.14	19.6
AC16	29.68	3.52	7.03	8.69	8.99
AC18	26.29	2.87	4.66	5.37	6.16
AC19	–	3.16	4.21	4.22	13.9
AC20	23.07	3.24	5.96	6.35	7.2
AC21	–	2.23	2.15	2.17	3.1
AC22	–	2.06	1.85	1.85	0.39
AC23	10.38	1.34	1.01	0.68	0.4
AC25	–	2.71	0.51	0.51	1.71
mean value	20.72				

measured gas permeabilities with the exception of two values from mobile NMR at the depth of about 3225 m.

For the next comparisons, the measured relaxation times were used for estimating permeability. Figure (13) presents the data sets which were calculated with the empirical Kenyon equation (5) using a factor of $a = 4$. There is a large disagreement between NMR and gas permeabilities, as the predicted values are too low. This can be explained by the use of a high porosity exponent equal to 4 in equation (5) and the low porosity values of the Allermoehe samples, between 2% and 11%. It should be noted that this equation is only suitable for formations of high porosity ($\phi > 15\%$).

Finally, the new method was applied by using equation (34) for calculating permeability from the relaxation data (Figure. 14). In this case, the data sets for the CMR log, the core scanner, and the gas flow measurements compare well.

7. Conclusion

NMR logging data recorded in a sandstone formation or NMR measurements performed with a mobile NMR core scanner on corresponding core plugs are useful to

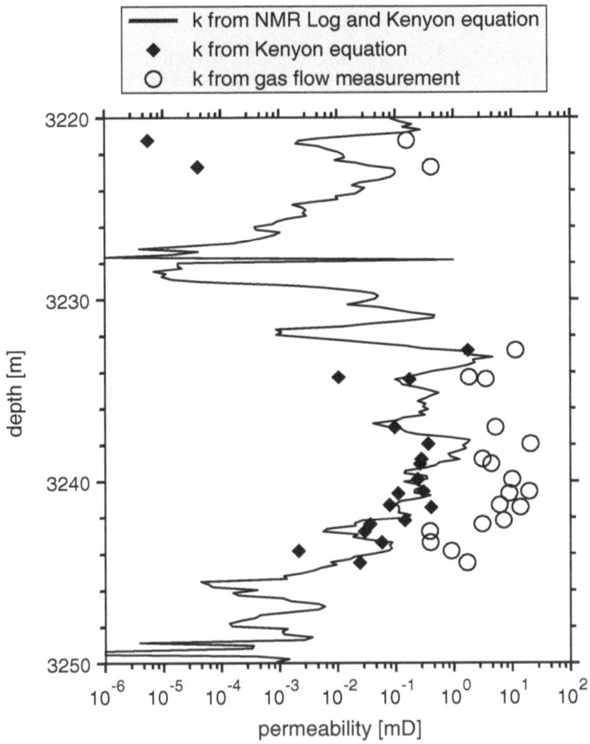

Figure 13
Comparison of NMR logging and mobile NMR sensor data: The CMR permeability (full line) is compared to NMR core permeability (diamonds) based on equation (5). Permeability values derived from gas flow measurements on core plugs (circles) are shown for comparison.

determine porosity directly and accurately. The empirical Kenyon equation with $T_{2,LM}$, which is commonly used in the oil industry for estimating permeability from NMR logging data, is acceptable for typical reservoir rocks with porosity >10%. A high exponent to porosity ϕ in this equation restricts the applicability to a small range of porosities. Other permeability equations contain the surface relaxivity ρ_2 explicitly. Then, the effective hydraulic pore radius has to be larger than about 10 μm for using a constant surface relaxivity ρ_2. In the case of the Rhaetian sandstone formation at Allermoehe with low porosity ($\leq$11%) and permeability, a calculation scheme with a constant ρ_2 yields large errors. Therefore, the relationship between ρ_2 and pore radius r has been studied. From T_1-T_2 correlation experiments, the relationship between the surface relaxivities ρ_1 and ρ_2 can be determined. A constant value for ρ_1 can be obtained from the comparison of T_1 distributions with mercury injection curves. The relaxation of transverse relaxations mainly depends on the dephasing of magnetic spins due the internal inhomogeneous magnetic field. This process can be described by a model of an interlayer with large magnetic field gradients

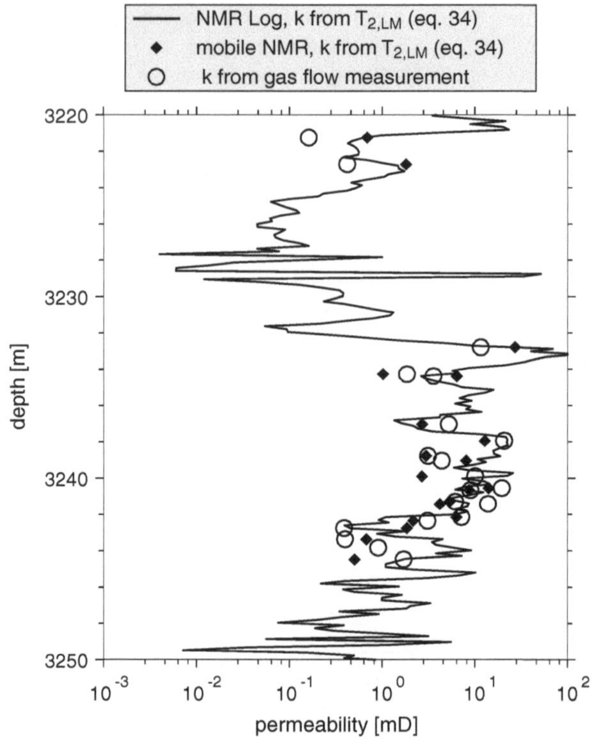

Figure 14
Results of the new model applied to NMR logging and mobile NMR sensor data for calculating permeability based on equation (34): the CMR permeability (full line) is compared to NMR core permeability (diamonds). Permeability values derived from gas flow measurements on core plugs (circles) are shown for comparison.

inside the pore walls. The thickness of this interlayer and a relaxation term ρ_{dephase} can be calculated. The surface relaxivity ρ_2 results as $(\rho_1 + \rho_{\text{dephase}})$ multiplied by a correction term, which is 1 for a pore throat radius larger than 10 μm and increases with decreasing pore radius. Based on this analysis it is possible to calculate an individual surface relaxivity ρ_2 for each T_2 value. On this basis, accurate permeability prediction from transverse relaxation measurements is possible. A hydraulic effective pore radius can be calculated for each sample from the logarithmic mean value of the T_2 distribution curves. Consequently, permeability can be predicted accurately from the effective pore radii with the physically based Kozeny-Carman equation. The new permeability model yielded equation (34), which can be conveniently used to calculate permeability from measured decay curves of transverse magnetization.

A comparison of NMR measurements with the core scanner to a CMR Log showed that both methods are compatible. For applying the new permeability model to NMR downhole measurements, one needs the value of ρ_2 for large pores. This can be determined by calibration on cores. Otherwise, the relationship between ρ_1 and ρ_2 can be studied by two dimensional T_1-T_2 measurements in the borehole. The surface relaxivity ρ_1 may be assumed on the basis of literature data. When the permeability prediction with equation (34) is based on relaxivity values from literature, it might be necessary to adjust them to the calibration radius $r_{\rm eff}$, if the values were originally calibrated to the pore body radius.

Acknowledgements

The authors would like to thank P. Glover of Université Laval for his constructive comments. The theoretical part of the paper was strengthened by his work on permeability prediction and transformation of length parameters of porous media. Yves Bernabé of MIT is gratefully acknowledged for his help to improve the paper substantially.

Appendix

Here, we explain equations (19) and (25) which relate T_1 and $T_{\rm dephase}$ to the pore radius r and a reduced radius (r-d) respectively. We use a spherical pore model.

According to SEEVERS (1966), inverse longitudinal relaxation time T_1^{-1} is described as the arithmetic mean of the free fluid relaxivity $\rho_{FF}[\mathrm{s}^{-1}]$ for the free water volume and the so-called interlayer relaxivity $\rho_{\rm interlayer1}$ $[\mathrm{s}^{-1}]$ for the interlayer$_1$ at the pore wall. The interlayer$_1$ with thickness δ_1 is different from the interlayer with large field gradients in Figure 4 with thickness d.

$$\frac{1}{T_1} = \frac{V_{total} - V_{interlayer1}}{V_{total}}\rho_{FF} + \frac{V_{interlayer1}}{V_{total}}\rho_{interlayer1}. \tag{36}$$

The free fluid relaxivity results from the interaction between water molecules themselves, whereas the interlayer$_1$ relaxivity is explained by the interaction of water molecules with the crystal lattice of the pore wall. As ρ_{FF} is much smaller than $\rho_{\rm interlayer1}$, the first term can be neglected for small pores. $V_{\rm interlayer1}/V_{\rm total}$ can be expressed by $S_{\rm por,NMR}$ and the thickness δ_1 of interlayer$_1$ which is in the range of nanometers:

$$\frac{1}{T_1} = \rho_{interlayer1}\delta_1 S_{por,NMR} = \rho_1\frac{3}{r}. \tag{37}$$

The value of longitudinal magnetization $M_{\rm longitudinal}$ starts with zero for $t = 0$ and approximates M_∞ for a prolonged period. The function $(M_\infty$-$M)_{\rm longitudinal}$ would be equal to transverse magnetization if dephasing did not appear. Figure 15 presents a schematic plot of $(M_\infty - M)_{\rm longitudinal}$ versus diametric extension from the pore wall to the center (dashed line). The local values of $(M_\infty - M)_{\rm longitudinal}$ correspond to the concentration of protons in

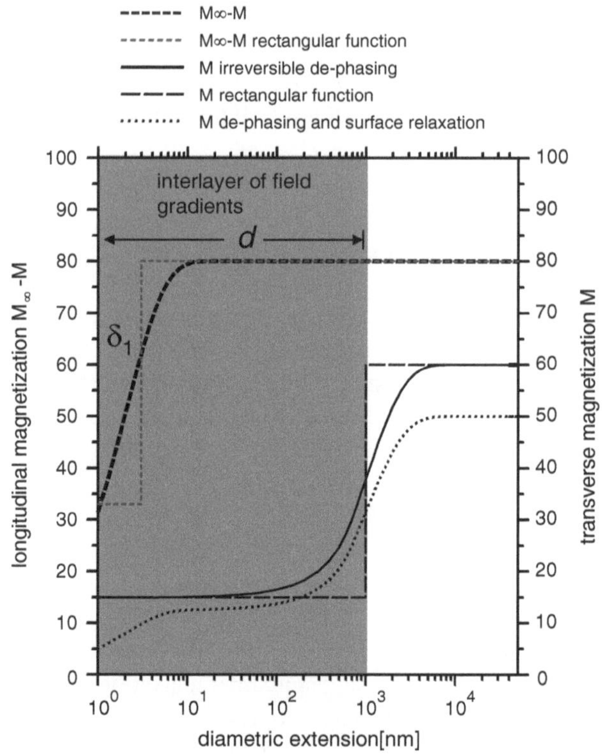

Figure 15
Schematic drawing for explaining the relationship between T_1 and r_1 and the relationship between $T_{dephase}$ and $(r\text{-}d)$. See text for more details.

the activated magnetic state, which is arbitrarily set to 80% at the center. The protons loose their energy at the pore wall. The shape of the curve results from diffusion of activated water molecules towards the pore wall. The thickness of interlayer δ_1 is constructed with the help of a step curve which confines the same area as the original curve.

The geometrical conditions of transverse magnetization $M_{transverse}$ are illustrated in the same figure. The local values of $M_{transverse}$ correspond to the concentration of protons in the activated magnetic state, which are not irreversibly dephased and therefore can be measured by the spin-echo-method. This fraction can be brought back into phase during echo time, in contrast to those that changed their precession velocity by diffusion in an inhomogeneous magnetic field which is caused by paramagnetic minerals at the pore walls. An interlayer of thickness d enveloping the pore wall contains this internal, inhomogeneous magnetic field (see Fig. 4). The second curve (continuous line in Fig. 15) represents the transverse magnetization due to irreversible dephasing without surface relaxation at the pore wall. The value in the pore center is set to 60% in order to show that the relaxation due to dephasing progresses faster than pure surface relaxation fore $(M_\infty\text{-}M)_{\text{longitudinal}}$. The curve decreases with a steep slope at the beginning of the interlayer of large magnetic field gradients. This interlayer is about one order of magnitude thicker than interlayer$_1$. The thickness d is again constructed with the help of a step curve which confines the same area as the original curve. Below this curve lays the third curve (dash-dotted line) which represents the measured transverse magnetization due to surface relaxation coupled with irreversible dephasing. How these two processes work together is explained by equation (24). Here, we want to justify equation (25), which relates the relaxation time T_{dephase} to the reduced radius (r-d).

The curve in the middle of Figure 15 shows that it is allowed to neglect the portion of activated protons within the interlayer with thickness d. Therefore, the relaxation process of dephasing can be described by the decrease of activated protons in the inner sphere with volume $V_{[r\text{-}d]} = (4/3)(r\text{-}d)^3$. Some activated protons pass the surface with area $S_{[r\text{-}d]} = 4(r\text{-}d)^2$ by diffusion and become dephased. As the amount of activated protons is proportional to M, we get:

$$\frac{dM}{dt} = -\rho_{\text{dephase}} S_{[r-d]} \frac{M}{V_{[r-d]}}. \tag{38}$$

Integration yields:

$$\begin{gathered} \log M = -\rho_{\text{dephase}} \frac{S_{[r-d]}}{V_{[r-d]}} t + \log M_0 \\ M = M_0 \exp\left(-\frac{t}{T_{\text{dephase}}}\right) \text{ with } T_{\text{dephase}} = \frac{1}{\rho_{\text{dephase}}}\left(\frac{V_{[r-d]}}{S_{[r-d]}}\right). \end{gathered} \tag{39}$$

$V_{[\text{r-d}]}$ and $S_{[\text{r-d}]}$ are expressed by the radius (r-d):

$$\frac{1}{T_{\text{dephase}}} = \rho_{\text{dephase}} \frac{3}{(r \text{ - } d)}. \tag{40}$$

References

Anferova, S., Anferov, V., Arnold, J., Talnishnikh, E., Voda, M., Kupferschläger, K., Blümler, P., Clauser, C., and Blümich, B. (2007), *Improved Halbach sensor for NMR scanning of drill cores*, Magnetic Resonance Imaging, *25*(4), 474–480.

API, Recommended Practice for Determining Permeability of Porous Media, RP 27, 1956.

Archie, G.E. (1942), *The electrical resistivity logs as an aid in determining some reservoir characteristics*, Transact. Am. Inst. Mining Engin. *146*, 54–62.

Arnold, J., Clauser, C., Pechnic, R., Anferova, S., Anferov, V., and Blümich, B., (2006), *Porosity and permeability from mobile NMR core-scanning*, Petrophysics, *47* (4), 306–314.

Baermann, A., Kröger, J. und Zarth, M. (2000a), *Anhydritzemente im Rhätsandstein Hamburgs - Röntgen- und Kernspintomographische Untersuchungen und Lösungsversuche*, Zeitschrift für Angewandte Geologie, *46* (3), 144–152.

Baermann, A., Kröger, J., Taugs, R., Wüstenhagen, K. und Zarth, M. (2000b), *Anhydritzemente im Rhätsandstein Hamburgs - Morphologie und Strukturen*, Zeitschrift für Angewandte Geologie, *46* (3), 138–143.

BERNABÉ, Y., and REVIL, A. (1995) *Pore-scale heterogeneity, energy dissipation and transport properties of rocks*, Geophy. Resea. Lett., *22*, 1529–1532.

BLÜMICH, B., ANFEROVA, S., PECHNIC, R., PAPE, H., ARNOLD, J., and CLAUSER, C., (2004), *Mobile NMR for porosity analysis of drill core sections*, J. of Geophy. and Engin *1* (3), 177–180.

BRUNAUER, S., EMMET, H., and TELLER, E. (1938), *Adsorption of gases in multimolecular layers*, J. Am. Chem. Soc. *60*(2), 309–319.

BUNDESANSTALT FÜR MATERIALFORSCHUNG UND PRÜFUNG, Germany. *Catalogue of Reference Procedures, Testing and Chemical Analysis* (http://www.bam.de/en/kompetenzen/fachabteilungen/abteilung_1/fg13/index.htm, 2005).

CARMAN, P.C., *Flow of Gases through Porous Media* (Butterworth, London 1956).

COATES, G.R., PEVERARO, C.A., HARDWICK, A., and ROBERTS, D. (1991), *The magnetic resonance imaging log characterized by comparison with petrophysical properties and laboratory core data*, Proceeding of the 667th Annual Technical Conference and Exhibition, Formation Evaluation and Reservoir Geology, SPE, SPE 22723, 627–635.

COATES, G.R., XIAO, L., and PRAMMER, M.G., *NMR Logging — Principles and Applications* (Halliburton Energy Services, Houston 1999).

DAVID, C. (1993), *Geometry of flow path for fluid transport in rocks*, J. Geophys. Res. *98*(B7) (12), 267–278.

DULLIEN, F.A.L., *Porous Media: Fluid Transport and Pore Structure*, 2nd edn (Academic Press, New York 1979).

DUNN, K-J., BERGMANN, D.J., and LATORACCA, G.A., *Handbook of Geophysical Exploration, Seismic Exploration*, Vol. 3. *Nuclear Magnetic Resonance. Petrophysical and Logging Applications* (Pergamon, Elsevier Science, Oxford 2002).

GEOTEK (2000) *Multi-Sensor Core Logger Manual* Geotek, Nene House Drayton Fields, (Daventry Northants NN11 5EA, United Kingdom 2000).

GLOVER, P.W.J., and WALKER, E. (revised 2008), *A grain size to effective pore size transformation derived from electrokinetic theory*, Geophy.

GLOVER, P.W.J., ZADJALI, I.I., and FREW, K.A. (2006), *Permeability prediction from MICP and NMR data using an electrokinetic approach*, Geophy. *71*, No. 4, F49–F60.

GODEFROY, S., and CALLAGHAN, P.T. (2003), *Two-dimensional relaxation/diffusion correlation in porous media*, Magnetic Resonance Imaging, *21*, 381–383.

HOWARD. J.J., KENYON. W.E., and STRALEY, C. (1990), *Proton magnetic resonance and pore size variations in reservoir sandstones*, SPE 2060, 65th Annual SPE meeting, New Orleans, LA.

HÜRLIMANN, M.D., HELMER, K.G., LATOUR, L.L., and SOTAK, C.H. (1994), *Restricted diffusion in sedimentary rocks. Determination of surface-area-to-volume ratio and surface relaxivity*, J. Magn. Res., Series A, *111*, 169–178.

HÜRLIMANN, M.D., and VENKATARAMANAN, L. (2002), *Quantitative measurement of two-dimensional distribution functions of diffusion and relaxation in grossly inhomogeneous fields*, J. Magn. Res. *157*, 31–42.

KENYON, W.E. (1992), *Nuclear magnetic resonance as a petrophysical measurement*, Nuclear Geophys. *6* (2), 153–171.

KENYON, W.E. (1997), *Petrophysical principles of applications of NMR logging*, The Log Analyst, *38* (2), 21–43.

KLEINBERG, R.L., and HORSFIELD, M.A. (1990), *Transverse relaxation process in porous sedimentary rock*, J. Magn. Res. *88*, 9–19.

KLEINBERG, R.L., KENYON, and W.E. MITRA, P.P. (1994), *Mechanism of NMR relaxation of fluids in rocks*, J. Magn. Res., *108*, 206–214.

KOSTEK, S., SCHWARTZ, L.M., and JOHNSON, D.L., (1992), *Fluid permeability in porous media: Comparison of electrical estimates with hydrodynamical calculations*, Physic. Review B, *45*, 186–195.

KOZENY, J. (1927), *Über die kapillare Leitung des Wassers im Boden (Aufstieg Versickerung und Anwendung auf die Bewässerung)*, Sitz Ber. Akad. Wiss. Wien Math. Nat. (Abt. IIa) *136a*, 271–306.

LONNES, S., GUZMAN-GARCIA, A., and HOLLAND, R. (2003), *NMR petrophysical predictions on cores*, Trans. SPWLA 44th Ann. Logg. Symp., paper DDD, 13 p.

MORRIS, C.E., DEUTSCH, P., FREEDMAN, R., MCKEON, D., and KLEINBERG, R.L. (1996), *Operating guide for the combinable magnetic resonance tool*, The Log Analyst, *37*(6), 53–60.

PAPE, H., and CLAUSER, C. (this issue), *Improved interpretation of Nuclear Magnetic Resonance T_1 and T_2 distributions for permeability prediction - simulation of diffusion coupling for a fractal cluster of pores*, Pure Appl Geophys.

PAPE, H., CLAUSER, C., IFFLAND, J. (1999a), *Permeability prediction for reservoir sandstones based on fractal pore space geometry*, Geophys. *64*(5), 1447–60.

PAPE, H., CLAUSER, C. und IFFLAND, J. (1999b), *Zusammenhang zwischen Permeabilität und Porosität bei Aquifer-Sandsteinen auf der Grundlage einer fraktalen Porenraumgeometrie* (Geothermische Vereinigung und Stadtwerke Straubing (Hrsg.): *Geothermie: Forschung, Entwicklung, Markt*, Tagungsband der 5. Geothermischen Fachtagung, 12.-15.5.98, Straubing 1999, ISBN 3-932570-14–6).

PAPE, H., CLAUSER, C., and IFFLAND, J. (2000), *Permeability–porosity relationship in sandstone based on fractal pore space geometry*, Pure Appl. Geophys. *157*, 603–19.

PAPE, H., CLAUSER, C., and IFFLAND, J., KRUG, R., and WAGNER. (2005), *Anhydrite cementation and compaction in geothermal reservoirs: Interaction of pore-space structure with flow, transport, P–T conditions, and chemical reactions*, Int. J. Rock Mech. Min. Sci. *42*, 1056–1069, doi: 10.1016/j.ijrmms.2005.05.007.

PAPE, H., RIEPE, L., and SCHOPPER, J.R. (1987a), *Theory of self-similar network structures in sedimentary and igneous rocks and their investigation with microscopical methods*, J. Microsc. *148*, 121–47.

PAPE, H, RIEPE, L, and SCHOPPER, L. (1987b), *Interlayer conductivity of rocks - a fractal model of interface irregularities for calculating interlayer conductivity of natural porous mineral systems*, Colloids and Surfaces, *27*, 97–122.

PAPE, H., TILLICH, J.E., and HOLZ, M. (2006), *Time-dependent diffusion coefficients in sedimentary rocks predicted by a fractal capillary model*, J. Appl. Geophys. *58*(3), 232–252.

SCHOPPER, J.R. (1966), *A theoretical investigation on the formation factor/permeability/porosity relationship using a network model*, Geophys. Prospect. *14*(3), 301–341.

SCHWARTZ, L.M., SEN, P.N., and JOHNSON, D.L., (1989), *Influence of rough surfaces on electrolytic conduction in porous media*, Physic Review B, *40*, 2450-2458.

SEEVERS, D.O. (1966), *A nuclear magnetic method for determining the permeability of sandstone*, Trans. SPWLA 7th Ann. Logg. Symp. 1966; paper L.

SERRA, O., and SERRA, L., *Well Logging – Data Acquisition and Applications* (Serralog, Méry Corbon, France, 2004).

TIAB, D., and DONALDSON, E.C., *Theory and Practice of Measuring Reservoir Rock and Fluid Transport Properties*, (Gulf Publishing 1996).

WEBB, P.A., *An Introduction to the Physical Characterization of Materials by Mercury Intrusion Porosimetry with Emphasis on Reduction and Presentation of Experimental Data* (Micromeritics Instrument Corporation, Norcross, Georgia 2001).

(Received March 18, 2008, revised July 28, 2008, accepted August 28, 2008)
Published Online First: June 4, 2009

MIX
Papier aus verantwortungsvollen Quellen
Paper from responsible sources
FSC® C105338

If you have any concerns about our products,
you can contact us on
ProductSafety@springernature.com

In case Publisher is established outside the EU,
the EU authorized representative is:
Springer Nature Customer Service Center GmbH
Europaplatz 3, 69115 Heidelberg, Germany

Printed by Libri Plureos GmbH
in Hamburg, Germany